AF658377

Springer Proceedings in Physics

Volume 166

More information about this series at http://www.springer.com/series/361

Vasudevan Lakshminarayanan
Indrani Bhattacharya
Editors

Advances in Optical Science and Engineering

Proceedings of the First International Conference, IEM OPTRONIX 2014

Editors
Vasudevan Lakshminarayanan
School of Optometry and Vision Science
University of Waterloo
Waterloo, ON
Canada

and

Department of Physics
University of Waterloo
Waterloo, ON
Canada

and

Department of Electrical and Computer Engineering
University of Waterloo
Waterloo, ON
Canada

Indrani Bhattacharya
Engineering Physics and Electronics and Communication Engineering
Institute of Engineering and Management
Kolkata, West Bengal
India

ISSN 0930-8989 ISSN 1867-4941 (electronic)
Springer Proceedings in Physics
ISBN 978-81-322-2366-5 ISBN 978-81-322-2367-2 (eBook)
DOI 10.1007/978-81-322-2367-2

Library of Congress Control Number: 2015936156

Springer New Delhi Heidelberg New York Dordrecht London

Printed on acid-free paper

Springer (India) Pvt. Ltd. is part of Springer Science+Business Media
(www.springer.com)

This volume is dedicated to the bright young students of today who have the potential to be major players in the future worldwide scientific community

Preface

Optics/Photonics and Opto-electronics are indispensable in almost all spheres of life, ranging from everyday life requirements such as communication and biomedicine to advanced science and technology. Starting from basic classical geometric and physical optics and quantum optics, there has been tremendous technological advancement as a result of which we have areas such as optical communication networks, optical information processing, space optics, electro-optics, Opto-mechatronics, Organic Photonics and so forth. Keeping the present scenario in mind, the First International Conference on Opto-Electronics and Applied Optics, IEM OPTRONIX 2014, was organized by the Department of Electronics and Communication of Institute of Engineering and Management, IEM, at D1, Management Campus, Salt Lake, Sector V, Kolkata 700091 on 16–17th December, 2014 in technical collaboration with Optical Society of India, OSI along with the SPIE Student Chapter at IEM/Kolkata, and was financially supported by Department of Science and Technology, DST, Government of India and Council of Scientific and Industrial Research, CSIR.

The conference was organized to commemorate the International Year of Light, IYL 2015 which as proclaimed by the United Nations and was intended to raise awareness amongst both academic and non-academics. The aim was to raise awareness of students, researchers, faculty members, industry and appropriate government personnel about the recent trends in research and development in the area of Opto-electronics, Photonics, Applied Optics, Bio-medical Optics and other relevant fields.

At this Conference, the following 18 areas were covered:

- Quantum Optics and Information Processing
- Optical and Digital Image Processing
- Opto-electronic Devices
- Optical Design
- Photonics for Space Applications
- Application for Solar Energy
- Fibre Optics and Devices

- Diffraction Tomography
- Opto-electronic Materials
- Non-linear Optics and PCF
- Non-linear Phenomenon and Chaos
- Nano-Photonics, Bio-Photonics and Bio-Medical Optics
- Micro-electronics and VLSI
- E.M. Radiation Theory and Antenna
- Optical Communications and Networks
- Opto-mechatronics
- Organic Photonics
- Terahertz Technologies

Plenary, invited and contribute talks covered all these areas and there was an overwhelming response to our meeting. Among 120 contributed submissions, there were two keynote addresses, 12 plenary talks and 67 accepted research papers.

The Conference started with national anthem "Jana Gana Mana" sung by the distinguished dignitaries and audience followed by inaugural speech delivered by Prof. Dr. Satyajit Chakrabarti, Founder Director, IEM Kolkata and VC, University of Engineering and Management, UEM, Jaipur via Skype from Canada. The welcome speech was delivered by Prof. Satyajit Chakrabarti, Director, IEM, Kolkata followed by Ajoy Kumar Chakraborty, Professor, IEM, Kehar Singh, Former Professor, IIT, Delhi, Bishnu Pal, President, Optical Society of India, and Vasudevan Lakshminarayanan, Professor, University of Waterloo, Canada. Keynote addresses were delivered by Phillip Russell, Director, Max Planck Institute for the Science of Light, Erlangen, Germany on "Enhancing Light-matter Interactions Using Microstructured Glass Fibres" and Lorenzo Pavesi, Professor, Department of Physics, University of Trento, Italyon "Silicon Photonics". The Plenary talks were delivered by distinguished scientists; the details of which are available in the Conference website:

http://www.iem.edu.in/iem_optronix/IEM%20OPTRONIX/speakers.html.

The Central Glass and Ceramics Research Institute, CGCRI, a CSIR Research Laboratory at Kolkata and Light Logics, Trivandrum, Kerala, India exhibited different products and highlighted project opportunities existing at their institutions. We express our heartfelt gratitude for their contribution.

We are extremely thankful to the International Advisory Committee and the Technical Program Committee for their valuable guidance and extended support in this regard. A list of the committee members is appended to this Preface. We express our sincere gratitude to the Student Organizing Committee which has consistently supported and taken part in the different activities as and when required. We express our heartfelt gratitude to the staff and faculty members of IEM, Salt Lake, Kolkata for making the Conference a success.

Waterloo — Vasudevan Lakshminarayanan
Kolkata — Indrani Bhattacharya

IEM OPTRONIX 2014 Organizing Committee

Patron	**Prof. Dr. Satyajit Chakrabarti** Vice-Chancellor, University of Engineering and Management, UEM, Jaipur and Kolkata Founder Director, Institute of Engineering and Management, IEM, Salt Lake, Kolkata
	Prof. Banani Chakrabarti Director, Institute of Engineering and Management, IEM, Salt Lake, Kolkata
	Prof. Gopa Goswami Director, Placement Cell Institute of Engineering and Management, IEM, Salt Lake, Kolkata
Chairman	**Prof. Satyajit Chakrabarti** Director, Institute of Engineering and Management, IEM, Salt Lake, Kolkata
Co-Chairman	**Prof. Vasudevan Lakshminarayanan** University of Waterloo, Canada

Advisory Committee

Ajoy Ghatak
SPIE Fellow, Formerly Professor,
Department of Physics, IIT, Delhi, India

A.K. Chakraborty
Professor, IEM, Kolkata, India

Amitava Roy
Department of Science and Technology (DST),
Delhi, India

B.M.A. Rahman
City University London,
Fellow OSA, Fellow SPIE

B.P. Pal
President, Optical Society of India, Formerly Professor,
Department of Physics, IIT, Delhi, India

Chandan Kumar Sarkar
Professor, Jadavpur University, Kolkata, India

Kehar Singh
Formerly Professor, Department of Physics,
IIT, Delhi, India

L.N. Hazra
Emeritus Professor, University of Calcutta, Kolkata, India

Lorenzo Pavesi
University of Trenty, Italy

Mourad Zghal
Vice-President, International Commission of Optics, ICO
Engineering School of Communications,
University of Carthage, Tunisia

Phillip Russell
President, Optical Society of America
Director,
Max Planck Institute for the Light Sciences
Erlangen, Germany

	Shyamal Bhadra Principal Scientist, CGCRI, Kolkata, India
	Yatagai President Elect, SPIE, Utsumiya University, Japan
Convener	**Indrani Bhattacharya** IEM, Salt Lake, Kolkata, India
Technical Chairs	**C.S. Narayana Murthy** IIST, Thiruvananthapuram, Kerala, India
	Kallol Bhattacharya Calcutta University, Kolkata, India.
	Partha Roy Chaudhari Department of Physics, IIT Kharagpur, India
	Senthil Kumar M. Head, Reflective Optics Division, SAC(ISRO), Ahmedabad
	Takeo Sasaki Tokyo University of Science, Japan
	Yukitoshi Otani Centre for Optical Research and Education, CORE Japan
	Yushiro Mizutani University of Tokushima, Japan

Technical Program Committee

K.K. Ghosh
IEM, Salt Lake
Kolkata, India

G.S. Taki
IEM, Salt Lake
Kolkata, India

Malay Ganguly
IEM, Salt Lake
Kolkata, India

Indrani Bhattacharya
Convener of the conference
IEM, Salt Lake
Kolkata, India

Ratna Chakraborty
Co-Convener of the Conference
IEM, Salt Lake
Kolkata, India

Mili Sarkar
Co-Convener of the Conference
IEM, Salt Lake
Kolkata, India

Soumyadip Banerjee
IEM, Salt Lake
Kolkata, India

Srijita Chakraborty
IEM, Salt Lake
Kolkata, India

Tuhin Utsab Paul
IEM, Salt Lake
Kolkata, India

Triparna Dutta
IEM, Salt Lake
Kolkata, India

Saptarshi Goswami
IEM, Salt Lake
Kolkata, India

Student Organizing Committee

Rajarshi Roy Chaudhuri
Subhajit Dutta Chowdhury
Archisman Saha
Kaustav Dutta
Sahana Lahiri
Nitish Kumar Thakur
Soham Mandal
Amartya Bose
Prosanta Ghosh
Suparna Chakrabarty
Sabarno Chowdhury
Kuntal Sengupta
Samanway Ghosh
Samanway Guha
Soumyadeep Chatterjee
Ritam Ghosh
Pritom Pal

Acknowledgment

The editors wish to acknowledge Professor Dr. Satyajit Chakrabarti, Vice-Chancellor, University of Engineering and Management, UEM, Jaipur and Kolkata and Founder Director, Institute of Engineering and Management, IEM, Salt Lake, Kolkata, Mrs. Banani Chakrabarti, Director Madam and Mrs. Gopa Goswami, Placement Director, IEM for patronizing the effort from the very beginning.

The editors also acknowledge Professor Satyajit Chakrabarti (Alex), Director, Institute of Engineering and Management, IEM, Salt Lake, Kolkata, India for his continuous support and encouragement in making the Conference possible and it is due to his inspiration publishing of the Proceedings of IEM OPTRONIX 2014 in collaboration with M/s Springer has become possible.

Waterloo — Vasudevan Lakshminarayanan
Kolkata — Indrani Bhattacharya

Contents

Part XIV Opto-Electronic Devices

Part XV Opto-Electronic Materials

Editors and Contributors

About the Editors

Vasudevan Lakshminarayanan (Ph.D. University of California at Berkeley) is currently a professor of vision science, physics and electrical and computer engineering at the University of Waterloo. He has held visiting and faculty appointments at the universities of California at Berkeley, at Irvine, University of Missouri and the University of Michigan. He has had numerous honors, including Fellow of OSA, SPIE, AAAS, APS, IoP, etc. and the recipient of many awards including most recently the Esther Hoffman Beller medal of OSA, and the Optics Educator award of SPIE. He has been a KITP Scholar at the Kavli Institute of Theoretical Physics at the University of California at Santa Barbara. He was a AAAS science an technology policy fellow finalist and serves on the optics advisory board of the Abdus Salam International Center for Theoretical Physics in Trieste, Italy. Research interests range from optics to neuroscience, bioengineering, applied math and ophthalmology/optometry.

University of Waterloo
Email: vengu@uwaterloo.ca
Phone: +1 519-888-4567, Extension: 38167

Indrani Bhattacharya obtained her B.Sc (Honours) in Physics and M.Sc (Tech) in Applied Physics (with Specialization in Optics and Opto-Electronics) degrees in 1984 and 1988 respectively from Serampore College and the Department of Applied Physics, University of Calcutta. After obtaining her Maters degree, she joined in M/s Hindustan Cables Limited, HCL, A Government of India Undertaking as Fibre Optic System Development Engineer in Company's Fibre Optics Project, Naini, Allahabad and worked there for two years. During those days, she was actively involved in the development of Fused Fibre Optics Taper Biconical Couplers, Optical Fibre Joint Boxes including manufacturing and testing of slotted core, loose tube fibre optics cables supplied to Department of Telecommunications, DOT, Government of India. In the year 1992, she joined the HCL, Corporate Office at Kolkata and actively involved as the Core Group Member for the Computerization of the Company. Ms. Bhattacharya was trained in UNIX, SYBASE, ORACLE with Developer 2000 required in the computerization process.

After serving 19 years 6 months in HCL, she planned to join academics and accordingly, she left HCL on 31st March 2010 and enrolled in Ph.D. (Tech) program as a part-time research scholar in the Department of Applied Optics and Photonics, Calcutta University, under the guidance of Emeritus Professor Lakshminarayan Hazra. She had joined as Head of the Department, Electronics and Communication Engineering, Camellia College of Engineering and Technology, Burdwan and taught there for 2 years.

On 16th July, 2012 she joined Institute of Engineering and Management, IEM, Salt Lake, Kolkata as Assistant Professor (Sr. Grade) where she is working till date. In IEM, she has motivated students to be involved in research and development activities and opened SPIE and OSA approved IEM Student Chapters, Brahmatejas and Alokadhyaya. Prof. Bhattacharya is the core group member of the Research Council of IEM for promoting research and development activities among the faculty members and students of the Institute.

Prof. Bhattacharya is also the Co-coordinator of the Department of Science and Technology (DST), Govt. of India, sponsored INnovation in Science Pursuit for Inspired Research (INSPIRE) Science Camp organized in IEM, "The Enchanting World of Science", to motivate the top 1 % board students to be involved in research and development activities in Science.

Prof. Bhattacharya has published seven research works in scientific journals including Conference publications starting from 2011. Her research areas include Optical Image Processing, Optical Design, Diffractive Optics, Super Resolution, Solar Concentrator Optics, Bio-Photonics, Silicon Photonics.

Institute of Engineering and Management, IEM, Salt Lake, Kolkata, India
Department of Applied Optics and Photonics, University of Calcutta, Salt Lake, Kolkata, India
Email: indrani.bhattacharya@iemcal.com
Phone: +91 8697804729

Contributors

I.D. Aggarwal Physics Department, University of North Carolina, Charlotte, NC, USA

P.T. Ajith Kumar Light Logics Holography and Optics Pvt. Ltd, Thiruvananthapuram, Kerala, India

Shyam Akashe Department of Electronics and Communication Engineering, ITM University, Gwalior, India

A. Almazroa School of Optometry and Vision Science, University of Waterloo, Waterloo, Canada

R.K. Bahl Optical Communication Division (OCD), SNPA, Space Applications Centre, ISRO, Ahmedabad, India

Arup Banerjee Space Applications Centre (SAC), Indian Space Research Organization (ISRO), Ahmedabad, India

Koushik Banerjee ITM University (Gwalior), Gwalior, Madhya Pradesh, India

Soumyadip Banerjee Institute of Engineering and Management, Salt Lake, West Bengal, India

A. Banik Optical and Digital Communication Group, SNPA, SAC, Ahmedabad, India

A. Barh Department of Physics, Indian Institute of Technology Delhi, Hauz Khas, New Delhi, India

Koushik Basak Optical Communication Division (OCD), SNPA, Space Applications Centre, ISRO, Ahmedabad, India

Amine Ben Salem University of Carthage, Engineering School of Communication of Tunis (Sup'Com), GreS'Com Laboratory, Ariana, Tunisia

Shyamal K. Bhadra Fiber Optics and Photonic Division, CSIR—Central Glass and Ceramics Research Institute, Kolkata, India

A.K. Bhattacharjee Department of ECE, NIT, Durgapur, West Bengal, India

I. Bhattacharya Department of ECE, Institute of Engineering and Management, Salt Lake, Kolkata, India

K. Bhattacharya Department of Applied Optics and Photonics, University of Calcutta, Kolkata, India

Kaustav Bhattacharyya Center for Optical Research and Education (CORE), Utsunomiya University, Utsunomiya, Tochigi, Japan

N.S. Bhattacharyya Applied Optics and Photonics Research Laboratory, Department of Physics, Tezpur University, Assam, India

Divya Billaiya ECE Department, ITM University, Gwalior, Madhya Pradesh, India

K.S. Bindra Solid State Laser Division, Raja Ramanna Centre for Advanced Technology, Indore, India

A. Biswas Mechanical Engineering Department, NIT Silchar, Silchar, Assam, India

Arindam Biswas Department of ECE, Hooghly Engineering and Technology College, Hooghly, India

R. Biswas Applied Optics and Photonics Research Laboratory, Department of Physics, Tezpur University, Assam, India

Tushar Biswas Fiber Optics and Photonic Division, CSIR—Central Glass and Ceramics Research Institute, Kolkata, India

William R. Bobier School of Optometry and Vision Science, University of Waterloo, Waterloo, Canada

Shanmugam Boomadevi Department of Physics, National Institute of Technology, Tiruchirappalli, Tamil Nadu, India

R. Burman Department of Electronics and Telecommunication Engineering, Jadavpur University, Kolkata, India

S. Chakrabarti Department of Electronics and Communication Engineering, Institute of Engineering and Management, Salt Lake, Kolkata, India

Ratna Chakrabarty Department of Electronics and Communication Engineering, Institute of Engineering and Management, Kolkata, India

A. Chakraborty Institute of Engineering and Management, Salt Lake, Kolkata, India

Mrinmoy Chakraborty Dr. B.C. Roy Engineering College, Durgapur, India

Srijita Chakraborty Institute of Engineering and Management, Kolkata, India

B. Chatterjee Department of Electronics and Communication Engineering, Institute of Engineering and Management, Salt Lake, Kolkata, India; Department of Basic Science and Humanities, Institute of Engineering and Managemant, Kolkata, West Bengal, India

Sanjib Chatterjee Raja Ramanna Centre for Advanced Technology, Indore, India

Rajarshi Roy Chaudhuri Department of Electronics and Communication Engineering, Institute of Engineering and Management, Kolkata, India

Brijesh Kumar Chaurasia Department of CSE, ITM University, Gwalior, India

Rim Cherif University of Carthage, Engineering School of Communication of Tunis (Sup'Com), GreS'Com Laboratory, Ariana, Tunisia

Mayur K. Chhipa Department of Electronics and Communication Engineering, Government Engineering College, Ajmer, Rajasthan, India

Sabarno Chowdhury Institute of Engineering and Management, Kolkata, India

Subhajit Dutta Chowdhury Institute of Engineering and Management, Kolkata, India

Mukul K. Das UGC SAP Research Laboratory, Department of Electronics Engineering, Indian School of Mines, Dhanbad, India

N.R. Das Institute of Radio Physics and Electronics, University of Calcutta, Kolkata, India

Nandan K. Das Indian Institute of Science Education and Research, Kolkata, India

Koustav Dasgupta Department of Applied Optics and Photonics, University of Calcutta, Kolkata, India

Triparna Datta Institute of Engineering and Management (IEM), Kolkata, India

Debashis De Department of Computer Science and Engineering, West Bengal University of Technology, Kolkata, India

Khemraj Deshmukh SSTC Bhilai, CSVTU, Bhilai, Chhattisgarh, India

Subhadip Dey Bidhan Chandra Krishi Viswa Vidyalaya, Kalyani, India

Arpan Deyasi Electronics and Communication Engineering, RCC Institute of Information Technology, Kolkata, India

Pranab J. Dihingia Department of Physics, Dibrugarh University, Dibrugarh, Assam, India

R.S. Dubey Advanced Research Laboratory for Nanomaterials and Devices, Department of Nanotechnology, Swarnandhra College of Engineering and Technology, Narsapur, AP, India

D. Dutta Mechanical Engineering Department, NIT Silchar, Silchar, Assam, India

Kaustav Dutta Department of Electronics and Communication Engineering, Institute of Engineering and Management, Kolkata, India

Sibasish Dutta Applied Photonics and Nanophotonics Laboratory, Department of Physics, Tezpur University, Napaam, Assam, India

U.P.S. Gahlaut School of Physical Sciences, ITM University, Gwalior, Madhya Pradesh, India

Tarun Kumar Gangopadhyay CSIR—Central Glass and Ceramic Research Institute (CGCRI), Kolkata, India

R. Ganguly Institute of Engineering and Management, Salt Lake, Kolkata, India

K.P. Ghatak Department of Basic Science and Humanities, Institute of Engineering and Managemant, Kolkata, West Bengal, India

A. Ghosh Instruments Research and Development Establishment, Dehradun, India

K.K. Ghosh Institute of Engineering and Management (IEM), Kolkata, India

Nirmalya Ghosh Indian Institute of Science Education and Research, Kolkata, India

S. Ghosh INSPIRE Faculty, Institute of Radio Physics and Electronics, University of Calcutta, Kolkata, India

Y.C. Goswami School of Physical Sciences, ITM University, Gwalior, MP, India

A.K. Gupta Instruments Research and Development Establishment (IRDE), Dehradun, India

Ankur Kumar Gupta ECE Department, ITM University, Gwalior, Madhya Pradesh, India

P.K. Gupta Solid State Laser Division, Raja Ramanna Centre for Advanced Technology, Indore, India

Sundararaman Hari Hara Subramani Centre for Nonlinear Science and Engineering (CeNSE), School of Electrical and Electronics Engineering, SASTRA University, Thanjavur, Tamil Nadu, India

L.N. Hazra Department of Applied Optics and Photonics, University of Calcutta, Salt Lake, Kolkata, India

Lakshminarayan Hazra Department of Applied Optics and Photonics, University of Calcutta, Kolkata, India

N. Hutchings School of Optometry and Vision Science, University of Waterloo, Waterloo, Canada

Jijo V. Ittiarah CSIR—Central Glass and Ceramic Research Institute (CGCRI), Kolkata, India

Atibhi Jadon Department of ECE, ITM, Gwalior, Madhya Pradesh, India

M. Jayaraju MES Institute of Technology and Management, Chathanoor, Kerala, India

Pooja Joshi ITM Universe, Gwalior, Madhya Pradesh, India

S. Kalainathan School of Advanced Sciences, VIT University, Vellore, TN, India

Sanjib Kr. Kalita Department of Computer Science, Gauhati University, Gauhati, India

Angshuman Khan Department of Electronics and Communication Engineering, University of Engineering and Management, Jaipur, India

Saurabh Khandelwal ITM, Gwalior, Madhya Pradesh, India

Anil Kumar Instruments Research and Development Establishment (IRDE), Dehradun, India; Guru Jambheshwar University of Science and Technology, Hisar, India

Chandan Jyoti Kumar Department of Computer Science, Gauhati University, Gauhati, India

D. Kumar PG Department of Chemistry, SMS Govt Science College, Gwalior, MP, India

Dharmendra Kumar Department of Electronics Engineering, Indian School of Mines, Dhanbad, Jharkhand, India

Dhirendra Kumar Department of Physics, Indian Institute of Technology Patna, Patna, India

Jitendra Kumar Department of Electronics Engineering, Indian School of Mines Dhanbad, Jharkhand, India

Nitin Kumar Gurukula Kangari Vishwavidyalaya, Haridwar, Uttrakand, India

Vijay Kumar School of Physical Sciences, ITM University, Gwalior, India

Vinod Kumar Instruments Research and Development Establishment, Dehradun, India

Archana Kumari Kalyani Government Engineering College, Kalyani, India

Ankit Singh Kushwah Department of Electronics and Communication Engineering, ITM University, Gwalior, India

Priyanka Kushwah ITM, Gwalior, Madhya Pradesh, India

Vasudevan Lakshminarayanan School of Optometry and Vision Science, University of Waterloo, Waterloo, Canada; Department of Electrical and Computer Engineering, University of Waterloo, Waterloo, Canada; Department of Physics, University of Waterloo, Waterloo, Canada

Madan Lal Madan Faculty of Engineering and Technology, Mewar University, Chitordgarh, Rajasthan, India

V.P. Mahadevan Pillai Department of Optoelectronics, University of Kerala, Thiruvananthapuram, Kerala, India

Goutam Kumar Maity Electronics and Communication Engineering, MCKV Institute of Engineering, Howrah, India

Subir Majumder Fiber Optics and Photonic Division, CSIR—Central Glass and Ceramics Research Institute, Kolkata, India

Ashis Kumar Mandal Department of Physics, Jadavpur University, Kolkata, West Bengal, India

Bijoy Mandal Department of CSE, NSHM Knowledge Campus, Durgapur, West Bengal, India

Soham Mandal Institute of Engineering and Management, Kolkata, India

Nabin Baran Manik Department of Physics, Jadavpur University, Kolkata, West Bengal, India

Toijam Sunder Meetei Centre for Nonlinear Science and Engineering (CeNSE), School of Electrical and Electronics Engineering, SASTRA University, Thanjavur, Tamil Nadu, India

Isha Mehra Department of Physics, Indian Institute of Technology Patna, Patna, Bihar, India

Vishwas Mishra ITM University, Gwalior, Madhya Pradesh, India

Asish Mitra College of Engineering and Management, Kolaghat, India

Devendra Mohan Guru Jambheshwar University of Science and Technology, Hisar, India

Sayan K. Moitra Institute of Engineering and Management, Kolkata, India

Kajol Mondal Department of Physics, IIT Kharagpur, Kharagpur, India

Mayukh Mondal Department of Electrical Engineering, University of Engineering and Management, Jaipur, India

Deepak Motwani Department of CSE, Mewar University, Chitordgarh, Rajasthan, India

Chiradeep Mukherjee Department of Electronics and Communication Engineering, University of Engineering and Management, Jaipur, India

Moumita Mukherjee Centre for Millimeter-Wave Semiconductor Devices and Systems (CMSDS), DRDO, Kolkata, India

P.K. Mukhopadhyay Solid State Laser Division, Raja Ramanna Centre for Advanced Technology, Indore, India

Sabyasachi Mukhopadhyay Indian Institute of Science Education and Research, Kolkata, India

Aparna A. Nair Department of Electronics and Communication Engineering, College of Engineering Trivandrum, Trivandrum, Kerala, India

Yumiko Naka Department of Chemistry, Faculty of Science, Tokyo University of Science, Shinjuku-ku, Tokyo, Japan

Pabitra Nath Applied Photonics and Nanophotonics Laboratory, Department of Physics, Tezpur University, Napaam, Assam, India

C.M.S. Negi Department of Electronics, Banasthali Vidyapith, Banasthali, Rajasthan, India

Naveen K. Nishchal Department of Physics, Indian Institute of Technology Patna, Patna, India

S.M. Oak Solid State Laser Division, Raja Ramanna Centre for Advanced Technology, Indore, India

S.P. Ojha Department of Physics, Indian Institute of Technology, Banaras Hindu University, Varanasi, India

A. Ommani School of Optometry and Vision Science, University of Waterloo, Waterloo, Canada

Yukitoshi Otani Center for Optical Research and Education (CORE), Utsunomiya University, Utsunomiya, Tochigi, Japan

Bishnu P. Pal School of Natural Science, Mahindra Ecole Centrale, Bahadurpally, Hyderabad, India

Mrinmay Pal Fiber Optics and Photonic Division, CSIR—Central Glass and Ceramics Research Institute, Kolkata, India

Alok Kumar Pandey Optical Society of India, Ranchi, Jharkhand, India

Neeraj Pandey Instruments Research and Development Establishment, Dehradun, India

R.K. Pandey NIIT University Alwar, Rajasthan, India

Krishnamoorthy Pandiyan Centre for Nonlinear Science and Engineering (CeNSE), School of Electrical and Electronics Engineering, SASTRA University, Thanjavur, Tamil Nadu, India

Prasanta K. Panigrahi Indian Institute of Science Education and Research, Kolkata, India

K.K. Pant Instruments Research and Development Establishment, Dehradun, India

Prakash Pareek UGC SAP Research Laboratory, Department of Electronics Engineering, Indian School of Mines, Dhanbad, India

Jignesh Patel Space Applications Centre (SAC), Indian Space Research Organization (ISRO), Ahmedabad, India

D. Paul Applied Optics and Photonics Research Laboratory, Department of Physics, Tezpur University, Assam, India

Lorenzo Pavesi Nanoscience Laboratory, Department of Physics, University of Trento, Trento, Italy

B. Podder Mechanical Engineering Department, NIT Silchar, Silchar, Assam, India

L.P. Purohit Gurukula Kangari Vishwavidyalaya, Haridwar, Uttrakand, India

Kaamran Raahemifar Department of Electrical and Computer Engineering, Ryerson University, Toronto, Canada

B.M.A. Rahman City University London, London, UK

S. Rai Department of Physics, Mizoram University, Aizawl, Mizoram, India

P. Rajaram School of Studies in Physics, Jiwaji University, Gwalior, MP, India

Partha Roy Chaudhuri Department of Physics, IIT Kharagpur, Kharagpur, India

Uttam Roychoudhury Department of Chemical Technology, University College of Science and Technology, University of Calcutta, Kolkata, India

P.St.J. Russell Max Planck Institute for the Science of Light, Guenther-Scharowsky, Erlangen, Germany

Varun Sable ITM University, Gwalior, Madhya Pradesh, India

A. Saha Department of ECE, B.P. Poddar Institute of Technology, Poddar Vihar, Kolkata, India

Archisman Saha Department of Electronics and Communication Engineering, Institute of Engineering and Management, Kolkata, India

Supriti Samanta Department of Physics, Chaipat Girls' High School, Daspur, West Bengal, India

Swarup Samanta Department of ECE, Hooghly Engineering and Technology College, Hooghly, India

J. Sanghera Naval Research Laboratory, Code 5620, Washington, DC, USA

S. Saravanan Advanced Research Laboratory for Nanomaterials and Devices, Department of Nanotechnology, Swarnandhra College of Engineering and Technology, Narsapur, AP, India

K. Sarkar Department of Basic Science and Humanities, Institute of Engineering and Managemant, Kolkata, West Bengal, India

Mili Sarkar Institute of Engineering and Management, Kolkata, India

Takeo Sasaki Department of Chemistry, Faculty of Science, Tokyo University of Science, Shinjuku-ku, Tokyo, Japan

Nikhil Saxena ITM, Gwalior, Madhya Pradesh, India

Anasuya Sengupta ITM University (Gwalior), Gwalior, Madhya Pradesh, India

Chandan Sengupta SAMEER Kolkata Center, Kolkata, India; Department of Applied Optics and Photonics, University of Calcutta, Kolkata, India

David Ignacio Serrano-García Center for Optical Research and Education (CORE), Utsunomiya University, Utsunomiya, Tochigi, Japan

Dhrupesh Shah Space Applications Centre (SAC), Indian Space Research Organization (ISRO), Ahmedabad, India

Anurag Sharma Physics Department, Indian Institute of Technology Delhi, New Delhi, India

Dinesh Kumar Sharma Physics Department, Indian Institute of Technology Delhi, New Delhi, India

Hemant Sharma ITM University (Gwalior), Gwalior, Madhya Pradesh, India

Joshika Sharma ITM, Gwalior, Madhya Pradesh, India

Kapil Sharma Department of CSE, ITM University Gwalior, Gwalior, India

Pawan Sharma Department of ECE, ITM Universe, Gwalior, Madhya Pradesh, India

Payal Sharma Optical Communication Division (OCD), SNPA, Space Applications Centre, ISRO, Ahmedabad, India

Ranjana Sharma Department of Chemistry and Environmental Science, ITM-GOI, Gwalior, MP, India

S.K. Sharma Solid State Laser Division, Raja Ramanna Centre for Advanced Technology, Indore, India

Sneha Sharma Department of Electronics Engineering, Indian School of Mines Dhanbad, Jharkhand, India

L.B. Shaw Naval Research Laboratory, Code 5620, Washington, DC, USA

B. Chandar Shekar Department of Physics, Nanotechnology Lab, KASA, Coimbatore, India

Yogesh Shinde Space Applications Centre (SAC), Indian Space Research Organization (ISRO), Ahmedabad, India

Vipin Shrotriya School of Studies in Physics, Jiwaji University, Gwalior, MP, India

Siraj Sidhik CSIR—Central Glass and Ceramic Research Institute (CGCRI), Kolkata, India

A.J. Singh Solid State Laser Division, Raja Ramanna Centre for Advanced Technology, Indore, India

Bhoop Singh Department of Chemistry and Environmental Science, ITM-GOI, Gwalior, MP, India

C.P. Singh Solid State Laser Division, Raja Ramanna Centre for Advanced Technology, Indore, India

Hukum Singh Department of Applied Sciences, ITM University, Gurgaon, Haryana, India

Kehar Singh Department of Applied Sciences, ITM University, Gurgaon, Haryana, India

L.S. Singh Department of Electronics and Communication Engineering, National Institute of Technology, Manipur, Imphal, India

Rajeev Singh Department of Chemistry and Environmental Science, ITM-GOI, Gwalior, MP, India

Neetu Soni SSTC Bhilai, CSVTU, Bhilai, Chhattisgarh, India

Sonam Soni Department of CSE, ITM University Gwalior, Gwalior, India

Julia Soram Department of ECE, ITM University, Gwalior, Madhya Pradesh, India

A.B. Sreebha Department of Optoelectronics, University of Kerala, Thiruvananthapuram, Kerala, India

Sanjay Srivastava Department of Physics, MMV, Banaras Hindu University, Varanasi, India

Sanjeev K. Srivastava Department of Physics, Amity Institute of Applied Sciences, Amity University, Noida, India

S.K. Sudheer Department of Optoelectronics, University of Kerala, Trivandrum, Kerala, India

Soham Tewary Institute of Engineering and Management, Kolkata, India

Nitish Kumar Thakur Institute of Engineering and Management, Kolkata, India

Damber Thapa School of Optometry and Vision Science, University of Waterloo, Waterloo, Canada

Vipul Tiwari Department of CSE, ITM University Gwalior, Gwalior, India

Raj Kumar Tomar Department of Physics, Amity Institute of Applied Sciences, Amity University, Noida, India

V. Vadivelan Research and Development Centre, Bharathiar University, Coimbatore, Tamilnadu, India

R.K. Varshney Department of Physics, Indian Institute of Technology Delhi, Hauz Khas, New Delhi, India

Sunanda Vashisth Department of Applied Sciences, ITM University, Gurgaon, Haryana, India

A.K. Yadav Department of Applied Sciences, ITM University, Gurgaon, Haryana, India

Mourad Zghal University of Carthage, Engineering School of Communication of Tunis (Sup'Com), GreS'Com Laboratory, Ariana, Tunisia

…al Kumar Tomar Department of Physics, Amity Institute of Applied Sciences, Amity University, Noida, India

V. Vadivelan Research and Development Centre, Bharathiar University, Coimbatore, Tamilnadu, India

…s. Varshney Department of Physics, National Institute of Technology, Delhi, …ar Khan, New Delhi, India

Sunanda Vashisth Department of Applied Sciences, ITM University, Gurgaon, Haryana, India

A. K. Yadav Department of Applied Sciences, ITM University, Gurgaon, Haryana, India

Mourad Zghal University of Carthage, Engineering School of Communication of Tunis (Sup'Com), GreS'Com Laboratory, Ariana, Tunisia

Part I
Keynote Address

Chapter 1
Gas, Glass and Light: Controlling Light-Matter Interactions in Microstructured Fibres

P.St.J. Russell

Abstract Micro/nano-structured and photonic crystal fibres have in recent years been used to enhance light-matter interactions in many different ways. I will review some recent examples from the work of my group.

It is now well established that microstructured fibres, in particular photonic crystal fibres, permit remarkable control of the propagation of guided light, including introducing a new theme—the guidance of light, in a low-loss single mode, in a microscopic hollow channel. This last represents one of the most exciting opportunities in recent years, for it allows one for the first time effectively to eliminate beam diffraction in empty space or in materials with low refractive indices such as gases, vapours and liquids. As a result a new generation of versatile and efficient gas-based systems, for example pulse compression devices, light sources tunable from the vacuum UV to the near IR and nonlinear devices based on alkali metal vapours, is emerging [1–4].

Unlike the capillaries conventionally used for gas-based spectral broadening of ultrashort multi-mJ pulses, which provide only normal dispersion at usable pressure levels, hollow-core photonic crystal fibre provides pressure-adjustable normal or anomalous dispersion. Together with low-loss guidance in a bore $\sim 10\times$ narrower and a $100\times$ greater effective nonlinearity, this has led to a number of dramatic results, including pulse compression down to few-cycle pulses [5, 6], widely-tunable deep-UV light sources [7] and multi-octave Raman frequency combs [8, 9]. A new generation of versatile and efficient pulse compression devices and gas-based light sources, tunable from the vacuum UV to the near IR, is emerging.

PCFs with solid glass cores continue to inspire applications beyond the well-established fields of soliton dynamics and supercontinuum generation. For example, the presence of a transverse microstructure has a remarkable effect when the fibre is twisted continuously (by thermal post-processing) along its axis. This creates

P.St.J. Russell (✉)
Max Planck Institute for the Science of Light, Guenther-Scharowsky, Str. 1, 91058 Erlangen, Germany
e-mail: philip.russell@mpl.mpg.de

V. Lakshminarayanan and I. Bhattacharya (eds.), *Advances in Optical Science and Engineering*, Springer Proceedings in Physics 166,
DOI 10.1007/978-81-322-2367-2_1

orbital angular momentum (OAM) states in the cladding that couple to the core light at certain resonant wavelengths, creating deep dips in the transmitted spectrum [10]. Twisted PCF with a propeller-shaped core has the property of preserving the OAM sign [11].

Another new field is that of opto-acoustic devices, where the light itself drives mechanical resonances in the core structure. In one experiment, dual-frequency laser light sources were used to drive GHz acoustic resonances in a small solid glass core, resulting in the generation of a GHz frequency comb [12]. This was recently used to stably mode-lock a fiber ring laser at its 260th harmonic (1.57 GHz) [13].

Giant optically-driven optomechanical nonlinearities have been measured in a unique capillary fibre containing two parallel nano-webs (i.e., membranes) of glass [14]. This structure, when evacuated to μbar pressures, was recently observed to undergo noise-seeded Raman-like oscillations when pumped with only a few mW of CW light, resulting in generation of a frequency comb spaced by ~6 MHz [15].

These examples illustrate how the nanostructuring of glass fibres allows one to precisely control and enhance the interaction of light and matter in gaseous, liquid or solid form. New previously hard-to-access wavelengths of light can be produced (e.g., in the ultraviolet using gas-filled hollow core PCF or ZBLAN-based solid core PCF [16, 17]), intense optoacoustic interactions can allow stable mode-locking of fibre lasers at a very high harmonic of the fundamental round-trip frequency, and helical fibres can impose a twist on light.

References

1. Travers JC, Chang W, Nold J, Joly NJ, Russell PStJ (2011) Ultrafast nonlinear optics in gas-filled hollow-core photonic crystal fibers. J Opt Soc Am B 28:A11–A26
2. Russell PStJ, Hölzer P, Chang W, Abdolvand A, Travers JC (2014) Hollow-core photonic crystal fibres for gas-based nonlinear optics. Nat Photonics 8:278
3. Venkataraman V, Londero P, Bhagwat AR, Slepkov AD, Gaeta AL (2010) All-optical modulation of four-wave mixing in an Rb-filled photonic bandgap fiber. Opt Lett 35:2287–2289
4. Sprague MR, Michelberger PS, Champion TFM, England DG, Nunn J, Jin XM, Kolthammer WS, Abdolvand A, Russell PStJ, Walmsley IA (2014) Broadband single-photon-level memory in a hollow-core photonic crystal fibre. Nat Photonics 8:287–291
5. Mak KF, Travers JC, Joly NY, Abdolvand A, Russell PStJ (2013) Two techniques for temporal pulse compression in gas-filled hollow-core kagomé photonic crystal fiber. Opt Lett 38:3592–3595
6. Emaury F, Dutin CF, Saraceno CJ, Trant M, Heckl OH, Wang YY, Schriber C, Gerome F, Sudmeyer T, Benabid F, Keller U (2013) Beam delivery and pulse compression to sub-50 fs of a modelocked thin-disk laser in a gas-filled Kagome-type HC-PCF fiber. Opt Express 21:4986–4994
7. Mak KF, Travers JC, Hölzer P, Joly NY, Russell PStJ (2013) Tunable vacuum-UV to visible ultrafast pulse source based on gas-filled kagome-PCF. Opt Express 21:10942–10953
8. Couny F, Benabid F, Roberts PJ, Light PS, Raymer MG (2007) Generation and photonic guidance of multi-octave optical-frequency combs. Science 318:1118–1121

9. Abdolvand A, Walser AM, Ziemienczuk M, Nguyen T, Russell PStJ (2012) Generation of a phase-locked Raman frequency comb in gas-filled hollow-core photonic crystal fiber. Opt Lett 37:4362–4364
10. Wong GKL, Kang MS, Lee HW, Biancalana F, Conti C, Weiss T, Russell PStJ (2012) Excitation of orbital angular momentum resonances in helically twisted photonic crystal fiber. Science 337:446–449
11. Xi XM, Wong GKL, Frosz MH, Babic F, Ahmed G, Jiang X, Euser TG, Russell PStJ (2014) Orbital-angular-momentum-preserving helical Bloch modes in twisted photonic crystal fiber. Optica 1:165–169
12. Kang MS, Nazarkin A, Brenn A, Russell PStJ (2009) Tightly trapped acoustic phonons in photonic crystal fibres as highly nonlinear artificial Raman oscillators. Nat Phys 5:276–280
13. Pang M, Jiang X, Wong GKL, Onishchukov G, Joly NY, Ahmed G, Russell PStJ (2014) Picosecond fiber laser mode-locked at 260th harmonic by GHz acoustic resonance in photonic crystal fiber core. Advanced photonics (OSA technical digest), paper NTh4A.5
14. Butsch A, Kang MS, Euser TG, Koehler JR, Rammler S, Keding R, Russell PStJ (2012) Optomechanical nonlinearity in dual-nanoweb structure suspended inside capillary fiber. Phys Rev Lett 109:183904
15. Butsch A, Koehler JR, Noskov RE, Russell PStJ (2014) CW-pumped single-pass frequency comb generation by resonant optomechanical nonlinearity in dual-nanoweb fiber. Optica 1:158–164
16. Jiang X, Joly NY, Finger MA, Wong GKL, Babic F, Saad M, Russell PStJ (2013) Close to three-octave-spanning supercontinuum generated in ZBLAN photonic crystal fiber. Advanced solid-state lasers congress (OSA, Paris), postdeadline paper JTh5A.6
17. Jiang X, Joly NY, Finger MA, Babic F, Wong GKL, Travers JC, Russell PStJ (2015) Deep ultraviolet to mid infrared supercontinuum generated in solid-core ZBLAN photonic crystal fibre. Nat Photonics (to appear)

Chapter 2
Silicon Photonics

Lorenzo Pavesi

Abstract Silicon Photonics is no more an emerging research topic but is an actual technology with commercial products already available on the market. Quantum confinement of carriers or spatial localization of photons allows enhancing dramatically and widening the scope and potential of silicon photonics.

The tremendous evolution of internet has urged the microelectronic and telecommunication industries to look at technologies to face the bandwidth, speed and power consumptions that internet requires. One technology which is emerging as a comprehensive solution to this is integrated photonics where complexity, reconfigurability and intelligence are thrusts. An ideal platform for such integration is to use the same material that has permitted the success of microelectronics in the last century, i.e., silicon. Silicon photonics is a technology where photonics devices are fabricated by using standard silicon technologies [1–4]. In this way it is possible to leverage on the capabilities of the silicon industry to mass manufacture cheap photonics devices. A key ingredient to achieve this goal is to exploit the high level of integration that parallel and planar processing allows. Still, integration in photonics is limited to few hundreds of different components in a single chip due to the complexity of integrated photonic circuits and to the needs of hybrid integration technologies to face all the different functionalities needed. Nowadays, all the different functions have been generated and integrated in silicon photonics by using silicon based materials or, when they are not possible, hybrid integration [4]. An example of this is the signal source which in silicon photonics is provided by a hybrid III–V on silicon laser.

L. Pavesi (✉)
Nanoscience Laboratory, Department of Physics,
University of Trento, Trento, Italy
e-mail: pavesi@science.unitn.it

V. Lakshminarayanan and I. Bhattacharya (eds.), *Advances in Optical Science and Engineering*, Springer Proceedings in Physics 166,
DOI 10.1007/978-81-322-2367-2_2

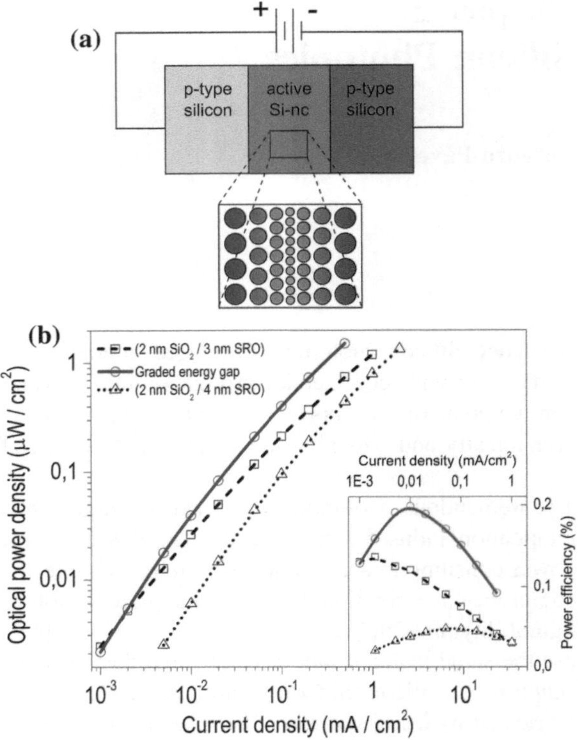

Fig. 2.1 **a** Scheme of graded band-gap LED. **b** Electro-optical characterization of LED with different Si-nc sizes (*black curves*) and multilayer structure (*red curve*). The *inset* shows the power efficiency

Alternatively one can try to change the properties of silicon: the use of nano-silicon is one of such an attempt [5]. The use of nano-Si in silicon photonics (waveguides, modulators, switches, sources and detectors) is reviewed and discussed in [6, 7]. Recent advances of nano-Si devices such as bio-imagers, optical resonators (linear, rings, and disks) are treated in [5, 8]. High efficiency light emitting diodes (Fig. 2.1, [9]) can be used for interchip bidirectional optical transceiver [10]. Silicon nanocrystals improve silicon solar cells [11]. In addition, third order nonlinearities in silicon nanocrystals allow fast all-optical switches (Fig. 2.2, [12]). On the other hand, confinement of photons to small microresonators allows tuning the photon properties [13]. Here also novel effects are found. Ultra high bandwidth robust optical switches for UDWDM [14], active suspended microdisk bistable devices [15], nonlinear optical generations [16] are only few applications where nanophotonics can be appreciated.

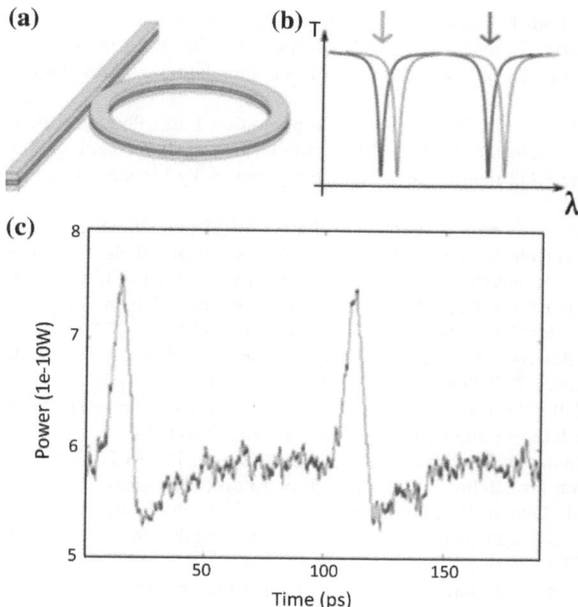

Fig. 2.2 **a** Scheme of the waveguide and the microring. The cyan layer indicates the nanosilicon layer. **b** Switching mechanism: *red arrow* indicates the pump beam, *cyan arrow* the CW signal beam. In the unperturbed state (*blue line*) the transmission of the probe is low, due to the intrinsic ring losses. While pumping (*red line*), a red shift is achieved and the probe beam is temporarily out of resonance and completely transmitted along the WG. **c** Experimental data of the optical switching achieved. FWHM of the pulses is about 20 ps

Acknowledgments This work was supported by ITPAR and the Italian Foreign Ministry.

References

1. Pavesi L, Lockwood D (eds) (2004) Silicon photonics. Topics in applied physics, vol 94. Springer, Berlin
2. Pavesi L, Guillot G (eds) (2006) Optical interconnects: the silicon approach. Springer series in optical sciences, vol 119. Springer, Berlin
3. Lockwood DJ, Pavesi L (eds) (2011) Silicon photonics II: components and integration series: topics in applied physics. vol 119, 1st edn. Springer, Berlin
4. Vivien L, Pavesi L (eds) (2013) Handbook of silicon photonics. CRC Press, Taylor and Francis Group, Florida, 849 pp. ISBN: 978-1-4398-3610-1
5. Pavesi L, Turan R (eds) (2010) Silicon nanocrystals; fundamentals, synthesis and applications. Wiley-VCH Verlag GmbH, Berlin. ISBN: 978-3-527-32160-5 (hardcover 688 pages Mar 2010)
6. Stefano O, Lorenzo P (2003) Priolo Francesco light emitting silicon for microphotonics. Springer Tracts Mod Phys 194:206

7. Gaburro Z, Bettotti P, Daldosso N, Ghulinyan M, Navarro-Urrios D, Melchiorri M, Riboli F, Saiani M, Sbrana F, Pavesi L (2006) Nanostructured silicon for photonics—from materials to devices, volumes 27–28 of materials science foundations. Trans Tech Publications Ltd., Zuerich
8. Daldosso N, Pavesi L (2009) NanoSilicon photonics. Laser Photonics Rev 3:508–534
9. Marconi A, Anopchenko A, Wang M, Pucker G, Bellutti P, Pavesi L (2009) High power efficiency in Si-nc/SiO2 multilayer light emitting devices by bipolar direct tunneling. Appl Phys Lett 94:221110
10. Marconi A, Anopchenko A, Pucker G, Pavesi L (2011) Silicon nanocrystal light emitting device as a bidirectional optical transceiver. Semicond Sci Technol 26:095019
11. Sgrignuoli F, Ingenhoven P, Pucker G, Mihailetchi VD, Froner E, Jestin Y, Moser E, Sánchez G, Pavesi L (2015) Purcell effect and luminescent downshifting in silicon nanocrystals coated back-contact solar Cells. Sol Energy Mater Sol Cells 132:267–274
12. Martínez A, Blasco J, Sanchis P, Galán JV, García J, Martínez JM, Jordana E, Gau-tier P, Lebour Y, Guider R, Pellegrino P, Hernández S, Daldosso N, Garrido B, Fedeli JM, Pavesi L, Martí J (2010) Ultrafast all-optical switching in a silicon-nanocrystal based silicon slot waveguide at telecom wavelengths. Nano Lett 10:1506–1511
13. Pitanti A, Ghulinyan M, Navarro-Urrios D, Pucker G, Pavesi L (2010) Probing the spontaneous emission probability in Si-nc based micro disk resonators. Phys Rev Lett 104:103901
14. Mancinelli M, Guider R, Bettotti P, Masi M, Vanacharla R, Pavesi L (2011) Coupled-resonator-induced transparency concept for wavelength routing applications. Opt Express 19:12227–12240
15. Ramiro-Manzano F, Prtljaga N, Pavesi L, Pucker G, Ghulinyan M (2013) Optical bistability with Si nanocrystals in a whispering gallery mode resonator. Opt Lett 38:3562
16. Cazzanelli M, Bianco F, Borga E, Pucker G, Ghulinyan M, Degoli E, Luppi E, Véniard V, Ossicini S, Modotto D, Wabnitz S, Pierobon R, Pavesi L (2012) Second harmonic generation in silicon waveguides strained by silicon nitride. Nat Mater 11:148–154

Part II
Plenary Talks

Chapter 3
Specialty Optical Fibers for Mid-IR Photonics

Bishnu P. Pal, A. Barh, S. Ghosh, R.K. Varshney, J. Sanghera, L.B. Shaw and I.D. Aggarwal

Abstract We present chalcogenide (Ch) glass-based all-fiber application-specific microstructured optical fibers (MOFs)-based devices like all-fiber efficient discrete and broad-band light sources as well as fibers for distortion free high power light guidance in the mid-IR spectral regime. For new source generation, nonlinearities and dispersion of index guided Ch-MOFs were suitably tailored through structural optimization, while for high power guidance, an ultra large mode area fiber design and a second design for distortion free parabolic pulse generation will be described.

3.1 Introduction

In recent years *eye safe* mid-infrared (mid-IR) spectral range (2–10 μm) has assumed great important owing to a host of potential applications ranging from non-destructive medical diagnostics to strategic applications in military/defense. This wavelength (λ) range is also aptly known as *molecular fingerprint* regime of various organic/inorganic molecules [1, 2].

B.P. Pal (✉)
School of Natural Science, Mahindra Ecole Centrale,
Bahadurpally, Hyderabad 500043, India
e-mail: bishnu.pal@mechyd.edu.in

A. Barh · R.K. Varshney
Department of Physics, Indian Institute of Technology Delhi,
Hauz Khas, New Delhi 110016, India

S. Ghosh
INSPIRE Faculty, Institute of Radio Physics and Electronics,
University of Calcutta, Kolkata 700009, India

J. Sanghera · L.B. Shaw
Naval Research Laboratory, Code 5620, Washington, DC 20375, USA

I.D. Aggarwal
Physics Department, University of North Carolina, Charlotte, NC 28223, USA

V. Lakshminarayanan and I. Bhattacharya (eds.), *Advances in Optical Science and Engineering*, Springer Proceedings in Physics 166,
DOI 10.1007/978-81-322-2367-2_3

Chalcogenide glass (S-Se-Te)-based microstructured optical fibers (MOFs) are potentially a versatile platform for mid-IR applications due to their high optical transparency, extraordinary linear and nonlinear (NL) properties. Studies on MOFs reveal that, their zero dispersion wavelength (λ_{ZD}) as well as dispersion slope could be tailored over a large wavelength (λ) range. This fascinating dispersion feature along with high Kerr nonlinearity of Ch-glasses ($n_2 \sim 100$ times higher than silica) make them imminently suitable for generating new wavelengths via NL effects, such as, four wave mixing (FWM), supercontinuum generation etc. In this talk, we would present some of our recent work on FWM-based mid-IR source generation within 3–6.5 μm λ-range [3–5]. At the other extreme, NL effects could be detrimental when guided wave high optical power delivery becomes necessary or propagation of high optical power is essential in high-power fiber lasers. These problems can be overcome either by distributing the powers across a large mode area (LMA) fiber or by reshaping a propagating high power Gaussian pulse to a parabolic pulse (PP), which can overcome NL-induced distortions like pulse breaking in a fiber laser. We theoretically demonstrate applicability of Ch-based MOF platform for designing ultra-LMA MOF for mid-IR [6] and an up-tapered MOF for generating PP at 2 μm [7].

3.2 Modelling and Key Results

3.2.1 Ch-MOF Based Mid-IR Source Generation

NL-FWM in a tailored Ch-MOF is exploited for generating an all-fiber mid-IR source [3] through choice of appropriate pump power level to suppress other NL effects. In highly NL single-mode (SM)-fibers, FWM-induced frequency shift depends on both the magnitude and sign of its GVD parameters, e.g. a positive β_4 leads to broad-band and flat gain, while a negative β_4 reduces the flatness and bandwidth (BW) of the FWM output [4]. For high efficiency, perfect phase matching, large mode overlap, and low fiber loss are essential. Further, a small idler as seed enhances its efficiency [3].

Arsenic sulphide (As_2S_3)-based MOF with hexagonally arranged air holes is designed and proposed (Fig. 3.1a) to attain a discrete mid-IR source at 4.36 μm by assuming a 2.04 μm Tm^{3+}-doped fiber laser of 5 W power as the pump. Structural optimization led to attainment of signal amplification factor (AF_s) ~20 dB with a BW of 16 nm (Fig. 3.1b). Further, this design can be tuned to get broad-band source, where thermally compatible borosilicate glass is assumed to fill the air holes in a suitable As_2S_3-MOF. With a pump (~2.8 μm), broad-band mid-IR 3–4.2 μm source is realizable (Fig. 3.1c). We have also designed arsenic selenide (As_2Se_3) and PES based-MOF structures which can generate both narrow (~6.46 μm) and broad-band (5–6.3 μm) sources with a CO laser as the pump (Fig. 3.1d, e) [5].

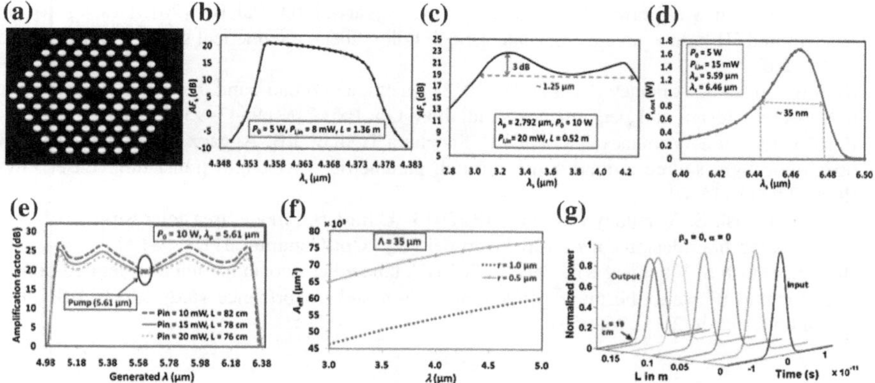

Fig. 3.1 **a** Typical cross-section of holey Ch-MOF. **b** AF_s of discrete source around 4.36 μm. **c** AF_s of broad-band source ranging from 3 to 4.2 μm. **d** AF_s of discrete source around 6.46 μm. **e** AF_s of broad-band source ranging from 5 to 6.3 μm. **f** Variation of effective area (A_{eff}) of MCOF. **g** Transformation of a Gaussian pulse to PP in a ~20 cm length of up-tapered MOF

3.2.2 High Power Mid-IR Light Guidance

Two thermally compatible, low index contrast Ch-glass based novel design of a microstructured core fiber (MCOF) is proposed as LMA fiber design [6], where ultra-LMA ≥ 75,000 μm^2 could be realized over 3–5 μm with effective SM operation (Fig. 3.1f). On the other hand, an up-tapered MOF design is proposed to achieve transformation of a high power Gaussian pulse to a PP at a wavelength of 2 μm within ~20 cm length of the tapered MOF (Fig. 3.1g) [7].

3.3 Conclusions and Acknowledgments

This talk will cover our recent theoretical work on Ch-MOF based new light source generation and high power light guidance at mid-IR spectral range.

This work relates to Department of the Navy Grant N62909-10-1-7141 issued by Office of Naval Research Global. The United States Government has royalty-free license throughout the world in all copyrightable material contained herein.

References

1. Jackson SD (2012) Towards high-power mid-infrared emission from a fibre laser. Nat Photon 6 (7):423–431
2. Serebryakov VA, Boĭko EV, Petrishchev NN, Yan AV (2010) Medical applications of mid-IR lasers. Problems and prospects. J Opt Technol 77(1):6–17

3. Barh A, Ghosh S, Agrawal GP, Varshney RK, Aggarwal ID, Pal BP (2013) Design of an efficient mid-IR light source using chalcogenide holey fibers: a numerical study. J Opt (IOP) 15 (3):035205
4. Barh A, Ghosh S, Varshney RK, Pal BP (2013) An efficient broad-band mid-wave IR fiber optic light source: design and performance simulation. Opt Exp 21(8):9547–9555
5. Barh A, Ghosh S, Varshney RK, Pal BP, Sanghera J, Shaw LB, Aggarwal ID (2014) Mid-IR fiber optic light source around 6 μm through parametric wavelength translation. Laser Phys (IOP) 24(11):115401
6. Barh A, Ghosh S, Varshney RK, Pal BP (2013) Ultra-large mode area microstructured core chalcogenide fiber design for mid-IR beam delivery. Opt Commun 311:129–133
7. Barh A, Ghosh S, Varshney RK, Pal BP (2014) A tapered chalcogenide microstructured optical fiber for mid-IR parabolic pulse generation: design and performance study. IEEE J Sel Top Quantum Electron 20(5):7500507

Chapter 4
Einstein's Photoemission, Magnetic Quantization and Heavily Doped III–V Quantum Well Superlattices with Graded Interfaces

S. Chakrabarti, L.S. Singh and K.P. Ghatak

Abstract In this paper an attempt is made to investigate, the Einstein's Photoemission (EP) from III–V heavily doped quantum well superlattices with graded interfaces in the presence of strong magnetic field on the basis of **newly formulated carrier energy spectrum** within the frame work of $\vec{k} \cdot \vec{p}$ formalism. Under magnetic quantization, the EP oscillates with inverse quantizing magnetic field. The EP increases with increasing photo energy in a step-like manner and the numerical values of EP with all the physical variables are totally band structure dependent for all the cases. The most striking features are that the presence of poles in the dispersion relation of the materials in the absence of band tails create the complex energy spectra in the corresponding HD constituent materials of such quantum confined superlattices and effective electron mass exists within the band gap which is impossible without the concept of band tails.

S. Chakrabarti
Department of Electronics and Communication Engineering,
Institute of Engineering and Management, Salt Lake, Kolkata, India

L.S. Singh
Department of Electronics and Communication Engineering,
National Institute of Technology, Manipur, Imphal 795001, India

K.P. Ghatak (✉)
Department of Basic Science and Humanities,
Institute of Engineering and Management, Kolkata 700091, West Bengal, India
e-mail: kamakhyaghatakcu@gmail.com

V. Lakshminarayanan and I. Bhattacharya (eds.), *Advances in Optical Science and Engineering*, Springer Proceedings in Physics 166,
DOI 10.1007/978-81-322-2367-2_4

4.1 Introduction

It is well known that the important concept of Einstein's Photoemission (EP), which is a physical phenomenon, occupies a singular position in the whole arena of modern opto-electronics in general and whose importance has already been established since the inception of Einstein's photoelectric effect which in recent years finds extensive applications in modern nano-photonics, characterization and investigation of condensed matter systems, photoemission spectroscopy and related aspects in connection with the investigations of the optical properties of nano-structures [1]. Interest in low dimensional silicon nanostructures also grew up and gained momentum, after the discovery of room temperature photoluminescence and electroluminescence of silicon nano-wires in porous silicon [2]. Work on ultrathin layers of $SiSiO_2$ superlattices resulting into visible light emission at room temperature clearly exhibited low dimensional quantum confinement effect [2] and one of the most popular techniques for analyzing the low dimensional structures is to employ photoemission techniques. Recent observation of room temperature photoluminescence and electro luminescence in porous silicon has stimulated vigorous research activities in silicon nanostructures [2]. The EP depends on the density-of-states (DOS) function which, in turn, is significantly affected by the different carrier energy spectra of different semiconductors having various band structures. In recent years, various energy wave vector dispersion relations of the carriers of different materials have been proposed [3] which have created the interest in studying the EP from HD materials and their quantized counterparts.

It is well known that heavy doping and carrier degeneracy are the keys to unlock the new properties of technologically important semiconductors and they are especially instrumental in dictating the characteristics of Ohomic contacts and Schottky contacts respectively. It is an amazing fact that although the heavily doped semiconductors (HDS) have been investigated in the literature but the study of the carrier transport in such materials through proper formulation of the Boltzmann transport equation which needs in turn, the corresponding heavily doped carrier energy spectra *is still one of the open research problems.*

With the advent of MBE, MOCVD, FLL and other experimental techniques and altogether new and technologically important artificial structure has been experimentally realized known as semiconductor superlattices (SLs) by growing two similar but different semiconductors in alternate layers with finite thicknesses. The compounds forming the alternate layers have the same kind of dispersion relation but having different band gaps. The concept of SL was developed for the first time by Keldysh [4] and was successfully fabricated by Esaki and Tsu finds extensive applications in modern nano-photonics.

It is worth remarking that the said SLs have been proposed with the fundamental assumptions that the interfaces between the layers having zero thickness devoid of any related interface effects. In reality, the potential distribution in superlattices, instead of 90° rotation will practically present a linear relationship and consequently a finite intermediate potential region will exist between the carriers in the wells and that in the barrier [5, 6]. The effect of finite thickness of the interfaces on the carrier energy spectrum is very important, since; the dispersion law of the carriers ultimately control the electron transport in such hetero-structures.

In this paper, the magneto EP from III–V, heavily doped superlattices with graded interfaces has been studied in theoretical background of Sect. 4.2. Section 4.3 contains the result and discussions pertinent to this paper.

4.2 Theoretical Background

The electron energy spectrum of the heavily doped constituent materials of III–V SLs whose energy band structures in the absence of band tailing effects are defined by three band model of Kane can be written as

$$\frac{\hbar^2 k^2}{2m_{cj}^*} = T_{1j}(E, \Delta_j, E_{gj}, \eta_{gj}) + iT_{2j}(E, \Delta_j, E_{gj}, \eta_{gj}) \tag{4.1}$$

where, $T_{ij}(E, \Delta_j, E_{gj}, \eta_{gj}) = \left(\frac{2}{1 + \mathrm{Erf}(E/\eta_{gj})}\right)\left[\frac{\alpha_j b_j}{c_j}\theta_0(E, \eta_{gj}) + \left[\frac{\alpha_j c_j + b_j c_j - \alpha_j b_j}{c_j^2}\right]\right.$

$$\gamma_0(E, \eta_{gj}) + \frac{1}{c_j}\left(1 - \frac{\alpha_j}{c_j}\right)\left(1 - \frac{b_j}{c_j}\right)\frac{1}{2}\left[1 + \mathrm{Erf}\left(\frac{E}{\eta_{gj}}\right)\right]$$
$$\left. - \frac{1}{c_j}\left(1 - \frac{\alpha_j}{c_j}\right)\left(1 - \frac{b_j}{c_j}\right)\frac{2}{c_j \eta_{gj}\sqrt{\pi}}\exp(-u_j^2)\left[\sum_{p=1}^{\infty}\frac{\exp(-p^2/4)}{p}\sinh(pu_j)\right]\right],$$

$b_j \equiv (E_{gi} + \Delta_j)^{-1}$, $c_j \equiv (E_{gi} + \frac{2}{3}\Delta_j)^{-1}$, $u_j \equiv \frac{1 + c_j E}{c_j \eta_{gi}}$ and

$T_{2j}(E, \Delta_j, E_{gi}, \eta_{gi}) \equiv \left(\frac{2}{1 + \mathrm{Erf}(E/\eta_{gi})}\right)\frac{1}{c_j}\left(1 - \frac{a_j}{c_j}\right)\left(1 - \frac{b_j}{c_j}\right)\frac{\sqrt{\pi}}{c_j \eta_{gi}}\exp\left(-u_j^2\right),$

η_{gi} is the scattering potential, m_{cj}^* is the effective electron mass at the band edge and $\mathrm{Erf}(E/\eta_{gj})$ is the error function. It is important to note that the presence of poles in the dispersion relation of the materials in the absence of band tails create the complex energy spectra in the corresponding HD constituent materials of such quantum confined superlattices and effective electron mass exists within the band gap which is impossible without the concept of band tails.

The electron energy spectrum in heavily doped quantum wells III–V SLs with graded interfaces can be written as

$$k_z^2 = G_8 + iH_8 \tag{4.2}$$

where the notations are defined in [3].

In the presence of strong magnetic field B along z direction, (4.2) assumes the form

$$k_z^2 = G_{8E,n} + iH_{8E,n} \tag{4.3}$$

where the notations are defined in [3].

Therefore the magnetic dispersion law in QW HD III–V SLs with graded interfaces can be written as

$$\left(\frac{n_z\pi}{d_z}\right)^2 = G_{8E_{41,n}} + iH_{8E_{41,n}} \tag{4.4}$$

where $E_{41,n}$ is the totally quantized energy in this case, n_z is the size quantum number and d_z is the nano thickness along z-direction respectively.

The electron concentration in this case can be expressed as

$$n_0(\pi\hbar/eBg_v) = \text{R.P of} \sum_{n=0}^{n_{\max}} \sum_{n_z=1}^{n_{z\max}} F_{-1}(\eta_{41}) \tag{4.5}$$

where g_v is valley degeneracy, $\eta_{41} = (k_BT)^{-1}(E_{\text{FSL}} - E_{41,n})$, E_{FSL} is the Fermi energy in this case and $F_j(\eta)$ is the well-known Fermi-Dirac integral of order j.

The EP in this case (J_{SL}) assumes the form

$$J_{\text{SL}}(hd_z/\alpha_0 e^2 Bg_v) = \text{R.P of} \sum_{n=0}^{n_{\max}} \sum_{n_{z\min}}^{n_{z\max}} F_{-1}(\eta_{41}) v_{z1}(E_{nz\text{SL}1}) \tag{4.6a}$$

where

$$v_{z1}(E_{nz\text{SL}1}) = 2n_z\pi \Big/ \left[\hbar d_z (G_8' + iH')\big|_{k_x=0,\, k_y=0,\, k_z=\frac{n_z\pi}{d_z} \text{ and } E=E_{nz\text{SL}1}}\right] \tag{4.6b}$$

$E_{nz\text{SL}1}$ is the quantized energy along z direction and is obtained by substituting $k_x = 0$, $k_y = 0$, $k_z = \frac{n_z\pi}{d_z}$ and $E = E_{nz\text{SL}1}$ in (4.4) and the primes denote the differentiation with respect to energy, α_0 is the probability of photon absorption.

4.3 Results and Discussion

Using the appropriate equations in Figs. 4.1, 4.2, 4.3 and 4.4, the normalized EP from QW HD III–V SLs (taking $GaAs/Ga_{1-x}Al_xAs$ and $In_xGa_{1-x}As/InP$ QW HD SLs) with graded interfaces under the magnetic quantization has been plotted as functions of the reciprocal quantizing magnetic field, normalized electron statistics, nano thickness and the normalized incident photon energy respectively. It appears from Fig. 4.1 that the EP in this case exhibits oscillatory dependence with inverse quantizing magnetic field due to Shubnikov-de Haas effect. Figure 4.2 exhibits the fact that the EP increases with increasing carrier degeneracy in an oscillatory way and the nature of oscillations is different as compared with Fig. 4.1. From Fig. 4.3, it can be inferred that the EP oscillates with film thickness and for certain values of

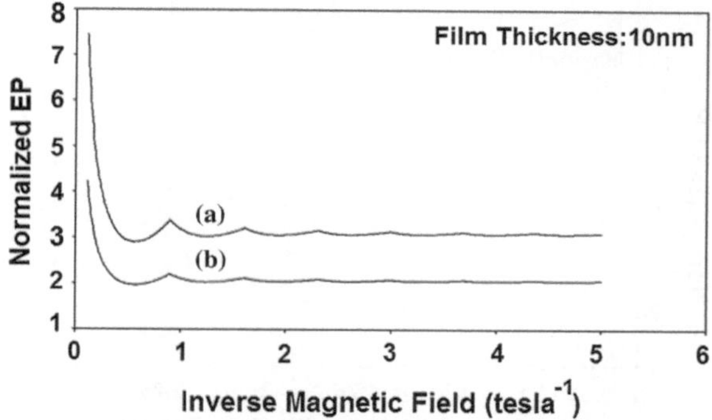

Fig. 4.1 Plot of the normalized EP from **a** $GaAs/Ga_{1-x}Al_xAs$ and **b** $In_xGa_{1-x}As/InP$ QWHD superlattices with graded interfaces as a function of inverse quantizing magnetic field

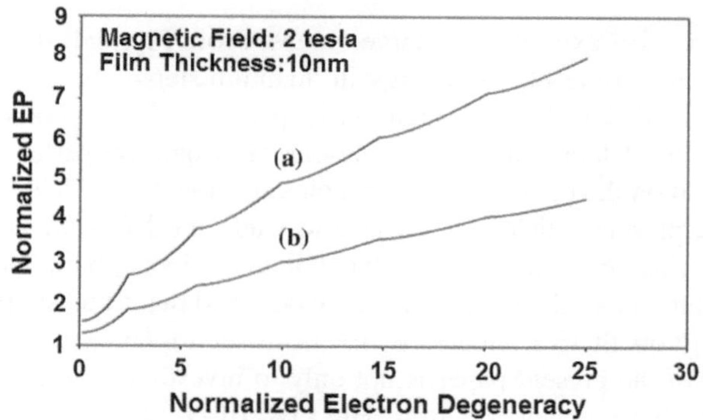

Fig. 4.2 Plot of the normalized EP from **a** $GaAs/Ga_{1-x}Al_xAs$ and **b** $In_xGa_{1-x}As/InP$ QWHD superlattices with graded interfaces under quantizing magnetic field as a function of normalized electron degeneracy

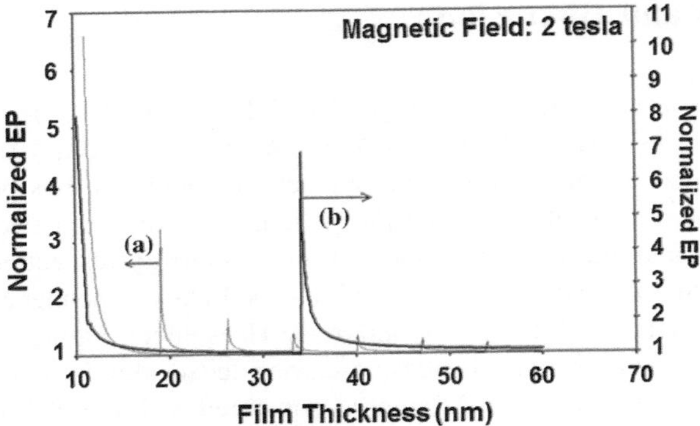

Fig. 4.3 Plot of the normalized EP from **a** $GaAs/Ga_{1-x}Al_xAs$ and **b** $In_xGa_{1-x}As/InP$ QWHD superlattices with graded interfaces under quantizing magnetic field as a function of film thickness

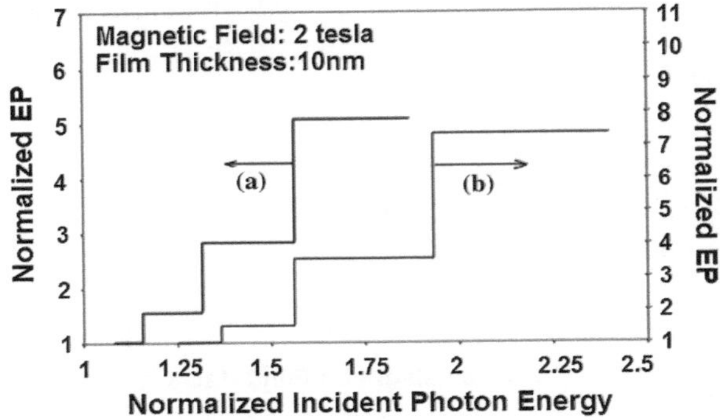

Fig. 4.4 Plot of the normalized EP from **a** $GaAs/Ga_{1-x}Al_xAs$ and **b** $In_xGa_{1-x}As/InP$ QWHD superlattices with graded interfaces under quantizing magnetic field as a function of normalized incident photon energy

film thickness the EP exhibits very large values. From Fig. 4.4, it appears that EP increases with increasing photon energy in quantum steps.

It may be noted that, the investigation of parallel and perpendicular transport properties and the detailed analysis of various electronic properties of HD superlattices are based on the band structure in their constituent materials. The theoretical analysis of the present article can be used to study the EP from various types of quantized superlattices and the constituent heavily-doped bulk materials in the absence of quantization phenomenon. It may be noted that t this simplified analysis shows the basic qualitative features of EP from such quantized hetero-structures. The basic aim of the present paper is not only to investigate the effect of quantum confinement on the magneto EP from HD QW III–V SLs but also to derive the temperature dependent carrier statistics since all the quantum processes in such hetero-structures are based on the electron density through Boltzmann transport equation.

Acknowledgments We are grateful to N. Debbarma, S. Debbarma and M. Debbarma for helpful discussion.

References

1. Canham LT (1990) Silicon quantum wire fabrication by electrochemical and chemical dissolution of wafers. Appl Phys Lett 57:1046–1048
2. Chakrabarti S, Sen SK, Chakraborty S, Singh LS, Ghatak KP (2014) Einstein relation in heavily doped quantum wires superlattices with graded interfaces: simplified theory and suggestion for experimental determination. Adv Sci Eng Med 6(9):1042–1057
3. Ghatak KP (2015) Dispersion relations in heavily doped nanostructures. Springer tracts in modern physics. Springer (in press)
4. Keldysh LV (1963) Effect of ultrasonics on the electron spectrum of crystals. Sov Phys Semicond 4:1658
5. Bhattacharya S, Ghatak KP (2013) Effective electron mass in low dimensional semiconductors. Springer series in materials sciences. Springer, Berlin
6. Jiang HX, Lin JY (1987) Band structure of superlattice with graded interfaces. J Appl Phys 61:624–628

Chapter 5
Optical Cryptography and Watermarking Using Some Fractional Canonical Transforms, and Structured Masks

A.K. Yadav, Sunanda Vashisth, Hukum Singh and Kehar Singh

Abstract We have carried out a few studies on image encryption and watermarking in various transform domains, for amplitude and phase images. We have used structured phase masks (SPM) based on devil's vortex Fresnel lens (DVFL) to enhance security by increasing the key-space. The proposed schemes have been validated by computer simulations and their efficacy is evaluated by computing mean-squared-error (MSE) between the original image and the decrypted image. They have been examined for their sensitivity to various encryption parameters, and robustness to noise and occlusion attacks.

5.1 Introduction

Due to ever increasing volume of information in recent years, its security has become extremely important. This is evident from a large body of literature on the subject. While the double random phase encoding (DRPE) [1] scheme for image encryption has been widely used, its modification and newer techniques have been proposed to overcome its inherent weaknesses. In the present paper we summarize the results of our investigations using structured masks and some of the fractional canonical transforms.

A.K. Yadav (✉) · S. Vashisth · H. Singh · K. Singh
Department of Applied Sciences, ITM University,
Sector 23A, Gurgaon 122017, Haryana, India
e-mail: akyadav@itmindia.edu

V. Lakshminarayanan and I. Bhattacharya (eds.), *Advances in Optical Science and Engineering*, Springer Proceedings in Physics 166,
DOI 10.1007/978-81-322-2367-2_5

5.2 Image Encryption

5.2.1 *Fully-Phase Image Encryption Using Double Random-Structured Phase Masks in Gyrator Domain [2]*

The phase-images are known to be more secure and resistant to noise as compared to their amplitude counterparts [3, 4]. The use of canonical transforms for encryption is well-documented in the literature. One such canonical transform, the gyrator transform (GT) was introduced in the field of optics and image processing by Rodrigo et al. [5].

5.2.1.1 The Gyrator Transform

The GT, similar to the fractional Fourier transform (FrFT), produces rotation in the twisted position-spatial frequency planes. The GT of a two-dimensional function $f(x, y)$ is written as [5],

$$G(u, v) = G^{\alpha}\{f(x, y)\}(u, v) = \iint f(x, y) K_{\alpha}(x, y; u, v) dxdy \tag{5.1}$$

where α is the transform angle and the kernel is defined as

$$K_{\alpha}(x, y; u, v) = \frac{1}{|\sin\alpha|} \exp\left[2i\pi \frac{(xy + uv)\cos\alpha - xv - yu}{\sin\alpha}\right] \tag{5.2}$$

The GT can be realized by an optimized optical system having plano-convex cylindrical lenses with fixed distance between them. The angle α is changed by proper rotation of these lenses. The GT has advantages such as computational ease and convenience in optical implementation since there is no need of changing the distances as in case of FrFT.

5.2.1.2 DVFL-Based Phase Function

A phase mask based on devil's lens can be described by one-dimensional Cantor function, a particular case of devil's staircase. A triadic Cantor set in the interval [0, 1] can be defined as [6]

$$F_S(x) = \begin{cases} \frac{l}{2^S} & \text{if } p_{S,l} \le x \le q_{S,l} \\ \frac{1}{2^S} \frac{x - q_{S,l}}{p_{S,l+1} - q_{S,l}} + \frac{l}{2^S} & \text{if } q_{S,l} \le x \le p_{S,l+1} \end{cases} \tag{5.3}$$

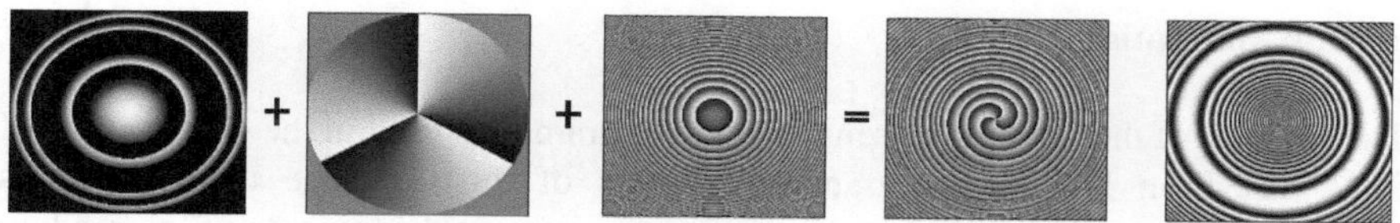

Fig. 5.1 Construction of DVFL: a combination of a devil's lens, a vortex lens and a Fresnel lens. Rightmost picture is of another SPM called toroidal zone plate (TZP)

where $F_S(0) = 0$, $F_S(1) = 1$, S is the order of Cantor function, l defines the number of horizontal sections of the function having a value from 0 to $2^S - 1$. Here q and p are the start and end points of each segment of the Cantor set. For some basic values of S, the values of $q_{S,l}$ and $p_{S,l+1}$ are provided in Mitry et al. [6]. A DVFL can be constructed by combining a devil's lens, a Fresnel lens and a vortex lens (Fig. 5.1) and has a phase function given by

$$\Omega_{S,m,f_o}(\zeta,\theta) = \exp\left[-i2^{S+1}\pi F_S(\zeta)\right]\exp(im\theta)\exp\left[-i\pi\zeta\frac{a^2}{\lambda f_o}\right] \tag{5.4}$$

where $\varsigma = (r/a)^2$ is the normalized quadratic radial coordinate, a is the lens radius, m is a non-zero integer called topological charge, θ is the azimuthal angle, f_O is focal length and λ is the wavelength of light used.

5.2.1.3 The Encryption Scheme (Fig. 5.2)

Mathematically, the encryption process can be written as

$$E(x,y) = G^{-\beta}\left\{\left[G^{\alpha}\left\{\exp[i\pi I(x,y)] \times \mathrm{RPM}_1 \times (\mathrm{DVFL})_{S=2}\right\}\right] \times \mathrm{RPM}_2 \times (\mathrm{DVFL})_{S=3}\right\} \tag{5.5}$$

Here, α and β represent orders of GT. The expression for decryption is given by

$$\exp[i\pi I(x,y)] = G^{-\alpha}[G^{\beta}\{E(x,y)\} \times \mathrm{RPM}_2^* \times (\mathrm{DVFL})^*_{S=3}] \times \mathrm{RPM}_1^* \times (\mathrm{DVFL})^*_{S=2} \tag{5.6}$$

where * indicates conjugate of the phase mask.

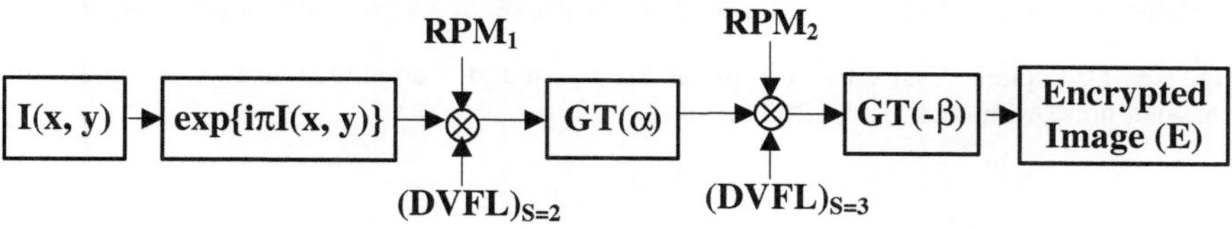

Fig. 5.2 Schematic diagram of the Encryption scheme

5.2.1.4 Simulation Results

The security of the system is strengthened by parameters used in the construction of SPM based on DVFL, and transform orders of the GT. The input image is recovered by using the correct encryption parameters, and conjugate of the random phase masks (RPMs). The use of a DVFL-based SPM enhances security by increasing the key space, and also overcomes the problem of axis-alignment associated with the optical set-up. The scheme is validated for grayscale and binary phase images (Fig. 5.3). We have also investigated the scheme's sensitivity to the encryption parameters such as GT angle, wavelength and focal length (Fig. 5.4), and robustness against occlusion and multiplicative Gaussian noise attacks (Fig. 5.5).

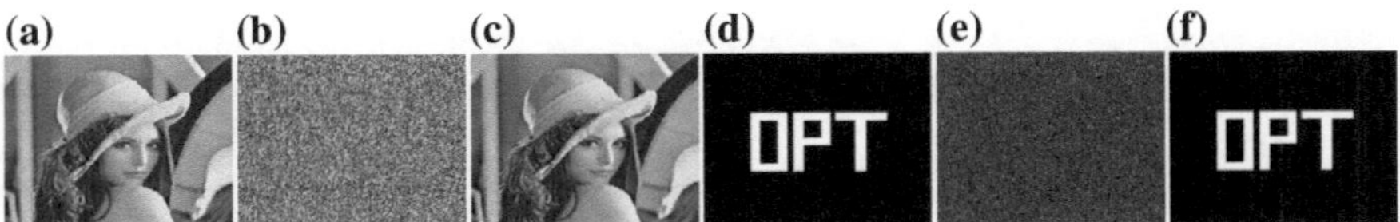

Fig. 5.3 **a** Input grayscale image, **b** encrypted image, **c** decrypted image and **d–f** corresponding results for binary image

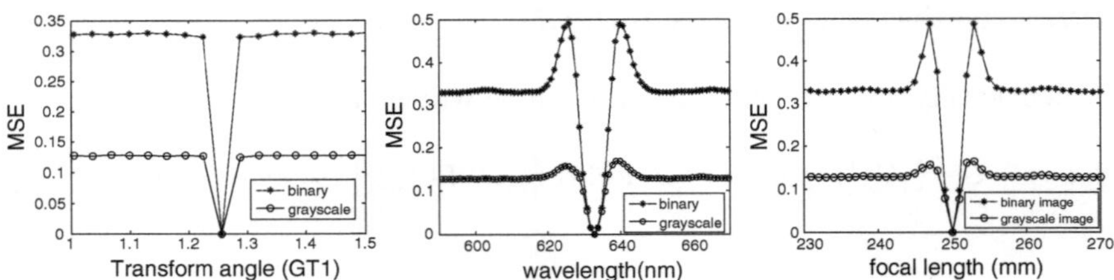

Fig. 5.4 Sensitivity of the proposed scheme against various encryption parameters

Fig. 5.5 Decryption results for **a–c** phase images and **d–f** amplitude images: multiplicative Gaussian noise with coeff. k = 0.2, 0.6, 1.8

5.2.2 Double Phase-Image Encryption Using Gyrator Transforms, and Structured Phase Mask in the Frequency Plane [7]

The phase-images are bonded with RPMs and gyrator transformed. The resulting images are then added and subtracted to give images which are bonded with a DVFL-based SPM. The images are again GT transformed to give the corresponding encrypted images (Fig. 5.6).

5.2.2.1 Simulation Results

The proposed scheme is validated (Fig. 5.7) and its performance is evaluated in terms of MSE between the input image and its decrypted image. In addition, the sensitivity to encryption keys such as SPM parameters, and transform angles of GT orders is investigated (Fig. 5.8). The technique is likely to provide enhanced

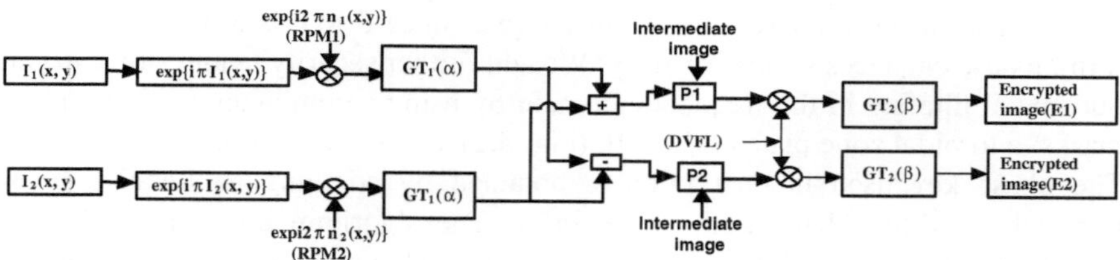

Fig. 5.6 Flowchart of the proposed encryption scheme

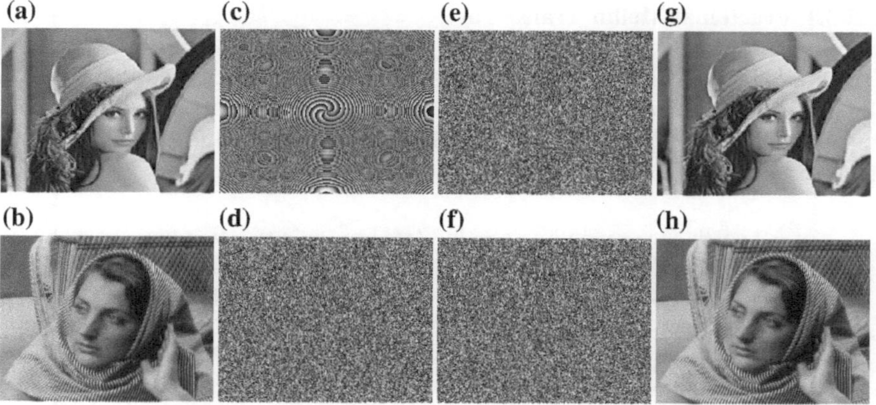

Fig. 5.7 **a–b** The double input phase-images, **c–d** the phase masks, **e–f** the encrypted images and **g–h** the decrypted images

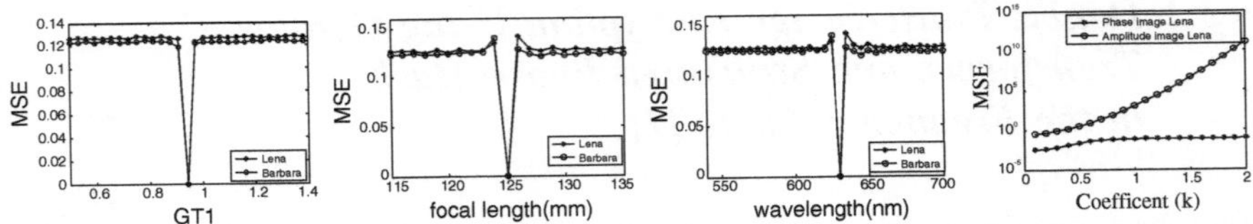

Fig. 5.8 Sensitivity plots showing MSE versus encryption parameters: GT order, focal length, and wavelength; effect of noise

security in view of the increased number of encryption parameters. Robustness of the system against occlusion and noise (Fig. 5.8) attacks has also been investigated.

5.2.3 Image Encryption Using Fractional Mellin Transform (FrMT), Structured Phase Filters, and Phase Retrieval [8]

An encryption scheme for amplitude images is proposed by using two SPMs in the FrMT plane, using a phase retrieval technique. Since FrMT is a non-linear transform, its use enhances system security. We add further security features by carrying out spatial filtering in the frequency domain by using a combination of two phase masks: a toroidal zone plate (TZP) [9], (Fig. 5.1) and a radial Hilbert mask (RHM). The phase key used in decryption is obtained by applying an iterative phase retrieval algorithm [10] based on the FrFT. The algorithm uses amplitude constraints of secret target image and the ciphertext obtained from multiplication of FrMT-transformed arbitrary input image and the masks TZP and RHM.

5.2.3.1 Fractional Mellin Transform

The FrMT is inspired by the FrFT. A two-dimensional FrMT of order (p_1, p_2) is the FrFT of the same order of a function in its log-polar transformation. In log-polar representation, the Cartesian space coordinates are converted to polar coordinates relative to the origin of coordinate system where:

$$x = r\cos\theta; \quad y = r\sin\theta; \quad \rho = \ln\sqrt{x^2+y^2} = \ln r; \quad \theta = \tan^{-1}(y/x).$$

In a Cartesian coordinate system, the two-dimensional FrMT of order (p_1, p_2) of an image $f(x,y)$ is given by

$$M^{p_1,p_2}(u,v) = \int_{-\infty}^{+\infty}\int_{-\infty}^{+\infty} f(x,y) \times x^{-\left(\frac{2iu\pi}{\sin\Phi_1}\right)-1} \times \exp\left[\frac{i\pi(u^2+\ln^2 x)}{\tan\Phi_1}\right] \times y^{-\left(\frac{2iv\pi}{\sin\Phi_2}\right)-1} \times \exp\left[\frac{i\pi(v^2+\ln^2 y)}{\tan\Phi_2}\right] dxdy \tag{5.7}$$

where $\Phi_1 = p_1\pi/2$ and $\Phi_2 = p_2\pi/2$. When the image is transformed to an annular domain by log-polar transformation, its FrMT of order (p_1, p_2) can be written as [11]

$$M^{p_1,p_2}(u,v) = C\int_{-\infty}^{+\infty}\int_{-\infty}^{+\infty} f(\rho,\theta) \times \exp[-2i\pi\left(\frac{u\rho}{\sin\Phi_1}+\frac{v\theta}{\sin\Phi_2}\right) + \frac{i\pi(u^2+\rho^2)}{\tan\Phi_1} + \frac{i\pi(v^2+\theta^2)}{\tan\Phi_2}]d\rho d\theta = F^{p_1,p_2}\{f(\rho,\theta)\} \tag{5.8}$$

where C is a constant.

5.2.3.2 Simulation Results

The encryption scheme has been validated for grayscale images, by numerical simulations (Fig. 5.9). The efficacy has been evaluated by computing MSE between the secret target image and the decrypted image. The sensitivity of the scheme has been studied for parameters such as phase keys, orders of FrMT and FrFT. The scheme is considered secure against most of the known attacks by virtue of its design being non-linear, asymmetric, and having a larger key-space.

5.2.4 Devil's Vortex Phase Structure as Frequency Plane Mask for Image Encryption Using the FrMT [12]

This study also uses the fractional Mellin transform for image encryption. A frequency plane phase mask based on Devil's vortex structure has been used. The phase key for decryption is obtained by an iterative phase retrieval algorithm (Fig. 5.10). The proposed scheme (Fig. 5.11) has been validated for grayscale secret target images, by numerical simulation.

5.2.4.1 Simulation Results

The efficacy of the scheme has been evaluated by computing mean-squared-error (MSE) between the secret target image and the decrypted image. Sensitivity

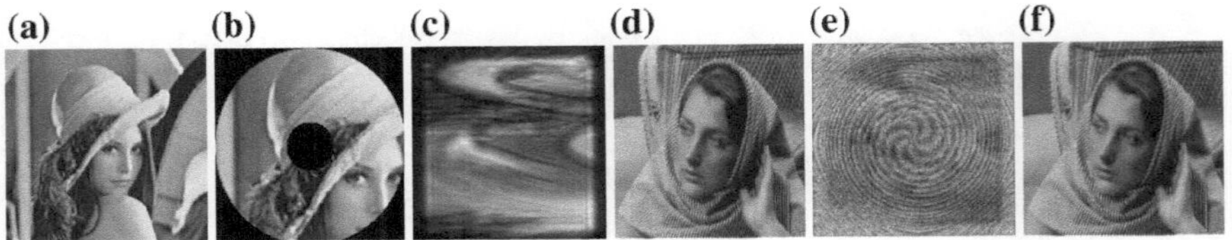

Fig. 5.9 Simulation results, **a** arbitrary input, **b** annular image, **c** FrMT image, **d** secret image, **e** phase key from phase retrieval algorithm, **f** decrypted image

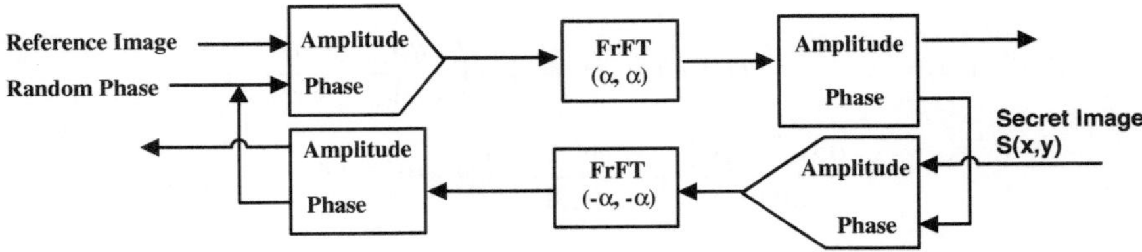

Fig. 5.10 Flowchart showing the phase retrieval algorithm. Adapted from Zhou et al. [11]

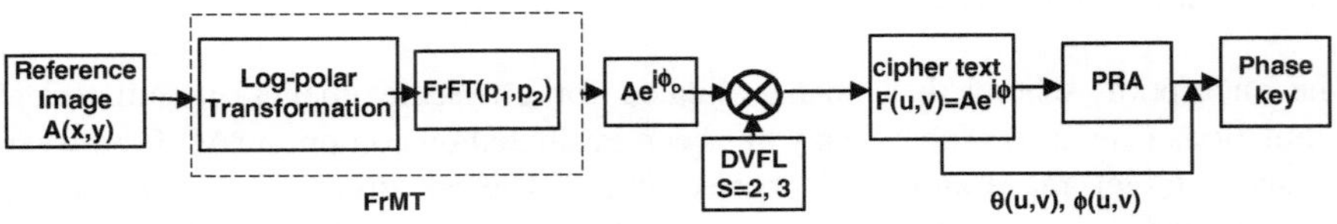

Fig. 5.11 Flowchart showing the proposed scheme

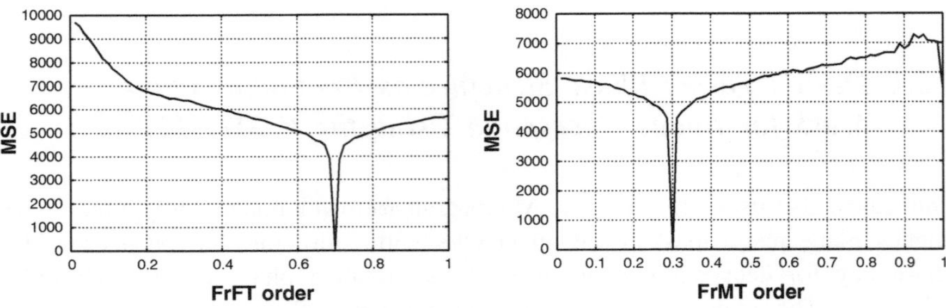

Fig. 5.12 Sensitivity plots showing MSE against orders of FrFT and FrMT

analysis of the decryption process to variations in various encryption parameters has been carried out (Fig. 5.12). The proposed encryption scheme has been seen to exhibit reasonable robustness against occlusion attack.

5.2.5 Double Phase-Image Encryption Using Gyrator Transform and FrFT with SPM in the Frequency Plane Followed by a Gyrator Transform [13]

This work is based on another encryption scheme for double phase-images. One of the phase-images is fractional Fourier transformed whereas the other one is gyrator transformed, after bonding them with RPMs. The two resulting images are then added and subtracted to give intermediate images which are bonded with a structured phase mask (SPM) in the frequency plane. Thereafter, the images are once again transformed using a GT to give the corresponding encrypted images. The performance of the scheme has been evaluated in terms of MSE between the input, and the decrypted image. In addition, the sensitivity to encryption keys such as SPM parameters, transform angles of GT and the FrFT orders has been investigated. The technique is likely to provide enhanced security in view of the increased number of encryption parameters. Robustness of the system for partial occlusion of the encrypted image has also been investigated.

5.3 Watermarking

5.3.1 A Phase-Image Watermarking Scheme in Gyrator Domain Using Devil's Vortex Fresnel Lens as a Phase Mask [14]

We have proposed a watermarking scheme (Fig. 5.13) for phase images, based on the use of devil's vortex Fresnel lens (DVFL) as a phase mask. The DVFL provides much-desired parameter-rich phase masks which contribute to the enhanced security of the scheme in addition to overcoming the problem of axis alignment in the optical setup.

Mathematically, the encryption of an input image $I(x, y)$ can be described as

$$E(\xi,\eta) = G^{-\alpha}[G^{\alpha}\{\exp(i\pi I(x,y)) \times \mathrm{DVFL2}\} \times \mathrm{DVFL3}] \tag{5.9}$$

The watermarked image $W(\xi,\eta)$ is obtained from the encrypted image $E(\xi,\eta)$ and a host image $H(\xi,\eta)$ using the following relation

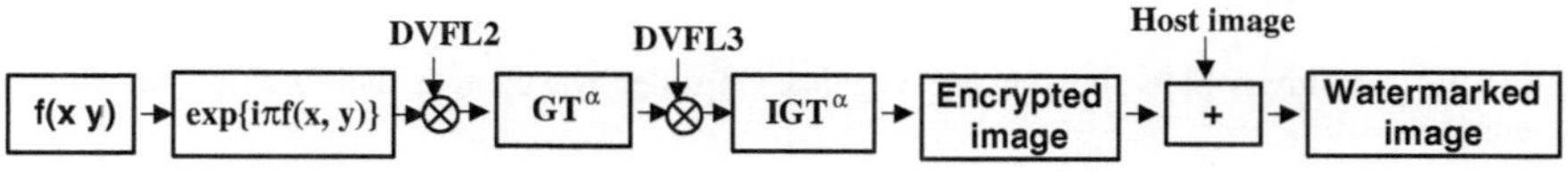

Fig. 5.13 Schematic diagram of the proposed watermarking scheme

$$W(\xi,\eta) = H(\xi,\eta) + \beta E(\xi,\eta) \tag{5.10}$$

where β is an attenuation factor chosen arbitrarily. However, the choice of β has to ensure the invisibility of the hidden image and its robustness against distortions.

5.3.1.1 Simulation Results

We have validated the scheme (Fig. 5.14) and evaluated its performance by studying its sensitivity to various encryption parameters including the order of the GT (Fig. 5.15). The scheme is sensitive to most of them. The robustness of the scheme is established against the occlusion (Fig. 5.15) attack and the multiplicative noise attack.

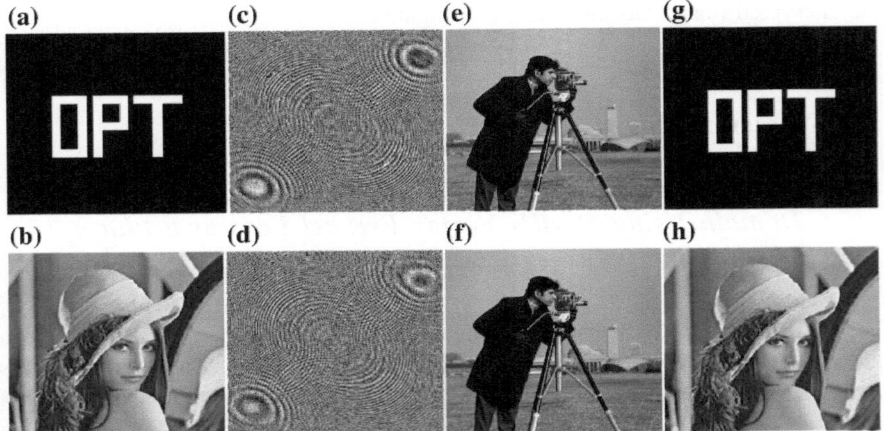

Fig. 5.14 Validation results for binary and grayscale images. **a–b** Input, **c–d** encrypted, **e–f** watermarked, **g–h** decrypted

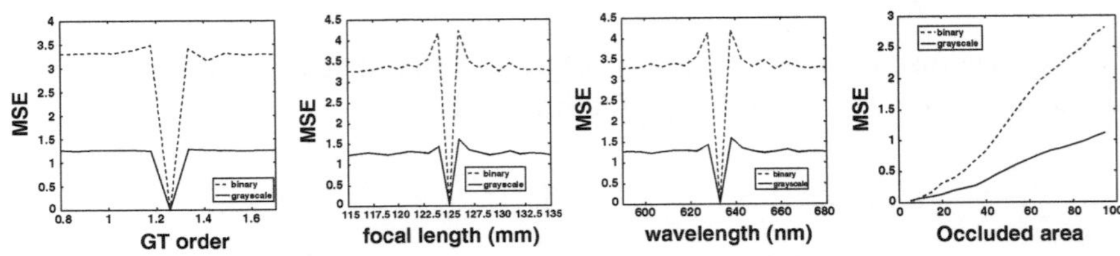

Fig. 5.15 Sensitivity plots showing MSE against GT order, wavelength, and focal length; effect of occlusion

5.3.2 *Watermarking Based on FrFT Using an RPM in the Input Plane and DVFL-Based Mask in the Frequency Plane [15]*

Another watermarking scheme has been proposed in the FrFT domain, using an RPM in the input plane and a DVFL-based phase mask in the frequency plane. The encrypted image resulting from the application of FrFT is attenuated by a factor and combined with a host image to provide a watermarked image. The validity of the proposed scheme has been established by comparing the results of decryption with the input images. The performance of the scheme has been evaluated in terms of MSE. The sensitivity to encryption keys such as DVFL parameters and the FrFT orders has been studied. The proposed technique provides enhanced security.

5.4 Conclusions

We have presented a brief review of a few encryption and watermarking studies for amplitude as well as phase-images. In most of the schemes we explored the use of structured or double random-structured phase masks to overcome the problem of axis-alignment and to enhance the security by increasing the key space. The schemes were validated through numerical simulations. The efficacy of the schemes has been investigated by carrying out sensitivity analysis against the encryption parameters. We have also examined the robustness through occlusion and noise attacks.

References

1. Javidi B (ed) (2005) Optical and digital techniques for information security. Springer, Berlin
2. Singh H, Yadav AK, Vashisth S, Singh K (2014) Fully-phase image encryption using double random-structured phase masks in gyrator domain. Appl Opt 53(28):6472–6481
3. Towghi N, Javidi B, Luo Z (1999) Fully phase encrypted image processor. J Opt Soc Am A: 16:1915–1927
4. Nishchal NK, Joseph J, Singh K (2003) Fully phase encryption using fractional Fourier transform. Opt Eng 42:1583–1588
5. Rodrigo JA, Alieva T, Calvo ML (2007) Gyrator transform: properties and applications. Opt Express 15:2190–2203
6. Mitry M, Doughty DC, Chaloupka JL, Anderson ME (2012) Experimental realization of the devil's vortex Fresnel lens with a programmable spatial light modulator. Appl Opt 51:4103–4108
7. Singh H, Yadav AK, Vashisth S, Singh K (2015) Double phase-image encryption using gyrator transforms, and structured phase mask in the frequency plane. Opt Lasers Eng 67:145–156
8. Vashisth S, Singh H, Yadav AK, Singh K (2014) Image encryption using fractional Mellin transform, structured phase filters, and phase retrieval. Optik 125:5309–5315

9. Barrera JF, Henao R, Torroba R (2005) Optical encryption method using toroidal zone plates. Opt Commun 248:35–40
10. Fienup JR (2013) Phase retrieval algorithms: a personal tour. Appl Opt 52:45–56
11. Zhou N, Liu X, Zhang Y, Yang Y (2013) Image encryption scheme based on fractional Mellin transform and phase retrieval technique in fractional Fourier domain. Opt Laser Technol 47:341–346
12. Vashisth S, Singh H, Yadav AK, Singh K (2014) Devil's vortex phase structure as frequency plane mask for image encryption using the fractional Mellin transform. Int J Opt 2014 (728056)
13. Singh H, Yadav AK, Vashisth S, Singh K (2015) Double phase-image encryption using gyrator, and fractional Fourier transforms with structured phase mask in the frequency plane followed by a gyrator transform. Asian J Phys 24(1):1–16
14. Yadav AK, Vashisth S, Singh H, Singh K (2015) A phase-image watermarking scheme in gyrator domain using devil's vortex Fresnel lens as a phase mask. Opt Commun 344:172–180
15. Singh H, Yadav AK, Vashisth S, Singh K (2014) A cryptosystem for watermarking based on fractional Fourier transform using a random phase mask in the input plane and structured phase mask in the frequency plane. Asian J Phys 23(4):597–612

Chapter 6
Nonlinear Fibre Optics: Application to Supercontinuum Generation

Amine Ben Salem, Rim Cherif and Mourad Zghal

Abstract Applications exploiting the mode confinement properties in specialty fibers take advantage of the extremely high intensity achieved in the minimum waist region yielding to high nonlinearities and supercontinuum (SC) generation. In fact, the generation of "white light" coined a supercontinuum, presents the result of the collection of nonlinear optical effects yielding remarkable spectral broadening of the original pump signal. By properly selecting the geometry and the material composition, the generation of broadband coherent SC from blue to mid-IR wavelengths regions is demonstrated in different highly nonlinear waveguides with low input pulse energy and in short fiber lengths.

6.1 Introduction

Driven by the potential applications in telecommunications including high bit-rate transmission, multi-wavelength, and wavelength division multiplexing (WDM) optical sources the development of low loss silica fibers led not only to a great revolution in the field of optical fiber communications but also to the advent of a new field of study known as nonlinear fiber optics. In fact, intensive investigations have been devoted to developing highly nonlinear optical waveguides which have great interest for compact, low power, and all optical nonlinear devices. During the last decade, the field of nonlinear fiber optics has not stopped growing due to the exploration of new highly nonlinear optical waveguides in addition to photonic crystal fibers (PCF). In fact, by increasing the strength of waveguiding when using a large contrast difference in the refractive index between the fiber core and the cladding materials, major and important effects on the nonlinear waveguide response have been observed. As a result of the reduction of fiber core size in order to maintain

A. Ben Salem · R. Cherif · M. Zghal (✉)
University of Carthage, Engineering School of Communication of Tunis (Sup'Com), GreS'Com Laboratory, GhazalaTechnopark, 2083 Ariana, Tunisia
e-mail: mourad.zghal@supcom.tn

V. Lakshminarayanan and I. Bhattacharya (eds.), *Advances in Optical Science and Engineering*, Springer Proceedings in Physics 166,
DOI 10.1007/978-81-322-2367-2_6

single-mode guidance, the nonlinearities are drastically enhanced due to the huge decrease in the mode effective area. Thus, a new family of waveguides is introduced and coined "photonic nanowires". This designation comes from the fact that their core size diameters are in the hundreds of nanometers such that they become sub-wavelength in scale. Moreover, as inherent features of nonlinear optics, the generation of new frequency components and the spectral broadening become very attractive and an interesting property in optical fibers and have been the subject of intensive studies since the early 1960s [1]. In fact, the particular process of spectral broadening known as supercontinuum (SC) generation occurs when narrowband incident pulses undergo extreme nonlinear broadening yielding a broadband (known as "white light") spectrally continuous spectrum [1–3].

In addition, by changing the material composition of the nanowire, new glasses can be exploited due to their nonlinear indices (n_2) higher than that of silica glass by more than one to three orders of magnitude. Among these glasses, chalcogenide and tellurite present excellent candidates and are very promising for highly nonlinear applications [4, 5]. Therefore, these materials offer a range of useful properties not possessed by silica, including good infrared transmittance.

In this paper, we recall the theoretical modeling of supercontinuum then we report on the generation of visible to mid-IR SC obtained in silica and non-silica fibers. More than one octave-spanning coherent single mode supercontinua extending are generated in different structures from visible to mid-IR wavelengths regions. The introduction of highly nonlinear specialty fibers with anomalous group velocity dispersion (GVD) has enormously contributed to the study of nonlinear optics, broadband SC generation, and soliton self-compression with few to single optical cycles. Thus, specialty highly nonlinear fibers present suitable waveguides and are very promising for mid-IR low-power devices and applications.

The remainder of this paper is organized as follows. Section 6.2 presents the theoretical modeling of SC generation in optical fibers. In Sect. 6.3, we demonstrate the generation of SC in different highly nonlinear fiber structures composed of silica and non-silica materials.

6.2 Theoretical Modeling of Supercontinuum Generation

In nonlinear optics, nonlinear propagation equations have been derived in both time and frequency domains. The time domain formulations have been often preferred because of analytic similarity to the nonlinear Schrodinger equation on which a large literature in many fields is based. However, the frequency domain formulation shows more directly the frequency dependence of effects such as dispersion, loss, and the effective mode area. Both approaches have been successfully used in the study of the supercontinuum generation.

By considering the electric field which is supposed to be linearly polarized along the $\hat{x}$ vector as $E(r,t) = 1/2.\hat{x}[E(x,y,z,t)\exp(-i\omega t) + c.c]$ and using the method of separation of variables, in the frequency domain, the Fourier transform of $E(x, y, z, t)$ is expressed as

$$\tilde{E}(x,y,z,\omega) = F(x,y,\omega)\tilde{A}(z,\omega-\omega_0)\exp(-i\beta_0 z) \tag{6.1}$$

where $\tilde{A}(z,\omega)$ is the complex spectral envelope, and β_0 is the propagation constant at the central frequency ω_0. $F(x, y, \omega)$ is the function corresponding to the transverse electric modes in the (x, y) plane where the z-axis is identical to the propagation direction. The time domain envelope $A(z, t)$ is obtained from applying the inverse Fourier transform to $\tilde{A}(z,\omega-\omega_0)$ where the amplitude is normalized such that $|A(z, t)|^2$ gives the instantaneous power in watts.

Using this notation, we obtain a time-domain generalized nonlinear Schrödinger equation (GNLSE) for the evolution of $A(z, t)$ [1]:

$$\frac{\partial A(z,t)}{\partial z} + \frac{\alpha}{2}A(z,t) - \sum_{m\geq 0}\frac{i^{m+1}\beta_m}{m!}\frac{\partial^m A(z,t)}{\partial t^m} = i\left(\gamma + i\frac{\alpha_2}{2A_{\text{eff}}}\right)\left(1 + i\tau_{\text{shock}}\frac{\partial}{\partial t}\right) \times \left(A(z,t)\int_0^{+\infty} R(t')|A(z,t-t')|^2 \mathrm{d}t'\right) \tag{6.2}$$

The left-hand side of this equation models linear propagation effects including the loss α and $\beta_m = (d^m\beta/d\omega^m)_{\omega=\omega_0}$ is the mth-order dispersion coefficient associated with the Taylor series expansion of the propagation constant $\beta(\omega)$ around ω_0.

The right-hand side models nonlinear effects including the nonlinear coefficient γ which takes the usual expression $\gamma = 2\pi n_2/(\lambda_p A_{\text{eff}})$, where $\lambda_p = 2\pi c/\omega_0$ is the pump wavelength. α_2 is the two-photon absorption (TPA) coefficient, A_{eff} is the effective mode area, and τ_{shock} is the term which models the dispersion of the nonlinearity. This is usually associated with effects such as self-steepening and optical shock formation forcing the group velocity to depend on the optical intensity, characterized by a timescale$\tau_0 = 1/\omega_0$. The shock term can be generalized to account for the frequency dependence of the effective area and to be as $\tau_{\text{shock}} = \tau_0 - \left\{\frac{\partial}{\partial\omega}[\ln A_{\text{eff}}(\omega)]\right\}_{\omega=\omega_0}$.

The nonlinear response function R(t) includes the instantaneous and the delayed Raman contributions. It is given by [1]:

$$R(t) = (1 - f_R)\delta(t) + f_R h_R(t) \tag{6.3}$$

f_R is the fractional contribution of the material, for instance, it is typically taken as $f_R = 0.18$ for the fused silica. h_R(t) is the delayed Raman response.

The dispersion coefficient β_1 gives the group velocity of the pulse envelope, while β_2 determines the dispersion of the group velocity of the pulse as a function of frequency and is responsible for pulse broadening. Therefore, β_2 is defined as the GVD parameter. For pulse propagation in optical fibers, it is much more common to describe the GVD dispersion coefficient referred to as chromatic dispersion $D_c(\lambda)$:

$$D_c = \frac{d\beta_1}{d\lambda} = -\frac{2\pi c}{\lambda^2}\beta_2 \tag{6.4}$$

c is the velocity of light in vacuum.

For instance, we show the chromatic dispersion of the fundamental mode HE_{11} as a function of the wavelength for different core diameter air-silica nanowires ranging from 400 to 900 nm, as seen in Fig. 6.1.

We find that silica photonic nanowires provide exotic dispersion profiles with the ability to exhibit two zero dispersion wavelengths. In fact, by adjusting the size of the nanowire and reducing its core diameter, the region of the anomalous dispersion exhibited between two zero dispersion wavelengths can be shifted toward the blue wavelength region. This behavior results from the dominance of the waveguide dispersion over the material dispersion. With the appropriate choice of core size, the overall GVD can be highly engineered and the second zero dispersion wavelength can be positioned near the pump wavelength to yield anomalous, normal, or zero GVD. We notice that the largest region of the anomalous GVD centered around a pump wavelength of λ_p = 800 nm is found for the 800 nm nanowire diameter and accompanied with low third-order dispersion (β_3) ($\beta_2 = -9.12 \times 10^{-2}$ ps^2 m^{-1} and $\beta_3 = -5.28 \times 10^{-6}$ ps^3 m^{-1}). In fact, pumping around that wavelength would be attractive for supercontinuum generation and soliton self-compression [4–8].

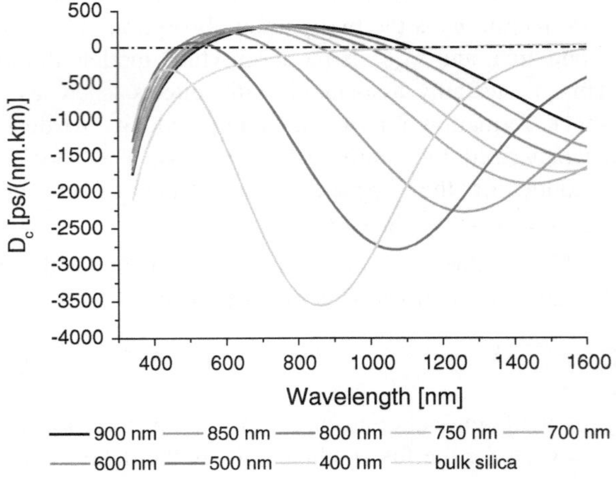

Fig. 6.1 Chromatic dispersion of air-silica nanowires with core diameters ranging from 400 to 900 nm

6.3 Supercontinuum Generation in Specialty Fibers

The introduction of new fiber geometries offers the possibility to shift and control the zero dispersion wavelength, which allows the optimization and enhancement of the nonlinearities. The anomalous dispersion regime is of considerable interest for the study of nonlinear effects because in this regime optical fibers support soliton propagation through a balance between the dispersive and nonlinear effects. In order to numerically investigate the nonlinear fiber effects, the GNLSE is resolved in the anomalous dispersion regime. The resolution of the GNLSE is performed by the symmetrized split step Fourier method where the injection of an input soliton is considered with an order N having a hyperbolic-secant pulse envelope field given by [1]: $A(0,t) = \sqrt{P_0}\,\sec h\left(\frac{t}{T_0}\right)\exp\left(-iC\frac{t^2}{2T_0^2}\right)$ where P_0 is the peak power and T_0 is the input soliton duration defined as $T_{FWHM}/1.763$. T_{FWHM} is the input pulse full width at half maximum (FWHM) duration. C is the chirp parameter controlling the initial chirp. The soliton order N is defined as $N^2 = L_D/L_{NL} = T_0^2\gamma P_0/|\beta_2|$ where $L_D = T_0^2/|\beta_2|$ is the dispersion length and $L_{NL} = (\gamma P_0)^{-1}$ is the nonlinear length.

6.3.1 Short Wavelength Extension of SC Generation

In the following, we investigate the interplay of nonlinear effects and experimentally demonstrate the generation of SC in a small core highly nonlinear silica PCF with zero dispersion wavelength λ_{ZDW} near to the pump wavelength of a Ti-Sapphire laser (800 nm). The small core silica PCF (see PCF cross section in Fig. 6.2 as inset) has zero dispersion wavelength of λ_{ZDW} = 670 nm, which allows to study the

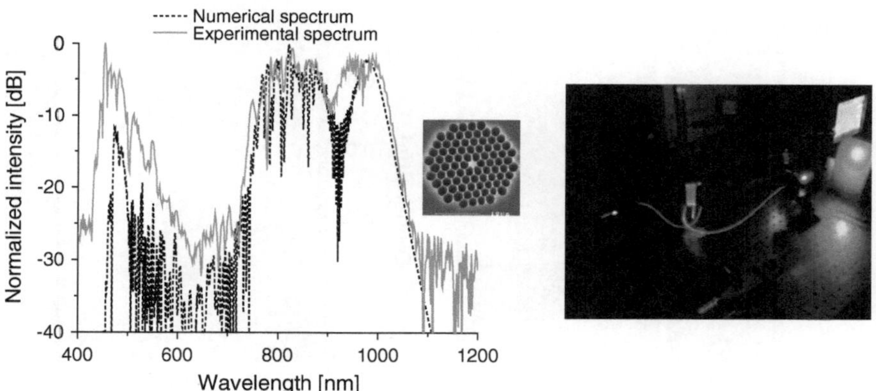

Fig. 6.2 *Left* Numerical and experimental supercontinuua generated in the small core PCF (PCF cross section as inset) of 150 fs pulses with an input energy of 625 pJ at λ = 800 nm. *Right* Blue-shifted light observed in the generated SC

nonlinear propagation in the anomalous GVD regime. The PCF has an effective mode area as $A_{\mathrm{eff}} = 1.78\ \mu\mathrm{m}^2$ and the nonlinear coefficient γ is evaluated to be $115\ (\mathrm{W\ km})^{-1}$ at $\lambda = 800$ nm. In fact, ultrashort pulses are generated by a Ti: Sapphire laser centered at 800 nm having duration of 150 fs with a petition rate of 80 MHz. By using a short length of 9.5 cm of the small core PCF, the supercontinuum is generated and collected with input pulse energy of 625 pJ corresponding to an average power of 50 mW.

Good agreement is found between numerical and experimental spectra, as shown in Fig. 6.2. We observe that the generated SC presents dominant blue-shifted wavelength components. Thus, small core PCFs present suitable waveguides for generating new frequency components in the very short wavelength region.

By analyzing the dynamics behind the supercontinuum generation in the anomalous dispersion regime, we find that the input pulses with an input soliton order of $N = 9.3$ undergo soliton fission which leads to the ejection of N fundamental solitons due to the Raman and high order dispersion effects. After the temporal pulse breakup, the generation of distinct spectral peaks in the anomalous GVD regime due to the soliton fission and the Raman self-frequency shift of ejected fundamental solitons are clearly observed in Fig. 6.3. Therefore, the presence of higher-order dispersions leads to the transfer of energy from the soliton to a narrowband resonance in the normal GVD regime. Then, dispersive waves (DW) are generated in the low wavelengths region.

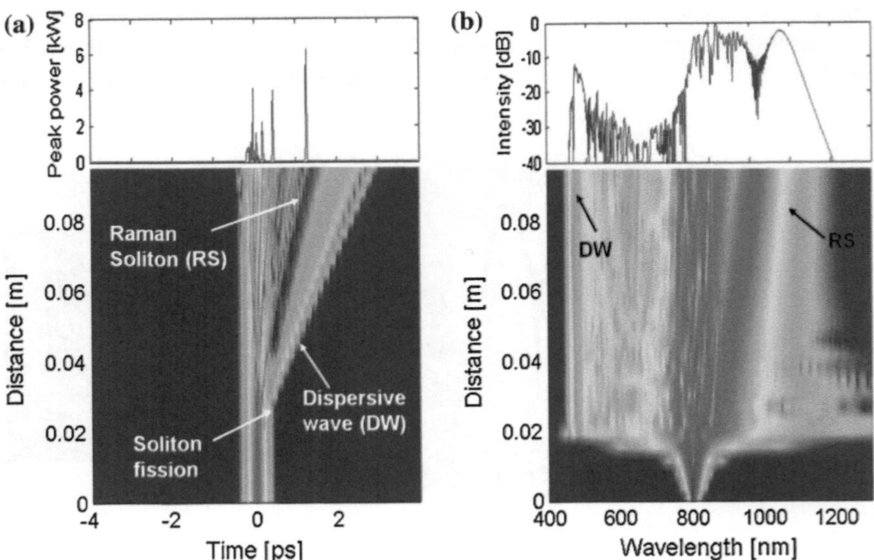

Fig. 6.3 Numerical simulations showing **a** temporal and **b** spectral evolution of the generated SC after 9.5 cm propagation distance

6.3.2 Pulse Compression and Visible Near-IR SC Generation

In this section, we study the temporal evolution in air-silica nanowires and we show that pulses could undergo temporal compression through soliton dynamics. Particularly, large anomalous GVD region with low third-order dispersion are two essential conditions required to achieve efficient broadband soliton self-compression [4].

In fact, we denote by z_{opt} the optimal length at which a maximum compressed pulse can be extracted ensuring high conserved soliton shape with lower pedestal energy [4, 6]. The compression efficiency is characterized by: $F_c = T_{\text{FWHM}}/T_{\text{comp}}$ the compression factor defined as the ratio of the input pulse (T_{FWHM}) and the output compressed pulse (T_{comp}) FWHM durations, and a quality factor $Q_c = P_{\text{comp}}/(P_0 F_c)$ defined as the ratio of the peak power (P_{comp}) of the compressed pulse with the input peak power P_0 and the compression factor F_c.

Figure 6.4a shows the temporal evolution as a function of the propagation distance of an initial 100 fs input chirped pulse with $C = -0.2$ delivered at 800 nm pump wavelength. A 1.4 fs compressed pulse is generated after 1.15 mm-long 800 nm air-silica nanowire corresponding to the excitation of an input soliton order $N = 19.8$. This interaction corresponds to a compression factor $F_c = 71.43$ and a

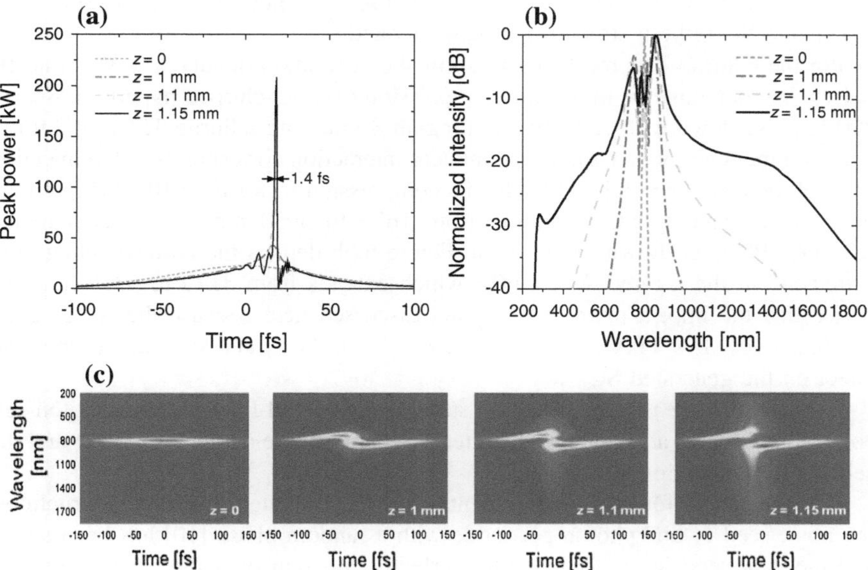

Fig. 6.4 **a** Temporal and **b** spectral evolution of 100 fs input pulse with 2.5 nJ input energy in the 800 nm air-silica nanowire as a function of the propagation distance at $\lambda p = 800$ nm. **c** Spectrogram representation of the pulse at each propagation distance. Maximum compression is obtained at $z_{\text{opt}} = 1.15$ mm

quality factor $Q_c = 0.13$. The corresponding spectral evolution and the spectrogram representations shown at each of the propagation distances are shown in Fig. 6.4b, c, respectively.

We notice that a broadband supercontinuum from ultraviolet to midinfrared regions extending from 260 to 1800 nm is generated in the 800 nm air-silica nanowire. These features reflect the growth of the symmetrical broadening when the propagation distance increases to 1 mm. This confirms that the effect of self-phase modulation is the predominant effect given such broadening.

Moreover, we should notice that by examining photonic nanowires, a large amount of the injected light is found to propagate outside the core boundaries which allow these subwavelength waveguides to be well suited for an efficient and controlled interaction of guided light with matter and perfect devices for sensing applications [9].

6.3.3 Mid-IR SC Generation

This section is dedicated to demonstrate that new designs of mid-IR sources operating beyond 2 μm are proposed by tapering PCFs yielding the generation of broadband mid-IR SC. We show that by tapering tellurite and As_2S_3 PCFs ZDW with large anomalous GVD can be exhibited beyond 2 μm since tapered SMFs exhibit only ZDW in the near-IR wavelengths [5]. Therefore, the generation of broadband SC in both structures is demonstrated.

For the tellurite Tapered PCF, we show the generation of sub-two-cycle mid-IR pulses. We demonstrate maximum compression of a pre-chirped hyperbolic-secant 200 fs pulse down to 19.1 fs by pumping an 8 mm-long tellurite TPCF at 2.9 μm with an input energy of 1 nJ. This nonlinear interaction corresponds to the injection of a soliton order $N = 3.4$ and leads to a compression factor $F_c = 10.47$. More than one octave-spanning SC extending from 1675 to 3950 nm is generated in the tellurite TPCF, as shown in Fig. 6.5a. Figure 6.5b depicts the corresponding generated SC in the tapered As_2S_3 PCF which extends from 3140 to 6330 nm. We notice that the impact of the TPA is not discussed here because the As_2S_3 glass exhibits a very low TPA coefficient (6.2×10^{-15} m W^{-1}) and simulations show no effect on the generated SC.

By analyzing the dynamics in both structures, this mid-IR broadband soliton-self compression SC is attributed to the interplay between the anomalous GVD and the self phase modulation.

We notice that for all the aforementioned SC generated in different structures, the noise seed of one photon per mode with a random phase [10] has been added and the generated SC is found to be perfectly coherent over the entire generated bandwidth. This proves that photonic nanowires are very promising for designing white light coherent sources with short lengths and low powers and are good candidates for generating mid-IR coherent supercontinuum.

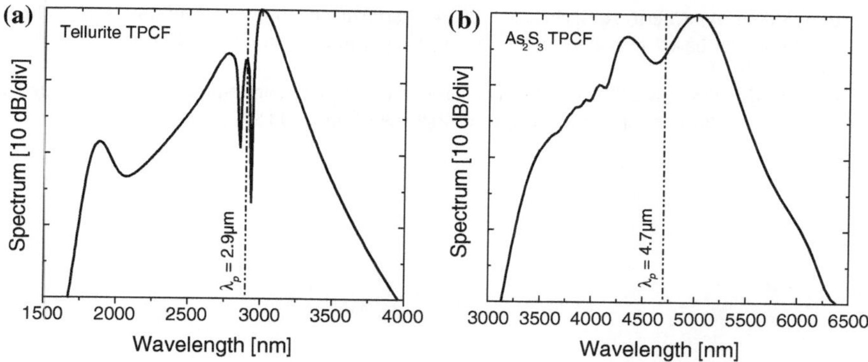

Fig. 6.5 Spectral broadening obtained after 8 mm-length in **a** the tellurite TPCF and **b** the As_2S_3 with 100 pJ input energy with 1 nJ input energy

6.4 Conclusion

The generation of supercontinuum has been reported in different highly nonlinear structures from blue to mid-IR wavelengths. By simply reducing the core size of nonlinear waveguides and generating tighter mode confinement in such specialty fibers including small core PCF and photonic nanowires with different material composition, tailoring the waveguide dispersion at any wavelength by changing the waveguide diameter is becoming possible which allows the optimization and enhancement of the nonlinearities.

References

1. Dudley JM, Taylor JR (2010) Supercontinuum generation in optical fibers. Cambridge University, Cambridge
2. Foster MA, Turner C, Lipson M, Gaeta AL (2008) Nonlinear optics in photonic nanowires. Opt Express 16:1300–1320
3. Agrawal GP (2007) Nonlinear fiber optics, 4th edn. Academic Press, London
4. Ben Salem A, Cherif R, Zghal M (2011) Soliton-self compression in highly nonlinear chalcogenide photonic nanowires with ultralow pulse energy. Opt Express 19:19955–19966
5. Ben Salem A, Cherif R, Zghal M (2012) Highly nonlinear tapered photonic crystal fibers for broadband mid-infrared supercontinuum generation in the few-optical-cycle regime. Opt Eng 51:105008
6. Ben Salem A, Cherif R, Zghal M (2011) Low-energy single-optical-cycle soliton self-compression in photonic nanowires. J Nanophotonics 5:059506
7. Ben Salem A, Cherif R, Zghal M (2012) Rigorous optical modeling of elliptical photonic nanowires. J Lightwave Technol 30:2176–2180
8. Ben Salem A, Cherif R, Zghal M (2011) Generation of few optical cycles in air-silica nanowires. Proc SPIE 8001:80011J

9. Ben Salem A, Cherif R, Zghal M (2014) Performance improvement in Mach-Zehnder interferometer-based refractive index sensor using elliptical photonic nanowires. J Mod Opt 61:263–269
10. Dudley JM, Coen S (2002) Coherence properties of supercontinuum spectra generated in photonic crystal fiber and tapered optical fiber. Opt Lett 27:1180–1182

Chapter 7
Light Propagation in Microstructured Optical Fibers and Designing High Gain Fiber Amplifier

Partha Roy Chaudhuri and Kajol Mondal

Abstract In this talk, the basic mechanism of light propagation in a triangular-lattice photonic crystal fiber (PCF) is first discussed with some key properties like, endlessly single-mode nature, controllable dispersion, high birefringence. Then a systematic study of a photonics crystal fiber design as a host of fiber amplifier is performed by varying all associated parameters towards utilizing controllable effective numerical aperture and tight modal confinement. A finite difference (FD) mode calculation analysis is used to determine the modal characteristics of the structure, which is then used to solve a standard *rate equation*. Results show that a spectral gain of the amplifier as high as 51 dB and that too over a short length ~2.5 m of the fiber is achievable. For field-deployment of the amplifier as inline component, the splicing/coupling loss (due to fundamental mode mismatch) of this all-fiber device is calculated. Notably, the coupling loss with standard telecom-grade SMF-28 fiber is reduced through an improved mode-matching of the structure-design. These results record a marked improvement in fiber amplifier performance in terms of realizing high-gain EDFA-PCF amplifiers.

7.1 Introduction

The PCF technology has offered a versatile platform to realize fibers and in-line components with many exciting modal characteristics which are far beyond the ones achievable through simple core-cladding structure [1]. The most useful attributes of PCFs are the high index-contrast between silica and air, and largely flexible geometry in terms of *size*, *shape* and *arrangement of air-holes* in the cladding. The high index-contrast enables to design fibers with large effective numerical aperture over a small core dimension and hence a very tight modal confinement as required by an amplifier host. Additionally, the average index-guiding effect in PCF with

P. Roy Chaudhuri (✉) · K. Mondal
Department of Physics, IIT Kharagpur, Kharagpur 721302, India
e-mail: roycp@phy.iitkgp.ernet.in

V. Lakshminarayanan and I. Bhattacharya (eds.), *Advances in Optical Science and Engineering*, Springer Proceedings in Physics 166,
DOI 10.1007/978-81-322-2367-2_7

makes it possible to achieve a largely tailorable mode-field area over a useful single-mode regime, which is a key requirement for controllable laser power delivery as well as for reducing splice/coupling loss. Also, the effective mode-field-diameter (MFD) for the pump (λ_p) and signal (λ_s) wavelengths can be made to control the associated overlap for ready positioning of the pump and signal fields across the active region. Erbium (Er) and Ytterbium (Yb) are the two most suitable active dopants in the silica host. A few works—both analysis and experimental demonstrations of Yb and Er-doped PCF based amplifiers have appeared in literature. However, a huge flexibility in designing with many unusual guiding features of PCF tempted us to revisit the exercise for a maximum utilization of these properties. In this work, the gain properties of erbium-doped triangular-lattice PCFs are investigated extensively by analyzing the effect of PCF geometry with the dopant radius in the core. By varying PCF geometry, we have optimized the pump and signal beam overlap, and a converged design has been proposed which exhibits an amplifier gain as high as 51 dB using a PCF length as short as 2.5 m. As an important requirement, the splice loss [2] of the PCF amplifier with standard telecommunication fiber (SMF-28) has been minimized through an adjustment of mode mismatch.

7.2 Design Analysis

Designing and analyzing the fiber-amplifier's characteristics primarily owe to a precise calculation of the propagating electromagnetic field distribution across the cross-section of the host fiber. The efficiency of the device is a direct outcome of the overlap of modal fields associated with the signal and pump beams within the doped area. Then, PCF is largely to look for the geometry where this overlap is strongest. In our analysis, we considered triangular-lattice PCF geometry, and for calculation of the mode-field distributions, a FD mode convergence recipe [3] has been implemented. Using the definitions of mode-effective area (A_{eff}) [4] for PCFs as well as considering an equivalent modal spot-size (r_{mod}), we evaluate the overlap integrals for the pump and signal beams. The mode-field distribution and mode effective index are the two key parameters in describing all important properties of guided modes. In PCF, these parameters largely depend on transverse geometry. We calculate the A_{eff} and hence r_{mod} following our mode-analysis method. The A_{eff} of the modal field is determined by taking surface integral of the field along the whole cross-section [4].

$$A_{eff} = \frac{\iint_s |E_t|^2 dxdy}{\iint_s |E_t|^4 dxdy} \tag{7.1}$$

The modal spot-size is then calculated ($A_{eff} = \pi r_{mod}^2$) from A_{eff} at the required wavelength which is the new input data to the *rate equations* [5].

7.3 Numerical Results

7.3.1 Gain Properties

The gain of a fiber amplifier depends on the emission and absorption cross-section of active-ion transitions and opto-geometric parameters. As an initial step, we consider the fiber studied in [6]. The fiber is a triangular-lattice PCF (the structure is shown in Fig. 7.1) and it has the geometrical parameters—air-hole diameter, $d = 1$ μm, pitch, $\Lambda = 2$ μm such that d/Λ of the PCF is 0.5. Following the work reported in [7], in our analysis, we take the erbium-ion concentration (ρ) as $\sim 2.6 \times 10^{25}$ ions/m^3 and assume that the erbium ions occupy a region of 0.5 μm radius at the center of the PCF core, pumped by a backward pump at 0.98 μm wavelength of power (P_p) 225 mW. Thus initially we attempt to reproduce the quoted results [7] and we notice a good agreement with the reported results. The results are plotted in Fig. 7.2, which clearly suggests that the obtainable maximum gain of EDFA-PCF amplifier can be made above 51 dB at 1.535 μm wavelength of signal light. The corresponding set of parameters that we consider in this plot is summarized in Table 7.1. Here, we detail our series studies on how the variations of d, Λ, doping radius (r_d), P_p and *input signal power* (P_s) of PCF amplifier influence the gain at the desired communication wavelength around 1.55 μm.

The mode-effective area i.e., field-confinement inside the core of PCF has a direct influence on the amplifier gain. The A_{eff} of the PCF can be tuned by changing the size and position of the air-holes. Figure 7.3 shows the gain as a function of active length of the PCF by varying d between 1.0 and 0.6 μm keeping r_d fixed at 1.0 μm and Λ at 2 μm (i.e., r_d/Λ is 0.5). The variation therefore shows d/Λ (marked in the inset) dependence of gain. As evident, the amplifier gain saturates beyond a

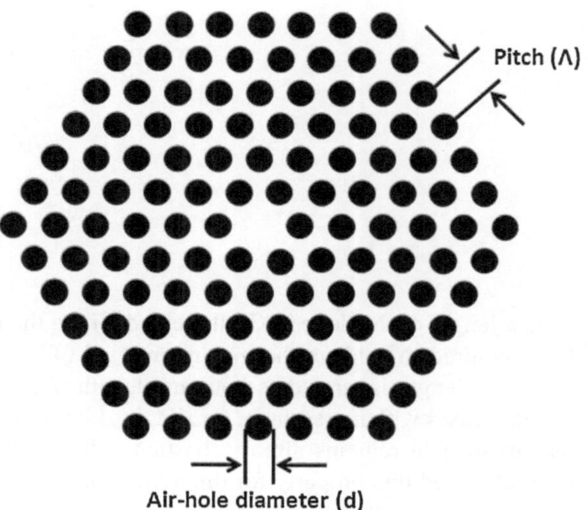

Fig. 7.1 Schematic of the considered PCF

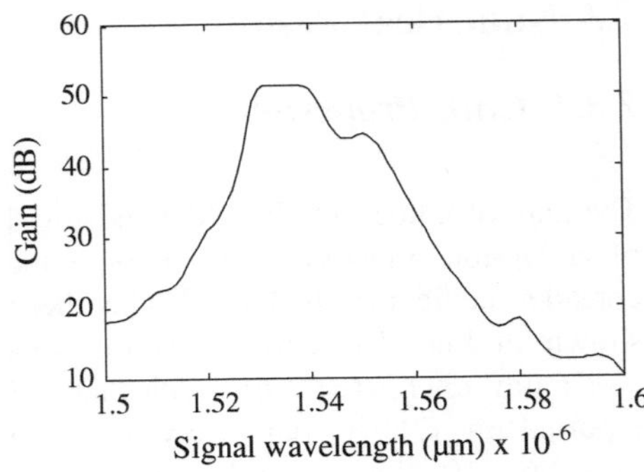

Fig. 7.2 Calculated spectral gain of the PCF-based erbium-doped fiber amplifier studied initially with parameters listed in Table 7.1

Table 7.1 PCF-EDFA parameters that have been used for the initial step of evolution

Parameters	Symbol	Value
Pump power	P_p	225 mW
Pump wavelength	λ_p	0.98 μm
Signal power	P_s	1 μW
Fiber length	L_F	7.8 m
Doping radius	r_d	0.5 μm
Doping concentration	ρ	2.6×10^{25} ions/m^3
Hole diameter	d	1 μm
Pitch	Λ	2 μm

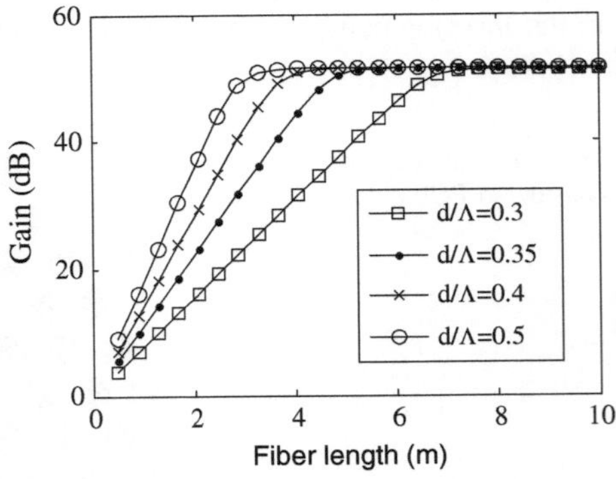

Fig. 7.3 Gain as a function of doped PCF length for different d values with $\Lambda = 2$ μm and $r_d/\Lambda = 0.5$

certain length of the doped PCF thereby defining the minimum active length of this PCF required to achieve the maximum level of amplification. The corresponding minimum length is termed as optimum length (L_{opt}). The figure also indicates that L_{opt} decreases with increasing d (or d/Λ, as Λ is fixed). So, in the above variation the maximum gain remains almost constant although the L_{opt} can be decreased. We have observed this property for different values of Λ. Also, dependence of gain on

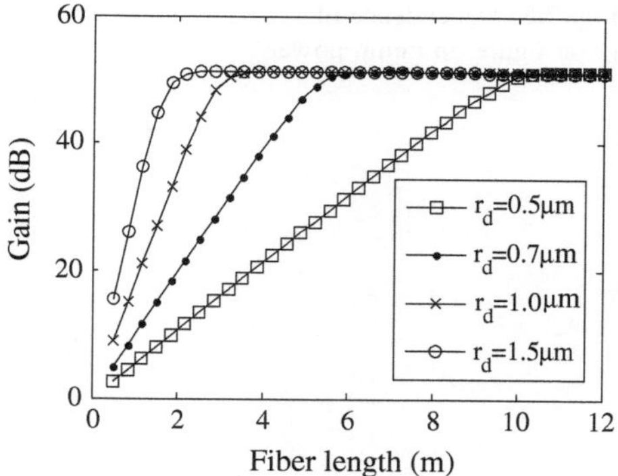

Fig. 7.4 Calculated amplifier gain as a function of fiber length for various doping radii at the core

d is seen to remain same for every value of Λ. The only difference observed is that for higher value of Λ, the maximum obtainable gain and L_{opt} decrease in comparison with that for lower value of Λ.

To study the variation of r_{d} with L_{F}, r_{d} is varied in the range of 0.5–1.5 μm, keeping the other parameters (e.g., P_{p}, λ_{p}, P_{s}, ρ, d and Λ) unchanged as considered in the initial step. The result of this study is plotted in Fig. 7.4. It shows that the L_{opt} decreases considerably with the increase of r_{d}. However, the maximum gain remains unchanged. For example, a gain of 51.4 dB is obtained with only 3.0 m long doped-PCF for $r_{\text{d}} = 1.5$ μm. In this optimization one should take care that increasing the r_{d} would cause a corresponding decrease of maximum obtainable gain. This resulting reduction in maximum obtainable gain is seen to decrease considerably for high Λ values. Then, we have estimated the effect of P_{p} and P_{s} on amplifier gain. This finding will be useful for final optimization of the amplifier design.

7.3.2 Noise Characteristics

The *amplified spontaneous emission* (ASE) results in the generation of noise within an amplifier. Here we consider the noise figure (NF) as ratio of signal to noise ratio $(\text{SNR})_{\text{i}}$ at the input to that at the output $(\text{SNR})_{\text{o}}$. For high-gain amplifier NF is defined as

$$\text{NF} = 2\left(\frac{n_{\text{s}}N_2}{n_{\text{s}}N_2 - N_1}\right) \tag{7.2}$$

where n_{s}, N_2 and N_1 are respectively the ratio of signal emission cross-section to signal absorption cross-section, upper state population and lower state population density. Figure 7.5 shows the calculated NF variation with P_{p} in the range of 5–300 mW for the considered configuration. Clearly, minimization of P_{p} also does not affect the noise figure.

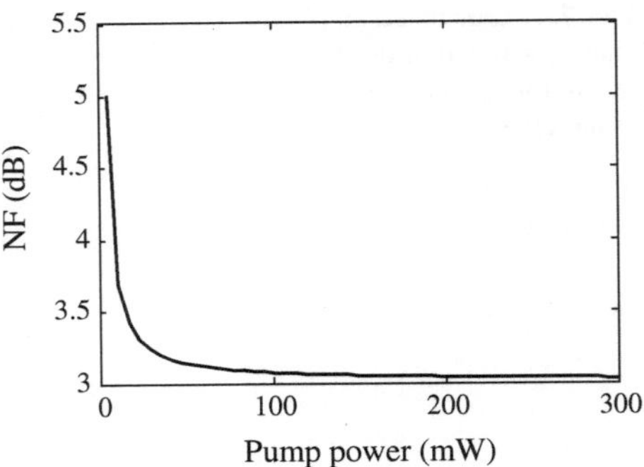

Fig. 7.5 Dependence of noise figure on pump power

7.3.3 Splice Loss

When deployed as communication/signal processing components, it is important to quantify splice loss. While optimizing amplifying features of the fiber, we also have evaluated the splice loss of the holey fiber with standard telecom fiber. In this phase, we have investigated the variation of splicing loss that occurs due to *fundamental mode mismatch* only. The splice loss has been calculated with the help of formulae proposed in [2] and the results are depicted in Fig. 7.6. Initially we analyzed a PCF with $d/\Lambda = 0.5$, and Λ between 2 and 10 μm. This result suggests that by adjusting the pitch ($\sim$7 μm) we can reduce the resulting splice loss. But a higher value of Λ, however, makes the PCF eventually multi-mode. Hence, we have reduced d/Λ from 0.5 to 0.3 through 0.4 and 0.35. According to the earlier discussion both the above d/Λ values (0.35 and 0.4) and Λ within the range of 2–7 μm exhibit almost same gain dependence on fiber length when r_d was varied from 0.25Λ to 0.75Λ.

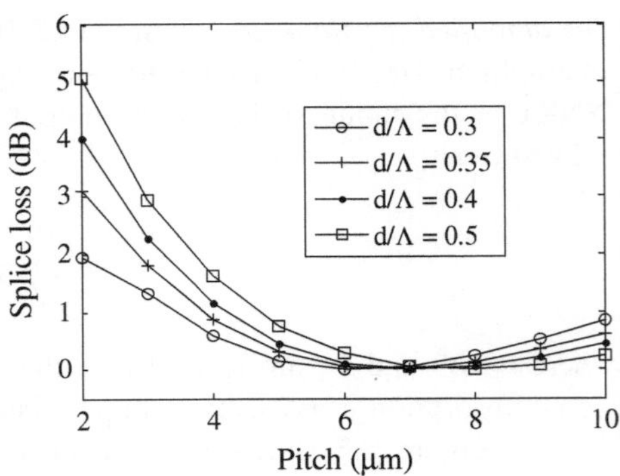

Fig. 7.6 Splicing loss between designed PCF and standard SMF

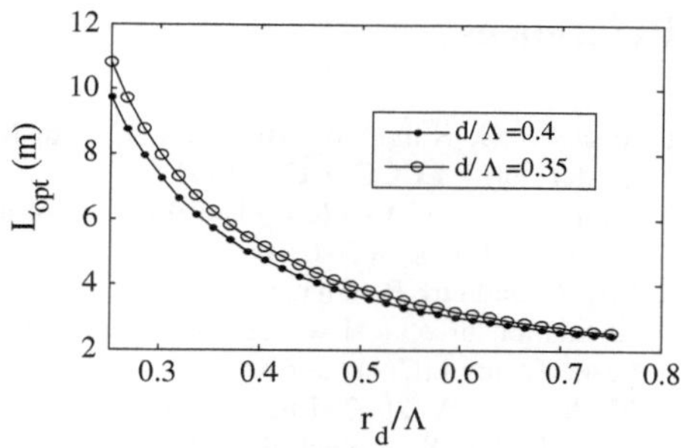

Fig. 7.7 Variation of optimum fiber length with r_d/Λ

Table 7.2 Converged set of optimized values of the amplifier parameters

Parameters	Symbol	Value
Pitch	Λ	7 µm
Hole-diameter	d	0.4Λ
Fiber length	L_F	2.5 m
Doping radius	r_d	0.75Λ

7.3.4 *Optimization*

Next, we go for final optimization of length of the PCF at the above-mentioned geometrical parameters, i.e., $d/\Lambda = 0.35$ and 0.4 having $\Lambda = 7$ where the fiber exhibits single-mode operation. As discussed earlier, these parameters are convenient in order to achieve low splice loss and less L_{opt} without affecting the maximum obtainable gain. Since the dopant radius controls the optimum length, converging to these parameters of the PCF towards a design, we then fine-tune our optimization by tailoring the effect of dopant radius on the amplifier gain. We studied this effect for the dopant radii r_d in the range of 0.25Λ–0.75Λ, beyond which the maximum obtainable gain tends to decrease. The results are shown in Fig. 7.7. Therefore, in summary, an amplifier gain of more than 51 dB can be obtained with 2.5 m long erbium-doped, single-mode PCF with considerably reduced splice loss. The converged set of parameters is given in Table 7.2.

7.4 Conclusion

In this report we have proposed a design of erbium doped fiber amplifier based on a systematic analysis of the host triangular lattice holey fiber. By utilizing the strong confinement properties of PCF, we have maximized the spectral gain to a value above 51 dB with corresponding minimized fiber length of 2.5 m. Splicing loss between the designed PCF based amplifier and a standard SMF-28 fiber is minimized at optimum fiber geometry.

References

1. Russell PSJ, Knight JC, Birks TA, Cregan RF, Mangan BJ, De Sandro JP (1997) Photonic crystal fibres. ECOC 97(3):63–64
2. Chong JH, Rao MK (2003) Development of a system for laser splicing photonic crystal fiber. Opt Exp 11(12):1365–1370
3. Roy Chaudhuri P, Ghatak AK, Pal BP, Lu C (2005) Fast convergence and higher-order calculation of optical waveguides: perturbation method with finite difference algorithm. Opt Laser Technol 37(1):61–67
4. Mortensen NA (2002) Effective area of photonic crystal fibers. Opt Exp 10(7):341–348
5. Chen J, Zhu X, Sibbett W (1992) Rate-equation studies of erbium-doped fiber lasers with common pump and laser energy bands. J Opt Soc Am B 9(10):1876–1882
6. Kogure T, Furusawa K, Lee JH, Monro TM, Richardson DJ (2003) An erbium doped holey fiber amplifier and ring laser. ECOC 03:1–3
7. Cucinotta A, Poli F, Selleri S (2004) Design of erbium-doped triangular photonic-crystal—fiber-based amplifiers. IEEE Photon Technol Lett 16(9):2027–2029

Chapter 8
Design and Development of Plasmonic Hollow Core Photonic Crystal Fiber for Sensing Applications

Tushar Biswas, Subir Majumder, Mrinmay Pal and Shyamal K Bhadra

Abstract We report design and fabrication of specialty hollow core photonic crystal fiber for utilizing them in surface plasmon resonance based material sensing. The refractive index of an unknown fluid can be determined by placing it into the hollow core of the fiber. Surface plasmons are generated by replacing air hole of the cladding selectively by silver or gold.

8.1 Introduction

Being inspired from the electronic band of a material, a clear conceptual import from the field of quantum mechanics was brought into the photonic crystal (PC) with a proposal of photonic band gap by John [1] and Yablonovitch [2] in late 1980s. The concept of photonic band gap is utilized successfully in optical guidance through hollow-core photonic crystal fiber (HCPCF), which is a periodic arrangement of air holes in silica matrix with a larger air defect. In 1991, Russell for the first time predicted the out-of-the plane photonic band gap with an idea that the light might be trapped inside a hollow core surrounded by PC throughout the length. First 2D photonic band gap [3] was theoretically predicted in 1995 and followed by fabrication [4] of first band gap fiber in 1996. Since the first development of HCPCF it has found various applications like gas laser, high power pulse transmission, pulse compression, low loss optical transmission, sensing etc. Recently, metallic PCFs have attracted tremendous research interest for their application in surface plasmon resonance (SPR) based refractive index (RI) sensing [5–14]. However, the existing PCF based sensor designs are tricky and might be difficult to fabricate. In this study we present the realistic design of HCPCF structure that can be utilized to obtain SPR

T. Biswas · S. Majumder · M. Pal · S.K. Bhadra (✉)
Fiber Optics and Photonic Division, CSIR—Central Glass and Ceramics Research Institute, Kolkata, India
e-mail: skbhadra@cgcri.res.in

V. Lakshminarayanan and I. Bhattacharya (eds.), *Advances in Optical Science and Engineering*, Springer Proceedings in Physics 166,
DOI 10.1007/978-81-322-2367-2_8

based RI sensor. The proposed sensor works on the principle of resonant coupling of surface plasmons (SPs) with the core guided mode. The air hole in PC cladding is replaced by silver and gold wire to generate SPs. The hollow-core of the proposed fiber can be filled up with the analyte whose RI is to be determined.

8.2 Design of HCPCF

In designing HCPCF one has to calculate the out of plane photonic band gap map, which is the variation of out of plane band gap with respect to out of plane wave vector. This map gives the operating wavelength region of an HCPCF. The electromagnetic wave propagation in a PC is governed by the following Maxwell's electromagnetic wave equations.

$$\nabla \times [\nabla \times E(\boldsymbol{x}|\omega)] = \frac{\omega^2}{c^2}\varepsilon(\boldsymbol{x}_{\|})E(\boldsymbol{x}|\omega) \tag{8.1}$$

$$\nabla \times [\nabla \times H(\boldsymbol{x}|\omega)] = \frac{\omega}{c^2}\varepsilon(\boldsymbol{x}_{\|})H(\boldsymbol{x}|\omega) \tag{8.2}$$

The photonic band gap map of a PC can be obtained by calculating the eigen frequencies using either (8.1) or (8.2). The real valued solutions of these equations are allowed to propagate through the PC whereas the complex valued solutions create photonic band gap regions. An electromagnetic mode having frequency value within the band gap region can be trapped into the defect core. The schematic diagram of cross section of an HCPCF is shown in Fig. 8.1a. Figure 8.1b shows the band gap map of such structure having pitch = 2.74 microns (centre to centre distance between two consecutive air holes) and d = 2.603 (diameter of air hole).

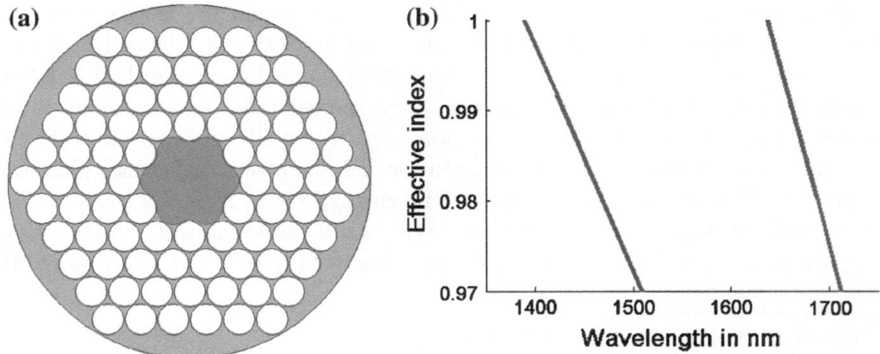

Fig. 8.1 **a** Schematic diagram of the cross section of an HCPCF. In this figure *white circles* and the region shaded in *blue* represent the air holes and the defect core respectively. **b** Photonic band gap of the HCPCF

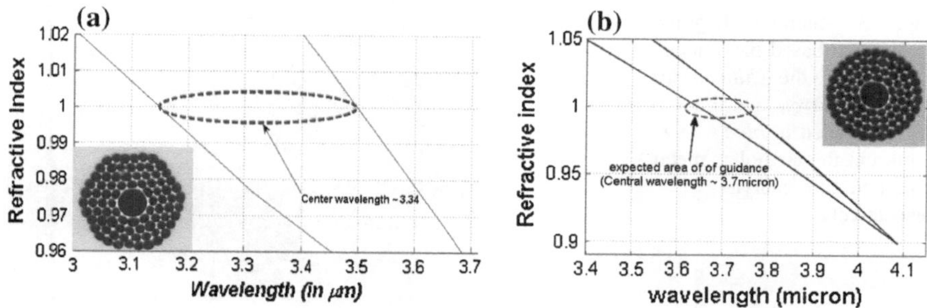

Fig. 8.2 Calculated band gap structure of **a** HCPCF5 and **b** HCPCF8 along with fiber microstructure (*inset*)

Numerical calculations reveal that the centre wavelength and band width of an HCPCF depends on pitch and diameter to pitch ratio respectively. We found that the structure has photonic band gap centered at 1550 nm.

8.3 Fabrication of HCPCF

We usually stack capillaries in a hexagonal geometry and put a central capillary of such a diameter as if seven capillaries have been removed. We make a 2 or a 3 mm diameter cane that are put into an 8/2 or a 10/3 mm tube and drawn into fiber of desired diameter. Usually the diameter of the fiber is 125 μm. Fabricated fiber HCPCF5 and HCPCF7 have been drawn from a cane inside 10/3 mm tube. HCPCF5 has an average pitch of 5.41 μm with filling fraction 94 % and shown central wavelength of guidance of 3.34 μm (theoretically), whereas for HCPCF7 respective values are 6.28 μm, 72 %, and 5.23 μm. Then we tried to fabricate the eighth fiber in the series (HCPCF8) in an 8/2 mm tube. The calculated guidance is at around 3.7 μm. Here pitch is 4.7 μm and the filling fraction is 82 % (Results of HCPCF 5 and 8 are in Fig. 8.2).

8.4 Design of HCPCF Based Plasmonic Sensor

So far we have restricted our attention only on the band gap guidance of HCPCF. In this section we show that an intelligently designed HCPCF can be employed to RI sensing by using index guidance property. The beauty of HCPCF lies in the fact that by filling the core with material of RI value greater than the effective index of the fundamental space filling mode (n_{FSM}) light can be guided through total internal reflection. The value of n_{FSM} of the HCPCF shown in figure remains below 1.3 in the wavelength region from 1300 to 1700 nm. Therefore, this fiber can guide light

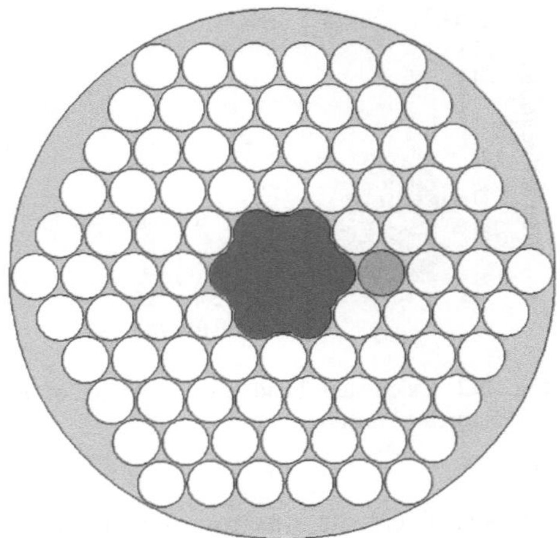

Fig. 8.3 Schematic diagram of HCPCF based bio sensor. In this figure the *white circles*, the *yellow circle* and the region shaded in *blue* represent the air holes, metal wire and core analyte respectively

in this wavelength region for core filling RI value greater than or equal to 1.3, which is desirable for designing bio sensor. The schematic diagram of the proposed sensing fiber is shown in Fig. 8.3.

We replace one of the air holes in the cladding by noble metal like gold and silver. SPs generate on the metal surface when light is launched into the core of the fiber. As a result of this the core guided mode gets attenuated. Attenuation reaches its maximum value for SPR condition, which is nothing but the establishment of phase matching between the SPP and the fundamental core guided mode. This occur for a particular wavelength (λ_{res}) for a fixed value of core RI. Now, if we vary the RI value of the core filling material the absorption peak, in other word the λ_{res}, gets shifted. This shifting of λ_{res} is a sensitive measure of change in the value of RI of analyte. We consider two cases. In the first case we replace the air hole by silver and by gold in the later case. Then, we vary the value of RI of the core and calculate the effective index of the fundamental core guided mode by using finite element method based COMSOL Multiphysics software. The absorption loss is calculated by using the imaginary part of the effective index. The variation of absorption loss peak with the change in the value of RI of analyte is shown in Fig. 8.4.

In Fig. 8.4 we can see for both cases three absorption peaks appear in the wavelength region from 1300 to 1700 nm. This happens due to cylindrical metal surface. There is only one absorption peak occurred for planar metal-dielectric surface because there is only one SPP mode. In case of cylindrical metal surface several absorption peaks can occur due to several couplings between core guided mode and various azimuthally polarized SPP modes. We designate three peaks as peak 1, peak 2 and peak 3 with respect to the ascending order of wavelengths of resonances.

Sensitivity is calculated from the sensitivity graph which is a plot of λ_{res} as a function of RI. The sensitivity graphs for the first peak of silver and gold wire based sensors are shown in Fig. 8.5a, b respectively.

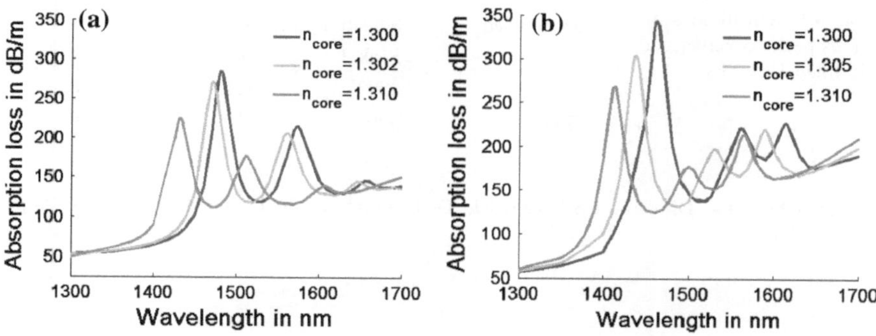

Fig. 8.4 Variation of absorption peaks with respect to RI for **a** silver and **b** gold wire

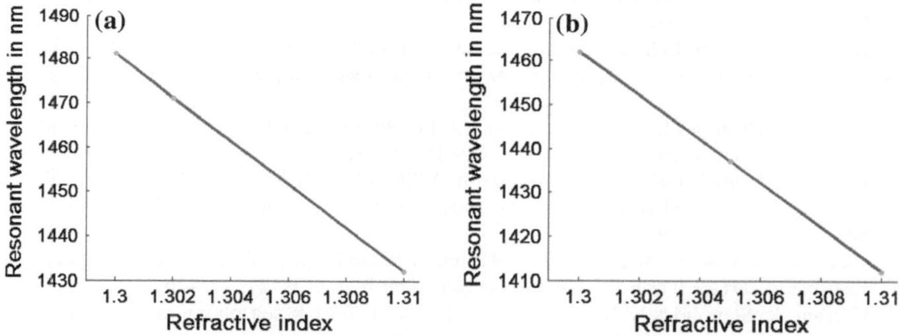

Fig. 8.5 Variation of λ_{res} with RI of **a** silver and **b** gold filled HCPCF

The spectral sensitivity is defined by the following equation

$$s_{\lambda}\left[nm\mathrm{RIU}^{-1}\right] = \frac{d(\lambda_{\mathrm{peak}}(n_a))}{dn_a} \tag{8.3}$$

Theoretically, sensitivity is obtained by calculating the slope of the linear fit of the sensitivity curve. The sensitivities of various peaks are shown in Table 8.1.

Numerical calculations of sensitivities shows that the first and second peaks of gold filled HCPCF are more sensitive than silver filled HCPCF. We have seen from the absorption loss spectra that in case of silver filled HCPCF the third absorption peaks are very flat. In case of gold filled HCPCF all the three peaks remain sharp. The sharpness of the absorption peak is desirable for faithful detection of RI.

In conclusion we have shown the designing of HCPCF to use them for developing surface plasmon resonance sensor. Then we discussed the fabrication procedure of HCPCF. We have considered silver as well as gold to obtain SPR based sensing effect in HCPCF. Gold filled HCPCF shows better sensitivity.

Table 8.1 Sensitivities of various peaks of silver and gold filled HCPCFs

Metal used	Sensitivity (*nm* RIU^{-1})		
	Peak 1	Peak 2	Peak 3
Silver	4893	6178	5000
Gold	5000	6200	4900

Acknowledgments Director, CSIR-CGCRI, DST, Govt. of India, CSIR 12th Plan Project (GLASSFIB).

References

1. John S (1987) Strong localization of photons in certain disordered dielectric superlattice. Phys Rev Lett 58:2486
2. Yablonovitch E (1987) Inhibited spontaneous emission in solid-state physics and electronics. Phys Rev Lett 58:2059
3. Birks T et al (1995) Full 2D photonic band gaps in Air/Silica structures. Electron Lett 31:1941
4. Knight J et al (1996) All silica single-mode optical fiber with photonic crystal cladding. Opt Lett 21:1547
5. Biswas T, Chattopadhyay R, Bhadra SK (2014) Plasmonic hollow-core photonic band gap fiber for efficient sensing of biofluids. J Opt 16:045001
6. Hassani A, Gauvreau B, Fehri MF, Kabashin A, Skorobogatiy M (2004) Photonic crystal fiber and waveguide based surface plasmon resonance sensors for application in the visible and near-IR. Electromagnetics 28:198–213
7. Gauvreau B, Hassani A, Fehri MF, Kabashin A, Skorobogatiy M (2007) Photonic bandgap fiber-based surface plasmon resonance sensors. Opt Expr 15:11413–11426
8. Hassani A, Skorobogatiy M (2009) Photonic crystal fiber based plasmonic sensors for the detection of biolayer thickness. J Opt Soc Am B 26:1550–1557
9. Yu X, Zhang Y, Pan S, Shum P, Yan M, Leviatan Y, Li C (2010) A selectively coated photonic crystal fiber based surface plasmon resonance sensor. J Opt 12:015005
10. Shuai B, Xia L, Zhang Y, Liu D (2012) A multi-core holey fiber based plasmonic sensor with large detection range and high linearity. Opt Expr 20:5975–5986
11. Town GE, Yuan W, McCosker R, Bang O (2010) Microstructured optical fiber refractive index sensor. Opt Lett 35(6):856–858
12. Pei-pei Z, Jian-quan Y, Hai-xia C, Ying L (2013) A surface plasmon resonance sensor based on a multicore photonic crystal fiber. Optoelectron Lett 9:0342
13. Lu Y, Hao CJ, Wu BQ, Musideke M, Duan LC, Wen WQ, Yao JQ (2013) Surface plasmon resonance sensor based on polymer photonic crystal fibers with metal nanolayers. Sensors 13:956–965
14. Fu X, Lu Y, Huang X, Hao C, Wu B, Yao J (2011) Surface plasmon resonance sensor based on photonic crystal fiber filled with silver nanowires. Opt Appl XLI:941–951

Chapter 9
Determination of Refractive Index In-Homogeneity of Transparent, Isotropic Optical Materials

Sanjib Chatterjee

Abstract Interferometric methods for the measurement of refractive index inhomogeneity of transparent, isotropic optical materials such as optical glass, fused silica, laser glass etc. have been discussed.

9.1 Introduction

In transparent, isotropic optical materials, the variation of refractive index (RI), i.e., inhomogeneity can occur, generally, in three different forms, namely striae, particulate, and extended inhomogeneity [1]. In high quality optical glass, fused silica, laser glass etc., the slow variation of RI, over an extended region, which is called extended inhomogeneity, is the most important factor, since it produces the wavefront distortion. The RI variation is measured from the optical path difference (OPD) variation suffered by an incident plane wavefront while passing through a slab of the test material. The OPD variation is obtained from the interferograms as the test beam interferes with a reference beam. Fizeau, Twyman-Green and Mach-Zehnder interferometers are most commonly used for measurement [2]. Chatterjee [3] discussed a Jamin interferometer based setup for the measurement of integrated wavefront distortion of a laser rod.

9.2 Principle

Single pass wavefront deformation suffered by a plane wavefront passing through a slab of isotropic optical material such as a laser glass slab (LGS) is illustrated in Fig. 9.1. An expanded, collimated He–Ne (632.8 nm) laser beam represented by

S. Chatterjee (✉)
Raja Ramanna Centre for Advanced Technology, Indore 452013, India
e-mail: schat@rrcat.gov.in

V. Lakshminarayanan and I. Bhattacharya (eds.), *Advances in Optical Science and Engineering*, Springer Proceedings in Physics 166,
DOI 10.1007/978-81-322-2367-2_9

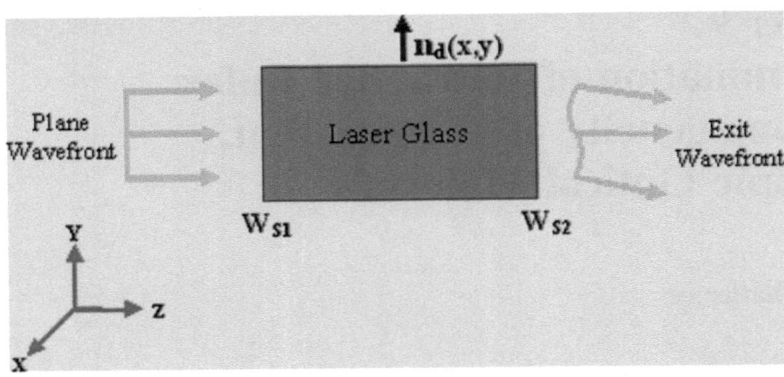

Fig. 9.1 Single pass wavefront deformation suffered by plane waves while passing through a LGS

a plane wavefront, which travels along the z-axis of a rectangular Cartesian co-ordinate system, passes through the LGS. The refractive index variations in directions normal to the propagation direction of the laser beam are represented by $n_d(x, y)$. The optical path difference (OPD) suffered by the plane wave front is given by

$$\Delta_0(x,y) = [n_d(x,y) - n_L][W_{S1}(x,y) + d + W_{S2}(x,y)] \tag{9.1}$$

where n_L is the refractive index of the surrounding medium in which the LGS is immersed. $W_{S1}(x, y)$ and $W_{S2}(x, y)$ are surface form errors of the entrance and exit planes of the LSG, respectively and d is the length/thickness of the LSG.

Figure 9.2 shows a Fizeau interferometer (FI) based setup for the measurement of refractive index homogeneity of isotropic optical materials. An expanded, collimated beam from a corrected telescope objective (TO) passes through a reference surface RS1, a sample optical material sandwiched between two high quality plane parallel optical glass/fused silica plates (HGP1) and (HGP2) using index matching liquid at the interfaces, and falls (normal incidence) on another reference surface RS2. The Fresnel reflected component of the beam from the RS2 double passes through the sandwiched sample and the RS1 and is made to interfere with the Fresnel reflected component of the incident beam from the RS1. It is evident that with a FI based setup one actually gets a double pass integrated wave front deformation, i.e., $2\Delta_0$, and contributions of the end surfaces need to be subtracted to get the refractive index inhomogeneity.

In the following, I discuss a Sagnac interferometer (SI) based setup for the determination of refractive index inhomogeneity of optical materials. The optical schematic of the SI based setup is shown in Fig. 9.3. A polarization SI [4] formed by a polarization beam splitter (PBS1) and plane mirrors M1, M1′; which are at 45° to each other, is used to produce a pair of laterally separated, mutually parallel, collimated beams with orthogonal planes of polarization, i.e., p and s polarizations. Both the beams pass through a liquid-filled cell (LFC), with plane parallel glass

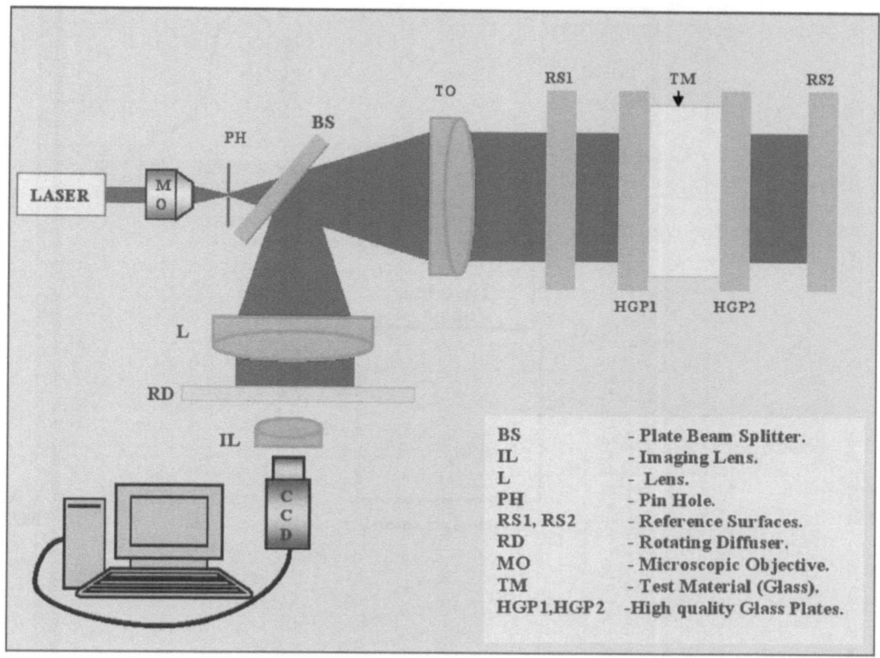

Fig. 9.2 FI based setup for measurement of index inhomogeneity

windows, kept in the path of the beams. The p-polarized beam serves as the probe/test beam as it passes through the LGS, with plane parallel end surfaces, kept in the p-polarized beam path in the index matching liquid while the s-polarized beam, which serves as the reference beam, travels a parallel path through the liquid. Another polarization SI formed by another polarization beam splitter (PBS2) along with plane mirrors M2, M2′, which are also at 45° to each other, recombines the test and reference beams by removing the lateral shear. A quarter wave plate (QWP) at 45° transforms the state of polarizations of the emergent probe and reference beams to opposite circular polarizations of which the linear components selected by a linear polarizer interfere. Interference fringes formed on a rotating diffuser screen (RDS) is grabbed by means of a CCD camera connected to a Personal computer. Polarization phase shift is applied by varying the angular orientation of the linear polarizer. The system aberration has been removed by subtracting the OPD variations without the test object from that with test object.

In Fig. 9.3, the expanded, collimated incident beam is split up at O1 on PBS1, and the split up beam components are recombined at O2 on PBS2. The OPD between the split up beam components, without the test LGS, assuming the beams are propagating along the z-axis of a Cartesian co-ordinate system in the LFC, can be written as

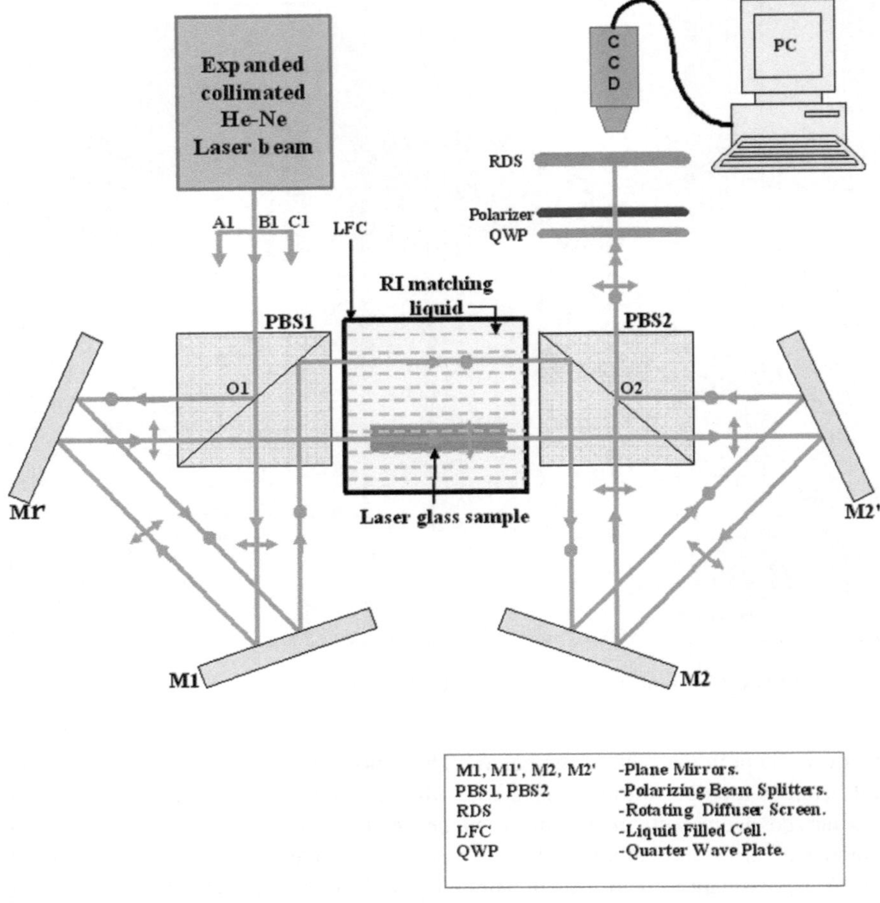

Fig. 9.3 SI based setup for the measurement of refractive index inhomogeneity

$$\Delta_1(x, y) = W_{\text{probe}}(x, y) - W_{\text{Ref}}(x, y) \tag{9.2}$$

where $W_{\text{probe}}(x, y)$ and $W_{\text{Ref}}(x, y)$ are the optical paths traversed by the probe and reference beams, respectively, without the test object. $\Delta_1(x, y)$ is the intrinsic OPD variations of the interferometer setup.

The OPD between the split up components with the probe beam passing through the test LGS can be written as follows

$$\Delta_2(x, y) = W_{\text{probe}}(x, y) + [n_{\text{d}}(x, y) - n_{\text{L}}][W_{S1}(x, y) + d + W_{S2}(x, y)] - W_{\text{Ref}}(x, y) \tag{9.3}$$

$n_d(x,y)$ is refractive index of the test object in a plane perpendicular to the direction of propagation of the probe beam, n_L is the refractive index of the liquid, $W_{S1}(x,y)$ and $W_{S2}(x,y)$ are the surface form errors of the end surfaces of the sample. d is the mean length of the sample in the direction of propagation of the probe beam.

We can eliminate the effect of system aberration by subtracting (9.2) from (9.3) as follows.

$$\begin{aligned}\Delta(x,y) &= \Delta_2(x,y) - \Delta_1(x,y)\\ &= [n_d(x,y) - n_L]d + \{[W_{S1}(x,y)\\ &\quad + W_{S2}(x,y)][n_d(x,y) - n_L]\}\end{aligned}$$

or

$$\left[\frac{\Delta(x,y)}{d}\right] = n_d(x,y) - n_L + \left[\frac{(W_{S1} + W_{S2})[n_d(x,y) - n_L]}{d}\right] \tag{9.4}$$

The contribution of the third term of (9.4) can be made negligible by (1) proper choice of matching liquid so as to make, $\Delta n = n_d - n_L$, a negligible quantity, (2) minimizing the surface form errors of the end faces, (3) and by increasing d. It is imperative to find out a tolerance for surface form error of the end faces of the sample for a chosen matching liquid and length of the sample. The maximum surface form error can be equated with an allowable peak to valley surface form deviation $\left(\frac{\lambda}{x}\right)$, where $\left(\frac{1}{x}\right)$ is a fraction of wave length. By taking contribution of the third term of (9.3) in the order of 10^{-7}, we can write

$$\left(\frac{n_d - n_L}{d}\right)\left(\frac{\lambda}{x}\right) = 10^{-7}$$

or

$$x = \Delta n\left(\frac{\lambda}{d}\right)10^{7} \tag{9.5}$$

Thus by proper choice of matching liquid and other parameters, the third term of (9.4) can be made negligible and we get.

$$\frac{\Delta(x,y)}{d} = n_d(x,y) - n_L(x,y) \tag{9.6}$$

We are basically interested in determining the variation of n_d, i.e., δn_d. Suppose $\delta\Delta(x,y)$ be the change in $\Delta(x,y)$ due to δn_d. Thus we can write from (9.6)

$$\left[\frac{\Delta(x,y) + \delta\Delta(x,y)}{d}\right] = n_d(x,y) + \delta n_d(x,y) - n_L \tag{9.7}$$

And finally, we obtain

$$\delta n_{\mathrm{d}}(x, y) = \left[\frac{\delta\Delta(x, y)}{d}\right] = \left[\frac{\delta\Delta_1(x, y) - \delta\Delta_2(x, y)}{d}\right] \tag{9.8}$$

Thus $\delta n_{\mathrm{d}}(x, y)$, i.e., the variation of refractive index of the test object, in a plane normal to the direction of propagation of the probe beam, can be determined by evaluating $\delta\Delta_1$, $\delta\Delta_2$ of (9.8). First, Δ_1, Δ_2 are determined by applying polarization phase shifting interferometry (PPSI) [4] and then applying least squared fitting each set of data is fitted to plane and the variations with respect to the fitted planes give us $\delta\Delta_1$, $\delta\Delta_2$.

In PPSI, a series of interferograms are captured as the phase difference between the interfering components is changed, in steps, by a constant amount by varying the angular orientation of the linear polarizer (Fig. 9.3) in steps of 45° for introducing phase shifts of $\pi/2$ between the frames [4].

Results obtained for a phosphate laser glass sample can be found in [4].

The important advantages of the SI based setup are: (1) Insensitivity to extraneous vibration. (2) Path compensated. (3) Applicable for un-polished test object. (4) The RI of the matching liquid can be slightly different. (5) High accuracy (10^{-6}) of measurement.

References

1. Reitmayer F, Schuster E (1972) Homogeneity of optical glasses. Appl Opt 11(5):1107–1111
2. Mantravadi MV, Malacara D (2007) Newton, Fizeau and Haidinger interferometers. In: Malacara D (ed) Optical shop testing. Wiley, New York, pp 17–32
3. Chatterjee S (2004) Measurement of single pass wavefront distortion of optical components with phase shifting Jamin interferometer. Opt Eng 43(4):872–879
4. Chatterjee S, Kumar YP (2013) Determination of the index inhomogeneity of transparent isotropic optical material with a dual Sagnac interferometer. Appl Opt 52:4820–4826

Chapter 10
Real Time Amplification of Moving Light Signals by Photorefractive Ferroelectric Liquid Crystal Mixtures

Takeo Sasaki and Yumiko Naka

Abstract The photorefractive effect in photoconductive ferroelectric liquid crystal blends containing photoconductive chiral compounds was investigated. Terthiophene compounds possessing chiral structures were mixed with an achiral smectic C liquid crystal mixture. The blends exhibit the ferroelectric chiral smectic C phase. The photorefractivity of the liquid crystal blends was investigated by two-beam coupling experiments. The photoconductive ferroelectric liquid crystal blends prepared in this study exhibited a large gain coefficient of over 1200 cm^{-1} and a fast response time shorter than 1 ms. Amplification of an moving optical image signal of over 30 fps using the photorefractive ferroelectric liquid crystal was demonstrated.

10.1 Introduction

The photorefractive effect is one of the phenomena that form holographic images within a material [1, 2]. The characteristic phenomenon of the photorefractive effect is asymmetric energy exchange in two-beam coupling that can be used to coherently amplify signal beams; therefore, the photorefractive effect has the potential to be used in a wide range of optical technologies similar to a transistor in electrical circuits. Optically transparent materials that show both photovoltaic and electro-optic effects can exhibit the photorefractive effect. Several photorefractive materials have been developed, including inorganic ferroelectric photoconductive crystals, organic crystals, photoconductive nonlinear optical organic polymers, amorphous organic photoconductive materials, photoconductive amphiphilic compounds, and photoconductive liquid crystals (LCs) [3–5]. Organic materials in particular have attracted significant interest in this context since 1994, because they exhibit large photorefractivity and shorter response times [6–9]. The photorefractive effect in

T. Sasaki (✉) · Y. Naka
Department of Chemistry, Faculty of Science, Tokyo University of Science,
1-3 Kagurazaka, Shinjuku-ku, Tokyo 162-8601, Japan
e-mail: sasaki@rs.kagu.tus.ac.jp

V. Lakshminarayanan and I. Bhattacharya (eds.), *Advances in Optical Science and Engineering*, Springer Proceedings in Physics 166,
DOI 10.1007/978-81-322-2367-2_10

LCs has been investigated previously [10]. LCs are basically liquid, so they can be easily driven by a low electric field. LCs are classified into several groups, the most well known of which are nematic and smectic LCs. Nematic LCs are used in liquid crystal displays (LCDs), whereas smectic LCs are very viscous and are thus seldom utilized in practical applications. Nematic LCs were first used as a photorefractive LC and large photorefractivity was obtained with the application of an electric field of only a few volts per micrometer [11]. The photorefractive effect in surface-stabilized ferroelectric liquid crystals (SS-FLCs) doped with a photoconductive compound has been reported [12, 13]. Ferroelectric liquid crystals (FLCs) belong to the class of smectic LCs that have a layered structure [14, 15]. FLCs belongs to a chiral smectic C (SmC*) liquid crystal in which the LC phase possesses a helical structure. In order to observe ferroelectricity in the FLCs, they must be formed into thin films [14]. The thickness of the film needs to be a few micrometers. When an FLC is injected between two glass plates in few micrometers gap, the helical structure of the LC phase uncoils and a surface-stabilized state (SS-state) is formed. In the SS-state, a spontaneous polarization (Ps) appears. For display applications, the thickness of the film is typically 2 μm. In the SS-state, FLC molecules can align in only two directions, and the direction of the FLC molecules changes according to the direction of the spontaneous polarization. When an alternating electric field is applied to the SS-FLC, the FLC molecules show a continuous switching motion. The electrical switching response time of FLCs is typically shorter than 1 ms. The direction of the spontaneous polarization is changed by the applied electric field, which gives rise to a change in the direction of the FLC molecules. Thus, when an internal electric field is created in an SS-FLC material, the direction of FLC molecules is changed by the field.

Figure 10.1 shows the mechanism of the photorefractive effect in FLCs. When laser beams interfere in a photoconductive FLC, charge separation occurs between the bright and dark positions of the interference fringe and internal electric fields are produced. The directions of spontaneous polarization in the area between the bright and dark positions are changed, that causes a periodic change in the orientation of the FLC molecules. Since the switching of FLC molecules is due to the response of bulk polarization, the switching is very fast. The photorefractive effect in commercially available FLCs mixed with a photoconductive dopant and that of FLC blends comprising liquid crystalline compounds and photoconductive chiral dopants have been investigated. The photoconductive FLC blend exhibited a large photorefractive effect. In this study, photoconductive FLC blends are prepared and the optical signal amplification was demonstrated.

10.1.1 Characteristics of the Photorefractive Effect

Since the change in the refractive index via the photorefractive effect occurs in the areas between the bright and dark positions of the interference fringes, the phase of the resulting index grating is shifted from that of the interference fringes. The phase

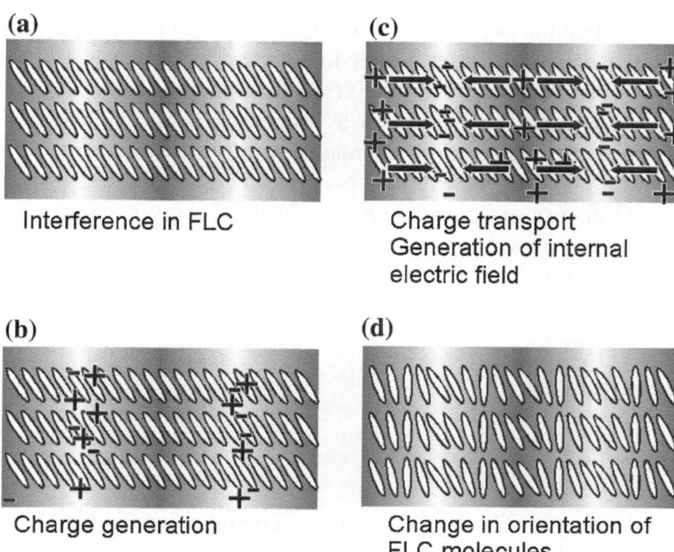

Fig. 10.1 The mechanism of the photorefractive effect in the ferroelectric liquid crystals. **a** Two laser beams interfere in the SS-state of the FLC/photoconductive compound mixture. **b** Charge generation occurs in the bright areas of the interference fringes. **c** While the electrons are trapped at trap sites in the bright areas, holes migrate to all parts of the material. As a consequence, the bright positions are charged negatively, and dark positions are charged positively. **d** The orientations of the spontaneous polarization and thus the orientations of the FLC molecules are changed by the internal electric field

of the refractive index grating is π/2-shifted from the interference fringes under conditions where one of the photogenerated charges does not move from the bright position. This phase shift is characteristic to the photorefractive effect. When the material is photochromic and is not photorefractive, then a photochemical reaction occurs at the bright areas, and a refractive index grating with the same phase as that of the interference fringes is formed in Fig. 10.2a.

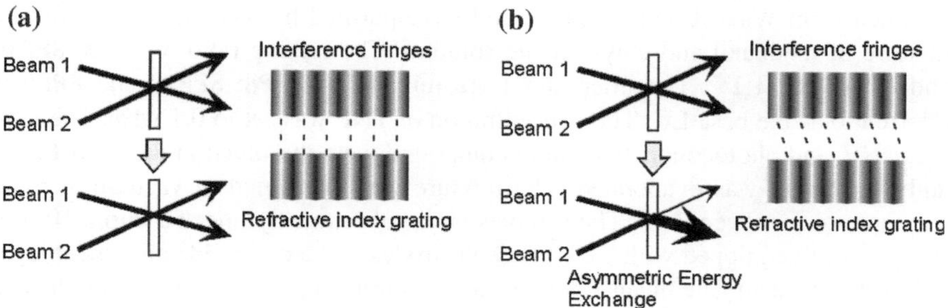

Fig. 10.2 **a** Photochromic grating and **b** photorefractive grating

In that case, when the intensities of the interfering beams are equal, the apparent transmitted intensities of the laser beams do not change. However, if the material is photorefractive, the phase of the refractive index grating is shifted from that of the interference fringes and this causes an asymmetric energy exchange [1, 2]. The transmitted intensity of beam 1 increases while that of beam 2 decreases (Fig. 10.2b). The photorefractivity of a material is confirmed by the occurrence of this asymmetric energy exchange. The asymmetric energy exchange is an amplification of one beam by another beam, so that this phenomenon can be utilized for an optical signal applications.

10.2 Experimental

10.2.1 Samples

Compared to nematic LCs, FLCs are more crystalline than liquid, so that the preparation of fine FLC films requires sophisticated techniques [14]. It is very difficult to obtain a uniformly aligned, defect-free SS-FLC using a single FLC compound. Mixtures of several LC compounds are usually used to obtain fine SS-FLC films. An FLC mixture is composed of the base LC, which is a blend of several LC compounds, and a chiral compound. The chiral compound introduces a helical structure to the LC phase through supramolecular interactions. To utilize a FLC as a photorefractive material, photoconductive compounds must be added to the FLC. However, the introduction of such non-LC compounds to the FLC often hinders the formation of a uniformly aligned SS state. Thus, appropriate design of the photoconductive compounds is crucial.

The structures of the LC compounds, the electron acceptor trinitrofluorenone (TNF) and the photoconductive chiral compounds used in this study are shown in Fig. 10.3. The photoconductive chiral compounds were synthesized as described in a previous paper. 13 TNF was obtained from Tokyo Kasei Co. The smectic C (SmC) liquid crystal used in this study was a 1:1:2 mixture of 5-Octyl-2-(4-octyloxyphenyl) pyrimidine (8PP8), 2-(4-Decyloxyphenyl)-5-octyl pyrimidine (8PP10), and 2-(4-hexyloxy)-5-octyl pyrimidine (8PP6). 8PP8, 8PP10 and 8PP6 were obtained from Wako Chemicals Co., and were purified by recrystallization from a mixture of methanol and ethyl acetate solution. The mixing ratio of 8PP8, 8PP10 and 8PP6 was 1:1:2. Hereafter, the 1:1:2 mixture of 8PP8, 8PP10 and 8PP6 is referred to as the base LC. The concentration of TNF was set to 0.1 wt%. The base LC, TNF, and photoconductive chiral compound were dissolved in dichloroethane, and the solvent was evaporated. The mixture was then dried in vacuum at room temperature for seven days. The samples were subsequently injected into a 10-μm-gap glass cell equipped with 1 cm^2 ITO electrodes and a polyimide alignment layer (LX-1400, Hitachi Chemicals Co.) for the measurements. The samples were heated to a temperature where the FLCs exhibit the isotropic phase. The samples were then

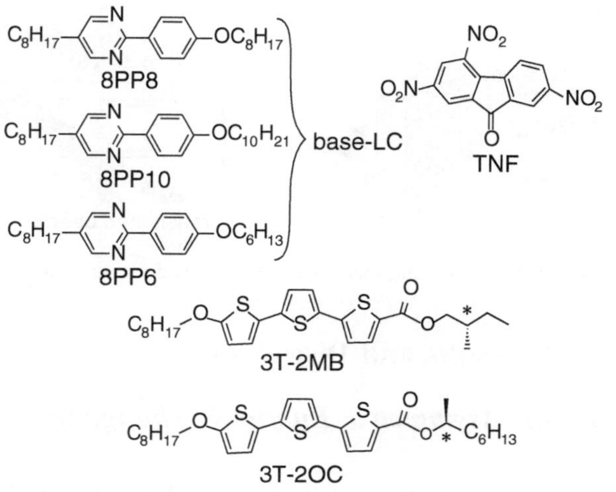

Fig. 10.3 Structures of the smectic LCs (8PP8, 8PP10 and 8PP6), photoconductive chiral dopants (3T-2 MB and 3T-2OC) and the sensitizer TNF

gradually cooled to ambient temperature at a rate of 0.1 °C/min. This process is necessary for preparing uniformly aligned FLC samples.

10.2.2 Measurements

Phase transition temperatures were determined using differential scanning calorimetry (DSC; DSC822, Mettler) and microscopic observations (FP-80, FP-82, Mettler; BX-50 polarizing microscope, Olympus). Spontaneous polarization (Ps) was measured by the triangular waveform voltage method (10 $V_{p\text{–}p}$, 100 Hz). The photorefractive effect was evaluated by a two-beam coupling experiment. A linearly polarized beam from a diode-pumped solid state laser (DPSS laser, Spectra Physics, Cyan-PC13689, 488 nm, continuous wave or Oxxious, 473S-50-COL-PP, 473 nm, continuous wave) was divided in two by a beam splitter; the two beams were then interfered in the sample film. A p-polarized beam was used in most of the experiments in this study. The laser intensity was 1 mW/cm^2 for each beam (1 mm diameter). The orientation of the rubbing direction and the beam incidence plane are shown in Fig. 10.4. The incident beam angles to the glass plane were 40° and 60°. The interval of the interference fringe was 1.87 μm. The sample was maintained at 25 °C using a thermo-controller (DB1000, Chino Co.). An electric field in the range of 0–10 V/μm was applied to the sample from a regulated DC power supply (Kenwood DW36-1), and the change in the transmitted beam intensity was monitored by photodiodes (ET-2040, Electro-Optics Technology, Inc.) and recorded by an oscilloscope. The formation time for the refractive index grating in the FLC was measured based on the simplest single-carrier model of photorefractivity [4, 5], in which the gain transient is exponential. The rising signal of the two-beam coupling was fitted by a single exponential function.

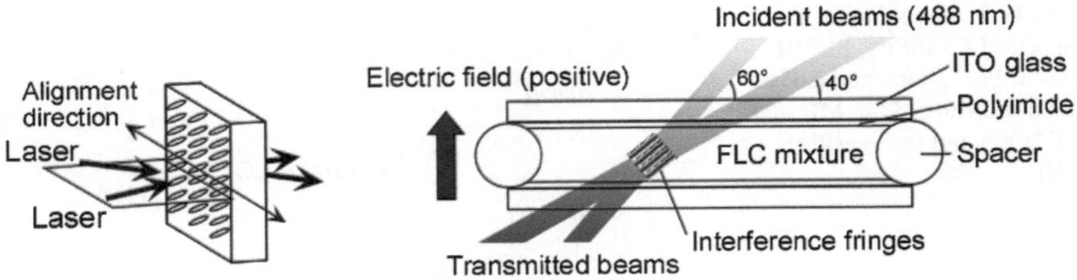

Fig. 10.4 Laser beam incidence condition and the structure of the LC cell

10.3 Results and Discussions

10.3.1 Asymmetric Energy Exchange in FLC Mixtures

The photorefractivity of the FLC mixtures was evaluated by two-beam coupling experiment. Figure 10.5 shows a typical example of the asymmetric energy exchange observed in a mixture of the base LC, 3T-2 MB and TNF at 25 °C with application of an electric field of 1 V/μm. Interference of the laser beams in the FLC mixture resulted in increased transmittance of one of the beams and decreased transmittance of the other beam. The increase and decrease of the transmittance intensities of these laser beams were reversed when the polarity of the applied electric field was reversed. Asymmetric energy exchange was observed only when an electric field was applied, that means that beam coupling was not caused by a thermal grating. As can be seen from Fig. 10.5, about 50 % of the intensity of the laser beam L_2 transferred to the L_1 beam.

In order to calculate the two-beam coupling gain coefficient, the diffraction condition needs to be correctly identified. The two possible diffraction conditions are the Bragg regime and the Raman-Nath regime, distinguished by the dimensionless parameter Q [2].

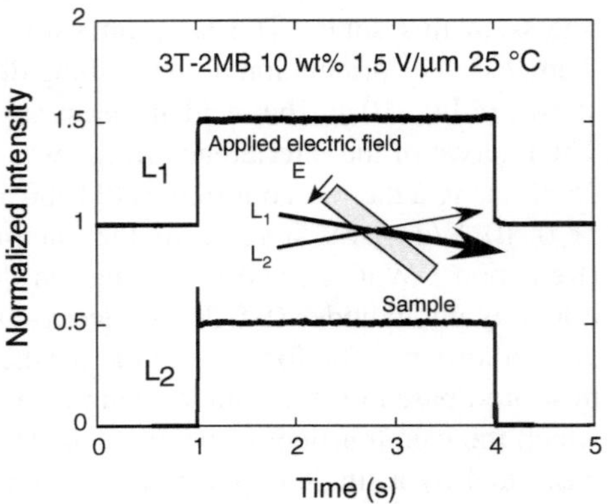

Fig. 10.5 Typical result of two-beam coupling experiments for a ternary mixture base LC, 3T-2 MB and TNF measured at 25 °C. The pump beam was incident at 1 s and closed at 4 s

$$Q = 2\pi\lambda L/n\Lambda^2, \tag{10.1}$$

where λ is the wavelength of the laser, L is the interaction path length, n is the refractive index, and Λ is the grating spacing. The Bragg regime of optical diffraction is defined as $Q > 1$, and excludes multiple scattering to produce only one order of diffraction of light. $Q < 1$ is defined as the Raman-Nath optical diffraction regime, in which many orders of diffraction can be observed. A Q value greater than 10 is usually required to guarantee that diffraction occurs entirely in the Bragg regime. Under the present experimental conditions, Q is calculated to be 6.2. Therefore, the diffraction observed in this experiment occurs predominantly, but not entirely, in the Bragg regime, with a small Raman-Nath component. However, because higher-order diffraction was not observed, the two-beam coupling gain coefficient Γ was calculated assuming Bragg diffraction [1–5].

$$\Gamma = 1/D\ln(gm/(1 + m - g)), \tag{10.2}$$

where $D = L/\cos(\theta)$ is the interaction path length for the signal beam (L = sample thickness, θ = propagation angle of the signal beam in the sample), g is the ratio of the signal beam intensities behind the sample with and without a pump beam, and m is the ratio of the beam intensities (pump/signal) in front of the sample.

The gain coefficients of the samples were measured as a function of the applied electric field strength (Fig. 10.6). The gain coefficient was calculated to be 1200 cm^{-1} in the 10 wt% sample with the application of only 1.5 V/μm. This gain coefficient is much higher than those of FLCs reported to date [8–13]. It was considered that a higher transparency of the LC mixture contributed to the high gain coefficient. The small electric field required to activate the photorefractive effect in FLCs is a great advantage for photorefractive devices. The response time decreased

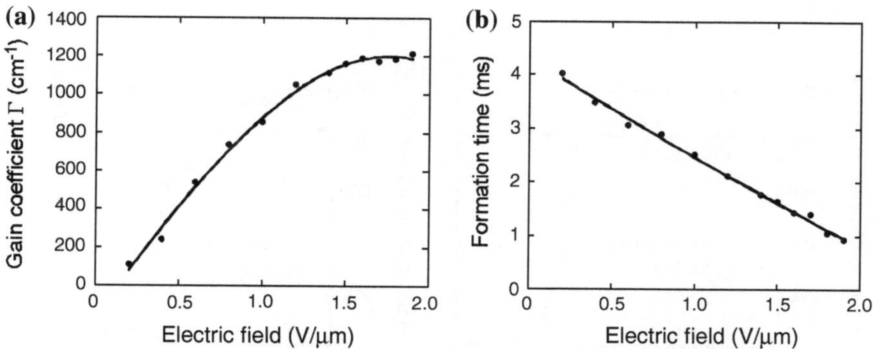

Fig. 10.6 **a** Electric field dependence of the gain coefficients for mixtures of the base LC, 3T-2 MB (10 wt%), and TNF (0.1 wt%) measured at 25 °C. **b** Refractive index grating formation times (response time) for mixtures of the base LC, 3T-2 MB (10 wt%), and TNF (0.1 wt%) measured at 25 °C

with increasing electric field strength due to the increased charge separation efficiency. The shortest formation time obtained was 0.9 ms for an external electric field of 1.9 V/μm. The large gain and fast response are advantageous for realizing optical devices such as real-time image amplifiers and accurate measurement devices [16].

10.3.2 Temperature Dependence of the Photorefractive Effect of FLCs

The temperature dependence of the gain coefficient of the 3T-2 MB samples is shown in Fig. 10.7. Asymmetric energy exchange was observed at temperatures below the SmC*-SmA phase transition temperature. The helical pitch of the 3T-2 MB samples observed in the polarizing microscope is plotted as a function of temperature in Fig. 10.8. Similarly, the helical pitch diverged when the temperature approached the phase transition temperature. In our previous study, the asymmetric energy exchange in an FLC sample was observed only in the temperature range in which the sample exhibits spontaneous polarization. Thus, the photorefractive effect appears only in the temperature range in which the sample exhibits ferroelectric properties.

10.3.3 Effect of the Concentration of Photoconductive Chiral Dopant

Figure 10.9 shows the gain coefficient of the samples with different 3T-2 MB concentrations plotted as a function of the magnitude of the external electric field. The gain coefficient increased with the strength of the external electric field up to

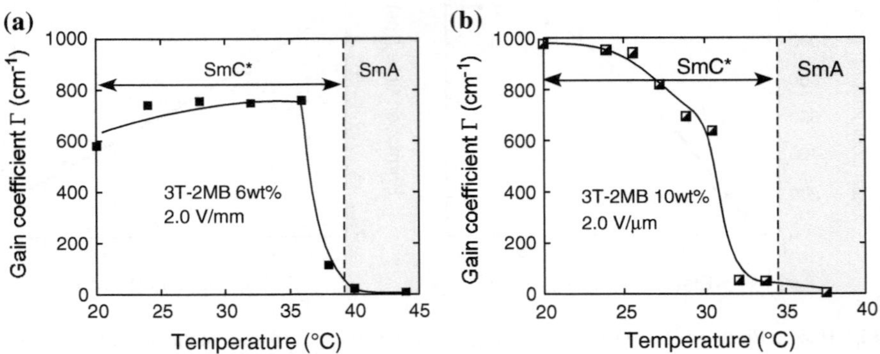

Fig. 10.7 Temperature dependence of gain coefficients and formation times. The concentration of 3T-2 MB was **a** 6 wt% and **b** 10 wt%

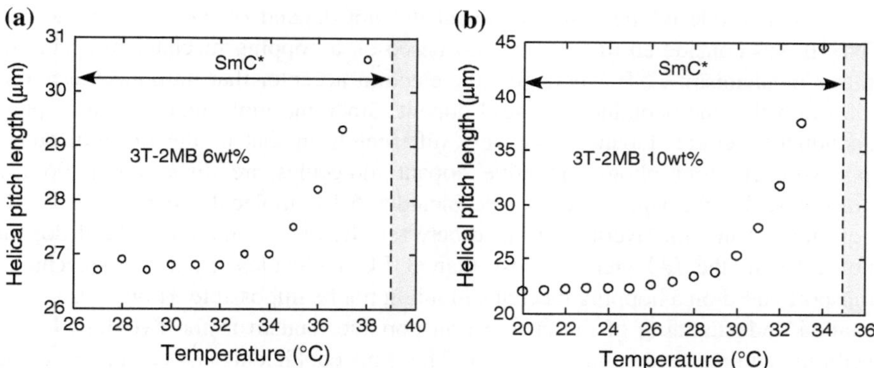

Fig. 10.8 Temperature dependence of helical pitch lengths. The concentration of 3T-2 MB was **a** 6 wt% and **b** 10 wt%

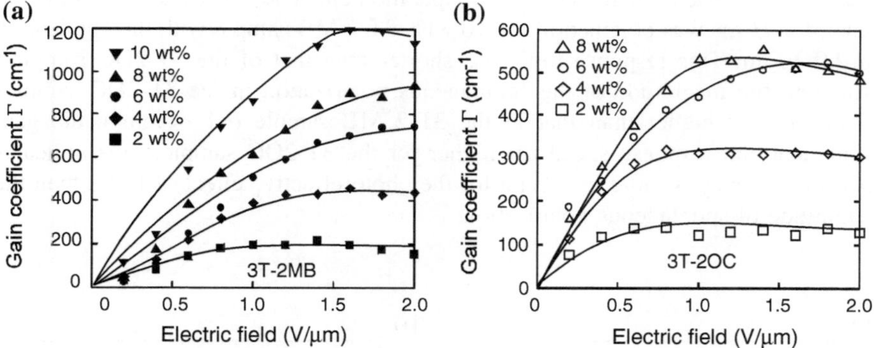

Fig. 10.9 **a** Magnitude of the gain coefficients of the mixtures of the base LC, 3T-2 MB and TNF (0.1 wt%) as a function of the strength of the applied external electric field measured at 25 °C. The concentrations of 3T-2 MB were in the range of 2–10 wt%. **b** Magnitude of the gain coefficients of the mixtures of the base LC, 3T-2OC and TNF (0.1 wt%) as a function of the strength of the applied external electric field measured at 25 °C. The concentrations of 3T-2OC were in the range of 2–10 wt%

0.2–0.6 V/μm and reached a constant value. It is considered that the decrease in the gain coefficient at a high external electric field is due to the realignment of the FLC molecules being restricted because the strength of the external electric field exceeded that of the internal electric field. As increasing the concentration of the photoconductive chiral dopants, the gain coefficient became larger. This may be due to increase in photoconductivity of the FLC medium and an increase in the magnitude of Ps.

The magnitude of the gain coefficient did not depend of the concentration of TNF. It was considered that conduction based on a hopping mechanism occurred. In this mechanism, TNF acts as just the electron acceptor that introduces electron holes into the photoconductive chiral dopant. Since the molecular weight of photoconductive chiral dopant is not very different from that of the LC molecules, approximately four photoconductive dopant molecules are dispersed in 96 LC molecules. A cube wherein each side includes 5 LC molecules contains 125 LC molecules. Thus, the average distance between the photoconductive chiral dopant molecules in the LC phase is less than 3 LC molecules. In this case, charge transport based on a hopping mechanism might not be impossible. However, a more detailed understanding of the charge transport mechanism in the dye-doped FLC medium requires further investigation. The gain coefficient was calculated to be 1200 cm^{-1} in the 10 wt% 3T-2 MB sample with the application of only 1.6 V/μm. This gain coefficient is twice as high as that of the 3T-2OC sample.

The electric field dependences of the response time of the samples is shown in Fig. 10.10. As increasing the strength of the external electric field, the response time was shortened due to increased charge separation efficiency. The shortest formation time of 0.93 ms was obtained in the 10 wt% 3T-2 MB sample with the application of 2.0 V/μm. This response time was shorter than that of the 3T-2OC samples. Although the magnitude of the spontaneous polarization in the 3T-2OC samples (5 nC/cm^2) is higher than that in the 3T-2 MB sample (<1 nC/cm^2), the gain coefficient and response speed are higher for the 3T-2OC sample. This indicates that transparency is more important for the photorefractive effect of FLCs than the magnitude of spontaneous polarization.

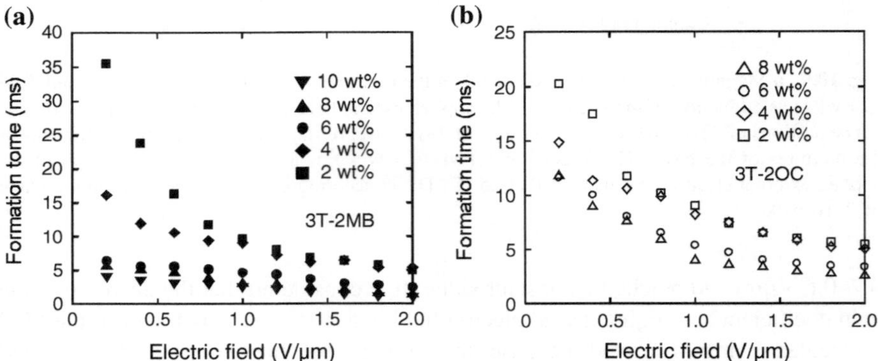

Fig. 10.10 **a** Electric field dependences of the refractive index grating formation time (response time) of the mixtures of the base LC, 3T-2 MB and TNF (0.1 wt%) measured at 25 °C. The 3T-2 MB concentrations were in the range of 2–10 wt%. **b** Electric field dependences of the refractive index grating formation time (response time) of the mixtures of the base LC, 3T-2OC and TNF (0.1 wt%) measured at 25 °C. The 3T-2OC concentrations were in the range of 2–10 wt%

10.3.4 Optical Image Amplification

Optical image amplification was demonstrated in this work (Fig. 10.11). A computer-generated image was displayed on a spatial light modulator (SLM) and a 473 nm DPSS laser beam was irradiated. The laser beam containing the image (signal beam) was transmitted through the FLC sample and monitored by a CCD camera. When the pump beam (a beam divided from the signal beam before SLM) was incident to the FLC sample and interfered with the signal beam, the signal was amplified through asymmetric energy exchange. The crossing angle of the beams at 15° yielded the largest amplification and the best contrast. The intensity of the signal beam was amplified 6-fold compared to that without the pump beam.

A computer-generated animation was displayed on the SLM. The frame rate was 30 fps. A 473 nm beam was irradiated on the SLM and the reflected beam was incident on the FLC sample. A pump beam interfered with the beam from the SLM in the FLC sample. A laser beam containing the moving image of the animation was amplified by the incident pump beam (Fig. 10.12). This result shows that the response of the photorefractive FLC was fast enough to amplify the optical image in real time. If a typical photorefractive polymer with a response time of ~100 ms is used in place of the FLC sample, the amplification would not occur. In that case, although a still image can be amplified, the intensity of the image would be weakened to the original magnitude when the image starts to move at a video rate.

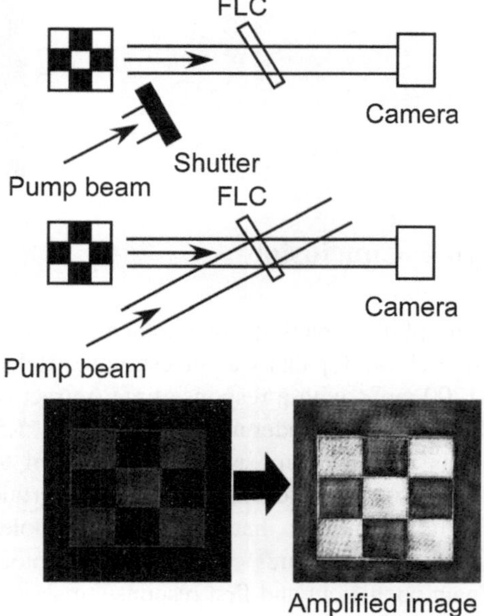

Fig. 10.11 Optical image amplification experiment. A computer-generated image was displayed on the SLM. The reflected beam form the SLM (473 nm) was irradiated on the FLC sample and interfered with the pump beam. The image transmitted through the FLC sample (3T-2 MB, 10 wt%) was monitored by a CCD camera

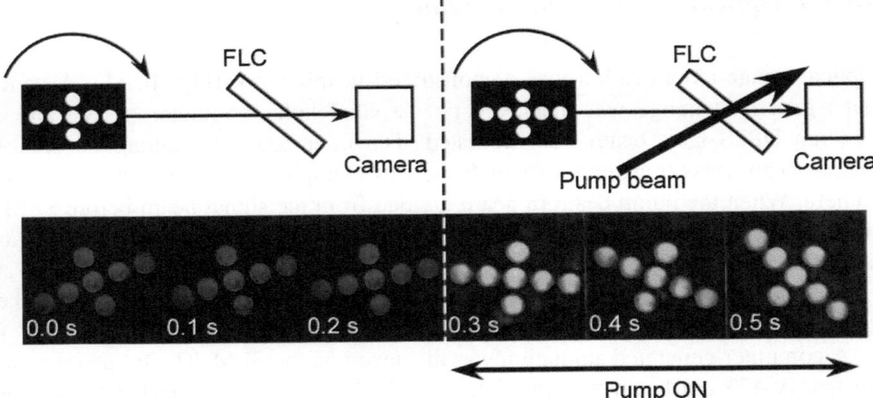

Fig. 10.12 Optical image amplification experiment. A computer-generated animation was displayed on the SLM. The reflected beam form the SLM (473 nm) was irradiated on the FLC sample and interfered with the reference beam. The image transmitted through the FLC sample (3T-2 MB, 10 wt%) was monitored by a CCD camera

Video 1 Optical image amplification experiment. http://www.rs.kagu.tus.ac.jp/photoref/Amplification_1_WMV.wmv

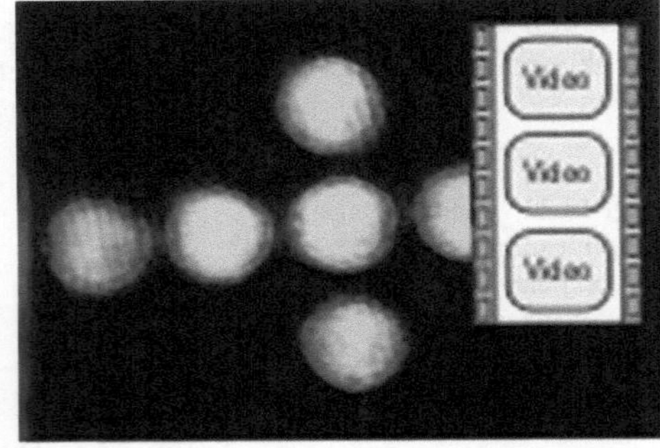

10.4 Conclusion

The photorefractivity of photoconductive FLC mixtures containing photoconductive chiral dopants was investigated. Relatively large gain coefficients higher than 1200 cm^{-1} with a response time (refractive index grating formation time) of 1 ms were obtained under application of only 1.5 V/μm in the SS-state of a FLC mixture. This response time is sufficiently short to realize real-time dynamic holograms. Because the molecular shape of the terthiophene-type photoconductive chiral dopants is similar to that of the base LC molecules, the miscibility with the base LC is high. FLC mixtures containing the photoconductive chiral dopants exhibited high gain coefficient and first response, making them useful for photorefractive devices. Real-time dynamic amplification of the optical image signal using the photorefractive ferroelectric liquid crystal was demonstrated.

References

1. Sasaki T, Ikegami M, Abe T, Miyazaki D, Kajikawa S, Naka Y (2013) Real-time dynamic hologram in photorefractive ferroelectric liquid crystal with two-beam coupling gain coefficient of over 800 cm^{-1} and response time of 8 ms. Appl Phys Lett 102:063306
2. Solymar L, Webb JD, Grunnet-Jepsen A (1996) The physics and applications of photorefractive materials. Oxford University Press, New York
3. Yeh P (1993) Introduction to photorefractive nonlinear optics. Wiley, New York
4. Moerner WE, Silence SM (1994) Polymeric photorefractive materials. Chem Rev 94:127–155
5. Kippelen B, Peyghambarian N (2002) Advances in polymer science, polymers for photonics applications II. Springer, Heidelberg, pp 87–156
6. Ostroverkhova O, Moerner WE (2004) Organic photorefractives: mechanisms, materials and applications. Chem Rev 104:3267–3314
7. Sasaki T (2005) Photorefractive effect of liquid crystalline materials. Polym J 37:797–812
8. Tay S, Blanche PA, Voorakaranam R, Tunc AV, Lin W, Rokutanda S, Gu T, Flores D, Wang P, Li G, Hilarie P, Thomas J, Norwood RA, Yamamoto M, Peyghambarian N (2008) An updatable holographic three-dimensional display. Nature 451:694–698
9. Blanche PA, Bablumian A, Voorakaranam R, Christenson C, Lin W, Gu T, Flores D, Wang P, Hsieh WY, Kathaperumal M, Rachwal B, Siddiqui O, Thomas J, Norwood RA, Yamamoto M, Peyghambarian N (2010) Holographic three-dimensional telepresence using large-area photorefractive polymer. Nature 468:80–83
10. Sasaki T, Miyazaki D, Akaike K, Ikegami M, Naka Y (2011) Photorefractive effect of photoconductive liquid crystalline mixtures composed of photoconductive chiral compounds and liquid crystal. J Mater Chem 21:8678–8686
11. Wiederrecht GP, Yoon BA, Wasielewski MR (2000) Photorefractivity in ferroelectric liquid crystal composites containing electron donor and acceptor molecules. Adv Mater 12: 1533–1536
12. Talarico M, Termine R, Prus P, Barberio G, Pucci D, Ghedini M, Goelemme A (2005) Photorefractive properties of undoped chiral smectic C phases of cyclopalladated complexes. Mol Cryst Liq Cryst 429:65–76
13. Talarico M, Goelemme A (2006) Optical control of orientational bistability in photorefractive liquid crystals. Nature Mater 5:185–188
14. Sasaki T (2006) Photorefractive effect of ferroelectric liquid crystals. Chem Rec 6:43–51
15. Skarp K, Handschy AA (1988) Ferroelectric liquid crystals. Material properties and applications. Mol Cryst Liq Cryst 165:439–509
16. Koukourakis N, Abdelwahab T, Li MY, Höpfner H, Lai YW, Darakis E, Brenner C, Gerhardt NC, Hofmann MR (2011) Photorefractive two-wave mixing for image amplification in digital holography. Opt Express 19:22004–22023

Chapter 11
Is Higher Order Aberration Associated with Reduced Visual Acuity in Children?

Damber Thapa, William R. Bobier, Kaamran Raahemifar and Vasudevan Lakshminarayanan

Abstract The purpose of this study is to examine whether the retinal image quality metrics calculated from the optical aberrations of the eye account for variation in visual acuity (VA) as measured by high contrast single letter chart. Strehl ratios and modulation transfer functions (MTF) were calculated from the ocular aberrations obtained from Hartmann-Shack images of pre-school children aged from 3 to 6 years. The Strehl ratios were grouped into three VA groups of 6/6, 6/9 and 6/12 and compared. No significant differences in Strehl ratios were observed. This finding suggests that optical aberrations alone do not completely describe the visual performance of the eye.

11.1 Introduction

The human visual process of seeing objects starts with the formation of a retinal image. The retinal image is formed by the optical components of the eye, such as cornea and lens, by focusing the light onto the retina. The photoreceptors convert the image into electrical impulses and that are transmitted to the brain through nerve fibers. Therefore, vision (commonly expressed in terms of visual acuity (VA)) depends on both optical and neural components. A reduced VA is most commonly associated with refractive error of the eye. After the invention of aberrometers, [1]

D. Thapa (✉) · W.R. Bobier · V. Lakshminarayanan
School of Optometry and Vision Science, University of Waterloo, Waterloo, Canada
e-mail: dthapa@uwaterloo.ca

V. Lakshminarayanan
Department of Electrical and Computer Engineering, University of Waterloo, Waterloo, Canada

V. Lakshminarayanan
Department of Physics, University of Waterloo, Waterloo, Canada

K. Raahemifar
Department of Electrical and Computer Engineering, Ryerson University, Toronto, Canada

V. Lakshminarayanan and I. Bhattacharya (eds.), *Advances in Optical Science and Engineering*, Springer Proceedings in Physics 166,
DOI 10.1007/978-81-322-2367-2_11

which provide complete refractive elements of the eye, the term refractive error is usually referred to lower order aberration (i.e. defocus and astigmatism). The other higher order refractive elements are called higher order aberration (HOA). Several studies have been conducted to access if HOAs are associated with a reduced VA. Some studies [2, 3] have found that the visual performance of the eye was improved by reducing HOAs. In contrast, others have shown a very small or no effect of HOAs on visual performance [4, 5]. Thus, the relationship between HOAs and visual performance is not clear; however, the relationship is very important from a clinical point of view and considering this relationship may help to further develop an algorithm which best describes the visual performance of the eye.

It has been seen that when VA of pre-school children are measured while wearing their best refractive correction, not all will end up with 6/6 acuity. While the majority are 6/6 but many of them have 6/9 and a few have 6/12 acuities. This arises the following question: what factors are responsible for reducing VA in such a condition. While it may not be possible to rule out all non-optical (i.e. neural) factors, this paper looked at whether the reduced VA could in anyway be due to HOAs. To address this we compare the HOAs between three different VA groups (6/6, 6/9 and 6/12) in a large sample of pre-school children. Many studies have been conducted to examine whether or not the retinal image quality metrics calculated from the aberration account for variations in visual performance (such as VA) in the adult population but it is still unclear that whether the same results hold for a pre-school population, when the optical structure of an individual is changed. To examine this we compare the retinal image quality metrics, such as Strehl ratio and area under the modulation transfer function (MTF) of young eye population (3–6 years) between three different acuity groups. The Strehl ratio that characterizes the retinal image quality can be computed by taking the ratio of the area under the MTF of the real eye and area under the MTF of diffraction limited eye.

11.2 Methods

11.2.1 Subjects

For this analysis, archived records were retrieved from investigations in which the Welch Allyn SureSight autorefractor (Wlech Allyn, NY) was used to screen for refractive errors in pre-school children in Oxford County, Ontario, Canada between 2000 and 2006. These investigations selected pre-school children who attended a vision screening program that was held as part of a county-wide health fair in Oxford County. The details of the study and refractive findings have been published elsewhere [6]. These records included cyclopleged refractive measures using the Welch Allyn SureSight Autorefractor, retinoscopy measurements and VA taken with the Cambridge Crowding Cards (Clement Clarke, Co, UK). A total of 781

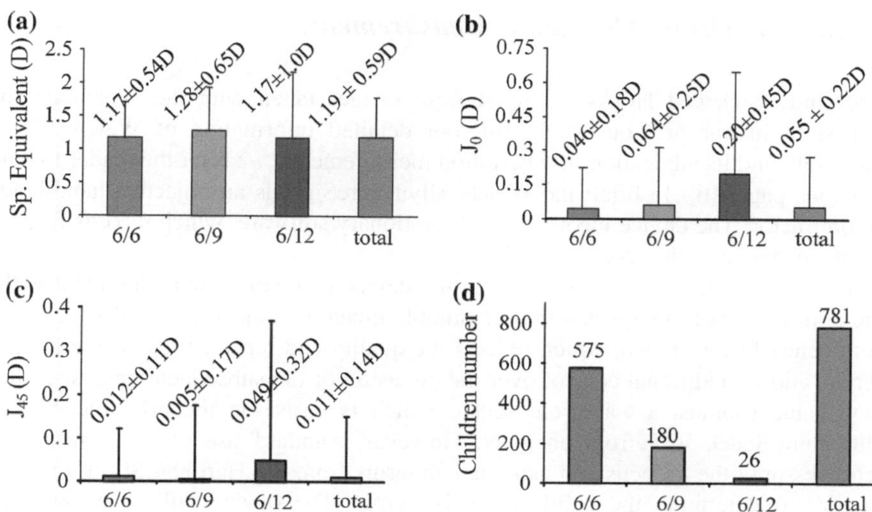

Fig. 11.1 Demographic summary of the subject divided in terms of visual acuity

eyes (right eyes 404 and left eyes 377) of 416 pre-school children (mean age 3.94 ± 0.94 years, range 3–6 years) were selected for this study. The sample showed a mean spherical equivalent of 1.19 ± 0.59D, a mean with-the-rule astigmatism of 0.055 ± 0.22D, and a mean oblique astigmatism of 0.011 ± 0.14D. Only visually normal records were considered, cases of binocular vision problems and amblyopia for example were omitted. The corrected VA of the subjects varied from 6/6 to 6/12. For comparisons, the subjects were divided into three groups (6/6, 6/9 and 6/12). A complete description of the subjects grouped by VAs is shown in Fig. 11.1.

11.2.2 Visual Acuity Measurements

The Cambridge Crowding Cards were used to measure the VA of pre-school children because they eliminate the need for children to name the letters. Children were asked to match the letters on the card to those on a matching board. VA test was conducted in accordance to strict methodological procedures outlined by the manufacturer. VA was measured monocularly while the fellow eye was occluded. Acuities were taken using single isolated letters while the child's refractive error was corrected and the eyes were cyclopleged. The examiners indicated the level of confidence for each acuity measure since children of this age are prone to losing their attention and cooperation as procedures become extended. Only visual acuities recorded with the highest level of co-operation and attention during testing were selected for this study.

11.2.3 Wavefront Aberration Measurements

The study reviewed HOAs extracted from images taken with the Welch Allyn SureSight in our previous study [6]. For detailed information of Welch Allyn SureSight and its adaptation for aberration measurements, we refer the reader to our previous paper [6]. In brief; the Welch Allyn SureSight is an objective hand-held autorefractor. The device incorporates discretionary software, which records five to eight images in rapid succession (2.4 s).

Images will be processed only when the device is in reasonable alignment with the optical axis of the eye and the measurable image is projected onto the CCD. A confidence factor (1–10), which reflects the quality and repeatability of the image series, allows additional control over the precision of the subsequent analysis. The device incorporates a wavefront sensor which is based on the principle of the Hartmann-Shack wavefront analyzer. However, standard use of the instrument retrieves only the defocus and astigmatism terms from the Hartman Shack images in rder to determine the childs refractive error. The device utilizes a 780 nm wavelength laser and a Sony ICX084AL CCD with unit cell size [7.4μm × 7.4μm].

11.2.4 Data Analysis

Zernike coefficients from 3rd to 8th orders (pupil diameter 5 mm) were used for the analysis. We calculated root mean square (RMS) values of total HOA, RMS of total coma (TC) ($C^{\pm1}{}_3$, $C^{\pm1}{}_5$ and $C^{\pm1}{}_7$), RMS of total trefoil (TT) ($C^{\pm3}{}_3$, $C^{\pm3}{}_5$ and $C^{\pm3}{}_7$), RMS of total spherical aberration (TSA) ($C_4{}^0$, $C_6{}^0$ and $C_8{}^0$) and RMS of total 3rd, 4th, 5th, 6th, 7th and 8th order aberrations. MTFs were calculated from the aberration structures of the children eye. The area under the MTF curve of each subject was calculated and the average area under the MTF curve was compared among the three different VA groups. Similarly, Strehl ratios calculated from the MTF curves were also compared between the VA groups. The data were analyzed using SPSS (version 17.0). Multivariate analysis of variance was used to examine the effect of VA on HOAs. Hotellings trace method calculated the significance levels with a significance level fixed at $p = 0.05$. If multivariate analysis of variance showed a significant difference in HOAs, then the Bonferonni post -hoc test was carried out to test multiple comparisons and the significance level was fixed at $p = 0.05/3 = 0.016$ to maintain the overall significance level of ($p < 0.05$).

11.3 Results

We calculated the mean ± standard deviation (STD) of HOAs for three different VA groups. The mean values are shown in Table 11.1. The multivariate analysis of variance calculated for total HOA, TC, TT and TSA showed that these aberrations were significantly different ($p = 0.021$) among VA groups. When individual aberrations were examined (Bonferonni post -hoc test) only the TT was significantly different ($p < 0.016$) among different acuity groups (Table 11.1). The TT of 6/6 VA group was significantly lower ($p < 0.016$) than that of the 6/12 VA group. The rest of the inter-acuity comparisons were not significantly different ($p > 0.016$). Mean values of total HOA, TC, TT and TSA are shown in Fig. 11.2.

The mean values of area under MTFs of different acuity groups were also compared. The area under the MTF varied across individuals with the STD of 5.99 arbitrary units (a. u.). The mean (±STD) area under the MTF for the 6/6 VA group was 21.57(±5.4) a. u., for 6/9 was 21.14(±5.8) a. u. and for 6/12 was 20.3(±6.0) a. u. One-way ANOVA was carried out to examine the differences in areas under the MTF curves across three different VA groups. The mean areas were not significantly different ($p = 0.381$). Similarly, the average (±STD) Strehl ratio of the 6/6 VA group was 0.516(±0.13), for 6/9 it was 0.506(±0.14), and for 6/12 it was 0.485 (±0.14). The average Strehl ratios were not statistically significant (One way ANOVA, $p = 0.381$) across VA groups. The average Strehl ratios are shown in Fig. 11.3.

Table 11.1 Mean values of higher order aberrations between different levels of visual acuity

Aberrations	6/6	6/9	6/12	*p*-value
HOA	0.1828 ± 0.0640	0.1922 ± 0.0755	0.2173 ± 0.1141	0.019
TC	0.1176 ± 0.064	0.1222 ± 0.0732	0.1329 ± 0.0922	0.422
TT	0.0848 ± 0.0431	0.0923 ± 0.0468	0.1197 ± 0.0803	0.000
TSA	0.0561 ± 0.0391	0.0552 ± 0.0394	0.0624 ± 0.0366	0.687
3rd order	0.1447 ± 0.0652	0.1536 ± 0.0739	0.1758 ± 0.1151	0.038
4th order	0.0820 ± 0.0381	0.0871 ± 0.0405	0.0959 ± 0.0349	0.077
5th order	0.0424 ± 0.0174	0.0434 ± 0.0195	0.0466 ± 0.0243	0.436
6th order	0.0300 ± 0.0123	0.0305 ± 0.0125	0.0351 ± 0.0171	0.136
7th order	0.0223 ± 0.0096	0.0224 ± 0.0097	0.0270 ± 0.0138	0.055
8th order	0.0185 ± 0.0076	0.0185 ± 0.0076	0.0213 ± 0.0087	0.186

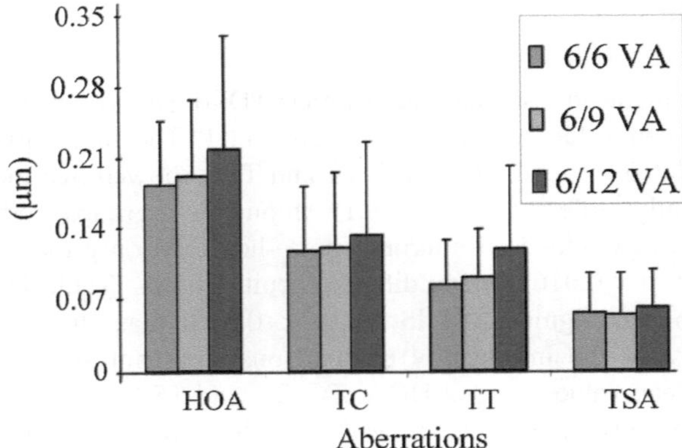

Fig. 11.2 Mean values of total higher order aberration (*HOA*), total coma (*TC*), total trefoil (*TT*) and total spherical aberration (*TSA*). Significant difference ($p < 0.016$) in TT between 6/6 and the 6/12 VA groups was found whereas the other comparisons were not significant

Fig. 11.3 Mean strehl ratios of 6/6, 6/9 and 6/12 visual acuity groups. No significant difference ($p = 0.381$) in strehl ratios was found

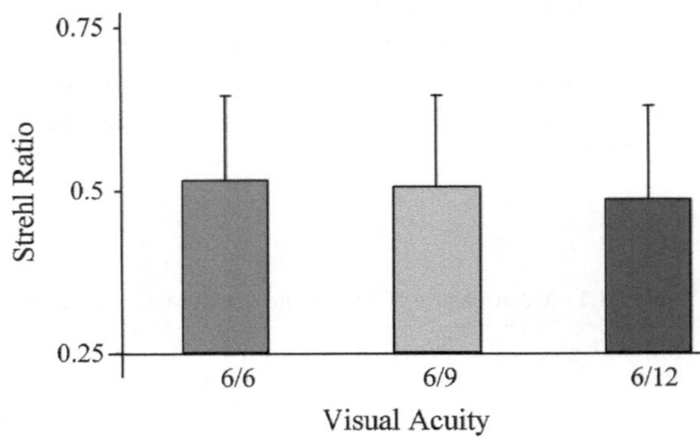

11.4 Discussion and Conclusion

In has been found that VA develops considerably from birth to the age of 1 year and after that it gradually increases; and around 3.5 years, the child usually acquire 6/6 level of VA for single letters [7]. Thus, the expected VA for these children tested with single letters would be 6/6 when eyes were cyclopleged and ametropia was corrected. However, the experimental data showed that VA of pre-school children varies widely between individuals ranging from 6/6 to 6/12. In this study, the data were retrieved from a large vision screening and follow up program that allowed us to group them into three different VA groups for comparative study. Although, there are many factors that can limit VA, such as cortical processing, photoreceptors density, ganglion cell density, intraocular scattering, neural noise and optical aberrations, this study only examined the effect of optical aberrations. The main objective was to examine whether or not measures of the monochromatic aberrations and resulting retinal image quality metrics, such as MTF and Strehl ratio, calculated from the aberration structure of pre-school children can account for variations in VA ranging from 6/6 to 6/12. Since, the sample included a small range of VA; the ocular aberrations were comparable across the different acuity groups. Although, RMS values of aberration were lower for better acuity groups, they were not significantly different amongst the different groups, except for TT. Our results are consistent with other studies carried out in adults [4, 5].

Area under the MTF curve and Strehl ratio characterize the retinal image quality due to the optical aberrations as they are not affected by the neural level and internal noise. We did not find significant differences in area under the MTF curves and Strehl ratios among different VA groups. This indicates that optical aberrations are not fully responsible for the visual performance of the eye. However, we measured VA using the high contrast single letter chart found in Cambridge Crowding Cards, the results could be different if the VA is measured in a way it is more sensitive to changes in optical quality because a study shows that the VA at low contrast level is more sensitive to HOA [8]. Similarly, new metrics that correlate better with visual performance may provide better prediction. A study has shown 31 optical quality metrics derived from optical aberrations to predict visual performance [9]. Inclusion of neural factors on the optical aberrations have also shown better prediction [10].

In summary, this study deals exclusively with the data which were collected retrospectively. We showed no significant difference in monochromatic HOAs and resulting retinal image quality metric calculated from the aberration structure of pre-school children among different VA groups (6/6, 6/9 and 6/12). The retinal quality metrics varied widely between individuals without showing a clear relationship with VA. In addition, subjects with reduced VA did not show any particular aberration pattern. This indicates that optical aberrations alone are not capable of perfectly describing the visual performance of the eye other factors, such as neural, are also responsible for the quality of vision.

Acknowledgments The authors acknowledge the industrial grant support from the Welch Allyn Company.

References

1. Rozema JJ, Van-Dyck DE, Tassignon MJ (2006) Clinical comparison of 6 aberrometers part 2: statistical comparison in a test group. J Cataract Refract Surg 32:33–44
2. Seiler T, Mrochen M, Kaemmerer M (2000) Operative correction of ocular aberrations to improve visual acuity. J Refract Surg 16:S619–S622
3. Williams D, Yoon G, Porter J, Guirao A, Hofer H, Cox I (2000) Visual benefit of correcting higher order aberrations of the eye. J Refract Surg 16:S554–S559
4. Levy Y, Segal O, Avni I, Zadok D (2005) Ocular higher-order aberrations in eyes with supernormal vision. Am J Ophthalmol 139:225–228
5. Villegas E, Alcn E, Artal P (2008) Optical quality of the eye in subjects with normal and excellent visual acuity. IOVS 49:4688–4696

6. Thapa D, Fleck A, Lakshminarayanan V, Bobier WR (2011) Ocular wavefront aberration and refractive error in pre-school children. J Mod Opt 58:1681–1689
7. Miller D (2003) Physiologic optics and refraction. In: Kaufman PL, Alm A (eds) Alders physiology of the eye. Mosby Inc. Missouri, pp 161–194
8. Applegate RA, Marsack JD (2005) Higher order optical aberrations predict visual acuity. IOVS 45:3613 E-Abstract
9. Marsack JD, Thibos LN, Applegate RA (2004) Metrics of optical quality derived from wave aberrations predict visual performance. J Vis 4:322–328
10. Lakshminarayanan V (2010) Model for predicting vision from a knowledge of wavefront aberrations. In: Khijwania S, Sharma AK (eds) [PHO-TONICS 2010] Proceedings of International Conference of Fiber Optics and Photonics, Viva Books, New Delhi, India, pp 165–170

Chapter 12
Design and Optimization of Silicon Photonic Devices

B.M.A. Rahman

Abstract Rigorous modal solutions of silicon nanowires are presented by using a full vectorial finite element method (FEM). Single mode operation range of these guides and full-vectorial modal field profiles for silicon based nanowires are presented. Results for novel vertical and horizontal slot waveguides are also presented showing TE and TM modes guidance through the low index dielectric regions including their design optimization. Designs of compact polarization rotators and highly sensitive biosensors are also presented.

12.1 Introduction

When the dimensions of an optical waveguide are much smaller than its operating wavelength, unique materials and structural dependent properties can be observed and these recently have been receiving much attention. In this regard, silicon (Si) has been particularly attractive as the low-cost and mature CMOS fabrication technology widely used in the electronics industry can be exploited. The high index contrast of Si allows light confinement in sub-micron size waveguides, along with the creation of very compact bends, to allow increased functionality of photonic integrated circuits.

12.2 Theory

A full-vectorial Finite Element Method (FEM) [1] is used in this paper to obtain the modal solutions of Si nanowire waveguides. The FEM, based on the vector-**H**-Field formulation, has been established as one of the most accurate and also numerically

B.M.A. Rahman (✉)
City University London, Northampton Sq, London EC1V 0HB, UK
e-mail: B.M.A.Rahman@city.ac.uk

V. Lakshminarayanan and I. Bhattacharya (eds.), *Advances in Optical Science and Engineering*, Springer Proceedings in Physics 166,
DOI 10.1007/978-81-322-2367-2_12

efficient approaches to obtaining the modal field profiles and propagation constants of the fundamental and higher-order quasi-TE and TM modes. The full-vectorial formulation is based on the minimization of the following functional in terms of the nodal values of the full **H**-field vector,

$$\omega^2 = \frac{\int [(\nabla \times \mathrm{H})^* \cdot \varepsilon^{-1}(\nabla \times \mathrm{H}) + p(\nabla \cdot \mathrm{H})^*(\nabla \cdot \mathrm{H})]\, dxdy}{\int \mathrm{H}^* \cdot \mu \mathrm{H}\, dxdy} \tag{12.1}$$

where **H** is the full vectorial magnetic field, * denotes the complex conjugate and transpose, ω^2 is the eigenvalue where ω is the angular frequency of the wave and ε and μ are the permittivity and permeability, respectively. A rigorous **H**-field based full-vectorial modal analysis has been carried out, which is able more accurately to characterize the abrupt dielectric discontinuity of a high index contrast optical waveguide. It should be noted that an equivalent **E**-field formulation would not be accurate as this cannot satisfy the boundary conditions at the dielectric interfaces. Arising from the results of the analysis, the characteristics of single mode operation, the vector field profiles, the modal ellipticity of circular, planar Si nanowires and Si horizontal and vertical slot waveguides can be obtained for both the quasi-TE and quasi-TM modes by using this **H**-field formulation.

12.3 Results

12.3.1 Strip Nanowires

First, a full-vectorial simulation of a typical Si strip optical waveguide or Si photonic nanowire waveguide [2] is presented. Such a waveguide consists in general of a Si core with a small rectangular cross-section, surrounded by either SiO_2 or air. The thickness of the core waveguide is taken as 260 nm and that of the lower SiO_2 buffer layer as 1.50 μm in this study. The refractive indices of the rectangular Si core and the SiO_2 cladding are taken as 3.50 and 1.5 respectively at a wavelength of 1.55 μm. The variations of the effective index, n_e, with the width, W, for the fundamental H_y^{11} and higher order modes with the SiO_2 cladding are shown in Fig. 12.1. It can be observed that when the width of the waveguide is large it can support many modes and n_e is closer to the effective index of a 260 nm thick slab waveguide. As W decreases, n_e also decreases and becomes closer to the refractive index (RI) of SiO_2 where the mode reaches its cut-off. It is shown here that the cut-off widths for the H_y^{41}, H_y^{31}, H_y^{21} and H_y^{11} modes are 1000, 700, 400 and 200 nm, respectively.

The H_y field profile of the fundamental H_y^{11} mode shown in Fig. 12.2 clearly identifies the maximum intensity occurring at the core centre. It is also shown that symmetry exists along the vertical, Y- and horizontal, X-axes, and the mode extends

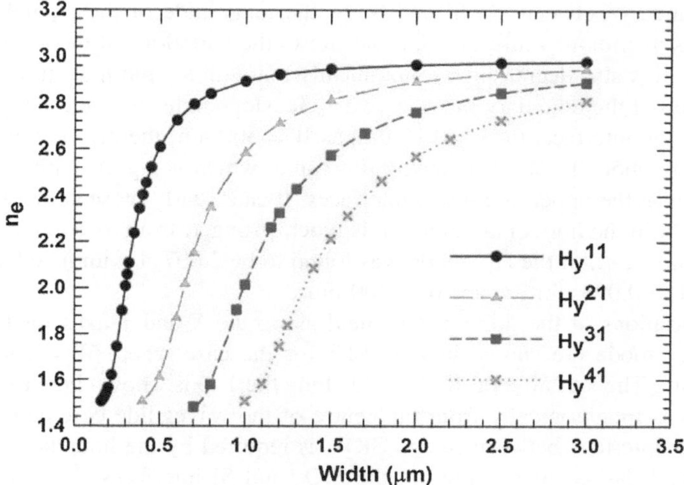

Fig. 12.1 Variations of the effective index, n_e with the waveguide width, for different quasi-TE modes for a rectangular strip waveguide with $H = 260$ nm

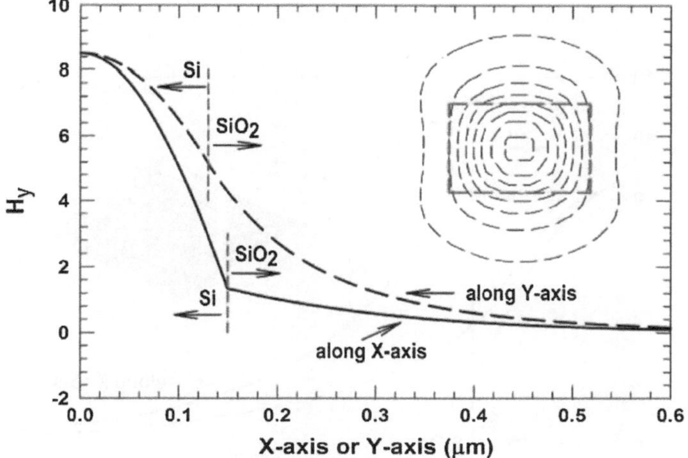

Fig. 12.2 Variations of the H_y field along X and Y-axes for the H_y^{11} mode with a SiO_2 buffer layer and cladding

considerably more into the top and bottom SiO_2 cladding region. To illustrate its variation more clearly, the variation of the H_y field along the Y-axis is also represented by a dashed line in Fig. 12.2. In this case, the H_y field is seen to be decreasing monotonically along the Y-axis and when $y = 0.13$ μm (as $H = 260$ nm), at the boundary between the Si and SiO_2, the magnitude of the H_y field reaches 60 % of its maximum value. A solid line shows the variations of the H_y field along the X-axis, this also decreasing monotonically but with a rapid reduction inside the Si core until at the boundary interface where its slope reduces in the SiO_2 cladding region. At the interface, (at $x = 0.15$ μm, as $W = 300$ nm) the magnitude of the H_y field is only about 16 % of its maximum value, which is significantly lower than field values at the upper and lower interfaces. It can clearly be seen that the modal confinement in the horizontal direction is much stronger, from its contour plot. The effective index, n_e, of the H_y^{11} mode was found to be 2.00711 with its effective area A_{eff}, equal to 0.098 μm^2, when $W = 300$ nm.

The variations of the dominant E_x field along the X and Y-axes for the fundamental H_y^{11} mode are shown in Fig. 12.3 for the case where $W = 300$ nm and $H = 260$ nm. The variation of the E_x field along the Y-axis, shown here as a dashed-line, reduces monotonically from the centre of the waveguide core and is continuous at the interface between Si and SiO_2, as required by the boundary condition. Two vertical dashed lines represent the SiO_2 and Si interfaces. However, the E_x field along the X-axis, shown here as a solid line, reduces more quickly in the core and at the Si/SiO_2 interface this increases abruptly in the SiO_2 region. Therefore the magnitude of the E_x field in the SiO_2 region can be significantly higher than that in the core region and this behavior is also demonstrated in as an inset shown in Fig. 12.3.

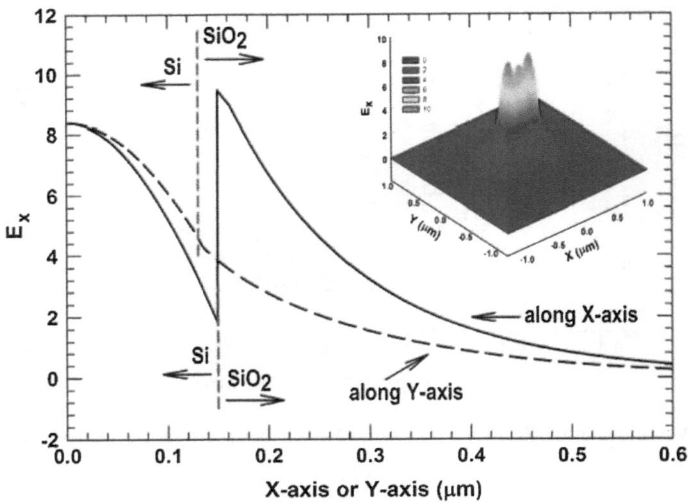

Fig. 12.3 Variation of the E_x field along the X and Y-axes for the H_y^{11} mode with SiO_2 buffer layer and cladding

12.3.2 Vertical Slot Nanowires

Recently, slot waveguide has emerged as a novel design where high field can be accessed in hollow or low-index region. In this case, the traditional Snell's law is no longer useful to explain the waveguiding in a low index region. The boundary condition at the dielectric interface demand that normal component of the **H**-field should be continuous. On the other hand, it also demands that the normal component of the **D**-vector should be continuous, which makes normal component of **E** vector not continuous. This rather increases the E_n field to be higher in the low index region. This fundamental concept has been exploited in optical slot waveguide where two narrow waveguides, both operating below their modal cut-off regions are brought close together to generate a high electric field in the low-index region of the waveguide. Next, such a Si slot optical waveguide is characterized. The waveguide can either be buried under a thick SiO_2 layer or surrounded by air. In this study the thickness of the Si waveguide is taken to be 300 nm where the SiO_2 (buffer) layer is of 1.50 μm dimension. The RI for the rectangular Si core and the SiO_2 are taken as 3.50 and 1.50, respectively at 1.55 μm wavelength.

The schematic cross section of the slot waveguide under investigation is shown as an inset in Fig. 12.4. The dominant H_y field component of the H_y^{11} mode is shown in Fig. 12.4 for the waveguide width, $W = 180$ nm and height, $H = 300$ nm with the slot width, $W_s = 50$ nm. Here, the outlined represents the boundary of the Si core. The H_y field profile shows that the maximum intensity occurs inside silicon core but close to the boundary between the Si waveguide and the SiO_2 slot and with significant H_y field magnitude extending into the slot region. It was noted that as W, increases, the dominant field component is shifted toward to the center of the Si core waveguide.

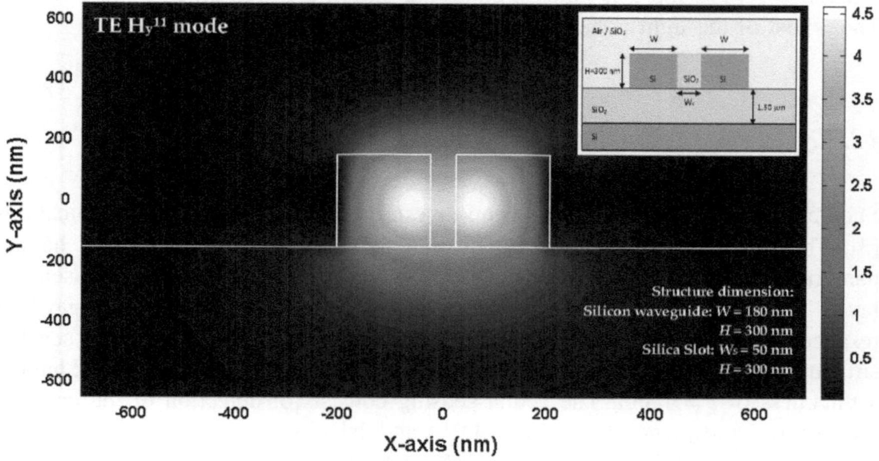

Fig. 12.4 The dominant H_y field profile of the H_y^{11} mode

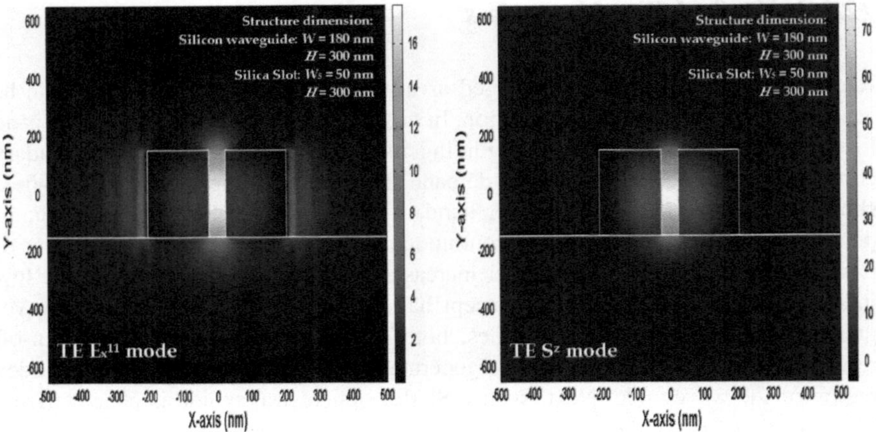

Fig. 12.5 The dominant E_x field profile of the H_y^{11} mode (*left*) and S^z field profile of the H_y^{11} mode (*right*)

As the width of the Si waveguide becomes significantly large, the field mode will be separated from the slot region and its maximum intensity will be concentrated more near the centre of the Si core waveguides. The fundamental quasi-TE (H_y^{11}) mode contains all the three components of the electric field, E_x, E_y and E_z. The contour E_x field profile of the H_y^{11} mode is shown in Fig. 12.5 (*left*), where the Si core waveguide is also outlined. The E_x field at the Si/SiO_2 interface increases abruptly with a step change in the SiO_2 region, thereby showing that there is a strong **E**-field in the slot region. From the full-vectorial **E** and **H** fields, the Poynting Vector (S^z) may be calculated and this is shown in Fig. 12.5 (*right*). It is indicated that the S^z intensity is primarily concentrated in the low index SiO_2 slot-separation region and it is much higher than that in the high index Si core region. Thus most of the light is being guided through the low-index slot region.

12.3.3 Vertical Slot Based Optical Sensors

Such a novel slot waveguide has been investigated for the biosensing applications [3]. The slot waveguide is formed by two Si wires close to each other having nanometer dimensions as shown in Fig. 12.6 with water as top cladding and sensing layer. The RI of silicon, silicon oxide and water are taken as 3.476, 1.444 and 1.31 respectively, at the operating wavelength of 1550 nm. In this design, the sensing structure is first coated with a linker layer (silanes) whose RI is taken as 1.42 having a thickness of t = 1 nm. The actual sensing consists of detection of the complementary DNA sequence. The RI of ssDNA and dsDNA are taken as 1.456 and 1.53 respectively. The thickness of the DNA probe layer is taken as n = 8 nm, and remains unchanged when binding of complementary DNA strands (targets) to DNA

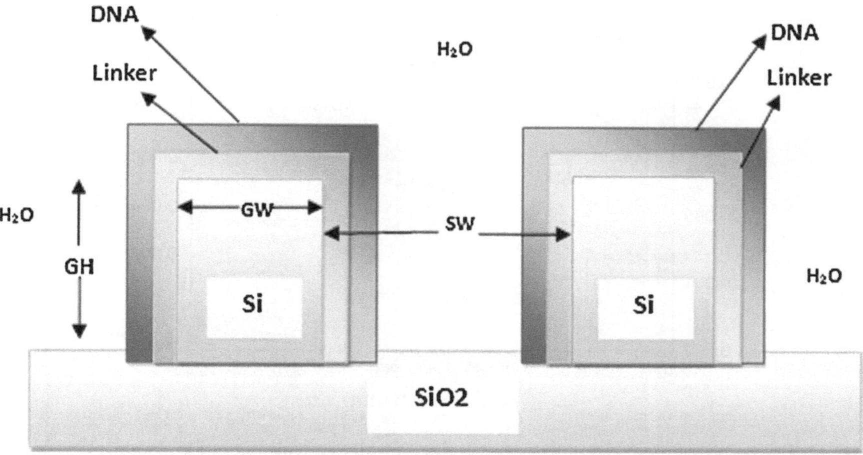

Fig. 12.6 Schematic diagram of a vertical slot-based bio-sensor

probes happens, i.e. only the RI changes from 1.456 (ssDNA) to 1.53 (dsDNA). A waveguide height, $GH = 320$ nm, and high index region width, $GW = 180$ nm, slot width, $SW = 100$ nm, linker layer thickness of $t = 1$ nm, and DNA probe thickness of $n = 8$ nm, are considered for the initial simulation study. In similar structures, by optimising design approaches, nearly 40–50 % power can be confined in the slot, but as the slot width is very small, $\sim$100 nm wide, it produces a very high power density, or intense electric field which can be exploited. The effective index change was produced either by a change of cover medium RI (homogeneous sensing) or by a change of thickness of DNA layer, which is immobilized on waveguide surface (surface sensing). The measurement sensitivity depends on optical field distribution in the sensing medium, so one of the most important design task is the waveguide optimization in order to maximize its sensitivity. The adlayer thickness and change of cover medium RI affects the effective index of propagating optical mode. The thickness of the DNA probe layer is taken as 8 nm and remains unchanged when binding of complementary DNA strands (targets) to DNA probes happens and the RI of ssDNA and dsDNA are taken as 1.456 and 1.53 respectively. The waveguide sensitivity can be written as

$$S = \Delta n_{\mathrm{eff}}/\mathrm{RI} \tag{12.2}$$

where RI is the DNA layer RI and Δn_{eff} is the effective index difference when ssDNA and dsDNA are present. The effective index difference is achieved by simulating first by adding ssDNA layer on top of linker layer and then replacing the ssDNA by a dsDNA layer.

A waveguide height, $GH = 320$ nm is fixed, and $SW = 60$, 100 and 140 nm is varied. The sensing layers are linker layer (thickness $t = 1$ nm and RI = 1.42), ssDNA (thickness $n = 8$ nm and RI = 1.456), and water respectively. The ssDNA

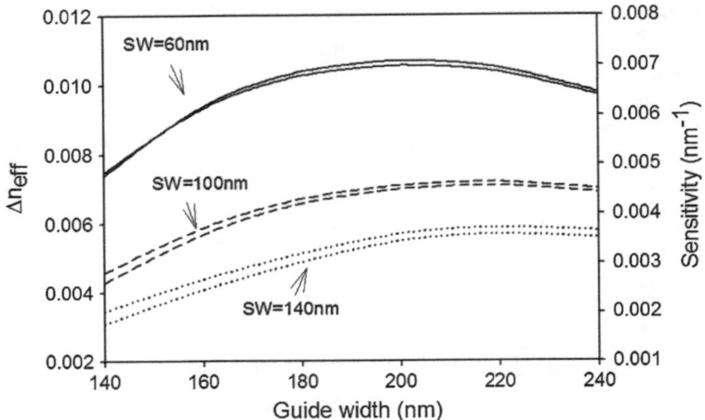

Fig. 12.7 Change of effective indices and sensor sensitivity with the guide width

layer is replaced with the dsDNA layer (thickness n = 8 nm and RI = 1.53) to achieve the difference between the two. Figure 12.7 shows larger effective index variation, and waveguide sensitivity is achieved at a guide width = 220 nm when the slot width is 60, 100 and 140 nm respectively. The greater the change in Δn_{eff}, the more sensitive the biosensor will be. Therefore, when the guide width is between 200 and 220 nm maximum index difference is achieved. Although a smaller slot width shows a more sensitive design, while considering the fabrication techniques available today, a 100 nm slot-width is seen as a suitable design.

12.3.4 Horizontal Slot

Next, the structure under investigation is a single horizontal slot optical waveguide. This type of waveguide is based on a low RI, SiO_2 slot sandwiched between two symmetric high-index Si core guides. Such waveguide with slot can easily be fabricated by using a SOI wafer on a Si substrate by layered deposition or thermal oxidation [4]. As they are grown layer by layer, it is expected to have better fabrication tolerances than the vertical slots waveguide. In this study the thickness of the Si waveguide, t_1 and t_2, taken as, silicon height, h, is taken to be 300 nm, the SiO_2 (buffer) layer is of a 1.50 μm and the width, w, of the waveguide is of 350 nm. The RI for the rectangular Si core at 1.55 μm wavelength is taken as 3.47. The RI for the SiO_2 cladding is taken as 1.45.

The quasi-TM (H_{11}^x) mode contains all the three components of the electric field, E^x, E^y and E^z: the E^y field is dominant, and the E^x and E^z field components are non-dominant. The E^y field profile of the H_{11}^x mode is shown as an inset in Fig. 12.8, where the Si horizontal waveguide is also outlined. It is clearly shown that due to the large index contrast at interfaces, the normal electric field undergoes a large

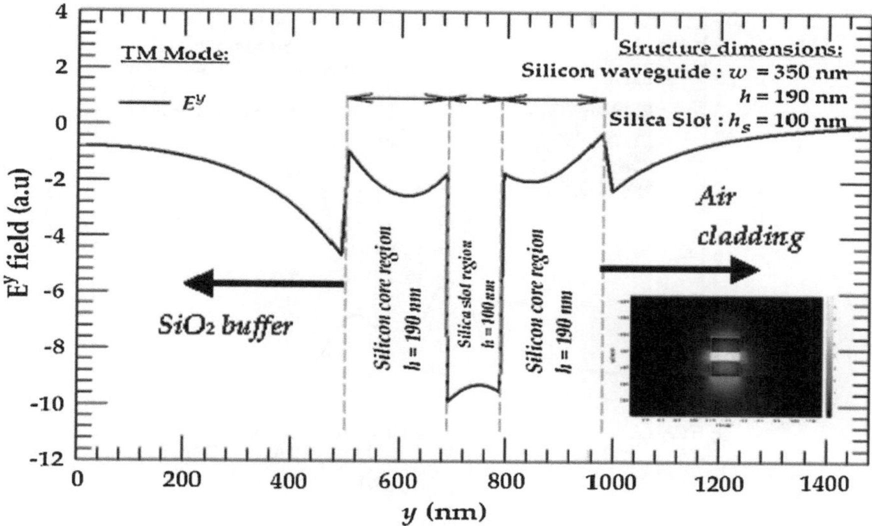

Fig. 12.8 Variations of E^y field along the y-axis for the quasi-TM H^x_{11} mode

discontinuity, which results in a field enhancement in the low-index region, where the E^y field is located.

In Fig. 12.8, the E^y field along the y-axis, shown that at the Si/SiO_2 interface the field increases abruptly with a step change in the SiO_2 region demonstrates that a strong **E**-field is present in the slot region. From the full vectorial **E** and **H** fields, the Poynting Vector (S^z) may be calculated. Figure 12.9 shows the S^z intensity along the y-axis, it shows that the S^z is at its maximum inside the slot area and it is significantly higher to the S^z intensity inside Si core 1 and Si core 2.

It is found that as the slot height, h_s, becomes large, the effective indexes decreases. The variation of the power confinement in the SiO_2 slot region with the core height, h, is shown in Fig. 12.10. For the TM mode of w = 500 nm and h_s = 100 nm, field is strongly confined when the core height is in the region of 150–200 nm and reaches its maximum at 170 nm when compared with two other waveguides dimensions. It is also shown that when Si core width is increased the power confinement in the slot increases. Figure 12.10 also shows that for the TM mode, a higher power confinement in the slot is achievable when the core height increases with the same width but of higher slot-height.

The high power confinement in silicon slot region can be exploited in the design sensitive biosensors (6 l). Here change in the effective indices with the presence of ssDNA is shown for a horizontal slot guide. Horizontal slot guide have lower modal loss and better fabrication tolerance but this guide only supports quasi-TM modes. The change in effective index values with the silicon guide width is shown in Fig. 12.11, indicating a very sensitive viable design. The high power confinement in silicon slot region can be exploited in the design sensitive biosensors (6 l). Here change in the effective indices with the presence of ssDNA is shown for a

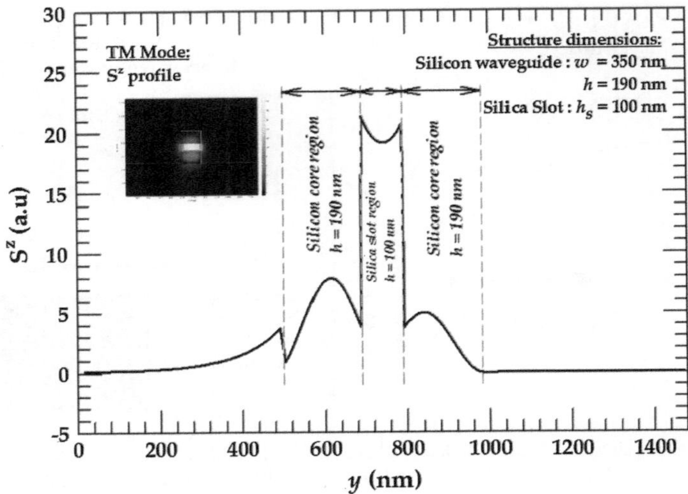

Fig. 12.9 Variations of S^z field along the y-axis for the quasi-TM H^x_{11} mode

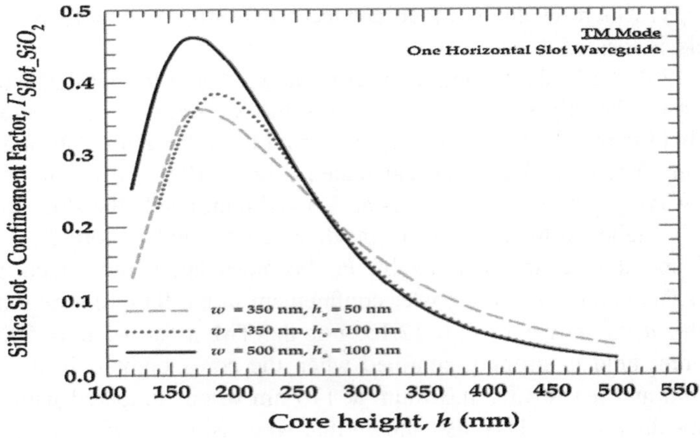

Fig. 12.10 Variations of power confinement in the slot region, $\Gamma_{Slot\_SiO_2}$, with the core height, h, for quasi-TE and TM modes

horizontal slot guide. Horizontal slot guide have lower modal loss and better fabrication tolerance but this guide only supports quasi-TM modes. The change in effective index values with the silicon guide width is also shown in Fig. 12.11, indicating a very sensitive viable design.

Polarization rotator is a key component in the implementation of a polarization diversity operation of modern optoelectronic system. Performance of a novel polarization rotator is shown, which only uses non-identical silicon nanowires [5].

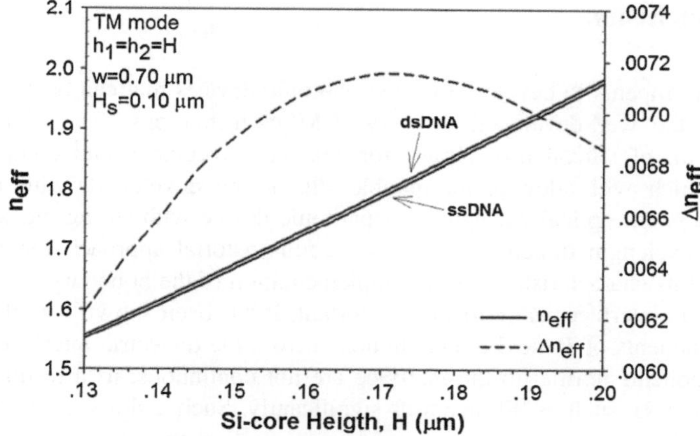

Fig. 12.11 Variations of power confinement in the slot region, $\Gamma_{Slot\_SiO_2}$, with the slot height, h_s, for quasi-TE and TM modes

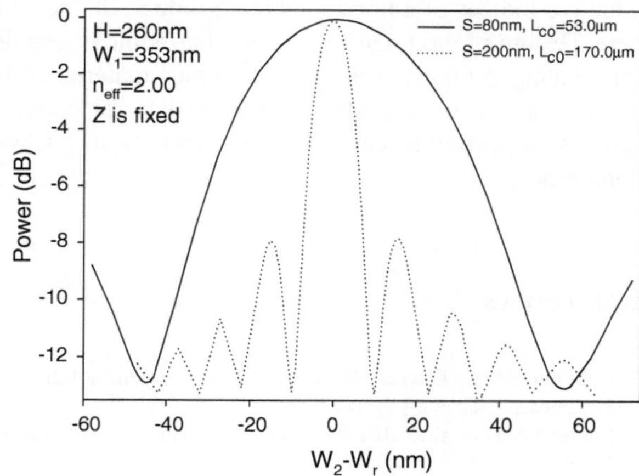

Fig. 12.12 Polarization conversion with the deviation in the waveguide width for simple polarization rotator exploiting TE and TM phase matching between two Si nanowires

In this case, height of the nanowires was 260 nm. The TE mode of a 353 nm wide Si nanowire is phase matched with the TM mode of a 595 nm wide Si nanowire. When they are separated by 200 nm gap, the resulting coupling length is 170 μm. However, if the separation distance is reduced to 80 nm then the coupling length is reduced to 53 μm. The performance of these compact polarization rotators with the change in the waveguide width is shown in Fig. 12.12. It can be observed, even when waveguide width, which might be changed due to fabrication tolerances, by a margin of ±20 nm, yet the performance remains better than 3 dB, indicating a viable design.

12.4 Conclusion

There is an impetus to develop low-cost photonic devices that can be furthered by exploiting the well-developed, low-cost CMOS technology. The higher index contrast, Δn, of silicon also allows for smaller waveguides and compact bend designs which will allow more reliable PIC to be developed, with increased functionality. An optical waveguide or photonic device with strong index contrast and sub-wavelength dimensions requires a full-vectorial approach for an understanding of its characteristics and the implementation of the boundary conditions at the dielectric interfaces are critically important. It has been shown here that the all three components of **H**-field are continuous across the dielectric interfaces, but **E**-field components normal to the interface are not continuous; as a result slot area with a lower RI, such as |**E**| increases significantly. Such a slot waveguide can be used to design very compact sensors where a small change in the RI in the slot region will have a significant measurable change in optical parameters for Mach-Zehnder or directional coupler based optical sensor system. Design of polarization rotator using polarized phase matching between two non-identical silicon nanowire have been exploited to achieve a very compact device. Change in the modal effective indices of a horizontal slot guide in the design of bio-sensor is also shown here. Design of modulator or second harmonic generation can also be considered by introducing a highly non-linear polymer material in the slot region. Some of the results for slot waveguides with potentials for creating nonlinear devices and for sensing applications and compact power splitters will also be presented in the conference.

References

1. Rahman BMA, Davies JB (1984) Finite element solution of integrated optical waveguides. J Lightwave Technol 2:682–688
2. Leung DMH et al (2010) Rigorous modal analysis of silicon strip nanoscale waveguides. Opt Express 18:8258–8539
3. Dar T, Homla J, Rahman BMA (2012) Label-free slot-waveguide biosensor for detection of DNA hybridization. Appl Opt 51:8195–8202
4. Sun R et al (2007) Horizontal single and multiple slot waveguides: optical transmission at λ = 1550 nm. Opt Express 15:17967–17972
5. Soudi S, Rahman BMA (2014) Design of compact polarization rotator using simple silicon nanowires. Appl Opt 53:8071–8077

Chapter 13
Global Synthesis of Optical Lens Systems

Lakshminarayan Hazra

Abstract The core optical functions of most optical/photonic systems are performed by a lens system. The design of such lens systems involves satisfying a set of objectives by means of a set of degrees of freedom provided by the construction variables; often each of the latter has specific physical and/or manufacturing constraints that need to be taken into account. Mathematically, the problem can be formulated as a problem of multi-objective nonlinear optimization in a constrained multivariate hyperspace. The large number of objectives as well as the large number of degrees of freedom even in case of a moderately complex problem precludes the direct use of any global optimization algorithm. In order to circumvent the limitations imposed by this 'Curse of Dimensionality', a prophylactic strategy for global synthesis of lens systems can be adopted. The strategy is likely to provide globally or quasiglobally optimal solutions for practical lens design problems. An outline of the strategy is reported.

13.1 Introduction

Most optical or photonic systems used in practice constitute a lens subsystem that carries out the core optical functions of the system. The term 'lens' is used in an extended sense to indicate all refracting, reflecting and diffractive elements used in the system. In practice, optimal design of these elements pertaining to specific requirements of different applications poses a major challenge in optical system design, except for the trivial cases. For the sake of illustration, Fig. 13.1 shows a typical multicomponent lens system that forms an image O′A′B′ of an object OAB. For the sake of simplicity the figure demonstrates the formation of a planar image on a transverse plane corresponding to a transverse planar object. In general, either

L. Hazra (✉)
Department of Applied Optics and Photonics, University of Calcutta,
JD 2 Sector III Salt Lake, Kolkata 700098, India
e-mail: lnhaphy@caluniv.ac.in; lnhazra@yahoo.com; lakshminarayanhazra@gmail.com

V. Lakshminarayanan and I. Bhattacharya (eds.), *Advances in Optical Science and Engineering*, Springer Proceedings in Physics 166,
DOI 10.1007/978-81-322-2367-2_13

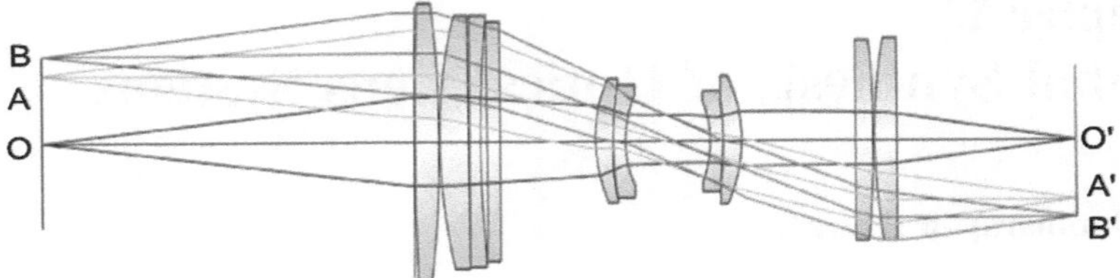

Fig. 13.1 A typical multicomponent optical lens system

one or both of the object or image may be located on a curved plane; also, either one or both of the object/image may be located at infinity. Again, some applications call for formation of a three-dimensional image of a three-dimensional object. Ideally there should be a one-to-one correspondence between the object and the image. For broad band systems, all rays of all wavelengths admitted by the optical lens system from an object point should meet at the corresponding image point in the image space. Real lens systems used in practice attempt to achieve these ideal goals as far as possible.

13.2 Designing a Lens System

The process of optical/lens design involves tackling an inverse problem (Fig. 13.2). In general, this involves determining an appropriate lens system which can perform the optical/imaging function in accordance with a set of specifications pertaining to the particular case.

Specifications made by the user usually consist of one or more from the following list:

- Object/Image size (Linear or Angular)
- Magnification (Transverse/Longitudinal/Angular)
- Object/Image location and/or object-to-image Throw

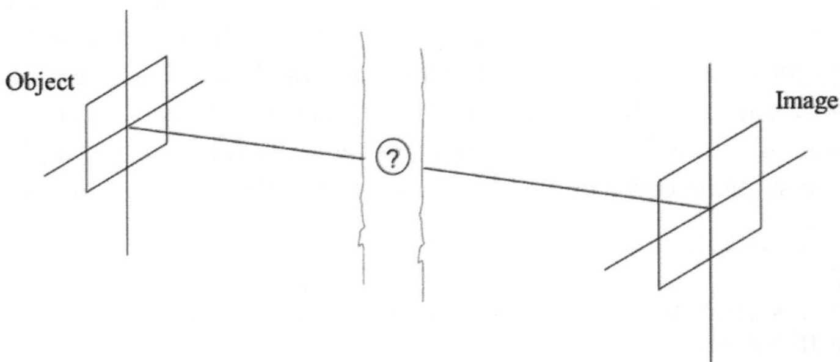

Fig. 13.2 Lens design is an inverse problem

- Nature of object: self-luminous/trans-illuminated/areal
- Image brightness
- Wavelength range of illumination (Broadband/Monochromatic/Narrowband)
- Detector: Eye/CCD/Film etc.
- Optical Material specifications

From these user specifications the designer has to work out the required optical specifications of the lens system that can perform the job. A list of optical specifications for conventional applications is given below:

1. Gaussian/paraxial requirements, e.g., focal length, F-no./Numerical Aperture etc.
2. One or more of the Image Quality Assessment Criteria. e.g. Optical Transfer Function (O.T.F.)/Point Spread Function (P.S.F.)/Edge Spread Function (E.S.F.)/Bar Spread Function (B.S.F.)
3. Image Brightness/Illumination
4. Depth of field/focus

The lens designer has also have to keep in view the following constraints:

(a) Physical realizability
(b) Manufacturability
(c) Material Availability
(d) Cost

The degrees of freedom available to the designer consist of a few or more from the following:

1. Number of lens elements
2. Curvatures
3. Glass thicknesses and Air separations
4. Stop position
5. Conic constants and Aspheric Coefficients
6. Optical (Glass) Materials.

13.3 Algorithm

The complexity of the problem of lens design arises out of the fact that, in general, the relations between the specifications and constraints with the degrees of freedom are highly nonlinear. Mathematically lens design can be formulated as a problem of nonlinear multi-objective optimization in a constrained, multivariate hyperspace.

Except for the trivial situations, all ventures for lens design take recourse to a recursive procedure, as illustrated in Fig. 13.3. Many commercial and in-house software packages are now routinely used for practical implementation of the methodology.

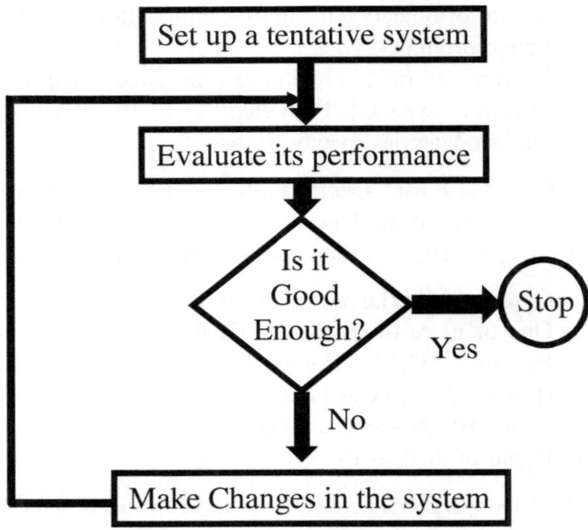

Fig. 13.3 A flowchart for methodology of lens design

13.4 Conclusion

In spite of ready availability of powerful optical design software, real success in optical system design calls for experience and ingenuity of the designer in working out the structural design layout for specific problems.

In order to facilitate the same, a prophylactic strategy to obtain globally or quasi-globally optimum solutions for the nondeterministic polynomial (NP) time hard problem of lens design optimization has been developed [1–4]. The inherent "curse of dimensionality" associated with this problem can be circumvented by taking recourse to a synergistic combination of top-down and bottom-up approaches. The strategy leads to a systematic approach for global synthesis of optical and electro-optical systems, obviating thereby the need for use of suboptimal makeshift solutions often used in practice.

The essential point in the whole approach is to realise that a "good" optical system can only be obtained by a combination of "bad" components. However, the defects of these bad components cannot be arbitrary defects; a good system for a given configuration calls for specific defects in its components or modules. The strategy advocated in this talk does not only lead to globally or quasi-globally optimum solutions for conventional optical systems, but it also provides cues for efficient use of unconventional optical elements in emerging applications.

For a complex system consisting of several modules, our strategy enables optimal modification of part/parts of the system or the system as a whole. Globally optimum values of desirable defects for each of the modules to be modified are determined first with the help of evolutionary programming. Next, a combination of semi-analytical procedures and global optimization techniques are utilized for synthesizing each of the defective modules. At each stage, appropriate measures are incorporated to prevent undesirable occurrence of higher order aberrations in the system. Lastly, the overall system obtained thereby can be re-optimized again with the help of available degrees of freedom.

Our prophylactic strategy, briefly enunciated as above, does not only reduce significantly the role of heuristics in optical design, but it also enables better understanding of the behaviour of modules comprising the optical system. The strategy is based on aberration theory, local optimization algorithms and evolutionary programming. Details of our prophylactic strategy has been enunciatedin [3].

References

1. Hazra LN (1984) Structural design of multicomponent lens systems. Appl Opt 23:4440
2. Banerjee S, Hazra LN (1998) Simulated annealing with constrained random walk in structural design of doublet lenses. Opt Eng 37(12):3260
3. Hazra LN, Chatterjee S (2005) A prophylactic strategy for global synthesis in lens design. Opt Rev (Japan) 12(3):247
4. Hazra LN, Pal S (2010) A novel approach for structural synthesis of zoom systems. Proc SPIE 7786:778607-1/11 (2010)

Chapter 14
Design and Analysis of Memristor Based Non-volatile Memories

Shyam Akashe

Abstract Emerging NVM devices like memristor, spin-transfer torque magnetic tunneling junction (STT-MTJ), phase-change memory has drawn great interests both from academia and industry. This article presents a new memory hierarchy system using next-generation non-volatile memory devices merging the characteristics of the conventional one. The utilization of non-volatile memory technology, has released new chances for low power, low cost, high density and higher endurance circuits which are very competitive and striving than DRAM as well as SRAM memory cells. By congregating the functions of DRAM, ROM, FLASH and SRAM, this non-volatile technology can lead to a new memory system which will simplify the architecture of the system and enhance its performance. There are different memory technologies of non-volatile nature in the literature such as FeRAM, STTRAM, RRAM, MRAM, PCRAM etc. These non-volatile memory technology acts as a backbone for providing required storage for nano-computing applications. Not a single technology can satisfy all the requirements of low cost, high density, low energy, low latency, and high endurance. So, Emerging Non-Volatile Memories were designed and analyzed.

14.1 Introduction

Realization of the ultra low energy, high density non-volatile memory devices is critically a major challenge for designers in a wide range of nano-computing applications. Scaling limits in complementary metal-oxide-semiconductor (CMOS) technology leads to various new models for information storage and processing [1]. The capacity of CMOS-based memories is so low that cannot be enhanced with the help of scaling of CMOS devices. The performance of the memory device will degrade as the data size increases. These limitations will force the designers to

S. Akashe (✉)
Department of ECE, ITM University, NH-75, Jhansi Road, Gwalior 474001, Madhya Pradesh, India
e-mail: shyam.akashe@itmuniversity.ac.in

V. Lakshminarayanan and I. Bhattacharya (eds.), *Advances in Optical Science and Engineering*, Springer Proceedings in Physics 166,
DOI 10.1007/978-81-322-2367-2_14

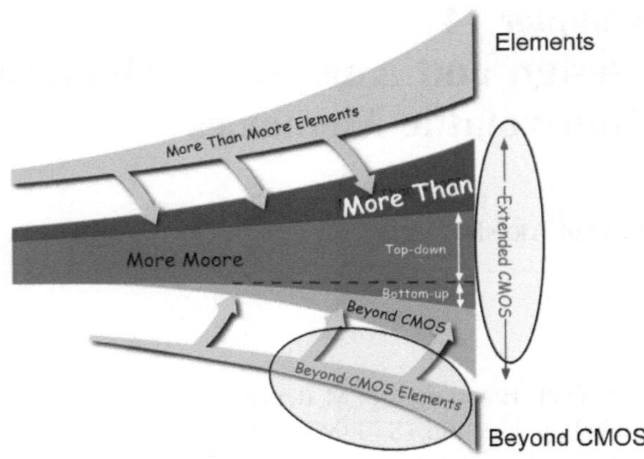

Fig. 14.1 Evolution of extended CMOS

accelerate their search for some advance memory technologies that can rectify these problems. The search for a universal memory storage device that combines rapid read and write speeds, high storage density and non-volatility is solved by the Emerging NVM technologies [2]. The performance gap between the main memory and disk oriented memory is fulfilled by these emerging non-volatile storage technologies like RRAM, PCM, MRAM, STT-RAM, etc. [3]. Figure 14.1 shows the evolution of the CMOS based technologies.

A high performance RRAM (resistive random access memory) with ultra low power is proposed in [4]. Magnetic random access memory (MRAM) recommends denser, fast, low-power, non-volatile memories as a substitute for CMOS based memories [5]. STT-RAM (spin torque transfer RAM) memory cells helped to lessen the high dynamic energy and slow write latencies [6]. Non-volatile memories act as the core of an Integrated Circuit because of their power to maintain their data even when the power is turned off [7]. The resistive based memories don't require many transistors or any other devices. They improve density, reduce power consumption, integrate with processors, and reduce total area.

The goal of this article is to explore the emerging non-volatile memories which are very effective than the existing CMOS based memories. Memristor [8, 9] is the fourth passive circuit element other than resistor, capacitor and inductor in the circuit theory. The Memristor-based memories are much denser and compact [10]. Also the Switching power consumption for Memristor is 20 times lesser than flash [11]. Memristor play an important role in improving the scalability and efficiency of existing memory technologies. Memristors exhibit a hysteretic relationship between current and voltage. Their nonvolatile property leads to fast access time, low energy and high integration density [12] (Fig. 14.2).

Thus, the level of interest in new kinds of RAM memory reflects the commercial potential of any successful product that can compete with both flash and DRAM. The ultimate ambition, according to Professor Alan Woodward of the University of Surrey, is to produce a type of universal memory "that you can take from one

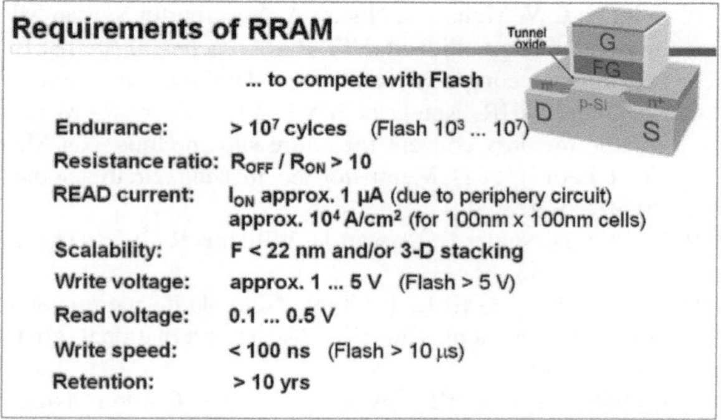

Fig. 14.2 Requirements of resistive RAM (or Memristor)

computing device to another". If ReRAM (Memristor) achieves dominance it promises faster running, more energy efficient, quicker-to-reboot PCs, tablets and handsets. It would also make its patent holders rich.

14.2 Conclusion

ReRAM technology appears to be a prime candidate to save energy in mobile devices and embedded systems. Memristors found its applications in input storage and processing, computational logic unit, latch or memory circuits etc. The evolution of these Emerging Non-Volatile Memories led to the emergence of new paradigms like resistive RAMs, non-volatile operation, multi-bit storage, and processing-in-memory.

References

1. Kudithipudi D, Merkel CE (2012) Reconfigurable memristor fabrics for heterogeneous computing. Adv Neuromorphic Memristor Sci Appl 4:89–106
2. Lee S-H, Jung Y, Agarwal R (2007) Highly scalable non-volatile and ultra-low-power phase-change nanowire memory. Nat Nanotechnol 2:626–630
3. Caulfield AM, Coburn J, Mollov T, De A, Akel A, He J, Jagatheesan A, Gupta RK, Snavely A, Swanson S (2010) Understanding the impact of emerging non-volatile memories on high-performance, IO-intensive computing. In: Proceedings of the ACM/IEEE international conference for high performance computing, networking, storage and analysis, Washington, DC, USA, pp 1–11
4. Cheng CH, Chin A, Yeh FS (2010) Novel ultra-low power RRAM with good endurance and retention. In: IEEE symposium on VLSI technology (VLSIT), Honolulu, pp 85–86
5. Scheuerlein R, Gallagher W, Parkin S, Lee A, Ray S, Robertazzi R, Reohr W (2000) A 10 ns read and write non-volatile memory array using a magnetic tunnel junction and FET switch in each cell. In: IEEE international on solid-state circuits conference (ISSCC), digest of technical papers, San Francisco, CA, USA, pp 128–129

6. Smullen CW, Mohan V, Nigam A, Gurumurthi S, Stan MR (2011) Relaxing non-volatility for fast and energy-efficient STT-RAM caches. 17th IEEE international symposium on high performance computer architecture (HPCA), San Antonio, TX, pp 50–61
7. Lankhorst MHR, Ketelaars BWSMM, Wolters RAM (2005) Low-cost and nano-scale non-volatile memory concept for future silicon chips. Nat Mater 4:347–352
8. Chua Leon (1971) Memristor-the missing circuit element. IEEE Trans Circuit Theory 18 (5):507–519
9. Strukov D, Snider G, Stewart D, Williams R (2008) The missing memristor found. Nature 453 (7191):80–83
10. Ho Y, Huang GM, Li P (2009) Nonvolatile memristor memory: device characteristics and design implications. In: IEEE/ACM international conference on computer-aided design, pp 485–490
11. Caulfield A, Grupp L, Swanson S (2009) Gordon: using flash memory to build fast, power-efficient clusters for data intensive applications. In: Proceeding of the 14th international Conference on Architectural Support For Programming Languages and Operating Systems , Washington, DC, USA, 7 March 2009. ASPLOS 09. ACM, New York 217–228
12. Shang Y, Fei W, Yu H (2011) Analysis and modeling of internal state variables for dynamic effects of nonvolatile memory devices. IEEE Trans Circuit Syst 159(9):1906–1918

Part III
Application of Solar Energy

Chapter 15
Recording and Optimization of Holographic Solar Concentrator in Ultra Fine Grain Visible Wavelength Sensitive Silver Halide Emulsion

V. Vadivelan and B. Chandar Shekar

Abstract Visible spectrum of solar light concentration on wavelength depended solar cell is to enhance the diffraction efficiency of the solar cell is our prime aim. Initial step of successful recording of high diffraction efficiency and good visible transmission holographic optical element were recorded. For this, we used ultra fine grain visible wavelength responsive silver halide holographic emulsion from Ultimate holography. The novelty of this work is three different laser sources of 442, 532 and 633 nm were used to record a single holographic transmission lens by using multiplex technique. The detail study of the holographic optical element recording is explained.

15.1 Introduction

Precise knowledge of solar irradiance components at the earth surface is leads to many solar energy applications. Specifically India's location in the equatorial Sun belt of the earth is its advantage for many solar related application. Theoretically India's solar power reception is about 5000 trillion kWh/year with about 300 clear sunny days in a year. The daily average solar energy incident over India varies from 4 to 7 kWh/m^2 with about 2300–3200 sunshine hours per year, depending upon location. This is far more than current total energy consumption. The annual global radiation varies from 1600 to 2200 kWh/m^2, which is comparable with radiation received in the tropical and sub-tropical regions. Conversion of solar energy into

V. Vadivelan (✉)
Research and Development Centre, Bharathiar University, Coimbatore 641046, Tamilnadu, India
e-mail: vvelan@gmail.com

B.C. Shekar
Department of Physics, Nanotechnology Lab, KASA, Coimbatore 641029, India

V. Lakshminarayanan and I. Bhattacharya (eds.), *Advances in Optical Science and Engineering*, Springer Proceedings in Physics 166,
DOI 10.1007/978-81-322-2367-2_15

electric energy, it is one of the most important and promising application of solar energy for the current world and future. Many conventional solar concentrators are involved to concentrate the solar light to enhance the conversion efficiency of the solar cells [1]. Current trends demand that the photovoltaic (PV) concentrators must achieve various goals of lowering costs at all levels and/or increasing the energy yield [2]. We used holographic optical element as solar concentrator. It has many advantages over the conventional concentrators, the ability of the spectrum splitting and off—axis solar concentration is the distinguished property of holographic solar concentrators. Its capability of focus the wanted wavelength of solar spectrum into the desired direction can concentrate light on the wavelengths dependable solar cells. One of a main advantage of a holographic solar concentrator as compared to a conventional one is seen in the overall reduction of investment cost and in the possibility to generate inexpensive solar electric power [3–5]. Multiplexed holograms have a great importance in a variety of applications, such as high density holographic data storage, optical computing, multiple beam combining, holographic color displays, head-up displays, filters and photonic crystals. Multi-color multiplexing has been recording in photopolymer and dichromated gelatins. Silver halide holographic emulsions exhibit a better sensitivity than dichromated gelatin and pan chromatic photopolymer. Nevertheless, although they have been used for recording multiplexed reflection holograms with different laser lines [6, 7].

15.2 Experimental Arrangement

Three laser sources of 442, 532 and 633 nm is used to record the holographic transmission lens, the laser beam is divided into two beams by using variable density beam splitter (BS) and the two divided beams named as reference and object beams. The reference beam is redirected by Aluminum front coated mirror (M1) to the recording plate (HP) with the desired angle. This beam is spatially cleaned by using spatial filter arrangement (SF2) and it is collimated by lens (L1). The other beam is expanded by spatial filter (SF1) and its placement is deciding the focal length of the holographic lens. The collimated beam and diverging beams are interfered at the ultra fine grain silver halide holographic emulsion. The schematic of the experimental arrangement is shown in Fig. 15.1.

The electronic shutter is controlled the interference of the two beams, not shown in the schematic. Again the same procedure followed for other two laser sources of wavelengths 442 and 532 nm. Wavelengths multiplexed at the same area of the plate. The exposure sensitivity is optimized by trial and error method for all the laser sources separately as well as combined. The whole experimental set up is arranged on the top of vibration isolation table and it is shown in Fig. 15.2.

The same experiment is followed for other two lasers too but the beams were interfered at the same area of the plate. The recorded interference pattern under the safe light condition with optimized laser exposures is processed by developer—bleach combination.

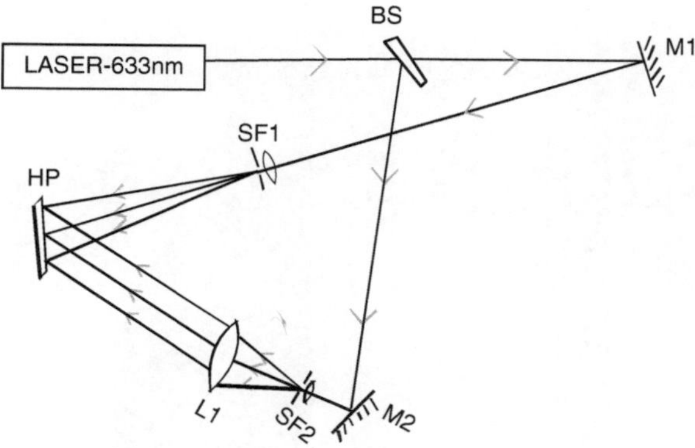

Fig. 15.1 Schematic representation of holographic lens recording experimental arrangement

Fig. 15.2 Experimental arrangement for holographic lens recording in *Blue*, *Green* and *Red* wavelengths

15.3 Result and Discussion

As a result, high diffraction efficiency of over 50 % was recorded and also high visible transmission was achieved by using Ultimate developer and modified rehalogenating bleach combination. High diffraction efficiency of hologram is achieved by using Silver Halide as a phase modulation material that is bleached holograms. We modification R10 bleach from the standard bleach by chemical concentration only, we used Pottasium Dichromate (1 g), Pottasium Bromide (35 g) and Sulphuric Acid (1 ml) for better transmission and diffraction efficiency. The recording of holographic solar concentrator in three laser wavelengths multiplexed on a single element is first time reported as per our knowledge. Exposure sensitivity of the plate for all three wavelengths was inspected by exposed the plate separately ranging from 100 to 2000 μJ/cm^2. Also exposure of the HSC was performed sequentially, starting with the blue wavelength of 442 nm, followed by the green wavelength of 532 nm and ending with the red wavelength of 633 nm. A set of HSC with different exposure energy combinations for each wavelength was obtained, ranging from 120 through 225 μJ/cm^2 for the He–Cd laser, 150 through

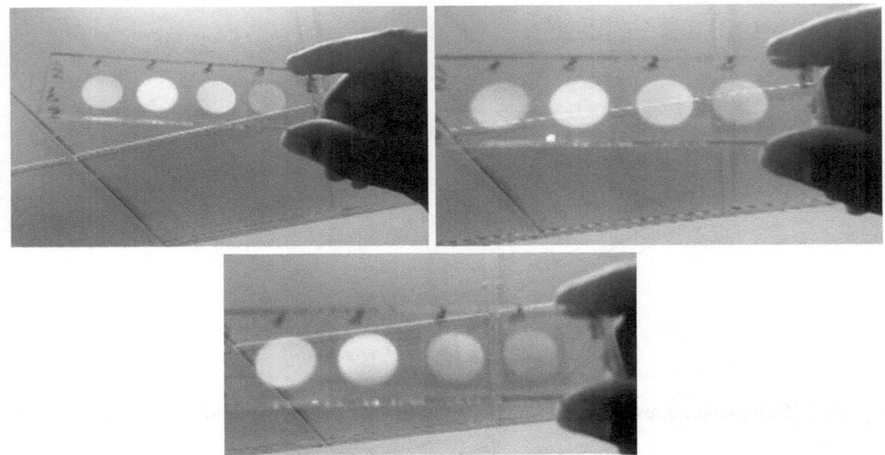

Fig. 15.3 Transmission holographic lens recorded by three visible laser wavelengths

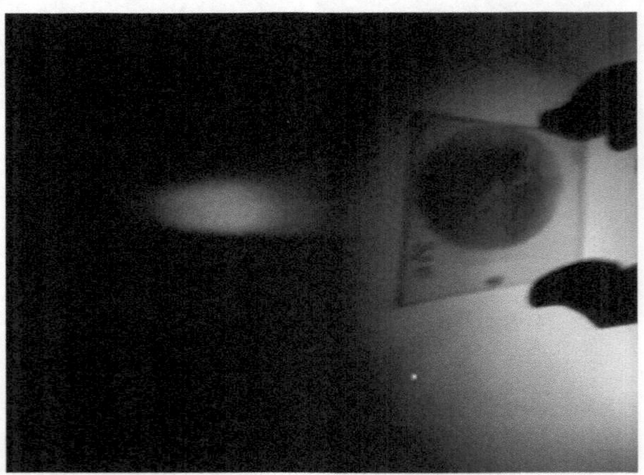

Fig. 15.4 Spectrum splitting of holographic solar concentrator

225 μJ/cm^2 for DPSS laser and 500 through 1000 μJ/cm^2 for the He–Ne laser. The recorded HSC is shown in Fig. 15.3.

HSC is used to concentrate the specific wavelength on the wavelength dependable solar cells. This will enhance the conversion system efficiency to 30–40 %. We proposed this wavelength multiplexed holographic optical element with high efficiency as well as its spectrum splitting and focusing property can be applied for HSC. Spectrum splitting of HSC is shown in Fig. 15.4.

15.4 Conclusion

Panchromatic silver halide holographic emulsion used for the fabrication of wavelength multiplexed holographic solar concentrator. We optimized the exposure sensitivity of the material separately for blue, green as well as red and also for combined wavelengths. Here we proposed the multiplexed element applied for concentrating the solar radiation to enhance the conversion efficiency of the wavelength dependable solar cells. Also this material can be used to manufacture cost effective holographic beam combiner as well as color display holograms and photonic crystals. Further progressing research is going on for the device engineering.

Acknowledgments I would like to thank Mr. T M Rajan, CMD of Ignetta Hologrpahic (P) Ltd for his encouragement and support.

References

1. Buljan M (2014) Recent trends in concentrated photovoltaics concentrators' architecture. J Photon Energy 4(1):040995
2. Nishikawa W, Horne S, Melia J (2008) LCOE for concentrating photovoltaics (CPV). In: International conference on solar concentrators for the generation of electricity (ISCS-5), (Institute of Electrical and Electronics Engineers, New York, 1900), Palm Desert, California
3. Ludman JE (1982) Holographic solar concentrator. Appl Opt 21(17):3057–3058
4. Castro JM, Zhang D, Myer B, Kostuk RK (2010) Energy collection efficiency of holographic planar solar concentrator. Appl Opt 49(5):858–870
5. Stefancich M, Zayan A, Chiesa M, Rampin S, Roncati D, Kierling L, Michel J (2012) Single element spectral splitting solar concentrator for multiple cells CPV system. Opt Express 20 (8):9004–9018
6. Lin LH, Pennington KS, Stroke GW, Lebeyrie AE (1966) Multicolor holographic image reconstruction with white-light illumination. Bell Syst Tech J 45:659–661
7. Upatnieks J, Marks J, Fedorowiez R (1996) Color holograms for white light reconstruction. Appl Phys Lett 8:286–287

15.4 Conclusion

References

Chapter 16
Solar Energy Potential of Silchar, Assam, India—A Resource Assessment

D. Dutta, B. Podder and A. Biswas

Abstract Solar energy is one of the promising renewable energy resources among all other non-conventional energy sources. Solar radiation data are required for resource assessment, model development, system design and collector testing along with other solar engineering and research. To optimise the solar technology and performances of various solar energy based equipment, solar energy assessment has to done beforehand. In our present work, our main objective is to assess solar intensity level of Silchar, Assam (India) using meteorological tools. The latitude of Silchar is 24.81°N. We have collected the solar radiation data of Silchar, Assam (India) from Solar Radiation Resource Assessment (SRRA) station which is installed in NIT Silchar by MNRE (Govt. of India). The data reported is based on the readings taken during the extreme summer in this region. A pyranometer and a Pyrheliometer are installed along with the solar tracker to measure global solar insolation and direct solar insolation respectively. Diffused Radiation is measured with the help of shaded pyranometer. On an average, the daily global and diffused solar radiations are 1024 and 756 W/m^2 respectively. The average solar insolation of Silchar is 4.29 kwh/m^2/day. It is compared with the National Renewable Energy Laboratory (NREL) satellite data for Silchar, which shows the same to be 4.66 kwh/m^2/day thereby making both the readings quite comparable. Therefore, intensity of solar radiation of this place is in the low range.

16.1 Introduction

With rapid increase in world population the global energy requirement is also increasing tremendously and it is estimated that by the end of 2050, global energy requirement will increase by five fold of the present energy status. To meet this huge energy demand, various researches are going on to use renewable energy

D. Dutta (✉) · B. Podder · A. Biswas
Mechanical Engineering Department, NIT Silchar, Silchar 788010, Assam, India
e-mail: debasish.ddutta@gmail.com

V. Lakshminarayanan and I. Bhattacharya (eds.), *Advances in Optical Science and Engineering*, Springer Proceedings in Physics 166,
DOI 10.1007/978-81-322-2367-2_16

source in a very efficient manner. Of all the renewable resources known to us, solar energy has been found a potential resource to satisfy our huge energy needs in the years to come.

In most part of the world, solar energy is found in abundance. The sun emits sufficient energy in an hour, compared to the annual global energy needs. It was observed that solar radiation of approximately 1.5×10^{18} kWh/year gets incident on earth's surface which is more or less 10,000 times the current annual global energy consumption. The total solar energy output or total solar irradiance (TSI) falling on earth's surface perpendicular to the solar radiation per unit time and per unit area is approximately 1366.1 w/m^2. This value is considered to be constant and is known as Solar constant.

16.1.1 Extraterrestrial Solar Radiation (Radiation Outside the Earth Surface)

The sun radiates throughout the entire electromagnetic spectrum from the shortest X-rays to long wave length radio waves. It was observed from the electromagnetic spectrum that maximum intensity of solar radiation falls under visible region in the wavelength range of around 480 nm and accounts for 40 % of the total energy. The solar energy contained in the ultraviolet region and the infrared regions are 8 and 52 % respectively.

Extraterrestrial Solar Radiation can be estimated by using the following relation between solar constant and the particular day of the year.

$$I_O = I_{SC}(1 + 0.033 Cos 360n/365)$$

where,

I_O Extraterrestrial radiation
I_{SC} Solar Constant
n day of the year

16.1.2 Terrestrial Solar Radiation (Radiation at the Earth Surface)

The total solar radiation incident on any horizontal or inclined surface consists of three types of solar radiation. These are: beam radiations (Direct radiations), Diffused radiation and Global radiation. A part of the solar radiation while passing through the atmosphere of the earth gets reflected by the atmosphere, a few portion of the solar energy gets scattered and the rest gets absorbed. The average flux density is nearly about 300 w/m^2 for the most preferred region of earth's surface.

The radiations which are coming from the sun directly without any reflection from any objects are known as Direct or Beam radiations. The portion of the solar radiations that are scattered and gets reflected from the various objects of the earth is called Diffused radiations. Total solar radiation or Global solar radiation is the combination of scattered, reflected and direct radiations. Solar radiation intensity does not remain constant throughout the year. It varies with the latitude, time of the day and atmospheric conditions of that place [1].

16.2 Literature Review

Izioman and Mayer [2] took 4 years data from 2001 to 2004 to estimate the global solar radiation of lowland and mountain site. They compared the observed global radiations with two types of models-cloud based (Kusten) model and the other is sunshine based (Angstrom-Prescott) model. They found that for Kusten model, the monthly men value of solar radiation and the observed value agrees within 2.5 % for lowland site and 13 % for mountain site. With Angstrom model, 2.5 % for lowland and 3.4 % for mountain site were observed.

Ali Rahoma [3] took the direct, global and diffused solar radiations for one year and analyzed the global index and diffused fraction for cloudless sky.

Rehman and Saleem [4] have used a geo statistical technique for the estimating the solar radiation intensity in Saudi Arabia.

Aksakal and Rehman [5] measured global solar radiation using real time solar radiation and meteorological data for one year in the Arabian Gulf Coast. It was found that the monthly mean solar radiation was 328 w/m^2 and highest measured daily solar radiation was 351 w/m^2.

Ramachandra and Subramanian [6] collected the solar radiation and meteorological data got from Indian Meterological Department (IMD), Pune and calculated the solar radiation intensity for various places of Karnataka districts. They found that in those regions, a substantial amount of solar radiation potential is obtainable round the year.

16.3 Measurement of Terrestrial Solar Radiation

In solar engineering and research, solar radiation data are very much necessary to asses various resources and to design and develop various solar energy utilization models. The solar radiation data is also used in testing the performance of various types of solar collectors.

In 1977, National Renewable Energy Laboratory (NREL) under Solar Energy Research Institute was established in the United States of America for extensive research and development of renewable energy [7]. The NREL has collaborated with Clean Power Research and the National Climatic Data Centre (NCDC) to update the National Solar Radiation Database (NSRDB). The NSRDB update includes data collected for 1454 locations in the US and its territories from 1991 to 2010.

There are basically three types of instrument for measuring the solar radiation falling in the earth's surface.

16.3.1 Pyrheliometer

This instrument is used for measuring only the direct radiation which is coming from the solar disc at normal incidence. This device is fitted with thermopile pyranometer which is situated at the lower portion of a long tube. This tube is also fitted with a two axis tracking mechanism to maintain the solar disc at an acceptance angle of 5° (Fig. 16.1) [8].

16.3.2 Pyranometer

This type of instrument is used to measure radiation from the entire sky. It consists of a black surface which heats up when it is exposed to solar radiation. On the black carbon disk, various thermocouples (thermopiles) are mounted. Thus one junction is connected to high temperature surface while the other is kept at relatively lower temperature. As a result an e.m.f is generated which measures global radiation. To safeguard the thermopiles from dust, air movements and rainfall, a dome is made from one or two layers of ground and polished optical glass or acrylic plastic (Fig. 16.2) [8].

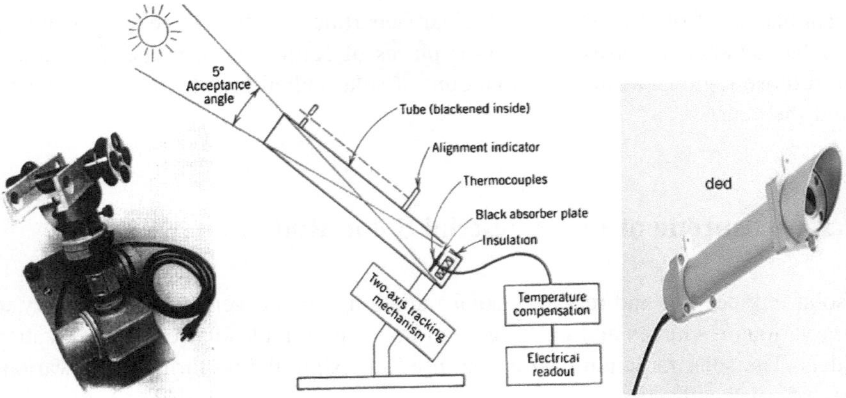

Fig. 16.1 A normal incidence pyrheliometer (NIP) used for measuring the direct component of solar radiation [8]

Fig. 16.2 Pyranometer[8]

16.3.3 Ring Type Shaded Pyranometer

This type of pyranometer shields the sensor from direct radiation from the sun. They are used to measure diffused radiation from the sky (Fig. 16.3).

Fig. 16.3 Ring type shaded Pyranometer[8]

16.4 Objective

The main objective of our present research is to analyze the solar energy potential of Cachar district, Silchar, Assam.

16.5 Information of Studied Area

The study was carried out in Cachar district (NIT Silchar) from Solar Radiation Data Centre (SRRA) which was installed by MNRE (Govt. of India). The Cachar district is located at a latitude of 24.8°N and a longitude of 92.7°E. It is situated in the southernmost part of Assam (Fig. 16.4), which is bounded by Jaintia hill on the north, on the south by Mizoram and on the east by the districts Hailakandi and Karimganj. It occupies a total land mass of 3786 km^2 and receives an annual average rainfall of more than 3000 mm (Fig. 16.5).

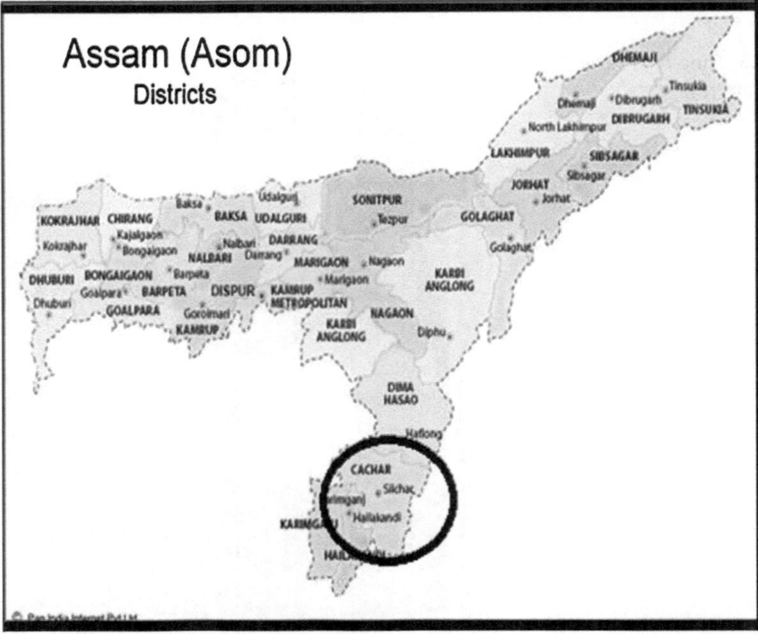

Fig. 16.4 Map of Assam with districts

Fig. 16.5 Study area: Silchar (Cachar district)

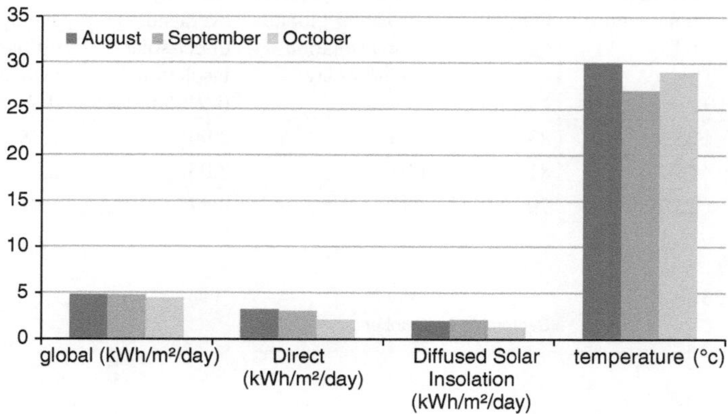

Fig. 16.6 Graph showing the variation of global, direct and diffused solar radiations along with average temperatures for three months obtained from SRRA station

16.6 Analysis of Solar Radiation Data of Silchar

Daily global and direct solar insolation data for the month of August, September and October is taken on houly basis from SRRA station which is installed at the NIT Silchar campus. Based on the above data, average solar insolation on the horizontal surface is estimated and is compared with the satellite data provided by NREL.

16.6.1 Data from SRRA station, Silchar

In the following table necessary atmosphere data is provided which is obtained from SRRA Station Silchar (Fig. 16.6).

Month	Average air temperature (°C)	Relative humidity (%)	Monthly averaged horizontal global solar insolation (kWh/m²/day)	Monthly averaged direct solar insolation (kWh/m²/day)	Monthly averaged diffused solar insolation (kWh/m²/day)
August	30	78	4.8	3.2	1.98
September	27	75	4.76	3.02	2.11
October	29	64	4.42	4.12	1.32

16.6.2 Atmospheric Data Taken from NREL[7]

See Fig. 16.7.

Month	Average air temperature (°C)	Relative humidity (%)	Monthly averaged horizontal global solar insolation (kWh/m²/day)	Monthly averaged direct solar insolation (kWh/m²/day)	Monthly averaged diffused solar insolation (kWh/m²/day)
August	25.9	83	4.38	2.99	2.28
September	25.1	81	4.07	3.06	2.02
October	23.7	70	4.42	4.47	1.41

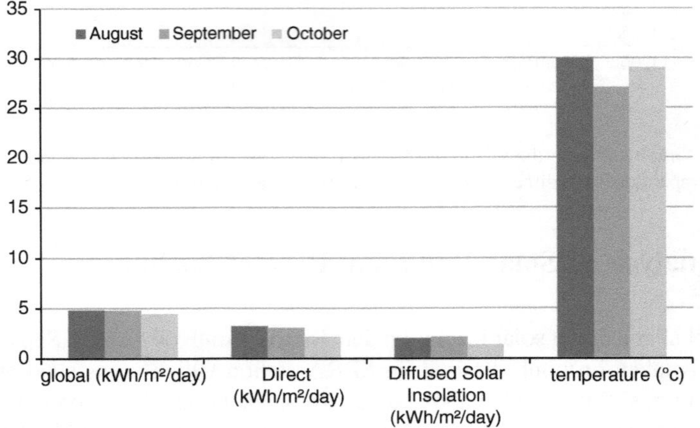

Fig. 16.7 Graph showing the variation of global, direct and diffused solar radiations along with average temperatures for three months obtained from NREL

16.7 Conclusion

By knowing the average annual solar insolation of any place we can be able to design and predict the performance of various solar energy devices.

Data collection and simulation is a very complex procedure and can have errors in the range varying from 3 to 20 %. There are also various limitations in using the satellite data of various agencies. Based on advantages and disadvantages of various data source it can be concluded that most reliable data is obtained from ground based weather station. So to asses the performance of various solar power system, data of IMD under MNRE is highly recommended.

From the above study it is observed that the annual average global solar radiation falling on Silchar (according to SRRA) is 4.29 kwh/m^2/day and from the NREL data it was found to be 4.66 kwh/m^2.

References

1. Rai GD (2004) Solar energy utilization. Khanna Publishers: New Delhi
2. Iziomon MG, Mayer H (2002) Assessment of some global solar radiation parameterizations. J Atmos Solar Terr Phys 64:1631–1643
3. Rahoma UA (2001) Clearness index estimation for spectral composition of direct and global solar radiations. Appl Energy 68:337–346
4. Rehman S, Saleem GG (2000) Spatial estimation of global solar radiation using geostatistics. Renew Energy 21:583–605
5. Aksakal A, Rehman S (1999) Global solar radiation in Northeastern Saudi Arabia. Renew Energy 17:461–472
6. Ramachandra TV, Subramanian DK (1997) Potential and prospects of solar energy in Uttara Kannada, district of Karnataka State, India. Energy Sources 19:945–988
7. http://rredc.nrel.gov/solar/old_data/nsrdb/
8. Sukhatme SP, Nayak JK (2008) Solar Energy, principles of thermal collection and storage. http://www.powerfromthesun.net/Book/chapter02/chapter02.html

Chapter 17
Design and Analysis of Thin Film Based Silicon Solar Cells for Efficient Light Trapping

S. Saravanan, R.S. Dubey and S. Kalainathan

Abstract Silicon based solar cells are still preferable due to its existing technology with inexpensive fabrication cost. However, silicon solar cells are having drawback of weak absorption in longer wavelength due its indirect band gap and needs efficient light trapping in active region. In this paper, we present a design of solar cells based on top anti-reflection coating (ARC) layer and back reflector which is composed of distributed Bragg reflector (DBR) and diffraction grating (GRA) using finite difference time domain (FDTD) method. Simulations show efficient trapping of photon in active region as comparison to reference solar cell. A relative enhancement of cell efficiency ~54 and 60 % is observed for designed solar cells C2 and C3 respectively. This enhancement in performance of solar cells is attributed to the increased absorption and quantum efficiency in red and infrared part of incident solar spectrum.

17.1 Introduction

The scientific community is intensively working to achieve optimal efficiency from thin film silicon solar cells. Among the various energy sources, solar energy is demanded as an alternative to existing one and has covered its broad applications from grid connected electricity production to consumer electronics. Silicon based thin films are cheaper and easy fabrication due to well developed silicon technology. Silicon based thin films have been the choice as active materials for solar cells in its various forms like crystalline silicon, amorphous silicon, micro-crystalline silicon etc. The efficiency of silicon based thin film solar cell is low because of their

S. Saravanan · R.S. Dubey (✉)
Advanced Research Laboratory for Nanomaterials and Devices, Department of Nanotechnology, Swarnandhra College of Engineering and Technology, Seetharampuram, Narsapur, AP, India
e-mail: rag_pcw@yahoo.co.in

S. Kalainathan
School of Advanced Sciences, VIT University, Vellore, TN, India

V. Lakshminarayanan and I. Bhattacharya (eds.), *Advances in Optical Science and Engineering*, Springer Proceedings in Physics 166,
DOI 10.1007/978-81-322-2367-2_17

weak absorption in longer wavelength range. Therefore, light trapping mechanism has become a vital role to enhance the overall conversion efficiency of silicon based solar cells. The Finite-Difference Time-Domain method (FDTD) is one of the well known techniques used to model advanced optoelectronic/photonic devices [1]. This computational method is demanded for the design of active and passive devices which involves application of boundary conditions to either an infinite or finite system. FDTD method is more suitable for solving of complex Maxwell's equations with high accuracy and more appropriate for metamaterials.

The thin film solar cell integrated with anti-reflection coating (ARC) and a combination of distributed Bragg reflector (DBR) and grating (GRA) as back reflector improves the performance of the device [2–5]. The purpose of ARC coating is to reduce the surface reflections with an increase in optical path length. A DBR also known as one-dimensional photonic crystal composed of two distinct dielectric layers are used as reflector for a designed wavelength range. Use of diffraction grating gives scattering and diffraction of light which enforces the light back in active region. In this way, once light is exposed to solar cell it should be diffused to active region after crossing ARC layer. The short wavelength light will be absorbed by active region while longer wavelength will reach to back reflector. Here, partial light will be scattered through grating and partial light will be interacted with DBR where it will decay evanescently and hence, enforce light to re-enter in the silicon-active region. This mechanism of light trapping plays a vital role for the enhancement of light absorption.

This paper deals with the design and analysis of solar cells using efficient light trapping back reflector composed of a DBR and diffraction grating. We have compared the performance of solar cells based on different types of back reflectors and found that back reflector of a DBR and diffraction grating is promising one due to its efficient trapping mechanism. In Sect. 17.2, design approach is presented while simulated results are discussed in Sect. 17.3. Finally, Sect. 17.4 concludes the paper.

17.2 Designing Approach

We have employed FDTD method for the design and analysis of solar cell performance. Figure 17.1 shows a schematic diagram of solar cell composed of ARC, diffraction grating (GRA) and one-dimensional photonic crystal/DBR. A DBR consists of alternate layers of amorphous-Silicon (a-Si) and silicon-di-oxide (SiO_2) with their refractive index 3.6 and 1.45 respectively. The thickness of each layer is determined by the quarter wave principle ($t = \lambda_C/4n$), where, n and λ_C are the refractive index and center wavelength of photonic band gap respectively. The ARC layer is made of silicon nitrate (Si_3N_4) with their thickness 80 nm. On top of the DBR, diffraction grating is embedded in c-Si region which is made of a-Si and SiO_2 with duty cycle (G_{dc} = 450 nm), period (G_p = 550 nm) with thickness (T_g = 115 nm). The silicon-active region thickness is restricted to 1.5 miro-meters for the proposed design.

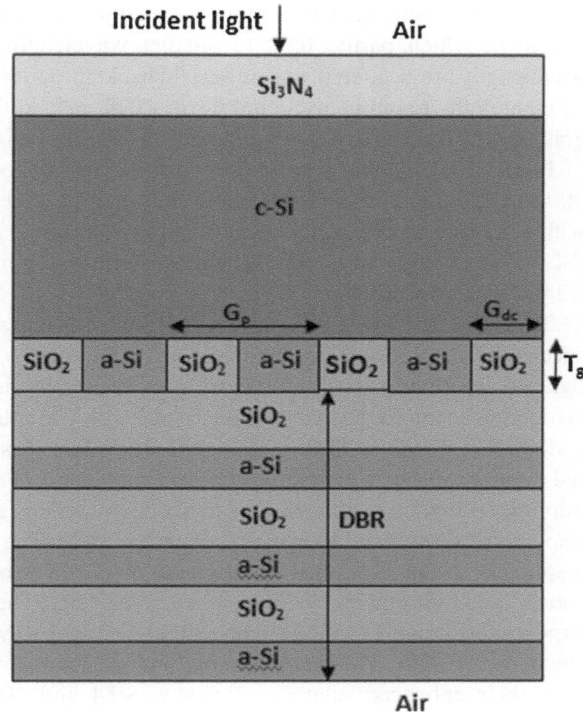

Fig. 17.1 The schematic diagram of designed solar cell structure

In this study, we have designed three solar cells: cell C1-Reference/planar, Cell C2-ARC + GRA + Metal and Cell C3-ARC + GRA + DBR using FDTD method. Perfectly matched layer (PML) is preferred as an absorbing boundary technique for the modeling of optoelectronic/photonic devices whereas periodic boundary conditions (PBC) is nothing but a set of boundary conditions used to approximate infinite system by considering a unit cell. For simulation x and y-axis boundary conditions are kept as PBC while z-axis boundary as PML. The planer solar cell C1 gives wave propagation in single way so lesser absorption of light is observed. However, solar cells C2 and C3 shows enhanced light interaction which is due to integrated back reflector of a DBR with grating and hence, more absorption of light. The purpose of light trapping is to prevent the entered photons into the solar cell with the probability to absorb within the active region.

17.3 Results and Discussion

The wave propagation in designed solar cells is shown in Fig. 17.2. We can perceive light interaction into the device when once light is made incident on the solar cell. Figure 17.2a shows absorption of shorter wavelength light while loss of longer wavelength light in bottom air medium. Figure 17.2b shows distinct field profile in which partial light of smaller wavelength are absorbed while longer wavelength are scattered and reflected back in active layer. Figure 17.2c shows efficient light trapping by coupling in DBR in addition light scattering through grating which enhances the interaction of light in active region.

Figure 17.3a shows absorption of solar cells in accordance to wavelength from 300 to 1200 nm. The performance of designed solar cells C2 and C3 are compared with reference solar cell C1 which mainly consists of ARC layer of silicon nitride (Si_3N_4). As compared to cell C1, enhanced light trapping can be observed for cells C2 and C3 respectively.

The solar cell C3 shows absorption peaks in red and infrared part of incident spectrum from 450 to 1160 nm while cell C2 indicates broad absorption peaks with reduced amplitudes. The light absorption of ARC, grating and metal based solar cell is found to be improved than reference solar cell. The solar cell based on ARC, grating and DBR (C3) shows further improved absorption. A sharp absorption peaks in the red and infrared part of solar spectrum for the case of cell C3 indicates an efficient trapping of light is sustained. This performance enhancement is due to the combined trapping of shorter and longer wavelength of light. The performance from cell C2 is lesser due to the skin effect induced by metal layer based reflector. Figure 17.3b shows quantum efficiency of designed solar cells. The quantum efficiency clearly exposes the light absorption enhancement in longer wavelength ranges in addition to small wavelength. The designed solar cell structures show improved performance with their enhanced relative efficiency ~54 and 60 % from cells C2 and C3 respectively. The solar cell C3 shows major contribution through enhanced absorption from 675 to 1160 nm due to an efficient trapping of light.

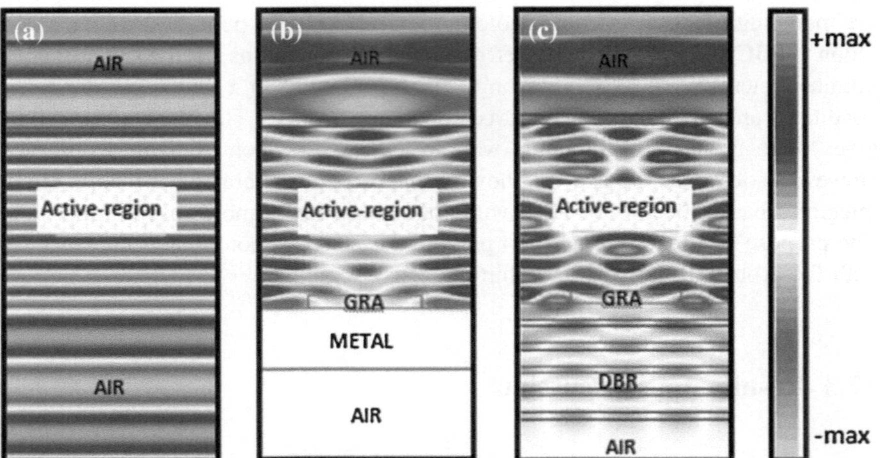

Fig. 17.2 The electric field distribution in different solar cell structures, Cell C1 (**a**), Cell C2 (**b**) and Cell C3 (**c**)

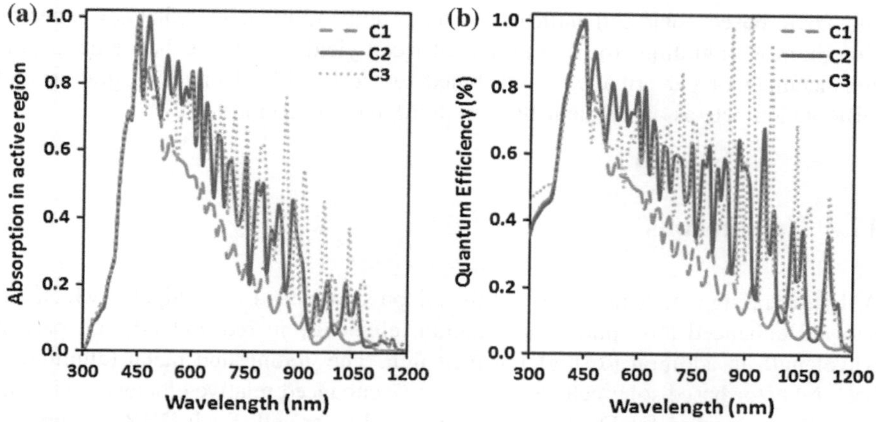

Fig. 17.3 Absorption spectra (**a**) and quantum efficiency (**b**) of solar cells C1, C2 and C3 over incident solar spectrum

The overall quantum efficiency is found to be increased from wavelength range 450–1160 nm. The quantum efficiency curves reveal enhanced conversion efficiency in a broad wavelength range. Solar cell structure C3 shows highest performance with their relative enhanced efficiency up to 60 %.

Figure 17.4 shows short-circuit current (mA/cm^2) versus voltage (V) characteristics of three solar cells C1, C2 and C3. An enhancement in short-circuit current can be observed for cells C2 and C3 respectively. The obtained short-circuit current of cells C2 and C3 are 18.3 and 19.1 mA/cm^2 with open circuit voltage 0.7 V and fill factor 84.5 % for all designed solar cells. The inset bar diagram in Fig. 17.4 shows the cell efficiency of cells C2 and C3. The combination of periodic grating

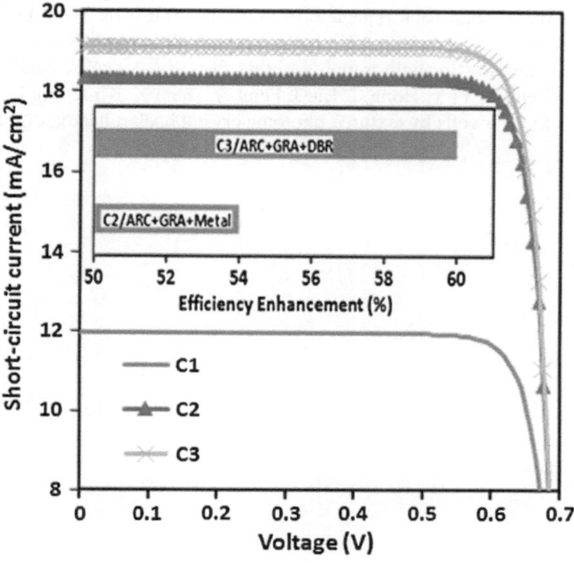

Fig. 17.4 J–V characteristics of designed solar cell structures C1, C2 and C3

and DBR based solar cell indicates ~60 % enhancement in efficiency whereas aluminum with grating structure based solar cell yields ~54 % enhancement. This enhancement of performance is attributed to the combined effect of grating and DBR which acts as an efficient light trapping back reflector.

17.4 Conclusions

A light trapping mechanism of thin film silicon solar cells is investigated which has showed enhanced absorption and quantum efficiency in red and infrared part of wavelength in addition to shorter wavelength. The grating and metal (aluminum) back reflector based solar cell showed ~54 % enhanced relative efficiency whereas ~60 % is obtained for DBR and grating based solar cell. Such DBR and grating based solar cell shows better performance but includes fabrication complexity too. Further, thickness of grating, duty cycle and period needs optimization to achieve better performance of the solar device.

Acknowledgment The financial support provided by Defence Research and Development Organisation (DRDO), New Delhi, INDIA is highly acknowledged.

References

1. Taflove A, Hagness SC (2000) Computational electrodynamics, 2nd edn. Artech House, Norwood, MA
2. Zhao L, Zuo YH, Zhou CL, Li HL, Diao HW, Wang WJ (2010) A highly efficient light trapping structure for thin film silicon solar cells. Sol Energy 84:110–115
3. Krc J, Zeman M, Luxembourg SL, Topic M (2009) Modulated photonic-crystal structures as broad band back reflectors in thin film solar cells. Appl Phys Lett 94(15):153501-1–153501-3
4. Sheng X, Johnson SG, Broderick LZ, Michel J, Kimerling LC (2012) Integrated photonic structures for light trapping in thin film Si solar cells. Appl Phys Lett 100:111110-1–111110-3
5. Zeng L, Yi Y, Hong C, Liu J, Feng N, Duan X, Kimerling LC (2006) Efficiency enhancement in Si solar cells by textured photonic crystal back reflector. Appl Phys Lett 89:111111-1–111111-3

Chapter 18
A Review Report on Solar Cell: Past Scenario, Recent Quantum Dot Solar Cell and Future Trends

Angshuman Khan, Mayukh Mondal, Chiradeep Mukherjee, Ratna Chakrabarty and Debashis De

Abstract Solar cell is the most promising renewable energy source in this modern era which converts light energy into electric energy. The solar cell has achieved a sharp growth as sustainable energy source in recent years. Solar cells can easily replace the fossil fuels as it is pollution free. The solar cell technology is also changing to find a new horizon. In this paper, the review of solar cell technology, starting from crystalline silicon solar cell to recent quantum dot solar cell is discussed. This paper also focused on the future trends of solar cells and its aspects.

18.1 Introduction

The most highly used sources for energy generation is fossil fuels. But the needs of energy increases day by day, hence the fossil fuels are consumed at a faster rate. This high rate of consumption make an early ends of the fossil fuels. So, there is a need for alternative energy sources. The renewable energy sources like solar, wind, sea wave are the best replacement of the fossil fuels which are used in a wide range of geographical areas. The solar energy is the most useful source of energy nowadays.

A. Khan (✉) · C. Mukherjee
Department of Electronics and Communication Engineering,
University of Engineering and Management, Jaipur, India
e-mail: angshumankhan2910@gmail.com

M. Mondal
Department of Electrical Engineering,
University of Engineering and Management, Jaipur, India

R. Chakrabarty
Department of Electronics and Communication Engineering,
Institute of Engineering and Management, Kolkata, India

D. De
Department of Computer Science and Engineering,
West Bengal University of Technology, Kolkata, India

V. Lakshminarayanan and I. Bhattacharya (eds.), *Advances in Optical Science and Engineering*, Springer Proceedings in Physics 166,
DOI 10.1007/978-81-322-2367-2_18

Photons are the main ingredient to convert the energy from light into electricity at the atomic level. The Solar cell device follows two functions first one is photo generation of charge carriers in a light absorbing semiconductor material, and second is the separation of charge carriers into a conductive contact which transmit the electricity. This type of conversion is known as photovoltaic effect. The solar cell related research field is known as photovoltaics.

The solar cells technology is a green approach and it doesn't generate any toxic material, but it is cost effective [1]. The recent photovoltaic technology based on the theory that, each cell must have two layer semiconductor materials. When photon strikes this combination of material, electrons gain energy from photon and moves from one layer to another, thus generate electricity.

In photovoltaics, the light absorbing material absorbs the certain energy from photon. As for example, for silicon, 1.14 eV can excite the electron into conduction band as silicon has a band gap of 1.14 eV. If the photon energy is more than 1.14 eV, then the excess energy will generate heat, not electron—hole pair. So this limits the performance of the solar cell. There are several approaches to increase the efficiency.

Different institute or organizations in the whole world are involving in research on solar cell. Some of the popular institution or organization or laboratories names are listed in Table 18.1.

Different types of solar cell technologies were used in the past century, some of them are costly, and some of them had high efficiency. Table 18.2 shows the different technologies of solar cells with their efficiency. Some of them were used in past and some them are currently in use and with some new concept which are yet to be implement.

From the past to recent technology of solar cell, different types of semiconductor materials have been used. Some of them have advantages and some of them have shortcomings. They have different band gap energy. The band gap energies along with conduction band (minimum) energies and valence band (maximum) energies of some important materials are shown in Table 18.3.

Table 18.1 Some of the popular institution or organization or laboratories names involving research on solar cell

Sl. No.	Name of institute/organization/laboratory
1	Fraunhofer Institute of Solar Energy System
2	Sanyo Corporation
3	Kaneka
4	National Renewable Energy Laboratory
5	University of New South Wales
6	Sharp
7	Radboud University Nijmegen

Table 18.2 Different technologies of solar cells with their efficiency

Sl. No.	Solar cell technology	Efficiency in %
1	Organic solar cell	6.1
2	Amorphous silicon	10.1
3	DSSC	11.1
4	$CuInS_2$	12.5
5	CdTe	16.5
6	Multi crystalline silicon	20.3
7	InP cell	21.9
8	HIT cell	23
9	Crystalline silicon	24.7
10	GaAs cell	26.1
11	Multi junction cell	40.8

Table 18.3 The band gap energies along with conduction band (minimum) energies and valence band (maximum) energies of some important materials

Sl. No.	Material	Band gap energy (eV)	C. B. minimum (eV)	V. B. maximum (eV)
1	PbSe	0.27	−4.93	−5.20
2	PbS	0.37	−4.74	−5.11
3	$CuInS_2$	1.50	−4.06	−5.56
4	CdSe	1.74	−4.10	−5.84
5	CdS	2.40	−3.98	−6.38
6	ZnSe	2.70	−3.40	−6.10
7	ZnO	3.20	−4.19	−7.39
8	TiO_2	3.20	−4.21	−7.41
9	SnO_2	3.20	−4.50	−8.00
10	ZnS	3.60	−3.46	−7.06

18.1.1 Generation of Solar Cell

Based on the generation the solar cell technology may be classified as follows:

First generation: The solar cell technology of this generation is completely silicon based. It has high efficiency. But due to high manufacturing cost and heavy weight it is obsolete.
Second generation: This generation is based on thin film solar cells. The materials like CdTe, InGaSe, and amorphous silicon are used as fabrication materials. They have less efficiency than first generation solar cells. Not only that, it has high cost than first generation. Light weight is the main characteristics of these types of solar cell which are mostly used in solar panel successfully.
Third generation: This generation solar cells include like quantum dot technology, multijunction cells technology, hot carrier cells technology etc. This generation

solar cells have higher efficiency, lesser weight but again it has high manufacturing cost [5–11].
Forth generation: This generation will be also multi junction solar cells. But number of junction will be more than third generation. Choosing of exact material for each layer is the most important factor in this generation. The design will be nano level. The size will be small; weight will be very less keeping the performance high.

18.2 Past Scenario

We cannot divide the research as past, present or future. Though some technologies are very old but research on the particular technology is still going on. In 1985, organic solar cell was first introduced using perylene pigment and CuPc. It had efficiency of 1 %. Next in 90s, another type of organic solar cell prepared by carbon and ZnPc confirmed the same efficiency. Recently introduce organic solar cell technology based on polymer fullerene composite with a higher efficiency [12–19].

The mesoscopic solar cell was introduced in 1990s. It was followed by nano-crystal dye sensitized solar cell, with efficiency of 10–12 %.

The tandem nanocrystalline dye sensitized solar cell with an efficiency of 15 % was introduced in 90s. The research is going on to develop a higher efficiency dye sensitized solar cells [20].

18.3 Solar Cell at Present

18.3.1 Multi Junction Solar Cell

The multi junction solar cells are manly uses in satellite and space applications. It consists of multiple thin films. GaAs solar cells are very popular for its efficiency of 44 %. The three layer GaAs solar cells consist of GaAs, Ge and $GaInP_2$ [21].

18.3.2 Quantum Dot Solar Cell

The quantum dot solar cells are based on dye sensitized solar cell. The low band gap semiconductor materials along with quantum dot are used to fabricate the cells. The popular materials to develop quantum dots are CdS, CdSe, Sb_2S_3, etc. It promises the low cost cells because of it can be made using simple chemical reactions. The quantum dots could be used to make thin film cells with better efficiency as it can emit more than one electron for every photon absorbed. It has efficiency of 63 %. But still, efficient quantum dot solar cell fabrication is very difficult [1–4, 22].

Table 18.4 Comparison between three stack and six stack quantum dot solar cells

Parameters	Six stack	Three stack
Short circuit current density (mA/cm^2)	8.00	4.00
Open circuit voltage (volt)	0.55	0.425
Normalized maximum power (mW/cm^2)	2.947	1.083
Normalized saturation current (nA/cm^2)	63.658	234.31

The popular solar cells are InAs quantum dots in GaAs substrate. There are three stack and six stack quantum dot solar cells. The comparison between them discussed in Table 18.4.

18.4 Future Improvements

To improve the efficiency, the disordered materials like GaInP have to use as the top layer in any multijunction cell. Another most hopeful design method is the increment of the number of junctions in solar cells. The four junction cell has already introduced, the research is going on five and six junctions solar cells. The solar cell using quantum dots, nano wires, carbon nano tubes, quantum wires can improve the performance of solar cell. The quantized states of quantum dot acts as intermediate states for better absorption of photons in solar cells, as a result the efficiency increases up to 63 %. So the multi stack quantum dot solar cell can improve the efficiency.

18.5 Conclusion

This paper mainly investigates the past and present scenario of solar cell technology. Table 18.2 shows the difference between the efficiency levels for different types of solar cell technology; Table 18.3 gives the idea about different band gap energy for different kind of semiconductor material used in solar cell. Table 18.4 shows the current density, open circuit voltage, maximum power and current of quantum dot solar cell. This review report shows the multi junction solar cells are mostly used and main focus for the future application in solar cell technology.

References

1. Afzaal M, O'Brien P (2006) Recent developments in II–VI and III–VI semiconductors and their applications in solar cells. J Mat Sci 16:1597–1602. doi:10.1039/B512182E
2. Chao CC et al (2007) Quantum dot solar cells, MATSCI316 term project. Standford University

3. Feteha MY, Ameen M (2013) CdHgTe quantum dots sensitized solar cell with using of titanium dioxide nanotubes. J Power Energy Eng 1(5):67–72
4. Laghumavarapu RB et al (2014) New quantum dot nano materials to boost solar energy harvesting. SPIE. doi:10.1117/2.1201401.005315
5. Choubey PC et al (2012) A review: solar cell current scenario and future trends. Recent Res Sci Technol 4(8):99–101
6. Yamaguchi M (2005) Super-high-efficiency multi-junction solar cells. Prog Photovoltaics Res Appl 13:125
7. Strobl GFX et al (2006) European roadmap of multijunction solar cells and qualification status. In: Proceedings of IEEE 4th conference, vol 2. doi:10.1109/WCPEC.2006.279839
8. Dimroth F, Kurtz S (2007) High-efficiency multijunction solar cells. MRS Bull 32 (3):230–235. doi:http://dx.doi.org/10.1557/mrs2007.27
9. Burnett B (2002) The basic physics and design of III–V multijunction solar cells. National Renewable Energy Laboratory, III–V research group, summer, 1–18
10. Martí A, Luque A (2004) Next generation photovoltaics: high efficiency through full spectrum utilization. Institute of Physics, Bristol
11. Chopra KL et al (2004) Thin film solar cells: an overview. Prog Photovoltaics Res Appl 12 (2–3):69–92. doi:10.1002/pip.541
12. Tang CW (1986) Two layer organic photovoltaic cell. Appl Phys Lett 48(2):183–185
13. Rostalski J, Meissner D (2000) Monochromatic versus solar efficiencies of organic solar cells. Elsevier Sol Energy Mater Sol Cells 61:87–95
14. Sariciftci NS, Smilowitz L, Braun D, Srdanov G, Srdanov V, Wudl F, Heeger AJ (1993) Observation of a photoinduced electron transfer from a conducting polymer (MEHPPV) onto C_{60}. Synth Met 56(2–3):3125–3130
15. Das B et al (2001) Multijunction solar cells based on nanostructure arrays. In: Electrochemical society meeting
16. Sariciftci NS, Braun D, Zhang C, Srdanov V, Heeger AJ, Stucky G, Wudl F (1993) Semiconducting polymer-backminsterfullerence heterojunctions:diodes,photodiodes, and photovoltaic cells. Appl Phys Lett 62(6):585–587
17. Sariciftci NS, Heeger AJ (1994) Reversible, metastable, ultrafast photoinduced electron transfer from semiconducting polymers to buckminsterfullerene and in the corresponding donar/accepter heterojunctions. Int J Mol Phys B8(3):237–274
18. Sariciftci NS (1995) Role of buckminsterfullerene C_{60} in organic photoelectric devices. Prog Quant Elec 19(2):131–159
19. Kim JY, Lee K, Coates NE, Moses D, Nguyen T, Dante M, Heeger AJ (2007) Efficient tendem polymer solar cells fabricated by all-solution processing. Science mag 317(5835):222–225
20. O'Regan B, Grätzel M (1991) A low cost, high-efficiency solar cell based on dye-sensitized colloidal TiO_2films. Nature 353:737–740
21. Sherif RA et al (2006) The multijunction solar cell: an enabler to lower cost electricity for concentrating photovoltaic systems. In: Proceedings of solar power conference
22. Sobolev NA et al (2001) Enhanced radiation hardness of InAs/GaAs quantum dot structures. Phys Stat Sol B 224(1):93

Chapter 19
Development of a Window Holographic Lens to Utilize Solar Energy

A.B. Sreebha, V.P. Mahadevan Pillai and P.T. Ajith Kumar

Abstract A transmission holographic optical element to perform like a lens was recorded successfully in a silver halide holographic recording material using off-axis geometry. The wavelength selectivity and the focusing property of holographic lens were used to design concentrator for solar radiation to enhance the power yield from photovoltaic systems. The fabricated holographic lens was coupled to a dye sensitized solar cell and an efficiency enhancement was found. The present paper explores a simple idea to use the holographic lens on a window pane of a house. The window pane can either accommodate a small holographic lens or act entirely as a holographic lens for solar concentration. The preliminary results obtained are reported.

19.1 Introduction

The emissionscreated by conventionalenergy-use are very high and there is a big requirement of non-polluting renewable energy sources. Solar energy is the major free source of inexhaustible obtainable energy among the various renewable energy sources. Solar energy is available in great intensity for a country like India. In most parts of India, clear sunny weather is experienced 250–300 days a year. The solar energy potential is about 6,000 million GWh per year [1]. But, so far the solar energy utilization in India is very small. The main constraints of solar power generation are low conversion efficiency and high cost. Across the globe much research work is

A.B. Sreebha (✉) · V.P. Mahadevan Pillai
Department of Optoelectronics, University of Kerala, Kariavattam,
Thiruvananthapuram 695 581, Kerala, India
e-mail: absreebha@gmail.com

V.P. Mahadevan Pillai
e-mail: vpmpillai9@gmail.com

P.T. Ajith Kumar
Light Logics Holography and Optics Pvt. Ltd,
Thiruvananthapuram 695 581, Kerala, India

V. Lakshminarayanan and I. Bhattacharya (eds.), *Advances in Optical Science and Engineering*, Springer Proceedings in Physics 166,
DOI 10.1007/978-81-322-2367-2_19

progressing to overcome the above and solar modules of different configurations are being tried. One approach is to reduce the total area of photovoltaics by concentrating sunlight by using optical components and focusing it on to high efficiency solar cells. The concentrators can be used as roof top concentrators or window concentrators to collect the impinging solar power. Traditional light concentrators are reflective or refractive elements and have limitations due to its bulky nature and concentration of heat on to the photovoltaic element. Holographic lens is a better alternative to conventional solar concentrators due to many reasons. Holographic lenses are recorded through laser interference and are diffractive elements. They are thin and inexpensive to mass produce. They are wavelength selective and can diffract away the heat radiation that can be detrimental to the photovoltaic element. Thus a holographic element can perform as a spectral splitter and concentrator. This facilitates effective energy coupling leading to increased conversion efficiency.

In passive solar building design, windows are made to collect solar energy [2]. The present paper explores a simple idea to use holographic optical elements on a window pane of a house. The window pane can either accommodate a small holographic lens or act entirely as a holographic lens for solar concentration. Specific windows are to be made sun following in order to have enhanced light collection.

19.2 Window Holographic Lens

We can depict holographic lens as a set of concentric rings which can diffract light waves in such a way that they add constructively at a preferred focal point. The lens hologram can also prevent the passage of light that interfere destructively at any desired point. A hologram can focus different wavelengths to different regions with high efficiency without any spectral overlap. White light when passes through a window hologram, different wave lengths diffract at different angles. The constructive components of the solar radiation strike solar cells to which their optical energies are compatible. The ultraviolet and infrared components of the solar spectrum are diffracted away from the solar cells. Here the hologram disperses sunlight so that each part of the solar spectrum is effectively coupled to the photovoltaic cell. The dispersion allows the addition of more solar cells to the module. Since holograms have both horizontal and vertical parallax, they offer varying three dimensional perspective when viewed from side to side or from top to bottom and vice versa [3].

The present paper describes the fabrication of a holographic lens in a PFG 01 silver halide material and an effort to couple it with a dye sensitized solar cell for the efficiency enhancement of the solar cell. The interference fringes are written on a holographic recording material as shown in Fig. 19.1a. The recorded holographic optical element (HOE) will act as a diffractive lens when illuminated by a reference beam. The HOE so made can be arranged in such a way to act on a window pane of a house (Fig. 19.1b). The window hologram splits and focuses the incident solar light to a focal line as shown in Fig. 19.1c. Suitable solar cells can be placed along the focal line in such a way that each wave length can be made to strike single cells

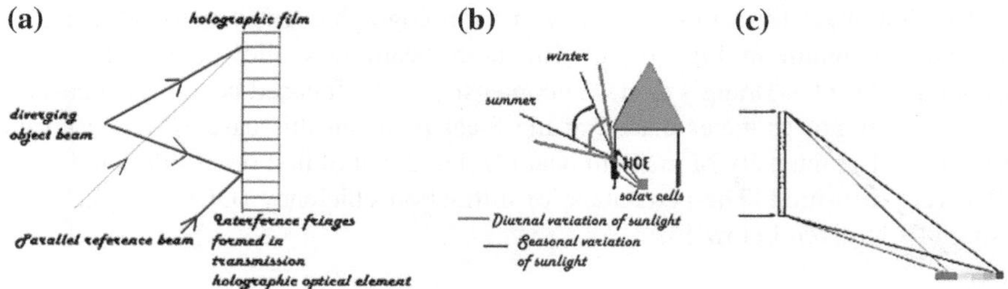

Fig. 19.1 **a** Transmission holographic lens. **b** Holographic lens as a window. **c** Spectral splitting and focusing of holographic lens

to which they are matched in terms of photon energies. The holographic window is designed in such a way that it can accommodate variation in the incident angle due to a variation of the Sun's position.

19.3 Recording Geometry and Diffraction Efficiency Measurements

The off-axis geometry as shown in Fig. 19.2a was used to record transmission holographic lens. A beam splitter was used to divide the beam from the laser into two beams and the beams were expanded using spatial frequency filters. The reference beam was made parallel using collimated lenses. The two beams were directed to the holographic film. The exposure time was controlled using a shutter placed in front of the laser.

The silver halide recording material PFG 01 was used for holographic recordings. For illumination, the laser radiation of wave length 632.8 nm from a He–Ne laser was used. The object beam was made to diverge from a point which was 10 cm away from the film. For recording HOE, the beam diameter of 1 cm was used and the inter beam angle between object and reference beams was kept at 27°. The samples were exposed to 80 μJ/cm^2 energy.

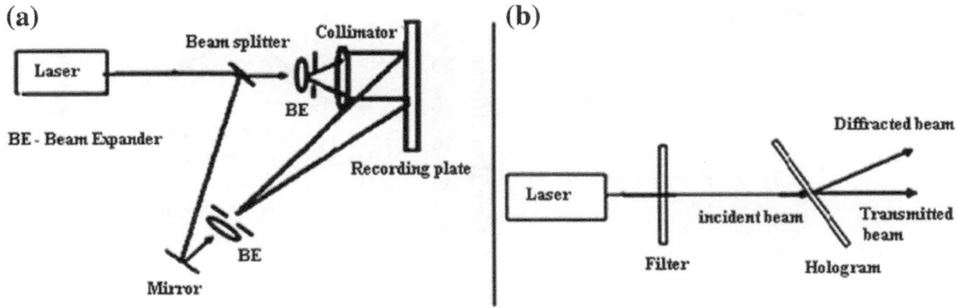

Fig. 19.2 **a** Transmission holographic optical element recording geometry and **b** schematic representation for the diffraction efficiency measurement of holographic optical elements

The first order diffraction efficiency of the holographic lens was found using the schematic diagram in Fig. 19.2b. The laser beam was allowed to fall on the hologram placed at Bragg's angle. The intensity of the focused beam was measured using a laser power meter placed at the focal point, in this case 10 cm from the hologram. The intensity of incident beam (P_i) and that of first order diffracted beam (P_1) were measured. The percentage of diffraction efficiency (DE) was calculated using (19.1) given below [4].

$$\mathrm{DE} = \left(\frac{P_1}{P_i}\right) \times 100 \tag{19.1}$$

19.4 Results

The first order DE of holographic lens for different beam ratio (ratio between object and reference beam) was measured with both direct and conjugate beams. Variation of diffraction efficiency with beam ratio is shown in Fig. 19.3. The lens showed a maximum DE for a beam ratio of 1:3. The maximum DE of 35 % was obtained for direct beam measurements and 33 % for conjugate beam measurements.

19.4.1 Coupling HOE with Solar Cell

The effect of coupling of the holographic lens with a standard yellow sensitive Dye Sensitized Solar Cell (DSSC) of area 0.25 cm^2 was studied at Department of Physics, IIT Chennai. The DSSC (fabricated by IIT, Chennai) used titanium dioxide as photoanode, platinum as counter electrode, N719 as sensitizing dye and tri-iodide/iodide redox couple as electrolyte. Simulations were made using Newport's ORIEL Class A solar simulator. The window hologram was used to direct the light

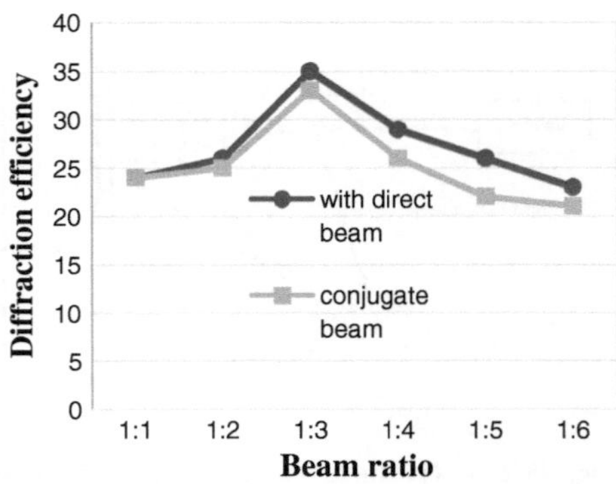

Fig. 19.3 Variation of diffraction efficiency with beam ratio

from the simulator to fall on DSSC. DSSC was initially placed at weakly illuminated region of the simulator and the efficiency measured at this position was 0.076. The HOE was placed vertically to the illuminating plane of simulator in such a way to make the yellow radiation falling on the DSSC and it was found that the efficiency of DSSC enhanced to 0.101 (32.9 % increment).

19.5 Conclusions

A transmission holographic off-axis concentrator lens was recorded and optimized in silver halide recording material. This fabricated holographic lens was coupled to a dye sensitized solar cell and an efficiency enhancement of 32.9 % was found. This result shows the possibility of using holographic lens as a concentrator element affixed on a window pane so as to focus and couple the incident solar light to a photovoltaic element. To achieve optimum energy coupling, it is required to track the window or more conveniently adjust the position of the photovoltaic element. More experiments with high efficiency recording materials and studies on efficiency variation with angle and concentration ratio are progressing.

Acknowledgments A.B. Sreebha is very much thankful for the financial support extended by *Kerala State Council for Science, Technology and Environment* (*KSCSTE*) under 'Back to Lab scheme'.

References

1. www.mnre.gov.in/information/solar-rpo/
2. Doerr T (2012) Passive solar simplified. 1st edn
3. Caulfield HJ, Ludman JE (2002) Holograms harness the rainbow for efficient solar power (8/1/2002)
4. Hariharan P (1987) Optical holography—principle, techniques and applications. Cambridge University Press, Cambridge

Part IV
Diffraction Tomography

Chapter 20
A New Approach to Diffraction Tomography Using Born Approximation

Soumyadip Banerjee

Abstract This paper mainly deals with an efficient method to solve the forward and the backward problem for 2D diffraction tomography. The method described here is a direct method which leads to faster reconstruction of the body's complex refractive index profile. The method is based upon Born approximation which holds for weakly scattering bodies. The method involves a new type of mapping from real space to K space. The reconstructed image of the body is studied for different noisy environment and for different resolution in K space. Also suitable forward and inverse formula is presented.

20.1 Introduction

Recently there has been an immense growth in the development of new algorithms and techniques in diffraction tomography problems. Though non-diffracting tomography is more popular in medical applications, still when it comes to reconstruction of the shape and the electrical properties (complex refractive index profile) of the body under test, diffraction tomography has an advantage. Early techniques employed Fourier diffraction theorem [1] to obtain data in frequency domain and that data is processed using an inverse algorithm [2] to get back the reconstructed image. Unlike the direct methods which work on far field data, there are methods which work on near field data and the reconstruction occurs through an iterative process and is based on moment method solution [2].

This paper is organized as follows. First the theoretical background of the work is provided along with the mathematical foundation for the new scanning method is explained and corresponding to the method the forward and the inverse formulas are provided. Lastly the results are discussed.

S. Banerjee (✉)
Institute of Engineering and Management, Saltlake, West Bengal, India
e-mail: soumyadip.banerjee@iemcal.com

V. Lakshminarayanan and I. Bhattacharya (eds.), *Advances in Optical Science and Engineering*, Springer Proceedings in Physics 166,
DOI 10.1007/978-81-322-2367-2_20

20.2 Theory

20.2.1 Wave Equation and Born Approximation

Considering an inhomogeneous medium for which the Helmholtz equation [3] is given by

$$[\nabla^2 + k^2]\mathbf{u}(\mathbf{r}) = 0 \tag{20.1}$$

where,

$k(r) = k_0 n(\mathbf{r})$
k_0 = average wave no. of the medium.

$n(\mathbf{r})$ is complex refractive index of the medium given by

$$n(\mathbf{r}) = \sqrt{\frac{\mu(\mathbf{r})\varepsilon(\mathbf{r})}{\mu_0\varepsilon_0}} \tag{20.2}$$

$u(\mathbf{r})$ = Complex amplitude of the total field at position given by $\mathbf{r}$.

The solution to (20.1) is given by [4]

$$\mathbf{u_s}(\mathbf{r}) = \int_V F(\mathbf{r}', \omega)\, g(\mathbf{r}|\mathbf{r}')\, u(\mathbf{r})\, \mathrm{d}v \tag{20.3}$$

where, green's function $g(\mathbf{r}|\mathbf{r}')$ is employed in the solution and $F(\mathbf{r})$ is the forcing function called the scattering potential [5].

There are rigorous methods to solve the above (20.3). But if the test body is weak in homogeneity then the above equation can be simplified using first order Born approximation [6]. In this approximation the total field in (20.3) is replaced by the incident field $u_0(\mathbf{r})$ and hence the simplified equation is given by

$$\mathbf{u_s}(\mathbf{r}) = \int_V F(\mathbf{r}', \omega)\, g(\mathbf{r}|\mathbf{r}')\, u_0(\mathbf{r})\, \mathrm{d}v \tag{20.4}$$

where, in 2D case the Green's function [7] is given by

$$g(\mathbf{r}|\mathbf{r}') = -\frac{j}{4} H_0^2(k_0|\mathbf{r} - \mathbf{r}'|)$$

20.2.2 Forward Problem

In 2D diffraction tomography the test object or objects are considered to be having refractive index variation along the cross-section and having no or gradual variation with height. The scattered field at the observation point in the far field region is given by

$$\mathbf{u}_s(\mathbf{r}) = A_0\sqrt{\frac{2j}{\pi \mathbf{k}_0 \mathbf{r}}} e^{-jk_0 r} \mathrm{f}(\mathbf{s}_0, \mathbf{s}) \tag{20.5}$$

where,

A_0 = amplitude of the incident field
$\mathbf{s}_0$ = unit vector directed towards the propagation direction of the incident field
$\mathbf{s}$ = unit vector directed towards the position of the receiver Rx.

$f(\mathbf{s}_0, \mathbf{s})$ is called the Scattering amplitude and is given as

$$f(\mathbf{s}_0, \mathbf{s}) = \mathrm{F}'[\mathbf{K}] = \iint F(\mathbf{r}', \omega) e^{-jK \cdot \mathbf{r}'}\, \mathrm{d}s \tag{20.6}$$

Here, $\mathbf{K} = k_0(\mathbf{s} - \mathbf{s}_0)$ is called the spatial frequency vector [8]. The significance of (20.6) is that the complex amplitude of the scattered field measured in the far field region depends entirely on only one Fourier component of the scattering potential $F(\mathbf{r})$ corresponding to a given K vector. The approach in this type of problem is to obtain Fourier component data for all K vectors within the K space. For that the scattered field is to be measured at all possible direction $\mathbf{s}$ and also for all possible direction of the incident wave $\mathbf{s}_0$. This is best diagrammatically described by Ewald's circle of reflection [8].

The method suggested here solves the forward problem. It is seen from (20.6) that the 2D Fourier transform value f depends mainly on $\mathbf{K}$. Generally, Fourier diffraction theorem is applied to obtain f data for different position of receiver and also for different angle of incidence. But this leads to a non-linear mapping from real space to $\mathbf{K}$ space. This makes the reconstruction procedure difficult. To circumvent the situation the sampled values of the scattered field is extracted at non-uniform spacing in real space so that the f data are obtained for uniform sampled values of $\mathbf{K}$. This makes the reconstruction problem easier to handle.

In this scanning method the $\mathbf{K}$ space is pictured with the help of a Cartesian co-ordinate system, where it is divided into uniform cells of size ΔK as shown in the Fig. 20.1.

In order to obtain f data corresponding to each cell the transmitting and the receiving antenna should be properly oriented with respect to the test body in the far field region. It is better to make the transmitter position fixed and change the angular position of the receiver accompanied by a proper rotation of the body to a particular angle.

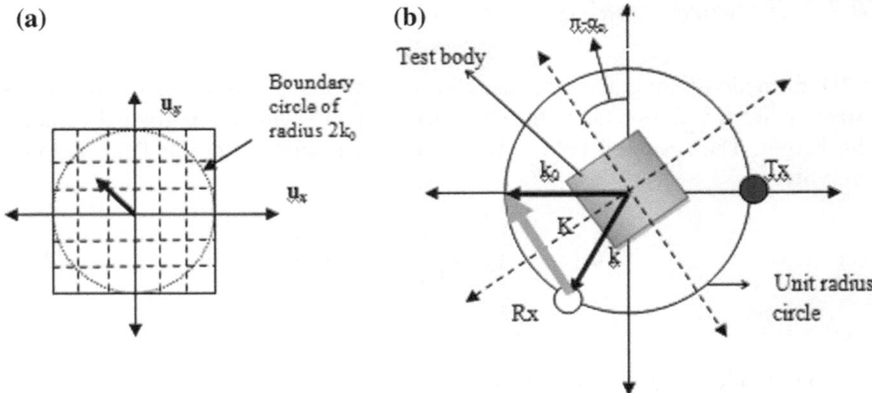

Fig. 20.1 **a** Showing **K** space in Cartesian co-ordinate system and **b** how Tx and Rx should be placed to obtain scattering amplitude data for a particular **K** vector

Let **K** in Cartesian co-ordinate system be represented as

$$K = K_x\mathbf{u_x} + K_y\mathbf{u_y} = m\Delta K\mathbf{u_x} + n\Delta K\mathbf{u_y}$$

and

$$\Delta \mathrm{K} = 2\mathrm{k}_0/\mathrm{M},$$

where m and n are rational no's and $(2M)^2$ is the total no. of cells the K space is divided into $\mathbf{u_x}$ and $\mathbf{u_y}$ are the unit vectors along two perpendicular directions in **K** space.

The co-ordinates of the transmitting antenna (Tx) on a unit radius circle is (1, 0) and that of the receiving antenna is $[-\cos(\alpha - \alpha_0), \sin(\alpha - \alpha_0)]$ where α_0 and α are given by the following (20.7) and (20.8)

$$\alpha_0 = \tan^{-1}\frac{n}{m} + \cos^{-1}\frac{\sqrt{m^2 + n^2}}{M} \tag{20.7}$$

$$\alpha = \tan^{-1}\frac{\sin\alpha_0 - 2\frac{n}{M}}{\cos\alpha_0 - 2\frac{m}{M}} \tag{20.8}$$

In order to obtain proper incident field direction the test body is rotated by an angle $\pi - \alpha_0$ in anti-clockwise direction.

The forward formula for obtaining the scattering amplitude [6] value is obtained by dividing the entire body or the scattering region into N no. of square cells of side length h. For a given K vector f is given by

$$f(\mathbf{K}) = h^2 \sin c\left(\frac{K_x h}{2}\right) \sin c\left(\frac{K_y h}{2}\right) \sum_{i=1}^{N} F(r_i')\mathrm{e}^{-j\mathbf{K}\cdot\mathbf{r}_i'} \tag{20.9}$$

where, $\mathbf{r}_i' = \mathrm{x}_i'\mathbf{u}_x + \mathrm{y}_i'\mathbf{u}_y$ specifies the position vector of the ith cell of the scattering region and h^2 is the cell area.

20.2.3 Backward Problem

As the K space is divided into uniform cells, so the low pass filtered approximation of the scattering potential [6] can be written as

$$F_{\mathrm{LP}}(\mathbf{r}) = \left(\frac{\Delta K}{2\pi}\right)^2 \sin c\left(\frac{\Delta Kx}{2}\right) \sin c\left(\frac{\Delta Ky}{2}\right) \sum_{i=1}^{(2M)^2} f(K_i)\mathrm{e}^{jK_i\cdot r} \tag{20.10}$$

ΔK^2 = area of each cell in K space.

Here, $\mathbf{K}_i$ specifies the position vector of the ith cell in K space. The above discrete sum is to be evaluated within the boundary circle.

20.3 Numerical Model

The real and imaginary part of the complex refractive index of the test body are shown in Fig. 20.2.

20.4 Results and Discussions

The reconstructed images of the body are obtained for high and low resolution in K space. The inverse method is tested with signals corrupted with different amount of noise. The type of noise considered over here is an Additive White Gaussian Noise (AWGN). The frequency of operation is 10 GHz and the incident field amplitude is 1 V/m for each case (Figs. 20.3 and 20.4).

The reconstructed images were also obtained considering low resolution in K space (for SNR-10 dB) is shown in Fig. 20.5. There is a marked degradation in the image quality due to insufficient sampling in K space.

When the reconstructed images are compared under the same noise condition then it is observed that the quality of the image is better when high resolution in K space is considered. This is more visible for images associated with low SNR values where it is hard to recover the profile of the body. Given that the forward

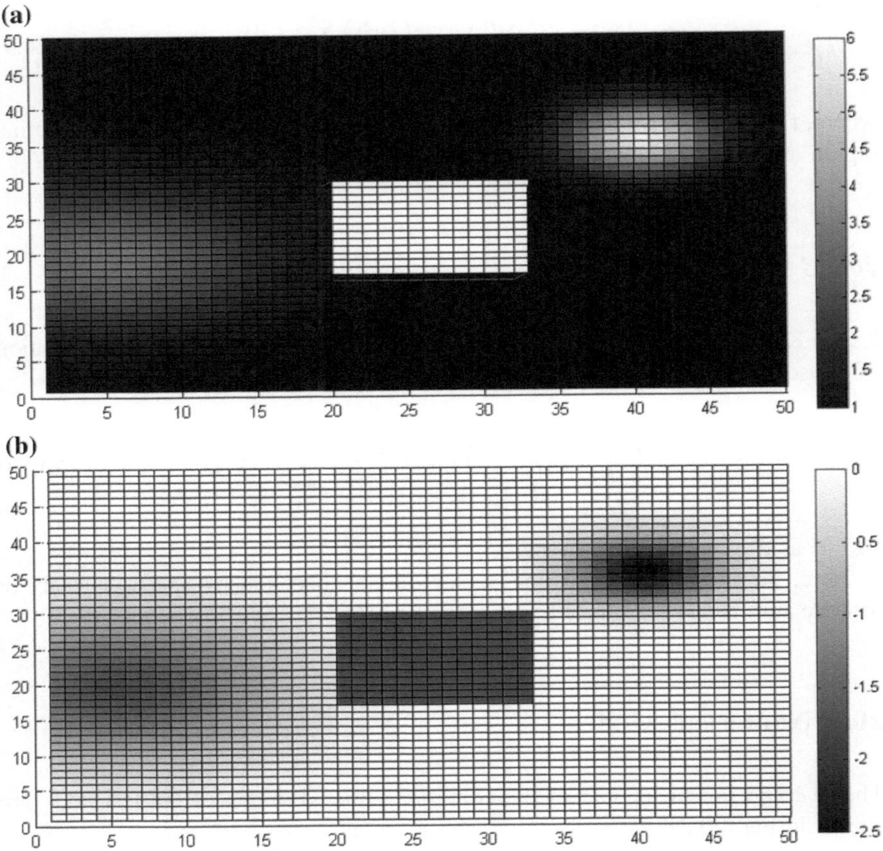

Fig. 20.2 Showing the position and the shape of the test body considered (dimension of the body is 10 cm × 10 cm) where **a** depicting the real part of the refractive index of the object (at 10 GHz) and **b** the imaginary part

data is available the inverse method took approximately 31 s to give back the reconstructed image of the object subjected to certain specifications (no. of cells K space is divided = 50 × 50, No. of cells the image frame is divided = 20 × 20). Execution time depends very much on the real and K space cell size.

The main practical challenge in these direct methods is to extract the phase information from the received signal. This requires sophisticated technologies like the use of vector analyzers. Other challenge is the proper positioning of Tx and Rx in the far field region.

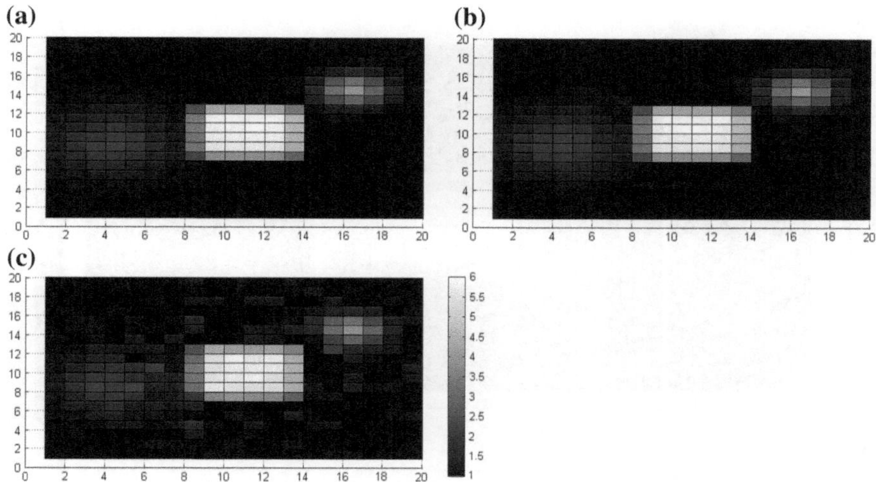

Fig. 20.3 Showing real part of the refractive index profile obtained after reconstruction process considering noise free (**a**) condition and also with noise corresponding to SNR = 30 dB (**b**) and SNR = 10 dB (**c**) at $f = 10$ GHz. No. of cells in K space = 50 × 50

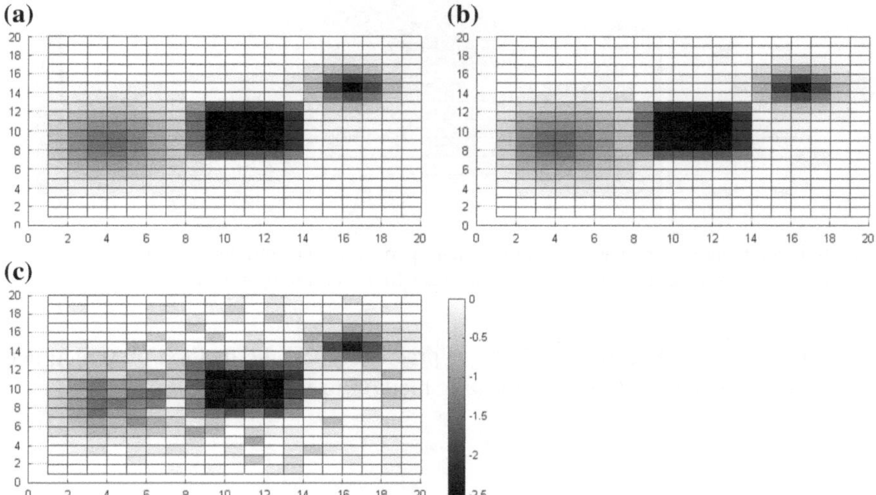

Fig. 20.4 Showing imaginary part of the refractive index profile obtained after reconstruction process considering noise free (**a**) condition and also with noise corresponding to SNR = 30 dB (**b**) and SNR = 10 dB (**c**) at $f = 10$ GHz. No. of cells in K space = 50 × 50

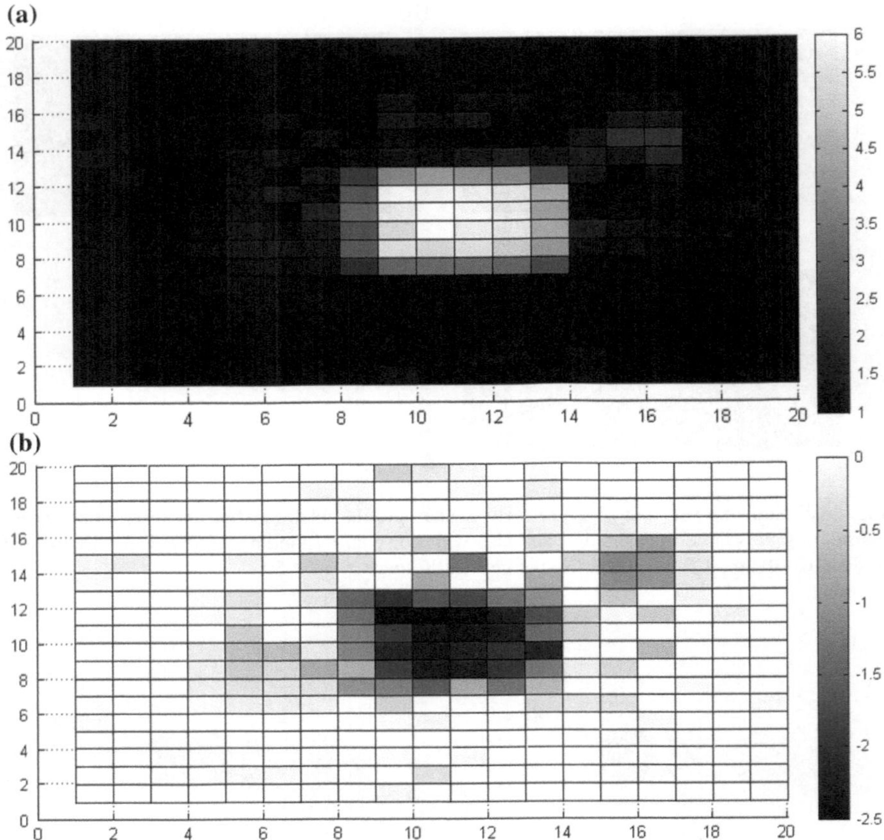

Fig. 20.5 Showing the real part (**a**) and imaginary part (**b**) of the complex refractive index profile for low resolution in K space (30 × 30 cells) under SNR condition of 10 dB

Further scope of research would be to practically test the method presented here with real test bodies and also extend this method to the 3D case.

References

1. Kak AC, Slaney M (1999) Principles of computerised tomographic imaging (Chap. 6). IEEE press, New Jersey, pp 218–220
2. Kundu AK, Bandyopadhyay B, Sanyal S (2008) An iterative algorithm for microwave tomography using modified Gauss-Newton method. In: IFMBE proceedings, vol 21, p 511
3. Born M, Wolf E (2005) Principles of optics-electromagnetic theory of propagation, interference and diffraction of light (Chap. 8). Cambridge University press, Cambridge, pp 695–703
4. Born M, Wolf E (2005) Principles of optics-electromagnetic theory of propagation, interference and diffraction of light (Chap. 8). Cambridge University press, Cambridge, pp 695–699

5. Born M, Wolf E (2005) Principles of optics-electromagnetic theory of propagation, interference and diffraction of light (Chap. 8). Cambridge University press, Cambridge, p 696
6. Kak AC, Slaney M (1999) Principles of computerised tomographic imaging (Chap. 6). IEEE press, New Jersey, pp 212–214
7. Green function for laplace and wave equation. http://www.springer.com/cda/content/.../cda.../9783319008783-c2.pdf
8. Born M, Wolf E (2005) Principles of optics-electromagnetic theory of propagation, interference and diffraction of light (Chap. 8). Cambridge University press, Cambridge, pp 700–703

Part V
E.M. Radiation Theory and Antenna

Chapter 21
Design and Analysis of Dual Band, DGS Integrated Compact Microstrip Antenna

Srijita Chakraborty, Sayan K. Moitra, Soham Tewary, Archana Kumari and Mrinmoy Chakraborty

Abstract A dual band rectangular microstrip patch antenna is proposed with novel defects introduced in the ground plane. Initially a plain rectangular microstrip antenna is designed which is resonant at 5.2 GHz i.e. the first WLAN band. With the introduction of two spiral angular defective ground structure, the microstrip antenna is found to be resonant simultaneously at 2.4 GHz i.e. the Bluetooth band and 5.8 GHz i.e. the second WLAN band. The different parameters of the microstrip antenna, such as the return loss, gain and impedance variations are extensively studied. Thus the proposed microstrip antenna behaves as compact, as well as a dual frequency band operated antenna.

21.1 Introduction

In the recent past, the performance of microstrip antenna is extensively developed and analyzed for various practical applications. The salient features of microstrip antenna which attracted the researchers are low profile, light weight, portable, ease in fabrication and cost effectiveness. Efforts are made to design compact microstrip antennas with higher percentage of miniaturization. In this regard, different techniques to achieve compact microtrip antennas, such as by cutting regular shaped slits, slots, defective ground structure etc., are explored in several papers. Due to these techniques a perturbation effect in the microstrip antenna is observed [1, 2], which makes the antenna operate at a lower application frequency band. Similarly efforts are made to design multiband microstrip antennas, which are simultaneously

S. Chakraborty (✉) · S.K. Moitra · S. Tewary
Institute of Engineering and Management, Kolkata, India
e-mail: srijitachakraborty2010@gmail.com

A. Kumari
Kalyani Government Engineering College, Kalyani, India

M. Chakraborty
Dr. B.C. Roy Engineering College, Durgapur, India

V. Lakshminarayanan and I. Bhattacharya (eds.), *Advances in Optical Science and Engineering*, Springer Proceedings in Physics 166,
DOI 10.1007/978-81-322-2367-2_21

resonant at more than one application frequency bands [3–7]. In the given paper, a compact dual-band rectangular microstrip antenna is proposed. At first the microstrip rectangular antenna is designed such that it is resonating at 5.2 GHz i.e. first WLAN band and then by the integration of two defects in the ground plane, the microstrip rectangular antenna is made to resonate simultaneously at 2.4 GHz i.e. the Bluetooth band and 5.8 GHz i.e. the second WLAN band. Thus a compact microtrip antenna with size reduction of 79.6 %, with dual band characteristics is obtained.

21.2 Design Principles

The detailed dimension of the structure of the proposed microstrip antenna is given in Fig. 21.1. The dielectric substrate FR4_epoxy with dielectric constant 4.4 and dielectric loss tangent of 0.002 is taken to design the proposed rectangular microstrip antenna and Zeland IE3D software tool is used to simulate the proposed antenna.

Originally the microstrip rectangular antenna is designed to be resonant at the first WLAN band i.e. 5.2 GHz. The dimensions of the rectangular patch for the microstrip antenna is taken to be 17.2 and 13.4 mm and the coaxial feed is placed at the coordinates of (0, 3.2) approximately. Then two identical angular spiral shaped defects with different length are introduced in the ground plane. Figure 21.2 shows the dimensions of the DGS integrated microstrip rectangular antenna and it can be noted that the length of the two DGSs are approximated to be 16 and 9.4 mm respectively, with the width of the DGS to be 0.5 mm. Both the defective ground structures are positioned at an optimal position with respect to each other, so that the effect of mutual coupling is minimized among the two defects. With the

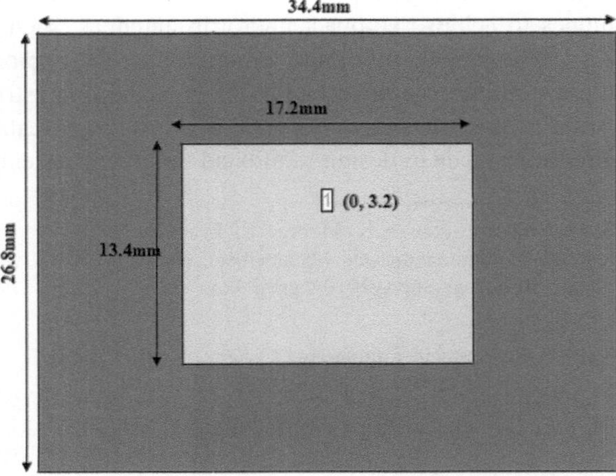

Fig. 21.1 Front view of the proposed rectangular microstrip antenna, resonant at 5.2 GHz

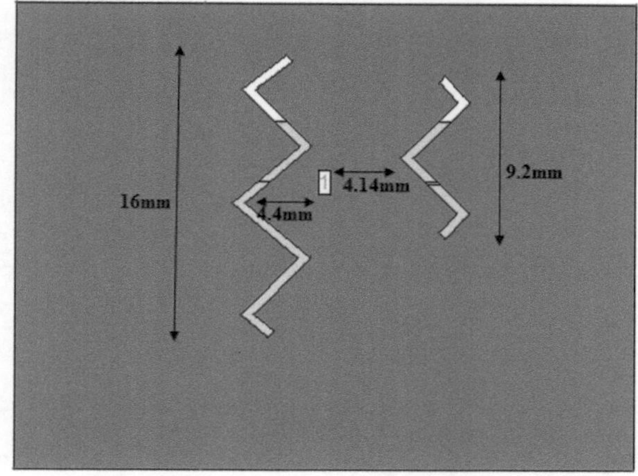

Fig. 21.2 Rear view of the proposed rectangular microstrip antenna, resonant at 2.4 and 5.8 GHz simultaneously

introduction of the defective ground structures, it is observed that the microstrip antenna is resonant simultaneously at 2.4 GHz i.e. the Bluetooth band and 5.8 GHz i.e. the second WLAN band.

21.3 Results

21.3.1 S11 Versus Frequency Graph

The S11 versus frequency graph of the rectangular microstrip antenna with defects and without defects are shown in Figs. 21.3 and 21.4 respectively. From the figures, it is observed that in absence of defective ground structure, the microtsrip rectangular antenna is resonating at 5.2 GHz (i.e. first WLAN frequency band) is −29 dB. When the defects are introduced in the ground plane the microstrip antenna is simultaneously resonant at 2.4 GHz (i.e. Bluetooth frequency band) and 5.8 GHz (i.e. second WLAN frequency band). The S11 measured in the case of the proposed

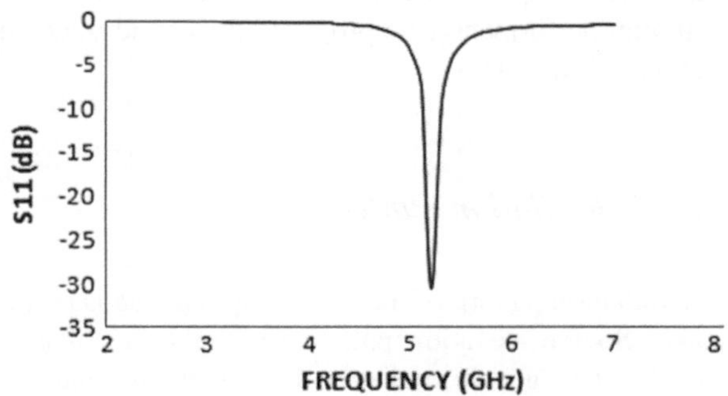

Fig. 21.3 Simulated return loss of the proposed antennas with antenna without DGS

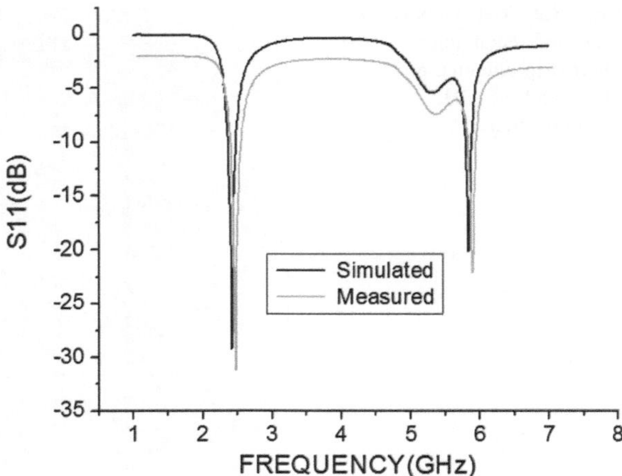

Fig. 21.4 Simulated and measured return loss of the proposed DGS, resonant at 2.4 and 5.8 GHz simultaneously

dual band antenna are −29 dB for 2.4 GHz and −20 dB for 5.8 GHz. Also from Fig. 21.4 it can be observed that the simulated and measured results are in sync with each other.

21.3.2 Impedance Versus Frequency Graph

The real part of impedance of the rectangular microstrip antenna at the resonant frequency should be approximately close to the characteristic impedance of the external transmission line i.e. 50 Ω. Similarly the microstrip antenna can be estimated with the help of an inductor–capacitor equivalent electric circuit and at resonance, as the net reactive component is zero, so the imaginary part of impedance should be close to 0 Ω. Figure 21.5 shows the impedance versus frequency plot for the rectangular microstrip antenna without the DGS and Fig. 21.6 shows the same plot for the proposed dual band microstrip antenna, resonating at 2.4 and 5.8 GHz respectively. From the figures it can be concluded that the real part of impedance for the proposed microstrip antennas are close to 50 Ω and the imaginary part of impedance at the resonating frequency is almost equal to 0 Ω. Thus for optimum performance the proposed dual band microstrip antenna is matched at the resonant frequencies.

21.3.3 Radiation Pattern

The radiation pattern of the microstrip antenna is perpendicular to the surface of the patch. So the elevation pattern for the value of $\varphi = 0°$ and $\varphi = 90°$ are to be calculated to get the radiation pattern of the antenna. The E-plane and H plane

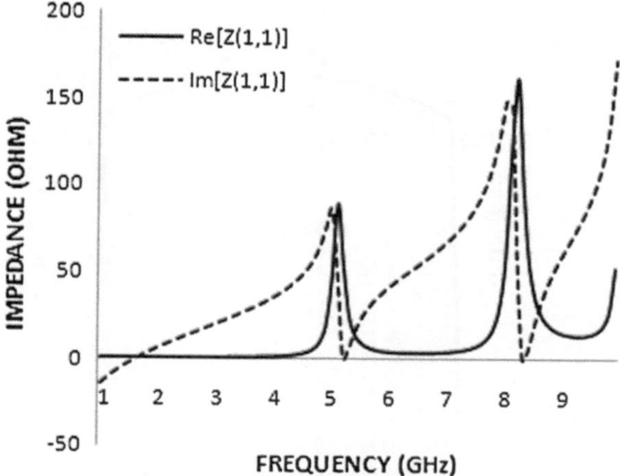

Fig. 21.5 Simulated impedance versus frequency plot for the microstrip antenna without DGS

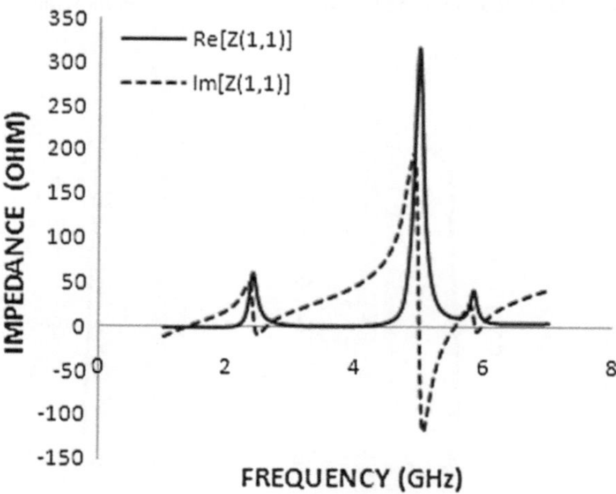

Fig. 21.6 Simulated impedance versus frequency plot for the microstrip antenna with DGS, resonant at 2.4 and 5.8 GHz simultaneously

radiation pattern at first WLAN frequency band (i.e. 5.2 GHz) is shown in Fig. 21.7. It can be observed that the gain is obtained at resonant frequency is 5.68 dBi. Figure 21.8 demonstrates the E-plane and H-plane radiation pattern at Bluetooth band (i.e. 2.4 GHz) and the maximum gain at 2.4 GHz is obtained as 2.35 dBi. Similarly Fig. 21.9 shows the E-plane and H plane radiation pattern at second WLAN band (i.e. 5.8 GHz) and the maximum gain at 5.8 GHz is obtained as 1.7 dBi. Figure 21.10 shows the total gain versus frequency plot for the proposed dual band rectangular microstrip antenna and from the gain variation over the frequency it can be concluded that the proposed antenna exhibits almost constant gain over the proposed application frequency bands.

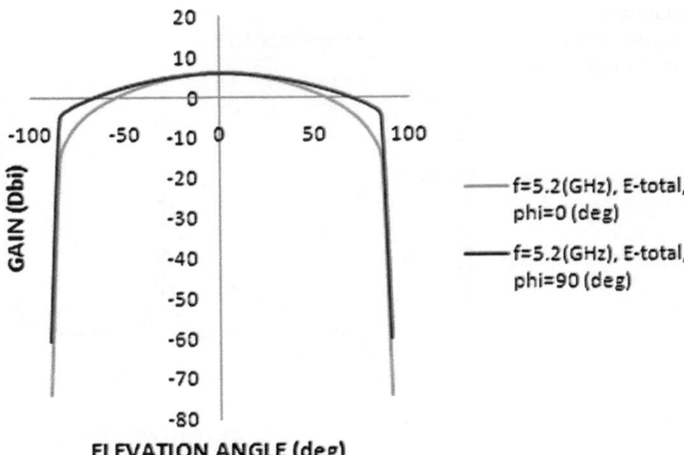

Fig. 21.7 Radiation pattern of the microstrip antenna without DGS, resonant at 5.2 GHz

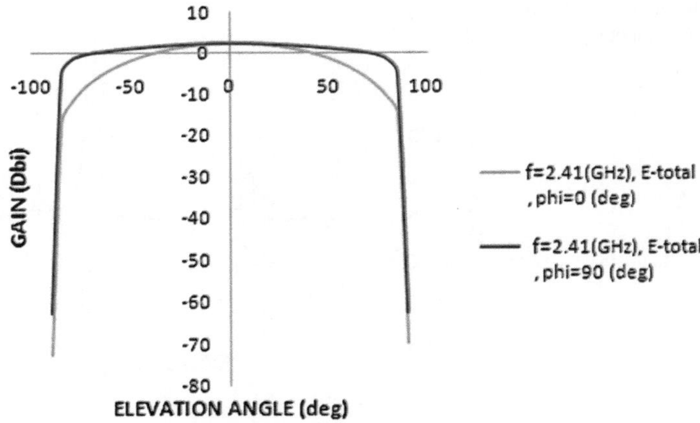

Fig. 21.8 Radiation pattern of the dual band microstrip antenna at 2.4 GHz

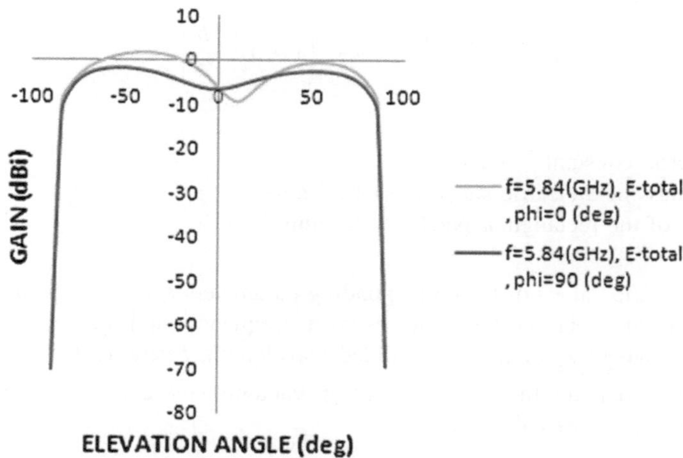

Fig. 21.9 Radiation pattern of the dual band microstrip antenna at 5.8 GHz

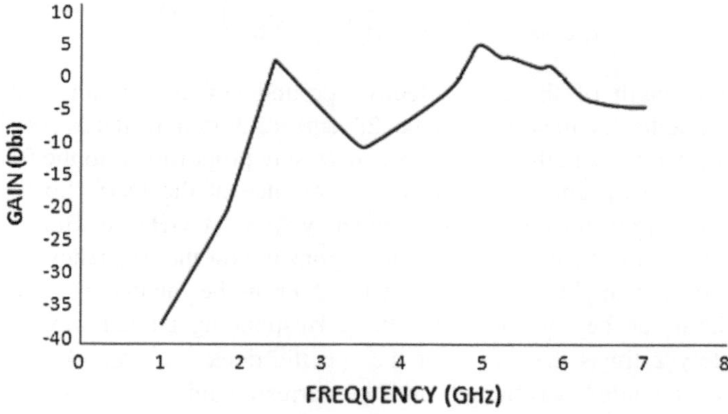

Fig. 21.10 Total gain versus frequency plot for the proposed dual band rectangular microstrip antenna

21.4 Mathematical Interpretation of the Dual Band Microtsrip Antenna

The proposed dual band microstrip antenna is found to resonate at frequency $f_1 = 2.4$ GHz i.e. Bluetooth application frequency band and at frequency $f_2 = 5.8$ GHz i.e. second WLAN application frequency band. To calculate the effective guided wavelength λ_g, the effective dielectric constant, ε_{reff} can be given as [7–9],

$$\varepsilon_{\mathrm{reff}} = \frac{\varepsilon_{\mathrm{r}} + 1}{2} + \frac{\varepsilon_{\mathrm{r}} - 1}{2}\left[1 + 12 \cdot \frac{h}{W}\right]^{-0.5}$$

where

ε_{r} Dielectric constant = 4.4

h Thickness of dielectric substrate = 1.57 mm

W Width of the rectangular patch = 17.2 mm

Thus putting the values of the corresponding parameters $\varepsilon_{\mathrm{reff}}$ is be calculated to be 3.87. From the relation between resonant frequency and guided wavelength ($\lambda_{\mathrm{g}} = \frac{c}{f\sqrt{\varepsilon_{\mathrm{reff}}}}$, where λ_{g} denotes the guided wavelength, f denotes the resonant frequency and c denotes the speed of light in vacuum), the corresponding value of wavelength λ_{g} can be calculated as,

$$\lambda_{\mathrm{g1}} = \frac{c}{f_1\sqrt{\varepsilon_{\mathrm{reff}}}} = \left(\frac{3 \times 10^{11}}{2.4 \times 10^9 \times \sqrt{3.87}}\right) = 63.5\,\mathrm{mm}$$
$$\lambda_{\mathrm{g2}} = \frac{c}{f_2\sqrt{\varepsilon_{\mathrm{reff}}}} = \left(\frac{3 \times 10^{11}}{5.8 \times 10^9 \times \sqrt{3.87}}\right) = 26.29\,\mathrm{mm}$$

The total length of the two defective ground structure implemented in the microtsrip antenna is measured to be 23 and 12.3 mm respectively. It can be observed [6] that the length of the DGS is inversely proportional to the frequency at which the microstrip antenna resonates in presence of the DGS. So the DGS of length 23 mm is responsible for the frequency f_1 = 2.4 GHz i.e. Bluetooth application and the DGS of length 12.3 mm is responsible for the frequency f_2 = 5.8 GHz i.e. second WLAN application frequency band. From the calculated value of guided wavelengths it can be concluded that the corresponding guided wavelength λ_{g} is approximately 2 times the length of the specific defective ground structure. The exact value of guided wavelength and the corresponding length of the defective ground structure are slightly altered due to the mutual coupling of the slots in the ground plane.

21.5 Conclusion

The spiral angular defective ground structures are integrated in the rectangular microstrip antenna which results in appreciable size miniaturization of about 79.6 %, along with dual frequency band operation of the microstrip antenna. The optimum gain reported is about 1.7 dBi for 2.4 GHz i.e. Bluetooth application frequency band and 2.7 dBi for 5.8 GHz i.e. second WLAN application frequency band. Thus the dual band microstrip antenna is suitable for practical wireless applications.

Acknowledgment The authors appreciate the support of the faculty members of Institute of Engineering and Management, Salt Lake and acknowledge the contribution of all concerned.

References

1. Chakraborty S, Pal S, Chakraborty M (2014) High performance DGS based compact microstrip patch antenna. In: Proceedings of 1st international science & technology congress 2014. Elsevier Publications 2014, Amsterdam, pp 404–409. ISBN 9789351072485
2. Panusa S, Kumar M (2014) MultiBand U-slot microstrip patch antenna with defective ground base. Int J Electron Electr Eng 7(10):1127–1132. ISSN 0974-2174
3. Singh G, Kaur J (2014) Design a multiband rectangular ring antenna using DGS for WLAN, WiMAX applications. Int J Adv Res Comput Commun Eng 3
4. Jadon BS, Vijay R, Nagar M (2014) Compact octagonal multiband slotted patch antenna with defective ground structure. PISER14 02(04/06):167–171
5. Shah SIH, Bashir S, Dildar Hussain Shah S (2014) Compact multiband microstrip patch antenna using defected ground structure (DGS). In: 8th European conference on antennas and propagation (EuCAP), April 2014, pp 2367–2370
6. Sharma PK, Jadaun VS (2012) Multi-band rectangular microstrip patch antenna with defected ground structure and a metallic strip. Int J Technol Explor Learn (IJTEL) 1(1): 29–31
7. Chakraborty M, Rana B, Sarkar PP, Das A (2012) Design and analysis of a compact rectangular microstrip antenna with slots using defective ground structure. Elsevier Sci Direct Proc Technol 4:411–416
8. Chakraborty M, Rana B, Sarkar PP, Das A (2012) Size reduction of a rectangular microstrip patch antenna with slots and defected ground structure. Int J Electron Eng 4(1):61–64
9. Balanis C (2012) Antenna theory analysis and design, 3rd edn. Wiley, New York

Part VI
Fibre Optics and Devices

Chapter 22
Design and Analysis of Chemically Etched and Biconically Tapered Fiber for Chemical Sensing Application

Siraj Sidhik, Jijo V. Ittiarah and Tarun Kumar Gangopadhyay

Abstract Evanescent field (EF) based sensor has been of great importance in chemical and gas sensing application. Chemical sensing using evanescent wave sensor occurs due to EF absorption, scattering or both. In this paper, the phenomenon of EF scattering for sensing is critically analyzed by modeling of (i) chemically etched fiber and (ii) tapered fiber using COMSOL. Based on the modeling, the tapered fiber is predicted to have better sensitivity to the external medium than the chemically etched fiber. Finally the tapered fiber is fabricated and tested for different concentration of sucrose and glucose solution. An effect of mode coupling is evident in the response indicated by a reversal of the spectrum after the cutoff frequency of single mode fiber.

22.1 Introduction

Fiber optic sensor (FOS) plays a significant role in various fields of application like automobiles, military, civil, medical and industries. FOS has numerous advantages e.g. high sensitivity, immune to electromagnetic interference, responsivity and reliability [1, 2]. Evanescent field (EF) interaction based sensing technology turns out to be a milestone in the field of chemical and gas detection [3, 4]. It is usually based on interaction between the external medium and the EF of light propagating in the fiber. It is mainly a sensor element that helps in extracting the exponentially decaying EF from the optical waveguide. An external medium with varying refractive index (RI), will have different impact on the penetration depth and energy of evanescent wave. It also changes the transmission intensity of light through the optical fiber. This variation in transmission intensity at the receiving end of the fiber helps in providing a judgment about the EF characteristics of different media.

S. Sidhik · J.V. Ittiarah · T.K. Gangopadhyay (✉)
CSIR—Central Glass and Ceramic Research Institute (CGCRI),
Kolkata 700032, India
e-mail: tkg@cgcri.res.in

V. Lakshminarayanan and I. Bhattacharya (eds.), *Advances in Optical Science and Engineering*, Springer Proceedings in Physics 166,
DOI 10.1007/978-81-322-2367-2_22

In this paper, the etched fiber [5] and tapered fiber [6, 7] as EF access block is modeled and fabricated using single mode fiber (SMF-28). Some experimental results based on RI sensing are also included.

22.2 Principle of Evanescent Access Block

The basic idea underlining the EF sensors is the EF energy. It is observed that when light is propagated through an optical fiber, the output light intensity depends on the EF distribution. On the other hand, it depends on the RI of the external medium. Light propagates through an optical fiber by total internal reflection. Even though total internal reflection is taking place, some portion of the light usually extend towards the cladding region and forms the standing wave according to the Maxwell's equation, which is referred to as the Evanescent wave [8]. If θ is the angle of incidence at the core-cladding interface with respect to the normal, the evanescent wave will travel in the vertical direction and will decay exponentially. The power of the evanescent wave field is given by the equation [9],

$$P = P_0 \exp(-x/\mathrm{d}p) \tag{22.1}$$

where P is the power of evanescent wave, P_0 represents the power of incident light, x is the perpendicular distance from the interface and dp is the penetration depth. The penetration depth dp is given by [9],

$$\mathrm{d}p = \frac{\lambda}{2\pi\sqrt{\left(n_1^2 \sin^2\theta - n_2^2\right)}} \tag{22.2}$$

where λ is the incident wavelength, n_1 is the RI of core, n_2 is the RI of cladding and θ is the angle of incidence at the core-cladding interface.

The impact of different media on the propagation and transmission of optical light through optical fiber depends on the RI. When light is usually propagating through the evanescent region, the energy is completely lost. Therefore as the amount of light propagating through the evanescent region increases, the energy of the light that is perceived at the output decreases. Based on this principle, an EF sensor can be designed to analyze the characteristics of EF by measuring the light intensity at the receiving end.

22.3 Modeling of Etched and Tapered Fiber Using Comsol

The 2D geometric model of etched and biconically tapered fiber created using the radio frequency (RF) module of comsol multiphysics is shown in Fig. 22.1.

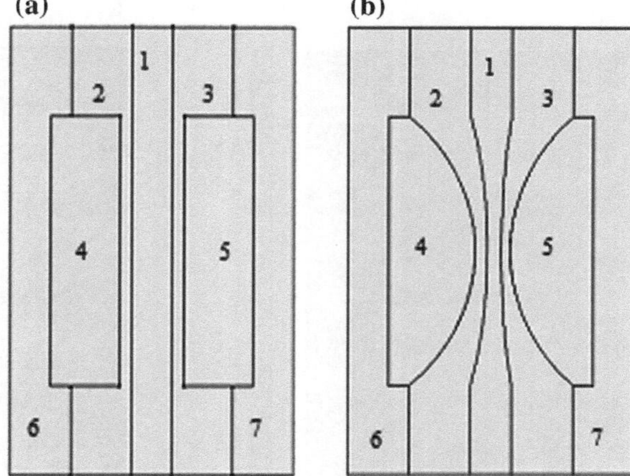

Fig. 22.1 2D geometric model of **a** etched fiber, **b** tapered fiber created using RF frequency module of comsol multiphysics

In the given model of etched fiber, region (1) forms the core having RI 1.457, region (2) and (3) forms the cladding having RI 1.45, region (4), (5) represents the etched portion of the optical fiber that forms the external medium, whose RI can be varied for performing the analysis and finally region (5), (6) is made as the air background. Similarly for the geometric model of tapered fiber, region 4 and 5 forms the tapered portion over which desired chemicals of varying RI are provided for sensing and all the other regions are same as that of the etched fiber. The bottom boundary of the core is made as the incoming port with an input power of about 1 W and the top boundary is defined as the receiving port. After that the model has been meshed using extra fine meshing and frequency domain, boundary domain and modal analysis has been set in the study subsection. The model is then simulated for a wavelength of 1550 nm to detect the transmission light intensity (I_0). The simulated result is shown in Fig. 22.2.

Figure 22.2a represents the visual map of the electric field along the etched fiber and Fig. 22.2b the tapered fiber when the RI of the external medium is equal to 1.35 for both the case. The figure shows that the major part of the light is concentrated in the core region (red color in online representation) but has certain loss in the form of EF (yellow color in online presentation).

22.3.1 Relationship Between RI of Evanescent Field and the Receiving Intensity

From Fig. 22.2, it is observed that the transmission decay through the optical fiber usually occurs due to the loss happening within the evanescent region. This loss within the optical fiber is a function of RI of the external medium at the etched and

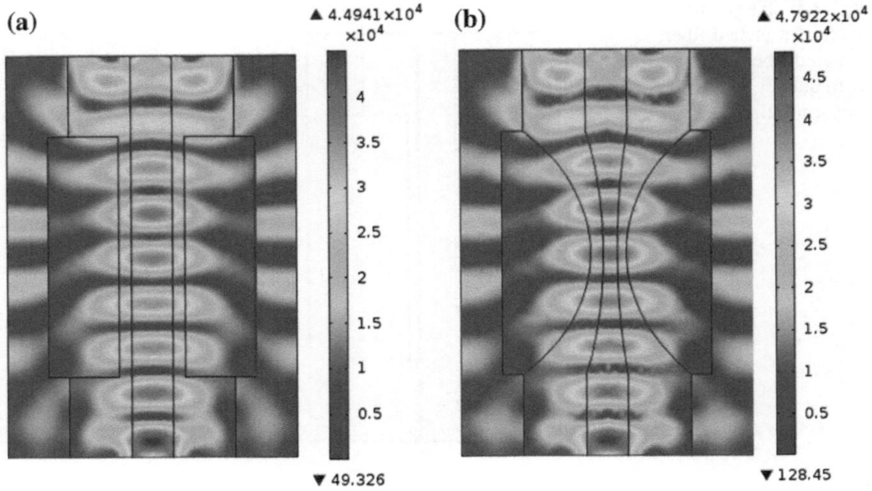

Fig. 22.2 Electric field propagating in **a** etched fiber, **b** tapered fiber when the RI of the evanescent field is 1.35

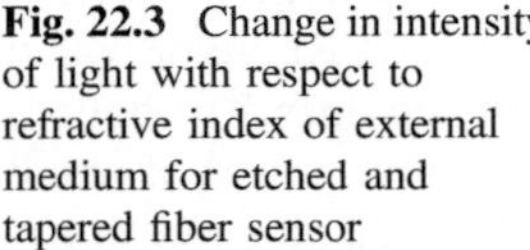

Fig. 22.3 Change in intensity of light with respect to refractive index of external medium for etched and tapered fiber sensor

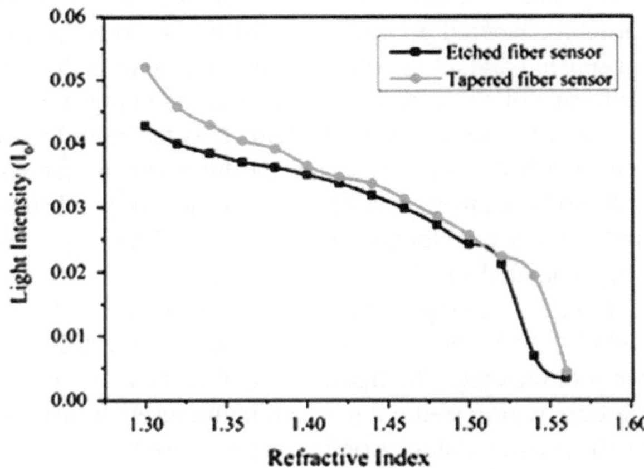

tapered portion of the optical fiber. So the RI of the external medium is varied and the effect on the transmission field intensity due to this change is observed and plotted. This variation of intensity with respect to the RI is plotted in Fig. 22.3.

It is observed from the result that, the biconically tapered fiber is having high sensitivity compared to that of the chemically etched fiber.

22.4 Experiment and Results

Two basic approaches are followed for the fabrication of EF sensor element by (i) chemical etching using hydro-fluoric (HF) acid and (ii) biconical tapering. In chemical etching process, a part of the optical fiber is placed in the HF solution which results in etching of desired portion of the cladding surface whereas in fused tapering process, the optical fiber is stretched precisely by controlling the arc interval of the fusion splicer. During stretching the diameter of core and cladding region is uniformly reduced for better evanescent interaction. The chemically etched and biconically tapered fiber has been fabricated at CSIR-CGCRI lab by keeping in account various design parameters on the basis of modeling done and is shown in Fig. 22.4.

The experimental setup for chemical sensing using tapered fiber is shown in Fig. 22.5. It comprises of a broadband light source, fiber axis alignment system, InGaAs detector and a data logging and acquisition system. The tapered fiber fabricated is properly aligned in between the broadband light source and the detector. Few drops of glucose ($C_6H_{12}O_6$) and sucrose ($C_{12}H_{22}O_{11}$) at different concentration are introduced into the sensor element with the help of a micro-injector. Further the wavelength of light source is scanned from 1100 to 1500 nm with a wavelength spacing of 1 nm each. The transmission spectrum obtained for different concentration of glucose and sucrose is shown in Fig. 22.6.

The power of the transmission spectrum is observed to have increased with the increase in concentration of glucose solution. For sucrose solution, the transmission power decreases with the increase of concentration. This happens due the EF scattering at the interface of the solution and the tapered region with receding index boundary. An interesting phenomenon of mode coupling is observed in which the spectral response is usually reversed after the cut-off frequency of the optical fiber (1260 nm) as indicated in the result.

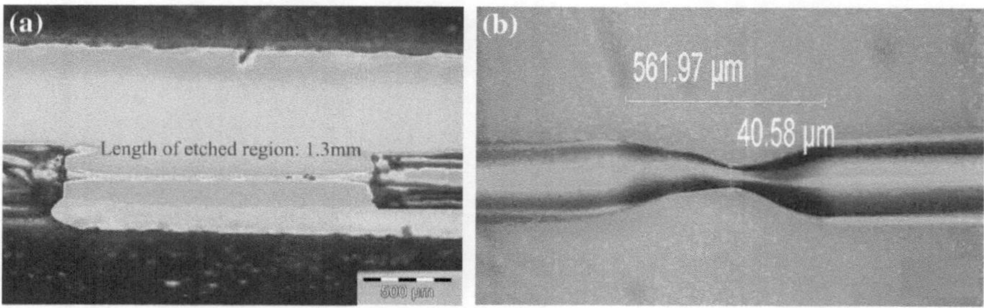

Fig. 22.4 **a** Photograph of chemically etched fiber fabricated at CGCRI [5], **b** photograph of tapered fiber fabricated at CGCRI

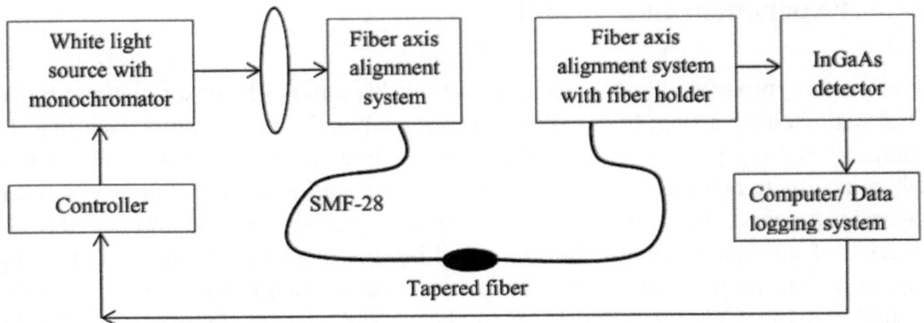

Fig. 22.5 Schematic representation of experimental setup for chemical sensing process

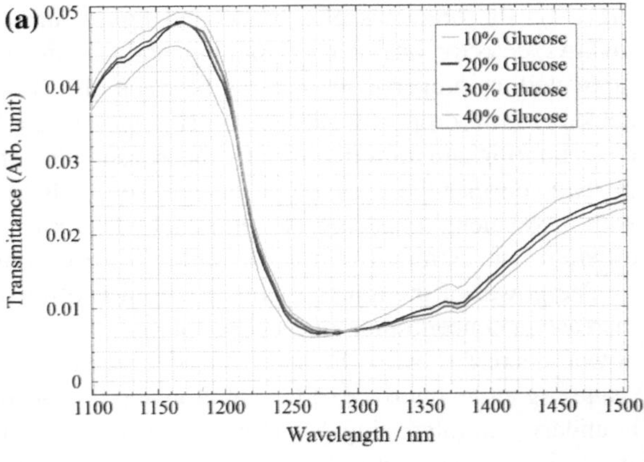

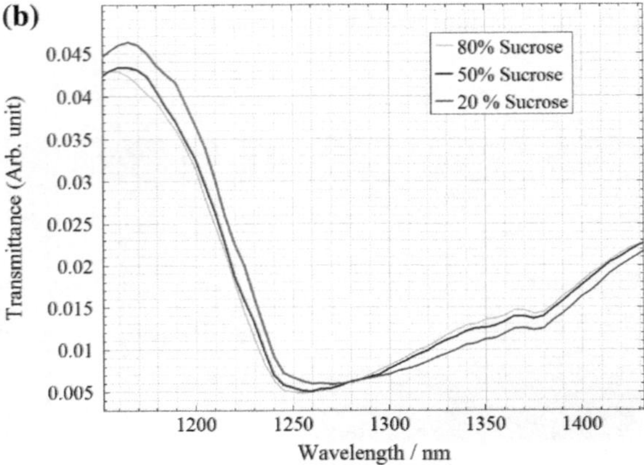

Fig. 22.6 Transmission spectrum for **a** glucose solution with different concentration using tapered fiber as the sensor element, **b** sucrose solution with different concentration using tapered fiber as the sensor element

22.5 Conclusion

A comparative analysis of chemical sensors based on biconically tapered fiber and chemically etched single mode fiber is performed. The response of the sensors with external medium having different RI is simulated. The results are plotted for further assessment. From the slope of the curve, it is concluded that biconically tapered fiber sensor is more sensitive to external RI than a chemically etched fiber sensor. Sensors are also found to be linear in the range of RI from 1.33 to 1.46. The tapered fiber is fabricated in-house using Fujikura fusion splicer by simultaneously controlling the arc time interval and stretching. Different concentration of glucose and sucrose solution is tested with the tapered fiber. The experimental results are found to be comparable with the theoretical predictions. Mode coupling is observed resulting in the reversal of spectral response after the cut off frequency of optical fiber.

References

1. Culshaw B (2005) Fiber-optic sensors. Opt Photonics News 16(11):24–29
2. Udd E (1993) Fiber optic sensors. Wiley Online Library, New York
3. Polynkin P, Polynkin A, Peyghambarian N, Mansuripur M (2005) Evanescent field-based optical fiber sensing device for measuring the refractive index of liquids in microfluidic channels. Opt Lett 30(11):1273–1275
4. Orghici R, Willer U, Gierszewska M, Waldvogel SR, Schade W (2008) Fiber optic evanescent field sensor for detection of explosives and CO_2 dissolved in water. Appl Phys B 90(2):355–360
5. Linslal CL, Syam Mohan PM, Halder A, Gangopadhyay TK (2013) Analysis and modeling of an optical fiber loop resonator and an evanescent field absorption sensor for the application for chemical detection. Sens Actuators Phys 194:160–168
6. Linslal CL, Mohan PMS, Halder A, Gangopadhyay TK (2012) Eigenvalue equation and core-mode cutoff of weakly guiding tapered fiber as three layer optical waveguide and used as biochemical sensor. Appl Opt 51(16):3445–3452
7. Linslal CL, Mohan PS, Halder A, Gangopadhyay TK (2012) Fabrication of single mode tapered fiber and used as biochemical sensor. In: Third international conference on sensors and related networks (SENNET-12), VIT University, Vellore, India. pp 18–21
8. Lezec H, Thio T (2004) Diffracted evanescent wave model for enhanced and suppressed optical transmission through subwavelength hole arrays. Opt Express 12(16):3629–3651
9. Rays AG, Fibers CS-M (1991) Fiber opticsTeich, M. C., & Saleh, B. E. A. (1991). Fundamentals of photonics. Canada, Wiley Interscience, 3

Chapter 23
Fluid Evaporation Monitoring of Volatile Organic Compound Using D-Shaped Fiber

Jijo V. Ittiarah, Siraj Sidhik and Tarun Kumar Gangopadhyay

Abstract A sensing configuration for monitoring of fluid evaporation using fiber loop ringdown spectroscopy is proposed. The sensing is induced by evanescent field scattering of the volatile liquid and the effect of optical light passing through the fiber cavity. A D-shaped fiber is fabricated in-house and deployed in the optical path for the evanescent wave interaction. The evaporation process of Ethylenediamine (EDA) is observed in terms of ringdown time variation. This configuration can be used to detect and study the evaporation rate of different volatile organic compound in time domain.

23.1 Introduction

The term evaporation is defined as the transition from liquid phase to gaseous phase, which occurs at a temperature below the boiling point of liquid and at ambient pressure. Evaporation purely depends on the properties of fluid, the environment conditions and the quality of the surface on which it is introduced [1]. Analysis of formation and evaporation of small volumes of liquid is thus required to understand numerous properties of fluids like surface tension, vapor pressure, viscosity, chemical composition and refractive index (RI) [2].

Fiber optic sensor (FOS) for liquid is usually limited to the measurement of concentration [3], RI [4] and absorption spectroscopy. But, the temporal study of evaporation of liquids proves to be vital in application of identification of fuel dilution, climatography, monitoring quality of water and so on [2]. FOS turns out to be an unpretentious and cost effective method for evaporation monitoring. Recently, Moura et al. [5] presented an evaporation monitoring system using

J.V. Ittiarah · S. Sidhik · T.K. Gangopadhyay (✉)
CSIR—Central Glass and Ceramic Research Institute (CGCRI),
Kolkata 700032, India
e-mail: tkg@cgcri.res.in

V. Lakshminarayanan and I. Bhattacharya (eds.), *Advances in Optical Science and Engineering*, Springer Proceedings in Physics 166,
DOI 10.1007/978-81-322-2367-2_23

suspended core fiber. Preter et al. [6] used a monolayer coated single mode fiber for monitoring droplet based evaporation.

In this work, monitoring of evaporation dynamics of Ethylenediamine (EDA) is demonstrated using fiber loop ringdown spectroscopy (FLRDS). D-shaped fiber which utilizes the principle of evanescent wave interaction between the sample and the optical light is used as the sensor element. FLRDS technique is preferred because of the enhanced sensitivity which is achieved due to the multiple interaction of light with the sensor element. Also it is insensitive to intensity and power fluctuation which makes it attractive among other FOS. The sensing is usually based on ringdown time variation induced due to the receding index boundary between the sensor element and liquid sample with evaporation.

23.2 Principle of Evaporation Monitoring Using FLRDS

The exponentially decaying pulse generated at the output of the FLRDS scheme is mathematically represented as [7],

$$P = P_0 \exp\left(-\frac{t}{\tau_0}\right) \tag{23.1}$$

where P is the power of optical pulse at time t, P_0 is the initial power of optical pulse and τ_0 denotes ringdown time given by,

$$\tau_0 = \frac{nL}{cT} \tag{23.2}$$

where τ_0 is the ringdown time without the chemical, n is the effective RI of fiber, L is the length of the fiber loop, c is the speed of light in vacuum and T represents the total loop transmission loss. Introduction of a volatile compound into the sensor element induces an additional loss T_1 which results in reduced ringdown time τ given by,

$$\tau = \frac{nL}{c(T + T_1)} \tag{23.3}$$

Solving (23.2) and (23.3)

$$T_1 = \frac{nL}{c}\left(\frac{1}{\tau} - \frac{1}{\tau_0}\right) \tag{23.4}$$

The desired loss T_1 is a result of EF attenuation due to EF scattering that occurs at the interface of volatile compound and the polished surface of fiber. The evaporation of volatile compound present on the sensor element, results in receding index difference at the boundary of interaction owing to the change in ringdown time.

23.3 Experimental Setup and Results

In this work, evaporation rate detection of EDA using EF interaction based FLRDS is illustrated. Initially a D-shaped fiber is fabricated at CGCRI with an insertion loss of 0.6 dB, polishing length 10 mm and polishing depth of 56 μm. Two couplers having low splitting ratio (99:1) with reduced insertion loss is developed. Now, both D-shaped fiber and couplers are implemented in the 30 m fiber loop to configure the FLRDS system. The experimental setup for evaporation monitoring using FLRDS scheme is shown in Fig. 23.1.

It comprises of a continuous wave (cw) laser source operating in the wavelength of 1550 nm, with a maximum power of 35 mW. The output of the laser source is directed to the high speed optical modulator for converting the cw signal into pulsed one. The modulator is driven by an electrical pulse having a power of 5 Vpp, frequency of 350 Hz and pulse width of 8 ns with the help of a pulse generator. The pulsed signal is then directed into the fiber loop of single mode fiber (SMF-28) and two identical 2 × 2 fiber couplers (99:1). The output of the fiber loop is then fed to a high frequency detector operating in the wavelength range of 850–1650 nm. The detected signal is further provided to a 2.5 GHz oscilloscope for observing the ringdown signal. The ringdown signal is observed to have large amount of intensity fluctuation arising due to the back reflection from the couplers and other devices. The total length of the fiber loop selected was 30 m.

A live monitoring and data acquisition system is designed using LabVIEW based program. It acquires the ringdown signal from the oscilloscope and simultaneously calculates the ringdown time for sensing purpose. The ringdown signal and the exponentially fitted curve obtained with the LabVIEW program is shown in Fig. 23.2.

Next, sub-microliter of EDA is introduced into the D-shaped fiber with the help of a micro-injector. The ringdown time variation with respect to evaporation of volatile EDA is shown in Fig. 23.3.

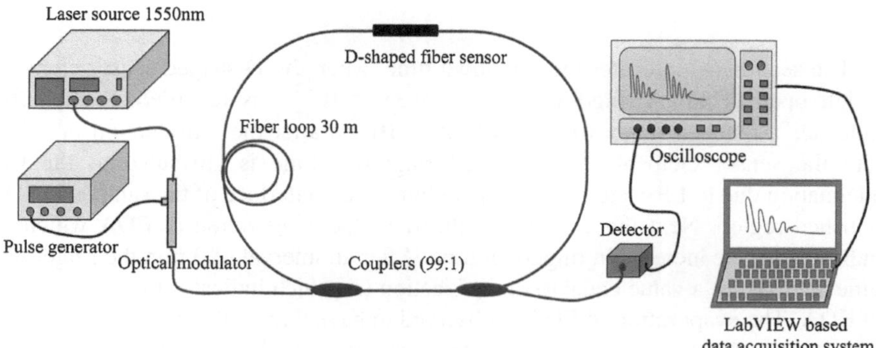

Fig. 23.1 Schematic diagram of FLRDS setup with 30 m fiber loop, two couplers and D-shaped fiber sensor for monitoring of evaporation

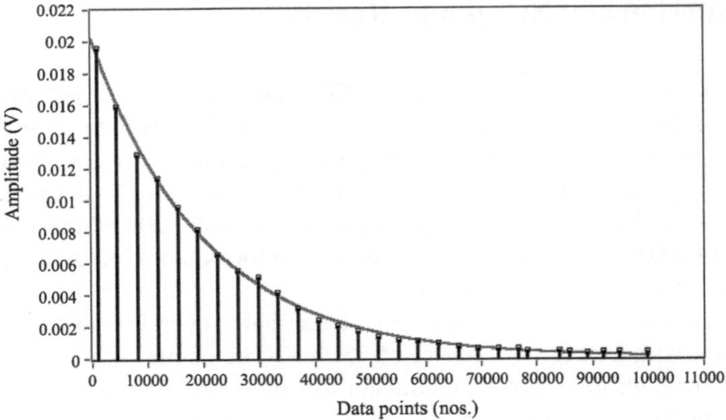

Fig. 23.2 The exponential curve fitted ringdown signal as represented in the front panel of the LabVIEW based program

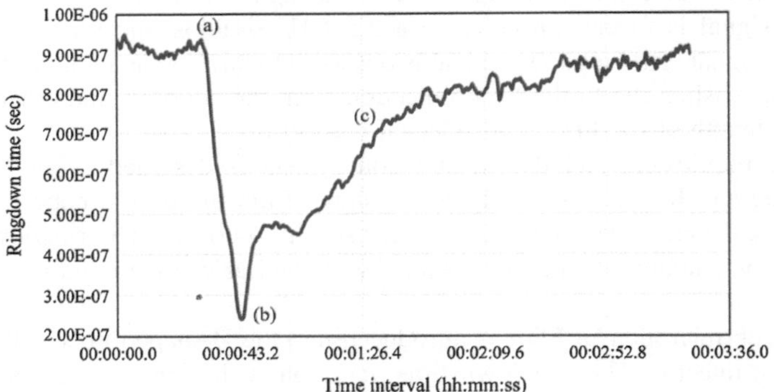

Fig. 23.3 Ringdown time variation showing the evaporation dynamics of Ethylenediamine as a function of time interval

The section (a) indicates the ringdown time, when the D-shaped sensor element is left open in air. A ringdown time of 9.38×10^{-7} s is recorded. Section (b) indicates a reduced ringdown time of 2.5×10^{-7} s after the introduction of EDA into the sensor element. This reduced ringdown time is attributed to the EF attenuation due to EF scattering of optical light at the interface of the sample and D-polished region. Next the section (c) illustrates the evaporation of EDA which is indicated by the increase in ringdown time. After an interval of 3 min the ringdown time approaches a value equal to that of section (a) which indicates full evaporation of EDA. The evaporation of EDA is observed to have three different phases. During the initial phase the evaporation rate as per ringdown time variation is 0.314×10^{-7} $\Delta\tau$/s. In the following phases the evaporation rate is found to be 0.078×10^{-7} and 0.011×10^{-7} $\Delta\tau$/s, respectively. Initial results are found to be promising in terms of evaporation rate detection of EDA. This work could be further explored for detection of different volatile organic compounds on the basis of evaporation rate.

23.4 Conclusion

We proposed and developed a technique for monitoring the evaporation of volatile organic compound using evanescent field based FLRDS. A D-shaped fiber with an insertion loss of 0.6 dB, polishing depth of 56 μm from cladding surface and polishing length of 10 mm is fabricated and integrated in the optical path. Evaporation rate detection of EDA is carried out using the FLRDS scheme. The evaporation of EDA is observed to have three different phases with an evaporation rate of 0.314×10^{-7}, 0.078×10^{-7} and 0.011×10^{-7} $\Delta\tau$/s.

Acknowledgment The research work is supported by the sponsored project (No. GAP0141) from DST, Govt. of India. The author would like to acknowledge the support of the Director, CGCRI, Dr. Mukul Paul, Dr. M. Pal and Dr. Shyamal Das of FOPD, CGCRI, Kolkata for their help.

References

1. Thomson BA, Iribarne JV (1979) Field induced ion evaporation from liquid surfaces at atmospheric pressure. J Chem Phys 71(11):4451
2. Birdi KS, Vu DT, Winter A (1989) A study of the evaporation rates of small water drops placed on a solid surface. J Phys Chem 93(9):3702–3703
3. Wang X-D, Wolfbeis OS (2013) Fiber-optic chemical sensors and biosensors (2008–2012). Anal Chem 85(2):487–508
4. Xia T-H, Zhang AP, Gu B, Zhu J-J (2010) Fiber-optic refractive-index sensors based on transmissive and reflective thin-core fiber modal interferometers. Opt Commun 283 (10):2136–2139
5. Moura JP, Baierl H, Auguste J-L, Jamier R, Roy P, Santos JL, Frazão O (2014) Evaporation of volatile compounds in suspended-core fibers. Opt Lett 39(13):3868–3871
6. Preter E, Katims R, Artel V, Sukenik CN, Donlagic D, Zadok A (2014) Monitoring and analysis of pendant droplets evaporation using bare and monolayer-coated optical fiber facets. Opt Mater Express 4(5):903
7. Tong Z, Jakubinek M, Wright A, Gillies A, Loock H-P (2003) Fiber-loop ring-down spectroscopy: a sensitive absorption technique for small liquid samples. Rev Sci Instrum 74 (11):4818

Chapter 24
Tellurite Glass Microstructured Optical Fibers: An Analytical Approach

Dinesh Kumar Sharma and Anurag Sharma

Abstract Microstructured optical fibers (MOFs) have generated broad interest due to their extraordinary optical properties which arises from the combination of wavelength scale features in the photonic crystal cladding and the high index contrast between the background material and the air. We analytically study the optical properties of nonsilica glasses such as tellurite (TeO_2) glass based MOFs by using our earlier developed analytical field model. We have compared the accuracy of our simulated results with numerical simulation results available in the literature.

24.1 Introduction

The potential of using photonic crystals both 2-dimensional and 3-dimensional in manipulating and controlling the electromagnetic waves have been reported by several researchers. Photonic crystals molded into optical fibers are very attractive from the scientific and technological point of view due to its broadband single-mode propagation characteristic. The combination of wavelength scale features in microstructured cladding and design flexibility offered by the microstructured optical fibers (MOFs) leads to a wider range of extraordinary optical properties [1–3]. The most common types of MOFs are made of waveguiding parts with a solid core and cladding layer consisting of spatially distributed periodic air-holes. The effective index of the holey cladding relative to the core is lower due to presence of these air-holes as a result the confinement and guidance of light in the MOFs with relatively high core index would occur via total internal reflection, and such type of MOFs are termed as index-guiding MOFs [3, 4]. The MOFs present greater advantages, such as tailorable effective modal areas, anomalous group velocity dispersion at visible and infrared (IR) band. Up to now, the majority of

D.K. Sharma (✉) · A. Sharma
Physics Department, Indian Institute of Technology Delhi, New Delhi 110016, India
e-mail: dk81.dineshkumar@gmail.com

V. Lakshminarayanan and I. Bhattacharya (eds.), *Advances in Optical Science and Engineering*, Springer Proceedings in Physics 166,
DOI 10.1007/978-81-322-2367-2_24

MOFs developed are made from pure silica glass and studied in the visible and the near-IR wavelengths [1–5].

Recently, compound glasses, soft glasses and polymer based MOFs have been fabricated. Following the work on silica glasses based MOFs the high-index non-silica glasses MOFs have developed rapidly offering advantages over silica MOFs in the application areas of high nonlinearity, low-loss mid-IR transmission and the compact active single-mode fiber devices with extremely large mode area. Nonsilica glasses such as tellurite (i.e., tellurite oxide TeO_2 based) having optical transparency in the wavelength range of 0.5–5.0 μm, fluoride with optical transparency in the wavelength range of 0.4–6.0 μm and chalcogenide glasses with optical transparency in the wavelength range of 1.0–16 μm have excellent optical transparency compared to silica [6–8]. Tellurite (TeO_2) glasses offers wide range of useful modal properties not possessed by silica glasses such as, good IR transmittance, high linear and nonlinear refractive indices moreover, they have high optical nonlinearity and relatively low phonon energy among oxide glasses [7–9]. Recently, IR supercontinuum (SC) from a TeO_2 fiber with a mode area as large as 30,000 μm^2 for the fundamental mode, pumped at 2.15 μm has been reported [6, 9]. Moreover, TeO_2 glass is much less toxic, more chemically and thermally stable. It is highly suitable fiber material used for IR nonlinear applications [6, 7]. The light guidance mechanisms in nonsilica MOFs is analogous to those silica based MOFs which arises due to periodic index modulation in the photonic crystal cladding [1, 10].

Using our earlier developed analytical field model [11–14] for index-guiding MOFs which is based on explicitly considering three circular rings of voids or air-holes in the fiber cross-section and one ring of lobes in the modal field, we have studied various optical properties [12–16] of solid-core MOFs which compared well with the experimental and numerical simulation results reported in the literature. Our simulated results compare well with the results that are available in the literature and hence, establish the validity of our analytical approach.

Here, we have extended our analytical field model further, for better accuracy. We have added more layers of circular air-holes in the model in order to take in account the actual picture of the fiber's cross-section. In this paper, our model is based on considering five circular rings of voids or air-holes around the central local defect (core) in the fiber cross-section and one ring of lobes in the modal field. We have used our slightly enhanced field model to study the modal properties of TeO_2 glasses MOFs with triangular lattice of circular air-holes in the photonic crystal cladding. Comparison with available numerical simulation results also has been included.

24.2 Tellurite Glass Microstructured Optical Fibers

The dielectric cross-section of a typical index-guiding MOF is composed of triangular lattice of circular voids or air-holes of refractive index unity in the TeO_2 glass matrix which repeats itself throughout the entire structure. In such type of array, each lattice point has six nearest neighbours at a distance of pitch, Λ

(center-to-center separation between two nearest voids or air-holes) and six next-nearest neighbours at a distance of $\Lambda\sqrt{3}$ followed by another six neighbours at a distance of 2Λ and so on [11, 12]. Thus, the central defect site (a missing air-hole at the center) can be considered to be surrounded by successive circular rings of radii Λ, $\Lambda\sqrt{3}, 2\Lambda, \Lambda\sqrt{7}\ldots$ etc., each ring containing circular voids placed symmetrically, as shown schematically in Fig. 24.1. Air-holes of first, third and fifth circular rings are arranged at the same angular position and it is assumed that the axis $\varphi = 0$ passes through the center of these rings. In second ring air-holes are shifted by an angle of $\pi/6$ while in fourth ring they are shifted by an angle of $\pi/12$.

The first circular ring has radius of $r_1 = \Lambda$ while next circular rings have radii of $r_2 = \Lambda\sqrt{3}, r_3 = 2\Lambda, r_4 = \Lambda\sqrt{7}$ and $r_5 = 3\Lambda$ respectively. We further assume that beyond the air-holes of fifth ring i.e., $r > (3\Lambda + a)$, where a is radius of each air-hole, we have a uniform medium of refractive index equal to the effective index of the fundamental space-filling mode (FSM). The FSM is also called fundamental cladding mode having largest value of the propagation constant and is defined as the effective index of the infinite photonic crystal cladding without any core (i.e., perfect cladding) [1, 3]. In our field model, we considered the field distribution mainly, consisting of two types of Gaussian terms: simple Gaussian that controls the field variation in radial (r) direction and the shifted Gaussian that controls the azimuthal variation of the field in angular (φ) direction [11–14].

To study the propagation properties of index-guiding TeO_2 glasses MOFs, we have used the following modal field for the fundamental mode, which is associated with the essential six-fold hexagonal symmetry of the structure [11–14];

$$\Psi(r,\varphi) = \exp\left(-\alpha r^2\right) - A\,\exp\left\{-\alpha_1 (r - \Lambda\sigma)^2\right\}(1 + \cos 6\varphi) \tag{24.1}$$

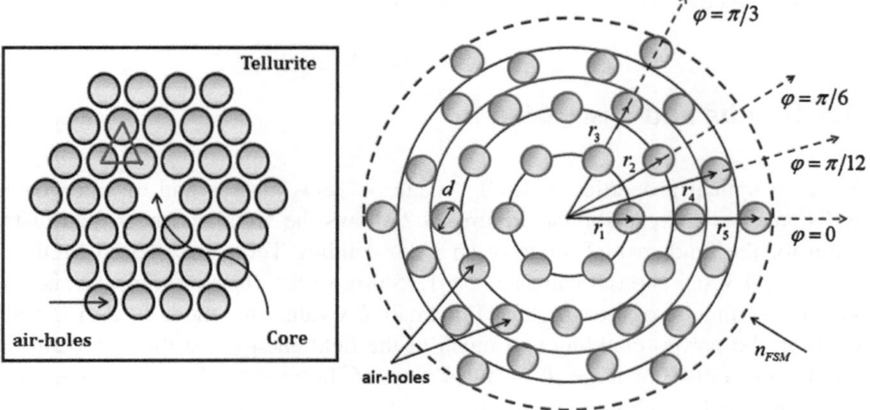

Fig. 24.1 Schematic diagram for the transverse cross-section of an index-guiding microstructured optical fiber with $N = 5$ circular rings

where α, A, α_1 and σ are the field parameters; Λ is the natural length of separation between the centers of two adjacent air-holes, known as pitch. We have used the variational techniques [17, 18] based on integral form of the scalar wave equation for the scalar guided modes, to obtain the optimized values of the propagation constant and the field parameters involved in (24.1), at a given wavelength and the waveguiding structure. In angular direction, our model possesses six-fold rotational symmetry which is shown explicitly in the Fig. 24.1. Using this symmetry feature we carried out calculations in a segment having a sector of $\pi/6$ which makes our calculations simpler and reduces the computation time.

Of all the TeO_2 glasses we considered a particular glass of the $0.8TeO_2$–$0.2WO_3$ chemical composition. The material dispersion of this glass is determined by the Sellmeier relation [10];

$$n(\lambda) = \left[2.4909866 + \frac{1.9515037}{1 - 5.6740339 \times 10^{-2}/\lambda^2} + \frac{3.0212592}{1 - 225/\lambda^2}\right]^{1/2} \tag{24.2}$$

where λ is measured in micrometers. Once we know the effective indices of the fundamental mode we can calculate the group velocity,

$$v_{gr} = c\left(n_{eff} - \lambda\frac{dn_{eff}}{d\lambda}\right)^{-1}, \tag{24.3}$$

and the dispersion parameter D,

$$D = -\frac{\lambda}{c}\frac{d^2 n_{eff}}{d\lambda^2} \tag{24.4}$$

where c is the speed of light in the vacuum, λ being the operating wavelength and n_{eff} is the effective index of the fundamental mode.

24.3 Results and Discussion

The main optical properties of MOFs made of TeO_2 glasses and of silica based MOFs are qualitatively similar. Figure 24.2 shows the simulated near-field distribution for the fundamental mode of an index-guiding TeO_2 glass based MOF for $d/\Lambda = 0.2$, 0.3, 0.5 and 0.8 values at $\lambda = 1.55$ μm, which clearly shows the six-fold rotational symmetry of the structure. For small d/Λ values the mode field distributes largely in the porous cladding and much of the field energy distributes around the air-holes or voids. For large d/Λ values most of the mode field energy is trapped inside the core due to high index contrast between the core and the porous cladding. In short, the distribution of the mode fields with smaller d/Λ values is apparently different from that of mode fields with larger d/Λ values. For larger d/Λ values the

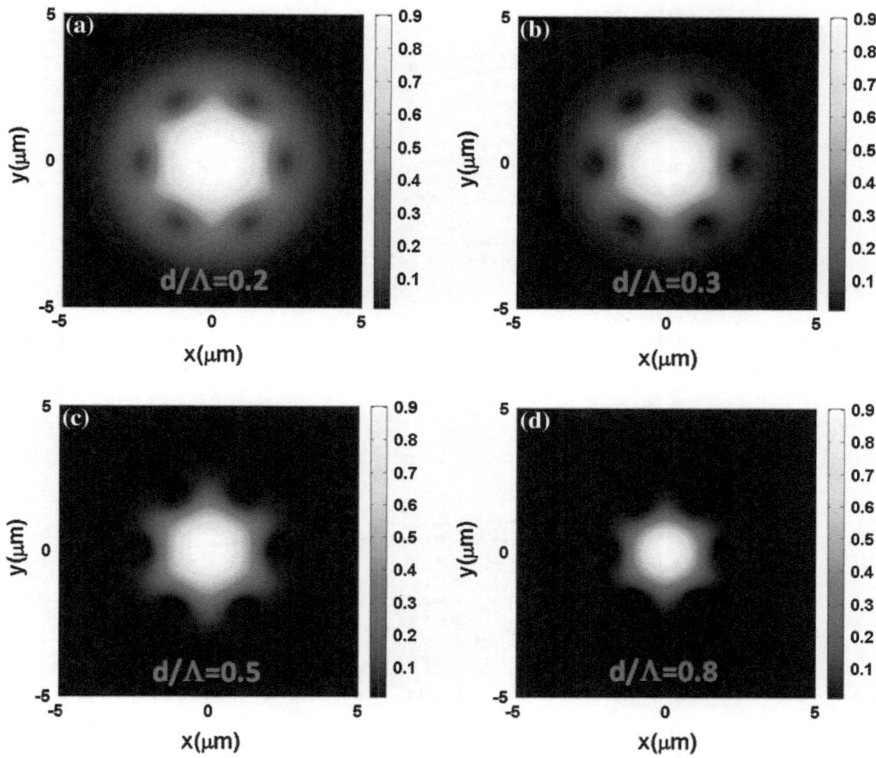

Fig. 24.2 Simulated near-field mode profiles of the fundamental mode for TeO_2 glass based microstructured optical fiber for Λ = 2.3 μm at λ = 1.55 μm

central part of the modal field becomes more compact and more circular in shape compared to smaller d/Λ values but the tail of field still has hexagonal features.

Figure 24.3 shows the spectral dependence of the effective index for the fundamental mode of the fiber with pitch, Λ = 2.3 μm for different normalized air-hole diameters: d/Λ = 0.2, 0.3, 0.5 and 0.8. We have compared the accuracy of our results with those obtained from the multipole method (MM) [10] as shown by circles, which are good in agreement. The MM is a rigorous numerical technique whereas our method involves only few numerical integrations and four parameters optimization which provides an analytical form to our modal field. Figure 24.4 shows the variation of group velocity to speed of light in the vacuum $\left(v_{gr}/c\right)$ as a function of wavelength in micron for different fiber geometries with pitch value of Λ = 2.3 μm. We have checked the accuracy of our results with those obtained using the MM [10] (as marked by circles). We have found good agreement between the results for small d/Λ values but this agreement deteriorates for large d/Λ ratios in longer wavelength regime.

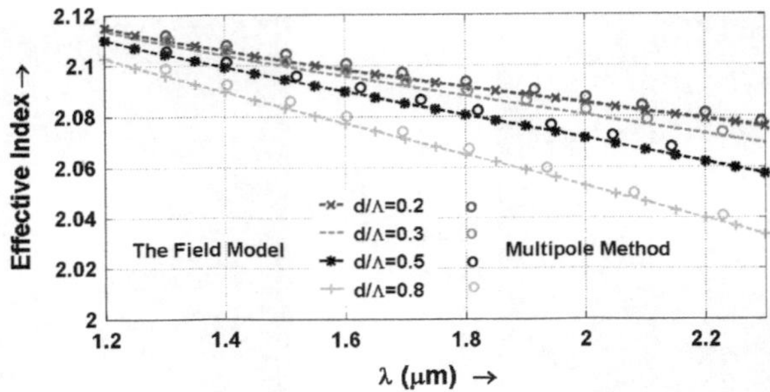

Fig. 24.3 Variation of the effective index of the fundamental mode as a function of λ for a given pitch of $\Lambda = 2.3$ μm

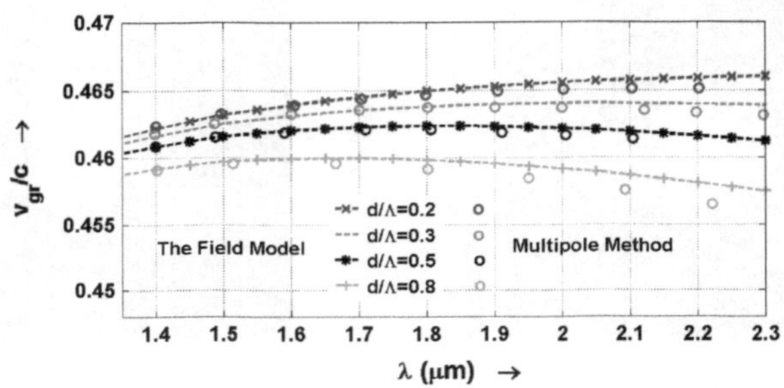

Fig. 24.4 Spectral dependence of the group velocity to speed of light in vacuum as a function of wavelength for different air-hole diameter to pitch ratios

We have also obtained the dispersion coefficient for a range of MOF parameters and have compared the accuracy with the MM [10]. The spectral dependence of the dispersion coefficient as a function of wavelength is shown in Fig. 24.5 for normalized air-hole diameters: *d*/Λ = 0.2, 0.3, 0.5 and 0.8 with Λ = 2.3 μm. These result matches very well with those obtained from MM [10] for smaller *d*/Λ values. But as the air-hole size increases our results deteriorates particularly at longer wavelength. This effect is more pronounced in the case of dispersion calculation compared to group velocity to speed of light ratio, (v_{gr}/c), as it depends on the second derivative of effective index of the fundamental core mode. In such cases the modal field spreads well into the porous cladding far away from the fiber's core and the field model based on five circular rings in the fiber geometry and one lobes (i.e., one shifted Gaussian term) of field is not sufficient. One must incorporate more circular rings of air-holes in the fiber structure as well as more lobes (i.e., shifted Gaussian terms) in the modal field. This would indeed increase the complexity of the numerical operations involved in our simulation as well as make computation somewhat tedious.

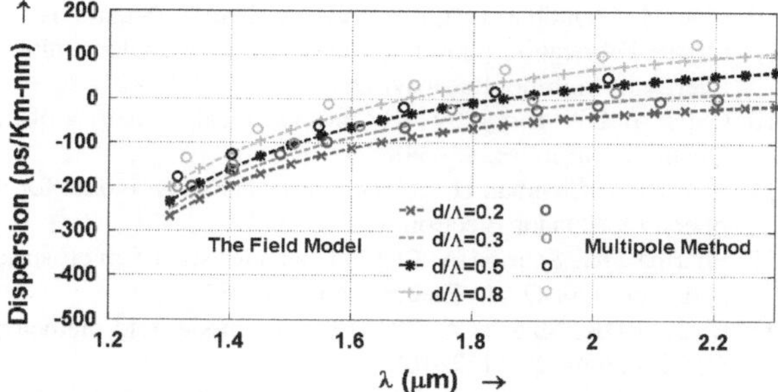

Fig. 24.5 Dispersion parameter as a function of wavelength for different *d*/Λ ratios

24.4 Conclusion

We have studied various optical properties of index-guided nonsilica glasses MOFs by using our earlier developed analytical field model in a slightly modified form which is based on explicitly considering five circular rings of air-holes in the fiber structure and one ring of dips in the modal field. We have considered TeO_2 glass based MOFs and have studied their modal characteristics. We have shown that our model predicts the effective indices appreciably well. The results of simulations based on the field model compares well with those numerical results obtained using the MM. Our model provides analytical form to the modal field therefore, it is easy to implement and it is well equipped to take in account field asymmetries around the air-holes. Our method consumes less computation time and memory compared to MM. For fibers with larger air-holes further improvements in the model would be required. Further, work is in progress.

References

1. Russell PStJ (2006) Photonic-crystal fibers. J Lightwave Technol 24:4729–4749
2. Roberts PJ, Mangan BJ, Sabert H, Couny F, Birks TA, Knight JC, Russell PStJ (2005) Control of dispersion in photonic crystal fibers. J Opt Fiber Commun Rep 2:435–461
3. Knight JC, Knight TA, Russell PStJ, Atkin DM (1996) All-silica single-mode optical fiber with photonic crystal cladding. Opt Lett 21:1547–1549
4. Birks TA, Knight JC, Russell PStJ (1997) Endlessely single-mode photonic crystal fiber. Opt Lett 22:961–963
5. Birks TA, Mogilevstev D, Knight JC, Russell PStJ (1999) Dispersion compensation using single material fibers. IEEE Photon Technol Lett 11:674–676
6. Feng X, Mairaj AK, Hewak DW, Monro TM (2005) Nonsilica glasses for holey fibers. J Lightwave Technol 23:2046–2054
7. Ravi Kanth Kumar VV, George AK, Knight JC, Russell PStJ (2003) Tellurite photonic crystal fiber. Opt Exp 11:2641–2645
8. Wang JS, Vogel EM, Snitzer E (1994) Tellurite glass: a new candidate for fiber devices. Opt Mat 3:187–203

9. Liao M, Chaudhari C, Qin G, Yan X, Kito C, Suzuki T, Ohishi Y, Matsumoto M, Misumi T (2009) Fabrication and characterization of a chalcogenide-tellurite composite microstructure fiber with high nonlinearity. Opt Exp 17:21608–21614
10. Gaponov DA, Biryukov AS (2006) Optical properties of microstructure tellurite glass fibers. Quant Electron 36:343–348
11. Sharma A, Chauhan H (2009) A new analytical model for the field of microstructured optical fibers. Opt Quant Electron 41:235–242
12. Sharma DK, Sharma A (2012) Characteristic of microstructured optical fibers: an analytical approach. Opt Quant Electron 44:415–424
13. Sharma DK, Sharma A (2013) On the mode field diameter of microstructured optical fibers. Opt Commun 291:162–168
14. Sharma DK, Sharma A (2014) Splicing of index-guiding microstructured optical fibers and single-mode fibers by controlled air-hole collapse: an analytical approach. Opt Quant Electron 46:409–422
15. Sharma DK, Sharma A (2013) Mode field expansion in index-guiding microstructured optical fibers. Proc SPIE 8794:87942A
16. Sharma DK, Sharma A (2013) Mode field diameters of index-guided microstructured optical fibers. Proc AIP 1536:701–702
17. Sharma A, Ghatak AK (1981) A variational analysis of single mode graded-index fibers. Opt Commun 36:22–24
18. Snyder AW, Love JD (1983) Optical waveguide theory. Chapman & Hall, London

Chapter 25
Design and Simulation of Octagonal Photonic Crystal Fiber for Supercontinuum Generation

Aparna A. Nair, S.K. Sudheer and M. Jayaraju

Abstract In this investigation, design of an octagonal ring photonic crystal fiber (OPCF) has been studied by introducing elliptical air holes at different rings. This structure provides improved nonlinearity and flattened dispersion of maximum ±12 ps/(nm.km). Using Finite Element Method (FEM) analysis has been employed to obtain each optical characteristics. By simulation the corresponding supercontinuum generation on each design is compared and also investigates the impact of pump pulse width on soliton order.

25.1 Introduction

Introduction of highly nonlinear photonic crystal fiber (PCF) has made possible of supercontinuum generation in recent years. PCF consists of solid or hollow core surrounded by microscopic air holes of different structure (circular, elliptical) arranged in different ordered array (hexagonal, octagonal etc.) [1]. Freedom of design has made PCF to obtain various features like high nonlinearity, dispersion compensation, low confinement loss, high birefringence, low effective area etc. Octagonal photonic crystal fiber (OPCF) has got many attractive features like smaller effective area low confinement loss and wideband single-mode operation than those of the hexagonal PCF [2, 3]. Whenever an intense optical pulse propagates through a nonlinear medium generates an extreme broad and continuous spectrum generally called Supercontinuum (SC) [4]. Generation of broad band SC spectrum sliced

A.A. Nair (✉)
Department of Electronics and Communication Engineering,
College of Engineering Trivandrum, Trivandrum, Kerala, India
e-mail: aparnayarphd@gmail.com

S.K. Sudheer
Department of Optoelectronics, University of Kerala, Trivandrum, Kerala, India

M. Jayaraju
MES Institute of Technology and Management, Chathanoor, Kerala, India

V. Lakshminarayanan and I. Bhattacharya (eds.), *Advances in Optical Science and Engineering*, Springer Proceedings in Physics 166,
DOI 10.1007/978-81-322-2367-2_25

sources has found applications in various fields like optical meteorology, Optical coherence Tomography (OCT), sources for wavelength division multiplexing (WDM), fiber optic sensing and so on. For ultrashort pulses like femtosecond pulses the spectral broadening of SC is dominated by soliton dynamics and self phase modulation (SPM) while for nanosecond pulses, the nonlinear Kerr effect and four wave mixing (FWM) dominates [5–7]. The interplay of dispersion and nonlinearity of the fiber leads to soliton generation. When a higher order soliton is perturbed; it splits into fundamental soliton through soliton fission process [8].

In the present study, two design has been proposed by introducing elliptical ring at inner and outer most ring of the OPCF. The parameters like nonlinear coefficient and dispersion characteristics of the each proposed design has been obtained. As the input pulse width $T_0 = 25$ fs, the soliton dynamics dominates and the soliton fission is clearly observed in the spectrum.

25.2 Theory and Design

Figure 25.1a, b shows the proposed design with elliptical air holes at inner ring (Type-1) and at outer ring of the OPCF (Type-2) respectively. The diameter of each circular hole, d = 1.2 μm, pitch $\Lambda = 2.27$ μm, $\Lambda_1 = 1.73$ μm and elliptical holes of major axis 1.2 μm and minor axis 0.8 μm were considered for each design. Once the modal effective refractive index (n_{eff}), obtained by solving an eigenvalue problem drawn from the Maxwell equations, then each optical properties like dispersion coefficient, nonlinear coefficient can be obtained. Dispersion plays a very important role in a nonlinear fiber which broadens the pulse while propagating through the fiber. Chromatic dispersion is the combined effect of material dispersion and wave guide dispersion. It is calculated using the relation as follows,

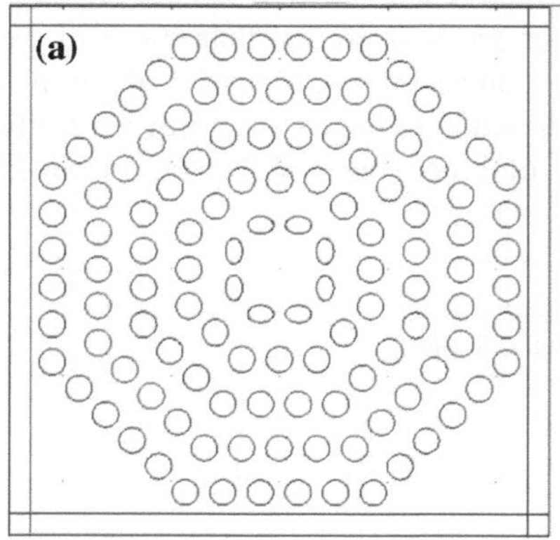

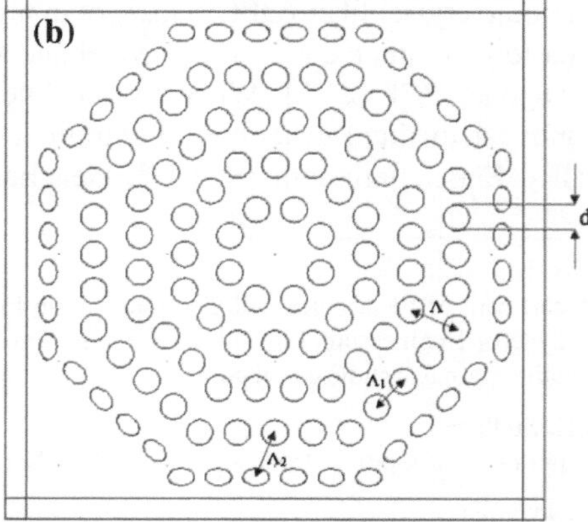

Fig. 25.1 Cross section of the proposed OPCF **a** with inner elliptical air hole ring (Type-1) and **b** with elliptical air hole ring at the end (Type-2)

$$D_c(\lambda) = D_{\mathrm{m}} + D_{\mathrm{w}} = \frac{-\lambda}{c}\frac{d^2Re(n_{\mathrm{eff}})}{d\lambda^2} \tag{25.1}$$

where D_m and D_w represents the material dispersion and waveguide dispersion respectively. $Re(n_{eff})$ is the real part of the complex modal refractive index. Nonlinear coefficient indicates the degree of occurrence of nonlinear effects (four wave mixing, stimulated Raman scattering etc.) in a fiber at higher optical intensities [9]. The nonlinear coefficient, γ of the PCF is calculated using the formula given by,

$$\gamma = \frac{2\pi * n_2}{\lambda * A_{\mathrm{eff}}} \tag{25.2}$$

where n_2 is the nonlinear refractive index of silica (2.66×10^{-20} m^2/W) and A_{eff} is the area covered by the light during propagation through PCF and it is calculated by,

$$A_{\mathrm{eff}} = \frac{\left[\int_{-\infty}^{\infty}\int_{-\infty}^{\infty}|E(x,y)|^2\mathrm{d}x\mathrm{d}y\right]^2}{\int_{-\infty}^{\infty}|E(x,y)|^4\mathrm{d}x\mathrm{d}y} \tag{25.3}$$

Generation of white light using high intensity pulses ranges from short to ultrashort pulses is technically termed as Supercontinuum. This process results due to various nonlinear phenomenon like four wave mixing, stimulated Raman or Brillouin scattering, self phase modulation (SPM) and Cross phase modulation (XPM) etc. Depending upon the characteristics like nonlinearity of the fiber, pulse width, power and center wavelength of the laser source, dispersion properties each physical processes dominates [10]. In the femtosecond regime optical soliton predominates for the SC generation [8]. The magnitude of the soliton self frequency shift is inversely proportional to the fourth power of the width of the soliton pulse [7]. In order to investigate generation of supercontinuum numerically, consider the generalized nonlinear Schrodinger equation (GNLSE) [11],

$$\begin{aligned}\frac{\partial}{\partial z}A(z,T) &= -\frac{\alpha(\omega)}{2}A(z,T) + \sum_{n\geq 2}\beta_n\frac{i^{n+1}}{n!}\frac{\partial^n}{\partial t^n}A(z,T) \\ &+ i\gamma\left(1+\frac{i}{\omega_0}\frac{\partial}{\partial T}\right)\int\limits_{-\infty}^{\infty} R(T')|A(z,T-T')|^2\mathrm{d}T'\end{aligned} \tag{25.4}$$

This equation represents the propagation of optical pulse through the fiber where $A(z, T)$ is the electric field envelop, α is the frequency dependent loss, β_n is the nth order dispersion at center frequency ω_0 and $R(T)$ is the Raman response function. This is solved using Split Step Fourier Transform method. This equation models nonlinear effects as well as dispersion effects during at propagation. The optical pulse is used is of hyperbolic secant pulse profile which is given as,

$$A(z, T) = \sqrt{P_0}\operatorname{sech}\left(\frac{T}{T_0}\right) \tag{25.5}$$

where P_0, T_0 are peak power and pulse width of the input pulse respectively. For ultrashort pulses like picosecond and femtosecond the SC generation is initially dominated by SPM and then leads to soliton dynamics at anomalous dispersion regime. When a pulse with sufficient peak power to constitute a higher-order soliton is perturbed and breaks up into a series of lower-amplitude sub pulses is termed as soliton fission [8]. During fission process the fundamental soliton experiences the red shift due to Intra pulse Raman scattering (IPRS) known as Raman soliton and higher order dispersion term transfers energy from soliton to narrow dispersive wave known as non soliton radiation shift towards shorter wavelength [5]. The Soliton order of the input pulse N is determined by both pulse and fiber parameters through

$$N^2 = L_D/L_{NL} \tag{25.6}$$

where L_D is the dispersive length and L_{NL} is the nonlinear length.

25.3 Result and Discussion

The proposed designs were implemented using COMSOL Multiphysics software. Figures 25.2 and 25.3 shows the dispersion coefficient and nonlinear coefficient for the wavelength ranges from 700 to 1700 nm. In the dispersion curve a good flattened dispersion approximately of ±12 ps/(nm km) has been obtained for Type-2

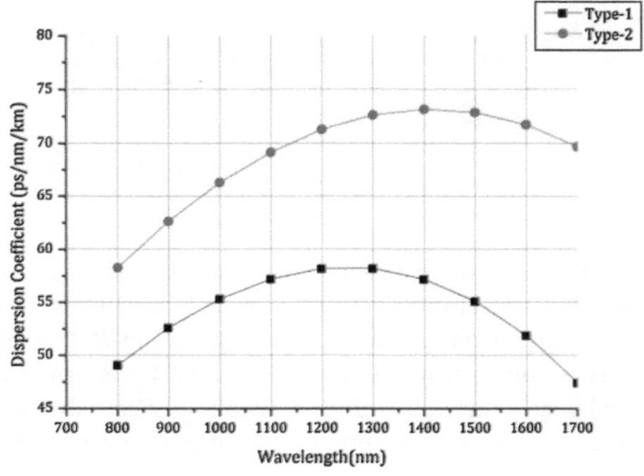

Fig. 25.2 Plot of dispersion coefficient versus wavelength for each proposed design

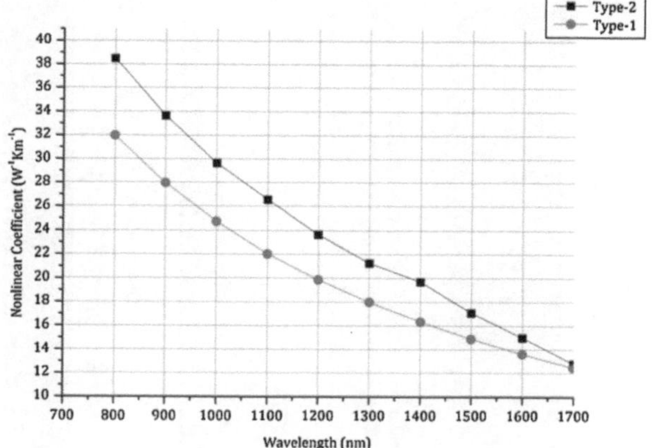

Fig. 25.3 Plot of nonlinear coefficient versus wavelength for each proposed design

and of ±8 ps/(nm km) for Type-1. By the introduction of elliptical holes provides more nonlinearity than the complete circular air hole rings [12]. Also it is shown that elliptical holes at outer ring (Type-2) shows more nonlinearity than the Type-1.

Supercontinuum generation using the proposed fiber designs has been shown at Fig. 25.4a, b. For the study of SC generation, the parameters considered are peak power P_0 = 7 kW, pulse width T_0 = 25 fs, γ = 38 W^{-1} km^{-1}. Using dispersion formula, $\beta_2 = \frac{-\lambda^2}{2\pi c} D$, Group Velocity Dispersion (GVD), β_2 and its higher order derivatives like β_3, β_4, β_5 etc are also calculated, where D is the dispersion coefficient. For each design the corresponding spectral evolution has been obtained is shown in Figs. 25.5 and 25.6 for Type-1 and Type-2 respectively. According to calculation the Soliton order, N = 4 for both designs. The characteristic length is $5L_D$ = 270 mm for Type-1 and 225 mm for Type-2. The Soliton break up (Soliton fission) becomes more visible only after the characteristic length. Greater numbers of soliton appears distinctly in both time as well as frequency domains. The Spectral broadening of the SC not only because of nonlinearity but also the effective of dispersion characteristics which determines the nature of the propagation dynamics as well as phase matching criteria [8]. The spectrum develops more towards higher wavelength due to Raman Soliton. This is clear from the bandwidth of the SC generated using Type-1 is approximately 450 nm while for Type-2 is of 325 nm. Figure 25.5 shows dependence of the spectra of SC with fiber length. At Fig. 25.6 the dependence of number of solitons (soliton order) to wavelength is obtained. As the pulse width increases the soliton is found to be increased as the spectral evolution increases. The result could be verified with the (25.6) for all other parameters kept constant.

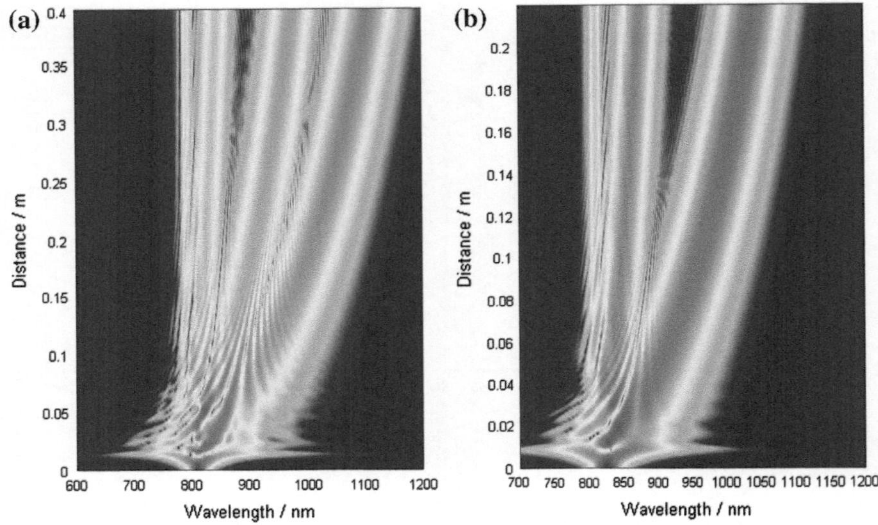

Fig. 25.4 Spectral evolution of supercontinuum generation **a** for soliton order of $N = 4$ for Type-1 and **b** for soliton order of $N = 4$ for Type-2

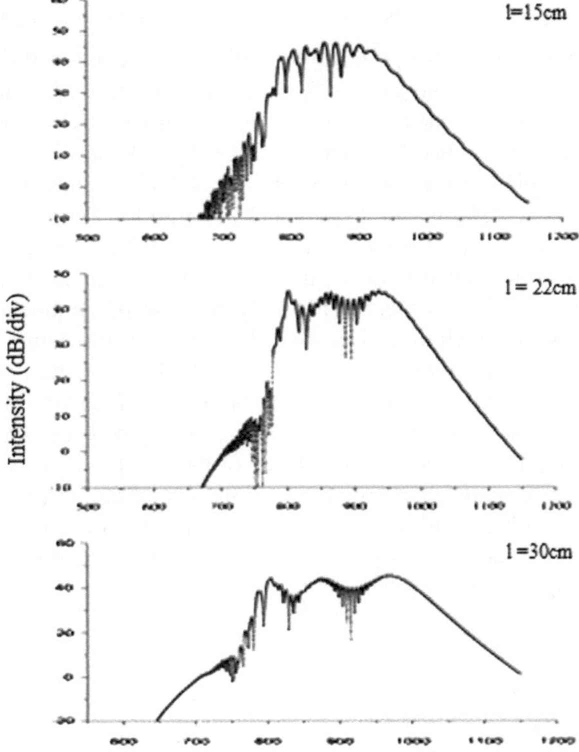

Fig. 25.5 Spectra of supercontinuum generation for soliton order of $N = 4$ for Type-2 at varying length

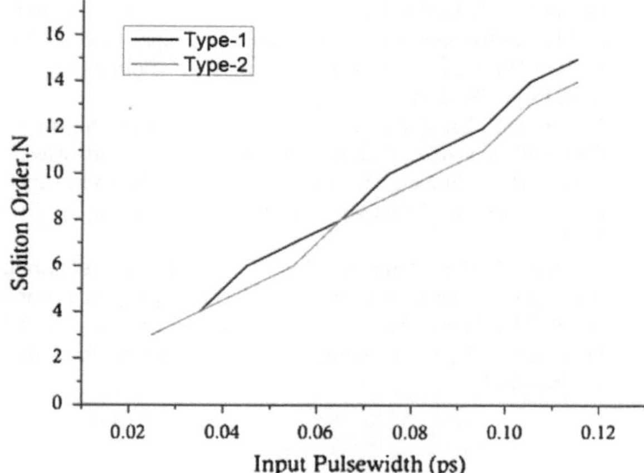

Fig. 25.6 Plot of number of solitons (soliton order) versus pulse width

25.4 Conclusion

In this paper, a novel design of an OPCF with introducing an elliptical air holes into the inner and outermost rings and studied its optical properties. A good flattened dispersion is obtained at telecommunication band and the elliptical holes enhanced the nonlinearity of the fiber. Such OPCF could provide good dispersion compensation over the entire telecommunication band. The nonlinear coefficient obtained at 800 nm is of 38 W^{-1} km^{-1} and at 1550 nm it is 16 W^{-1} km^{-1}. Here the OPCF of Type-1 provides more bandwidth i.e. 450 nm than the Type-2 with bandwidth 350 nm. Such SC sources are good for application like OCT. Also the red shifted pulses (Raman Soliton) are more than the blue shifted pulses (Dispersive Wave) in the spectrum. The spectrum broadening moves towards higher wavelength as the propagation length increases.

References

1. Russell PSJ (2006) Photonic-crystal fibers. J Lightwave Technol 24:4729–4749
2. Habib MS, Habib MS, Hasan MI, Razzak SA (2013) Tailoring polarization maintaining broadband residual dispersion compensating octagonal photonic crystal fibers. Opt Eng 52 (11):116111
3. Taghipour F, Olyaee S (2012) Doped-core octagonal photonic crystal fiber with ultra-flattened nearly zero dispersion and low confinement loss in a wide wavelength range. Fiber Integr Optics 31:178–185
4. Birks TA, Wadsworth WJ, Russell PSJ (2000) Supercontinuum generation in tapered fibers. Opt Lett 25:19
5. Ghosh D, Roy S et al (2009) Generation of supercontinuum and its theoretical study in three-ring silica microstructured optical fibers. Appl Opt 48:31
6. Liao M, Gao W, Cheng T, Duan Z, Suzuki T, Ohishi Y (2012) Supercontinuum generation in picosecond regime in a highly nonlinear tapered tellurite microstructured optical fiber. OSA 2012

7. Lehtonen M, Genty G, Ludvigsen H, Kaivola M (2003) Supercontinuum generation in a highly birefringent microstructured fiber. Appl Phys Lett 82:14
8. Dudley JM, Genty G, Coen S (2006) Supercontinuum generation in photonic crystal fiber. Rev Mod Phys 78:1135
9. Namihira Y, Liu J, Koga T et al (2011) Design of highly nonlinear octagonal photonic crystal fiber with near-zero flattened dispersion at 1.31 μm waveband. Opt Rev 18(6):436–440
10. Khan KR, Mahmood MF, Biswas A (2014) Coherent super continuum generation in photonic crystal fibers at visible and near infrared wavelengths. IEEE J Sel Topics Quantum Electron 20:5
11. Sharma M, Borgohain N, Konar S (2014) Supercontinuum generation in photonic crystal fibers possessing high birefringence and large optical nonlinearity. Phys Expr 4:26
12. Habib MS, Habib MS, Razzak SMA, Hossain MA (2013) Proposal for highly birefringent broadband dispersion compensating octagonal photonic crystal fiber. Opt Fiber Technol 19:461–467

Chapter 26
Highly Birefringent Fluoride Photonic Crystal Fiber with Low Confinement Loss

Sneha Sharma and Jitendra Kumar

Abstract A high birefringent and low confinement loss fluoride photonic crystal fiber (PCF) is proposed with two elliptical air holes in the cladding. The effective refractive index, birefringence, nonlinear coefficient and confinement loss are numerically simulated using the COMSOL Multiphysics software based on finite element method (FEM) with perfectly matched layer boundary (PML) condition. Its birefringence value reaches the magnitude of 1.636×10^{-2} and confinement loss is 3.7×10^{-2} dB/m at 2 μm wavelength.

26.1 Introduction

Photonic crystal fibers (PCFs) are single material optical fiber with periodic arrangement of air holes running parallel along its axis [1]. The internal periodic structures of PCFs are made of capillaries and filled with air. To propagate light along the core of the fiber a central solid region is realized by removing one or more central capillaries [2]. PCF provides enhanced design flexibility compared to standard optical fibers by changing the size and distance between the air holes and it has unique optical properties such as endlessly single-mode operation, controlled dispersion, large birefringence and high nonlinearity which are impossible to obtain in conventional fibers [3–7]. Among the properties of PCFs, one of the most interesting characteristics is birefringence. Circularly symmetric arrangement of air holes in PCF does not maintain the polarization state of the optical field along their length. In high birefringent PCF the polarization state is preserved along the propagation to achieve high precision measurements of physical quantities. Bire-

S. Sharma (✉) · J. Kumar
Department of Electronics Engineering, Indian School of Mines Dhanbad, Jharkhand, India
e-mail: sneha.ece@gmail.com

J. Kumar
e-mail: jitenkg@rediffmail.com

V. Lakshminarayanan and I. Bhattacharya (eds.), *Advances in Optical Science and Engineering*, Springer Proceedings in Physics 166,
DOI 10.1007/978-81-322-2367-2_26

fringence can be induced in PCF by breaking the symmetry of the structure and increasing the effective index difference between the two orthogonal polarization states. High birefringence fibers are mostly developed using the silica material for optical communication application. For mid-infrared application much attention has been paid to soft glasses such as tellurite, chalcogenide and fluoride [8–10]. Among these, fluoride fiber (ZrF_4-BaF_2-LaF_3-AlF_3-NaF, ZBLAN) is considered to be the most stable and technological advance for high power applications [11]. ZBLAN fiber has many unique characteristics such as lower absorption loss in the near infrared region, wide operating wavelength range, high emission efficiency and higher damage threshold [12, 13]. Compared to silica fiber, the loss of ZBLAN fiber is much smaller at around 2 μm wavelength [14].

In this paper, we designed a high birefringent and low confinement loss fluoride PCF. The various parameters of the proposed PCF are simulated using COMSOL Multiphysics software based on finite element method (FEM) with perfectly matched layer boundary (PML) condition.

26.2 PCF Design and Simulation Results

A fluoride PCF with square lattice having circular air holes and two elliptical air holes in the cladding is designed as shown in Fig. 26.1a. The diameter of the circular air holes is 0.96 μm, the distance between the air holes is 1.2 μm and the major and minor axis of elliptical air holes is 0.6 and 0.2 μm respectively.

The refractive index of air hole is taken as 1 and the refractive index of fluoride glass [12] can be expressed as

$$n_\lambda^2 - 1 = \frac{f_1\lambda^2}{\lambda^2 - \lambda_1^2} + \frac{f_2\lambda^2}{\lambda^2 - \lambda_2^2} \tag{26.1}$$

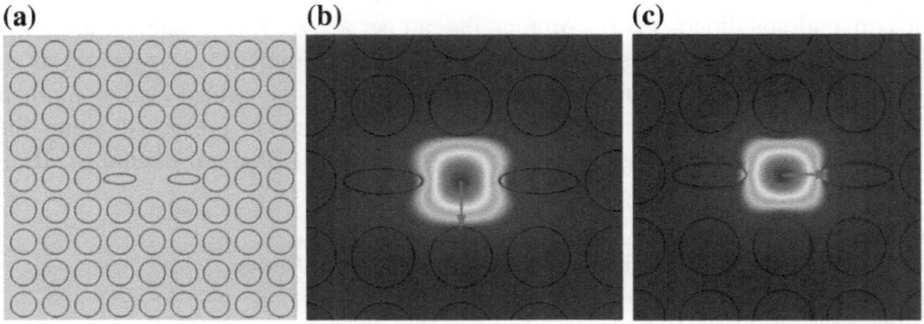

Fig. 26.1 **a** PCF structure, **b** electric field intensity for *y* and **c** *x* polarized mode

where $\lambda_1 = 0.08969$ μm and $\lambda_2 = 21.3825$ μm are the eigen absorption wavelengths in the UV and IR, respectively and $f_1 = 1.22514$ and $f_2 = 1.52898$ are the corresponding oscillator strengths for ZBLAN fiber. Figure 26.1b, c shows the electric field intensity for y and x polarized mode at 2 μm wavelength, respectively. The effective mode index of the fundamental x and y polarized modes are calculated by FEM with PML. Figure 26.2a shows the effective refractive index of the fundamental x and y polarized modes of the proposed PCF.

The birefringence of the PCF is defined as the difference between the effective indices of the two orthogonal polarization modes and is denoted by

$$B = \left| n_{\text{eff}}^{y} - n_{\text{eff}}^{x} \right| \tag{26.2}$$

where n_{eff}^{y} and n_{eff}^{x} are the real parts of the refractive indices of the y and x polarized modes of the PCF, respectively. Figure 26.2b shows the birefringence of the PCF as a function of wavelength. The birefringence value of the proposed fluoride PCF is 1.636×10^{-2} at 2 μm wavelength. The effective refractive index of the PCF decreases and the birefringence increases with the wavelength. The effective refractive index $\left(n_{\text{eff}}^{x}\right)$ of the x polarized mode is smaller than its corresponding y polarized mode $\left(n_{\text{eff}}^{y}\right)$.

The nonlinear coefficient (γ) is defined as

$$\gamma = \frac{n_2 \omega_0}{c A_{\text{eff}}} \tag{26.3}$$

where $n_2 \approx 2.1 \times 10^{-20}\,\text{m}^2/\text{W}$ is the nonlinear refractive index of fluoride glass, c is the speed of light in vacuum, ω_0 is the center frequency and A_{eff} is the effective mode area given by

$$A_{\text{eff}} = \frac{\left(\int_{-\infty}^{\infty} |E(x,y)|^2 \text{d}x\,\text{d}y\right)^2}{\int_{-\infty}^{\infty} |E(x,y)|^4 \text{d}x\,\text{d}y} \tag{26.4}$$

where E is the electric field in the medium.

Figure 26.2c, d shows the effective area and nonlinear coefficient of the proposed PCF for both the x and y polarized modes. It can be seen that the effective area of y polarized mode is greater than the x polarized mode because the spacing between the air holes along the y polarized mode is larger than the x polarized mode in the core. The effective area increases and nonlinear coefficient decreases with wavelength. As the cladding is periodic in PCFs the light confinement ability decreases in the core region and is called confinement loss (CL) given by the equation

$$\text{CL} = \frac{20}{\ln 10} k_0 \text{Im}(n_{\text{eff}}) \tag{26.5}$$

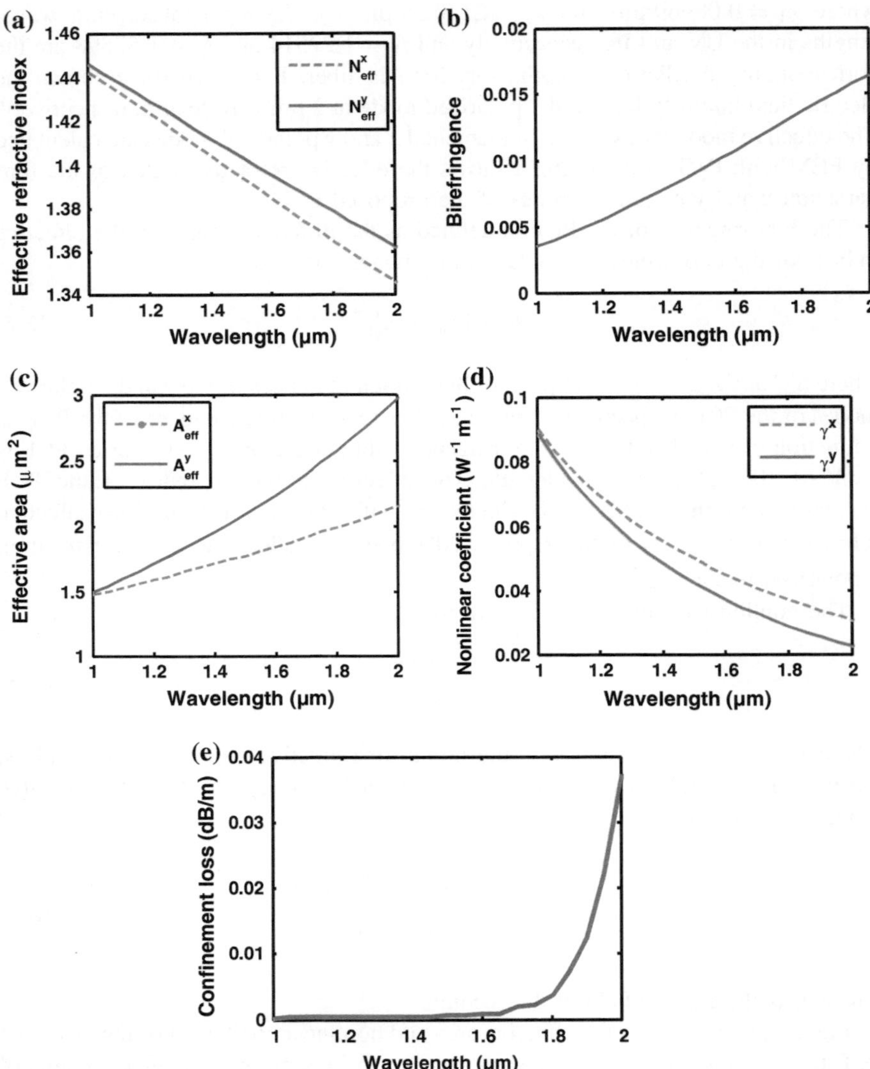

Fig. 26.2 **a** Effective refractive index, **b** Birefringence, **c** Effective area, **d** Nonlinear coefficient and **e** Confinement loss as a function of wavelength

where $\mathrm{Im}(n_{\mathrm{eff}})$ is the imaginary part of the effective refractive index and $k_0 = \frac{2\pi}{\lambda}$ is the wavenumber in free space. Figure 26.2e shows the calculated confinement loss of the proposed fluoride PCF. The confinement loss is 3.7×10^{-2} dB/m at 2 μm. wavelength. The proposed fluoride PCF with high birefringence and low confinement loss can be used for mid- infrared application.

26.3 Conclusion

In this paper, we design a square lattice fluoride PCF with circular air holes and two elliptical air holes in the cladding to induce high birefringence and low confinement loss. The birefringence value of the proposed PCF is 1.636×10^{-2}, confinement loss is 3.7×10^{-2} dB/m and nonlinear coefficient is 0.03 and 0.02 W^{-1} m^{-1} for both the x and y polarized modes, respectively.

References

1. Russell P (2003) Photonic crystal fibers. Science 299:358–362
2. Russell P (2006) Photonic-crystal fibers. J Lightwave Technol 24:4729–4749
3. Reeves WH, Skryabin DV et al (2003) Transformation and control of ultra-short pulses in dispersion-engineered photonic crystal fibres. Nature 424:511–515
4. Birks TA, Knight JC, Russell PSJ (1997) Endlessly single-mode photonic crystal fiber. Opt Lett 22:961–963
5. Sinha RK, Varshney SK (2003) Dispersion properties of photonic crystal fibers. Microwave Opt Technol Lett 37:129–132
6. Ju J, Jin W et al (2003) Properties of a highly birefringent photonic crystal fiber. IEEE Photonics Technol Lett 15:1375–1377
7. Blanch AO, Knight JC et al (2000) Highly birefringent photonic crystal fibers. Opt Lett 25:1325–1327
8. Price J et al (2012) Supercontinuum generation in non-silica fibers. Opt Fiber Technol 18:327–344
9. Gao W et al (2014) Mid-infrared supercontinuum generation in a four-hole As_2S_5 chalcogenide microstructured optical fiber. Appl Phys B 116:847–853
10. Liu L et al (2014) Numerical investigation of mid-infrared Raman Soliton source generation in endless single mode fluoride fibers. J Appl Phys 115:163102
11. Swiderski J et al (2013) Mid-IR supercontinuum generation in a ZBLAN fiber pumped by a gain-switched mode-locked Tm doped fiber laser and amplifier system. Opt Express 21:7851–7857
12. Gan F (1995) Optical properties of fluoride glasses: a review. J Non-Cryst Solids 184:9–20
13. Su W et al (2014) Highly birefringent ZBLAN photonic quasi-crystal fiber with four circular air holes in the core. Infrared Phys Technol 66:97–102
14. http://www.fiberlabsinc.com/fiber_technology.htm

Chapter 27
Splicing Hetero-core Fibers in Perspective of Different Material Compositions

D. Paul, R. Biswas and N.S. Bhattacharyya

Abstract A comprehensive study has been done utilizing effective index method. We analyze splice losses of Photonic crystal fiber (PCF) with single mode fiber (SMF) pertaining to different materials (phosphate, crown). We observe that losses with respect to splicing between SMF-PCF of dissimilar material composites, viz., phosphate and crown, are also influenced by modal parameters such as diameter and pitch of the micro-structured fiber. Results suggest dependence of losses on modal parameters with less magnitude of losses in crown compared to phosphate.

27.1 Introduction

Photonic crystal fibers (PCFs) has periodic array of air holes all throughout the length of the fiber. They provide two typical cross sections: One is an air-material cladding surrounding a solid material core, and the other is an air-material cladding surrounding a hollow core [1, 2].

In structural point of view PCF has completely different configuration compared to single mode fiber (SMF). Basic structural parameters in PCF are diameter (*d*), pitch (Λ). We intend to explore the losses based on the splicing between PCF and SMF with the aid of effective index approach (EIM) [3, 4]. Because of easy tenability of EIM approach with variation to modal parameters, the losses can be effectively investigated. Further, we extend our analysis to incorporation of different material composites of micro-structured fiber which enable to bring out prominent loss features in the context of splicing of PCFs w.r.t mono-mode fiber [1, 3].

In our approach, we have reflected on the fact that by the change of material composition in fibers we will have certain variation regarding losses. Maintaining this approach we consider splice loss between PCF-SMF in perspective of different material composites viz., phosphate and crown. Splicing has been carried out

D. Paul (✉) · R. Biswas · N.S. Bhattacharyya
Applied Optics and Photonics Research Laboratory, Department of Physics, Tezpur University, Assam 784028, India
e-mail: dmppaul@gmail.com; dimpi@tezu.ernet

V. Lakshminarayanan and I. Bhattacharya (eds.), *Advances in Optical Science and Engineering*, Springer Proceedings in Physics 166,
DOI 10.1007/978-81-322-2367-2_27

between non-identical PCF and step index SMF and the losses thereafter have been analytically established. We evaluate splice loss due to lateral, longitudinal in addition to angular offset. We observe good controlling features corresponding to alteration of structural parameters of micro-structured fibers spliced with SMFs.

27.2 Theory

In our study we have utilized spot sizes for SMF and PCFs of different material (phosphate and crown) to observe the variation of splice loss versus different offsets such as transverse, longitudinal and angular. We have used Peterman's and Marcuse formula to execute our analytical observation which are given as follows [5]:

$$w_{\mathrm{PCF}} = \left(0.065 + 1.619V^{-\frac{3}{2}} + 2.879V^{-6}\right)R \tag{27.1a}$$

And spot size equation for SMF is given as follows [5]:

$$w_{\mathrm{SMF}} = w_{\mathrm{PCF}} - \left(0.016 + 1.5161V^{-7}\right)R \tag{27.1b}$$

In (27.1b) R is considered as the radius of the fibers.

During splicing between two fibers, the main important phenomenon which has to be measured is the splice loss which depends on several factors like offsets due to which give rise to losses [6, 7]. In this paper we have utilized certain factors like transverse, longitudinal and angular offset.

Splice loss for transverse offset can be conveyed as follows [8, 9]:

$$\eta_{\mathrm{offset}} = \frac{4w_1^2 w_2^2}{(w_1^2 + w_2^2)} \mathrm{e}^{-\frac{2r_d^2}{w_1^2 + w_2^2}} \tag{27.2}$$

where, r_d is the transverse offset and w_1, w_2 are the spot sizes two fibers.

When longitudinal offset (z) arises between two fibers, then the splice loss is given by [6, 8, 10]:

$$\eta_z = \frac{4w_1^2 w_2^2}{\frac{z^2}{k_0^2} + (w_1^2 + w_2^2)^2} \tag{27.3}$$

If θ is the angular misalignment between two fibers then splice loss is given as follows [6, 8, 10]:

$$\eta_\theta = \frac{4w_1^2 w_2^2}{\left(w_1^2 + w_2^2\right)} \mathrm{e}^{-\frac{k^2 w_1^2 w_2^2}{2\left(w_1^2 + w_2^2\right)}} \tag{27.4}$$

27.3 Results and Discussions

Splicing between non-identical SMF-PCF fibers brings out a number of features concerning losses; with respect to various offsets such as transverse offset, longitudinal offset and angular offset and corresponding splice loss variation can be observed. In Fig. 27.1 we analyse the variation of splice loss versus transverse offset by varying structural parameter (d/Λ). As splice loss has a direct dependence on structural parameter for PCF and correspondingly with that of SMF. Here we see that there is an increase in splice loss with the increase in structural parameter. During splicing between two waveguide with increase in modal parameter give rise to more offsets and more the offsets more will be the loss. There is a difference of ~0.16 dB splice loss between higher and lower value of loss. Non-identical fibers and composite is also showing the exact pattern for splice loss versus transverse offset i.e.; with the increase in structural parameter there is an increase in the value of loss.

In Fig. 27.2, it is shown that with increase in air distances splice loss decreases and there are ~0.034 dB differences in losses. Increase in pitch means there will be an increase in air to air holes distance and hence lower offsets and hence there will a decrement in loss. Non-identical fibers and composite is also showing the exact pattern for splice loss versus λ i.e.; decrement of the values of loss is observed while increasing the values of Λ.

In Fig. 27.3 it is shown that with the increase in structural parameter splice loss versus longitudinal offset (z) increases and it is to be noted that the value of splice loss is very small ~0.009 dB it will hardly affect on transmission. In this case also we can see that with the increase in d/Λ during splicing as mentioned above there will be an increment in offsets and loss increase. Non-identical fibers and composite

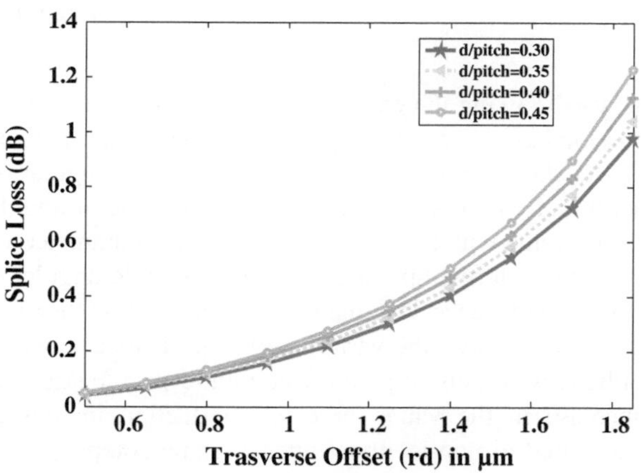

Fig. 27.1 Splice loss versus transverse offset (r_d) in µm for different structural parameter (d/Λ) at the intervals of 0.05 between non-identical fibers (SMF and PCF) and materials

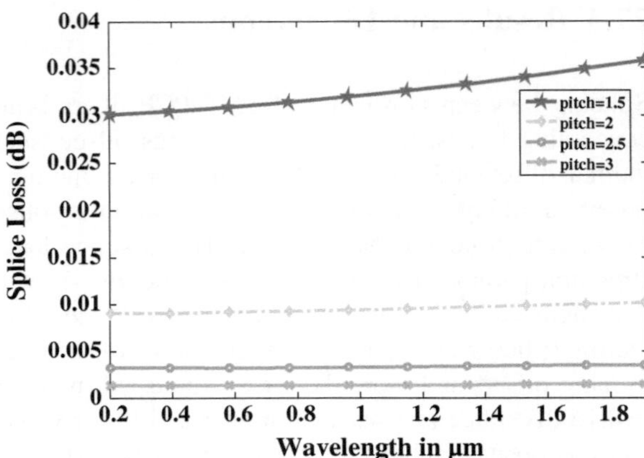

Fig. 27.2 Splice loss versus λ for different pitch (Λ) for constant transverse offset (r_d = 0.25 μm) hetero-core fibers (SMF and PCF) and non-identical materials

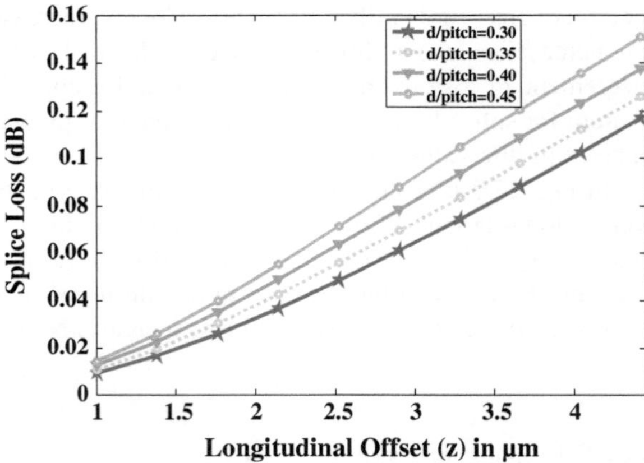

Fig. 27.3 Splice loss versus longitudinal offset (z) between SMF and PCF for non-identical materials

is also showing the exact pattern for splice loss versus longitudinal offset i.e.; with the increase in structural parameter there is an increase in the value of loss.

As shown in Fig. 27.4, splice loss versus angular offset exhibits declining trend with increase in structural parameter. In case of angular offset, there is an increase in the value with the decrease in structural parameter as more the value of the modal parameter during splicing there will be little angular deviation between the waveguides and hence loss increases with the decrease in structural parameter. In this case we consider the variation of non-identical fiber and non-identical composites. splice loss versus angular offset with respect to structural parameter shows that with increase in the value there is a reduction in loss and it shows a lower value compared to non-identical fiber of same composite.

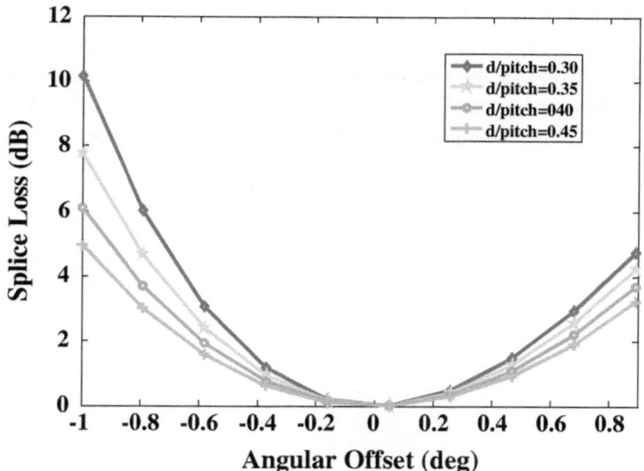

Fig. 27.4 Splice versus angular offset between hetero-core fibers (SMF-PCF) with different structural parameter

27.4 Conclusion

With the respective transmission behavior of PCFs is modeled with a vector effective refractive index approach, and then, the splice losses of photonic crystal fiber and SMF are analyzed. It is found that the splice loss is very sensitive to transverse offset and angular misalignment, the most prominent factor effecting splice loss is mismatching of mode and field. Eventually, the most important structural parameter affecting splice loss is the pitch Λ of PCFs, the splice loss of PCF-SMF is relatively low if the pitch of PCF is larger than the radius of SMF core.

References

1. Mortensen NA (2002) Effective area of photonic crystal fibers. Opt Express 10(7):341–348
2. Hansen TP, Broeng J, Jakobsen C et al (2004) Air-guiding photonic bandgap fibers: spectral properties, macrobending loss, and practical handling. J Lightwave Technol 22(1):11–15
3. Xiao L, Demokan MS, Jin W et al (2007) Fusion splicing photonic crystal fibers and conventional single-mode fibers: microhole collapse effect. J Lightwave Technol 25 (11):3563–3574
4. Li HB, Mafi A, Schulzgen A et al (2007) Analysis and design of photonic crystal fibers based on an improved effective-index method. J Lightwave Technol 25(5):1224–1230
5. Marcuse D (1978) Gaussian approximation of the fundamental modes of graded-index fibers. J Opt Soc Am 68(1):103–109
6. Kliros GS, Konstantinidis J, Thraskias C (2014) Prediction of macro-bending and splice losses for photonic crystal fibers based on the effective index method. http://arxiv.org/ftp/arxiv/papers/0705/0705.2875.pdf. Accessed on 14 May 2014
7. Meunier JP, Wang ZH, Hosain SI (1994) Evaluation of splice loss between two non identical single-mode graded-index fiber. IEEE Photonic Technol Lett 6(8):998–1000
8. Zendehnam A, Mirzaei M, Farashiani A et al (2010) Investigation of bending loss in a single-mode optical fibre. Pramana-J Phys 74(4):591–603

9. Aghaie KZ, Digonnet MJF, Fan SH (2010) Optimization of the splice loss between photonic-bandgap fibers and conventional single-mode fibers. Opt Lett 35(12):1938–1940
10. Kliros GS, Konstantinidis J, Thraskias C (2006) Splice loss calculations for index-guiding photonic crystal fibers. In: Proceedings of the 10th WSEAS international conference on communications, Vouliagmeni, Athens, Greece, pp 319–324

Part VII
Photonics for Space Applications

Chapter 28
Demonstration of Active Laser Beam Stabilization in Closed Loop for Free Space Optical Receiver

Koushik Basak, R.K. Bahl, Payal Sharma and A. Banik

Abstract Active beam stabilization is decisive in numerous applications such as free space optical (FSO) communication, military, biomedical etc. FSO communication receivers require very precise and stable pointing mechanism to maintain optimal operation. Any vibration, micro positional drift can affect efficiency and performance of active optical component and introduce an error which propagates throughout optical line [1]. Beam wandering and atmospheric effect also causes the contravention of FSO communication. In this paper we have demonstrated active beam stabilization technique using fast steering mirror (FSM) and Quad detector in closed loop for FSO receiver in laboratory. The quad-detector senses the error voltage according to the position of laser beacon. The error voltage is fed to FSM controller which is used in an external sensor feedback control mode with position feedback from quad photo detector and the FSM rectifies in the desired position.

28.1 Introduction

Free space optical (FSO) communication is becoming inevitable in the future communication era. FSO communication has advantages of narrow beam width compared with conventional radio link and hence low laser transmitter power is suitable to maintain the typical BER of 10^{-9}. The transmitter and receiver antenna

K. Basak (✉) · R.K. Bahl · P. Sharma
Optical Communication Division (OCD), SNPA, Space Applications Centre, ISRO, Ahmedabad, India
e-mail: koushik@sac.isro.gov.in

R.K. Bahl
e-mail: rkb@sac.isro.gov.in

P. Sharma
e-mail: payal@sac.isro.gov.in

A. Banik
Optical and Digital Communication Group, SNPA, SAC, Ahmedabad, India
e-mail: alak@sac.isro.gov.in

V. Lakshminarayanan and I. Bhattacharya (eds.), *Advances in Optical Science and Engineering*, Springer Proceedings in Physics 166,
DOI 10.1007/978-81-322-2367-2_28

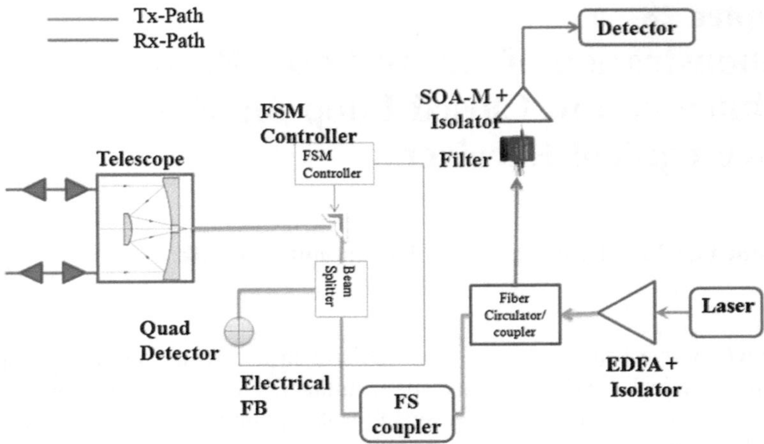

Fig. 28.1 Typical optical terminal incorporating active laser beam stabilization

dimensions are quite less (typically 20 times) compared to microwave antenna and the data rate is very high.

For FSO communication, atmospheric effects, e.g. turbulence, scattering degrades the optical link.

Figure 28.1 shows the block diagram of the typical optical terminal. In fiber based high speed optical receiver, the communication laser has to focus on the free space to fiber interface after the telescope. The multi mode fiber has the core dimension of 60 μm, so the alignment is very critical. The vibration of the satellite may cause the misalignment. In the experiment, with 10 Hz and 100 mV disturbance, 55 % accuracy is achieved around 60 μm fiber.

In this paper the fine pointing mechanism is defined to track the communication laser on the focal plane of a quad detector. The commercial quad photo diode array consists of four individual photo diodes build on single silicon substrate and is available with current-to-voltage amplifier and it provides top minus bottom and left minus right difference signal. Figure 28.2 shows block diagram of Quad sum and difference amplifier. The fast steering mirror (FSM) has a built-in controller along with it and used in external sensor feedback control mode to control the FSM depending upon the error voltage from quad cell according to the position of Laser. The FSM system has angular range of ±26.2 mrad from ±10 V and resolution of 1 μrad and hence suitable for our application.

28.2 External Mode of FSM Operation

The possible technology of Fast Steering Mechanism is available as: (i) FSM, (ii) Acousto-optic deflector, (iii) Deformable mirrors etc. Angular beam steering mechanism controls the precise rotation of two axes of mirror according to the error

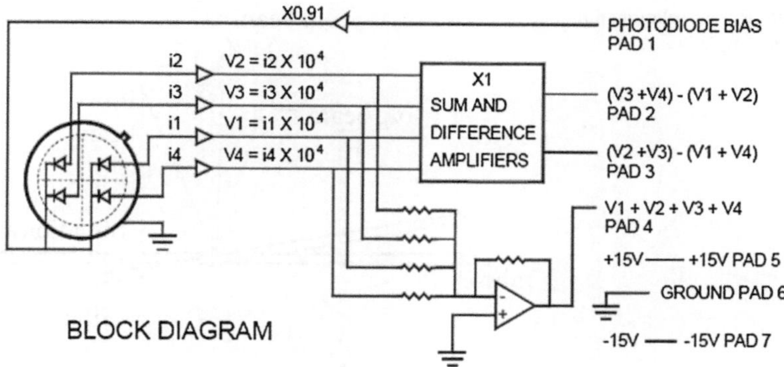

Fig. 28.2 Schematic block diagram of quad detector [2, 10]

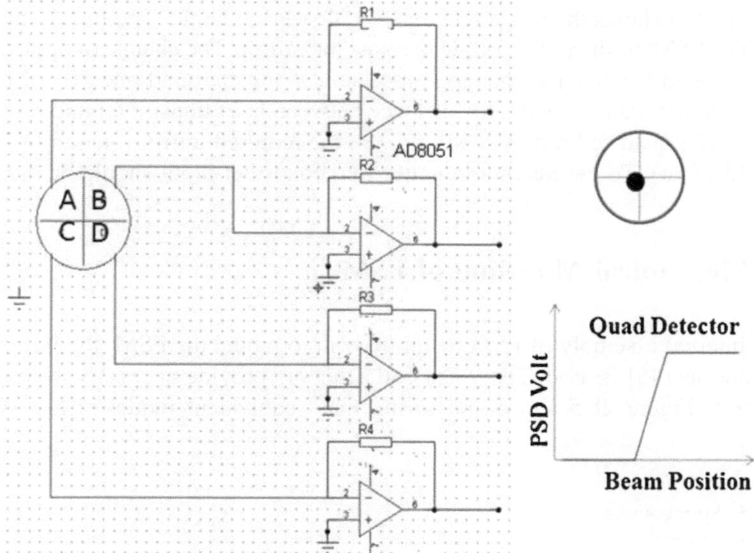

Fig. 28.3 Principle of quad cell

signal. In the FSM close loop, the error voltage is detected by Quad cell [2–4]. The four sectors of quad cell are represented by letters A, B, C and D as shown in Fig. 28.3.

The x and y displacement of the beam is given by:

$$x = \frac{(\mathrm{A}+\mathrm{C})-(\mathrm{B}+\mathrm{D})}{\mathrm{A}+\mathrm{B}+\mathrm{C}+\mathrm{D}} \quad y = \frac{(\mathrm{A}+\mathrm{B})-(\mathrm{C}+\mathrm{D})}{\mathrm{A}+\mathrm{B}+\mathrm{C}+\mathrm{D}} \tag{28.1}$$

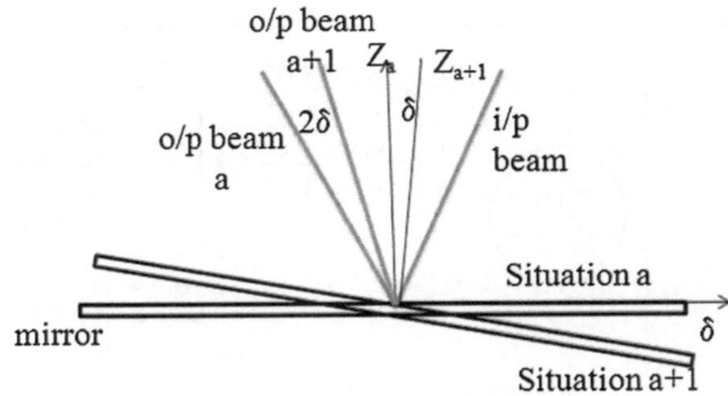

Fig. 28.4 Range relationship of steering angle

The radius of laser spot must be less than the diameter of quad detector. To reduce the spot size, a lens is used in front of quad cell. The error signal is corrected by the FSM controller. The FSM has full scale deflection of ±1.5° on either axis corresponding to ±10 V output swing.

Figure 28.4 shows the range relationship of steering angle. The linear voice coil actuator of FSM produces the torque to rotate the mirror. The change of optical angle of output beam is twice the change in mechanical angle around one axis of the XY plane. In the laboratory set up, the distance between laser transmitter and receiver is 132 mm. The quad cell radius is 3.9 mm. So the required mirror rotation of FSM is 0.86° (15 mrad). To get maximum deflection, controller input should be 5.7 V.

28.3 Mechanical Modeling of FSM

As the internal assembly of FSM is not known, to model the FSM, the information from datasheet [5] is used. The voice coil actuators provide necessary torque to tilt the mirror. Figure 28.5 shows the spring mass equivalent model of fast steering model.

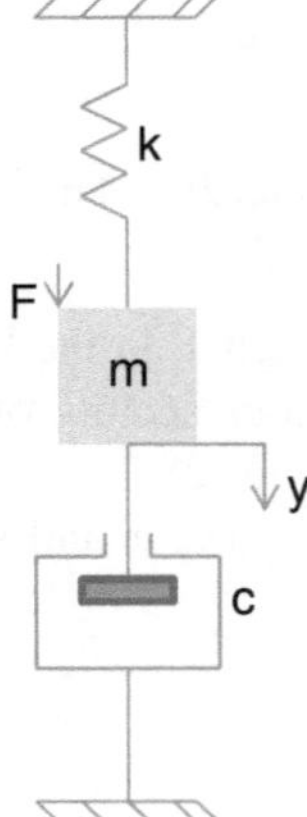

Fig. 28.5 Spring–mass equivalent model

This can be represented by [2, 6–9]:

$$\mu\ddot{y} = \sum F \quad (28.2a)$$

$$\mu\ddot{y} + c\dot{y} + ky = \sum F \quad (28.2b)$$

$$G(s) = 1/(\mu s^\wedge 2 + cs + k) \quad (28.2c)$$

Substituting the lateral movement to angular movement as follows: μ—I, F—T and x—φ;

Four actuators are mounted behind the mirror at each quadrant. The transfer function of FSM is given by:

$$G(s) = r/(Is^\wedge 2 + cs + k) \quad (28.3)$$

where r is the mirror radius, k is the spring constant, c is the friction coefficient and I is given my moment of Inertia of mirror.

The calculated mirror mass is:

$$\mu = \rho V = \rho\left(\pi r^2 h\right) = 2.23 \times 10^3\left(\pi \times 0.0127^2 \times 6 \times 10^{-3}\right) = 6.78\ \text{g}$$

where ρ denote the Pyrex mirror density, r and h denote the radius and thickness of mirror.

c denotes the friction coefficient. In FSM, the friction is caused by air drag only. The standard value of air viscosity

$$\varsigma = 1.73 \times 10^{-5}\ \text{N s/m}^2$$

So the friction coefficient is:

$$\begin{aligned} c &= \varsigma A = \mu\left(\pi r^2\right) \\ &= 1.73 \times 10^{-5}\left(\pi \times 0.0127^2\right) = 8.8 \times 10^{-9}\ \text{N s} \end{aligned}$$

where A is the surface area of mirror.

The radial stiffness is given by L_c = 0.00017 in./Lb and L_t = 0.00025 in./Lb. In our model, we used the average value. So stiffness is given by:

$$K = (L_c + L_t)/2 = 0.0002\ \text{in./Lb} = 0.0368\ \text{N/m}$$

Using the values of I, c and k, the transfer function is given by:

$$\begin{aligned} G(s) &= r/(Is^\wedge 2 + cs + k) \\ &= \frac{0.07}{0.0368 + 8.8 \times 10^{-9}\ \text{s} + 0.33 \times 10^{-3}\ \text{s}^2} \end{aligned}$$

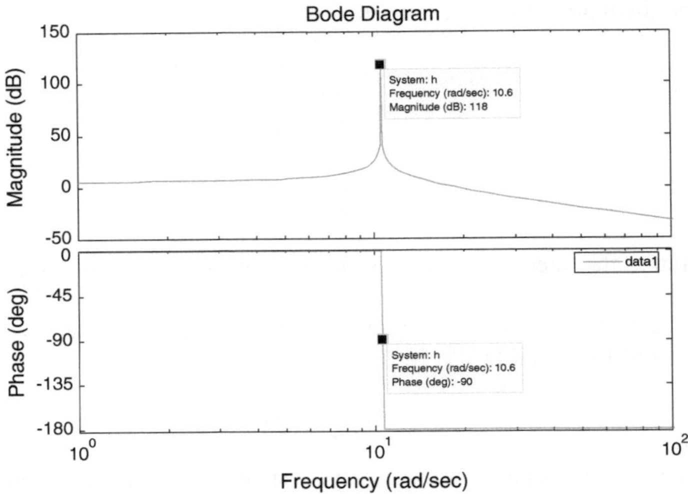

Fig. 28.6 FSM dynamics analysis

This equation is used to simulate the FSM model. The FSM is modeled by the above equation and plotted in MATLAB and shown in Fig. 28.6. The Bode plot shows the resonance frequency at 10.6 Hz.

28.4 Laboratory Set-up of Fine Pointing Mechanism

The block diagram of the set up of fine pointing mechanism is shown in Fig. 28.7a. The Laser of wavelength 650 nm and power of 1.5 mW is directed towards the FSM which is in external control feedback mode. This mode allows the FSM to lock the laser beam on the centre of quad cell. Using the neutral density (ND) filter, the beam is splitted (10:90) in two paths. The lower intense light is directed to the quad cell and the other is directed to communication receiver. Here we have used another FSM as a deformed mirror. The transmitted laser is initially directed to this deformed mirror. The actual laboratory set up is shown in Fig. 28.7b. Low frequency signal generator is used to move the FSM sinusoidally with different combinations of amplitude and frequency and at the receiver end a beam profiler is used to check the movement of the beam. In absence of the quad cell, the beam also moves sinusoidally as of deformed mirror moves. But in presence of quad cell, it controls the beam movement by steering the mirror and at the receiver end minimizes the angular error. When taking account the focused beam on the quad detector, the situation is shown in Fig. 28.8. The diameter of quad cell is 7.8 mm. An achromatic lens of 15 mm focal length is used in front of quad cell. The angular

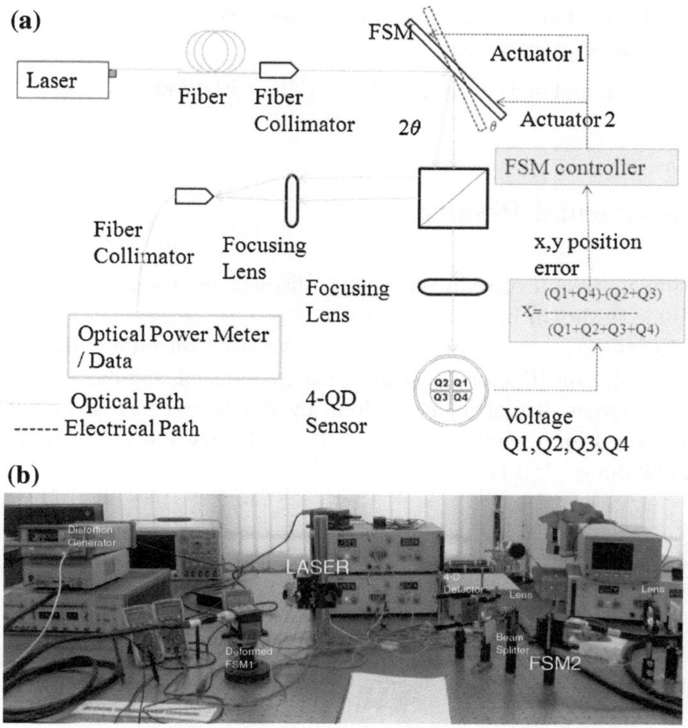

Fig. 28.7 **a** Block diagram of fine pointing mechanism. **b** Laboratory set up of fine pointing mechanism

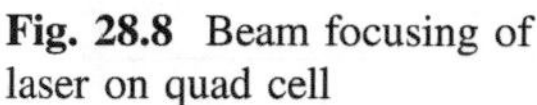

Fig. 28.8 Beam focusing of laser on quad cell

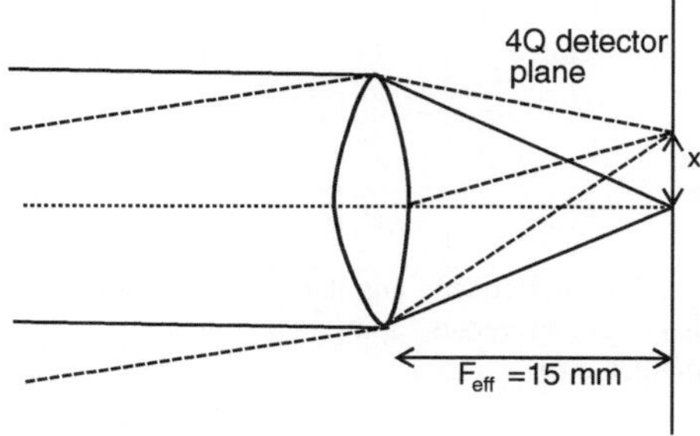

change of optical signal converts into lateral displacement after the lens and given by:

$$\delta = \arctan\left(\frac{x}{f_{\text{eff}}}\right) \tag{28.4}$$

where f_{eff} is the effective focal length of lens. The FOV volume defines the region of space that the sensor can detect.

$$\text{FOV} = 2\delta = 2 \text{ arctan } \left(\frac{x}{f_{\text{eff}}}\right) = 2 \text{ arctan } \left(\frac{0.39}{15}\right) = 30 \text{ mrad.}$$

28.5 Experimental Results

The experimental result of reduced angular displacement with variation of distortion amplitude is shown in Fig. 28.9.

Figure 28.10 shows the input angular error versus reduced angular error. As the figure shows, almost 10 times improvement in reduced angular error. The measurement of angular displacement is done by the laser beam profiler. The FSM system has closed loop bandwidth of 800 Hz at 10 mV but the quad detector has −3 dB bandwidth of 250 Hz.

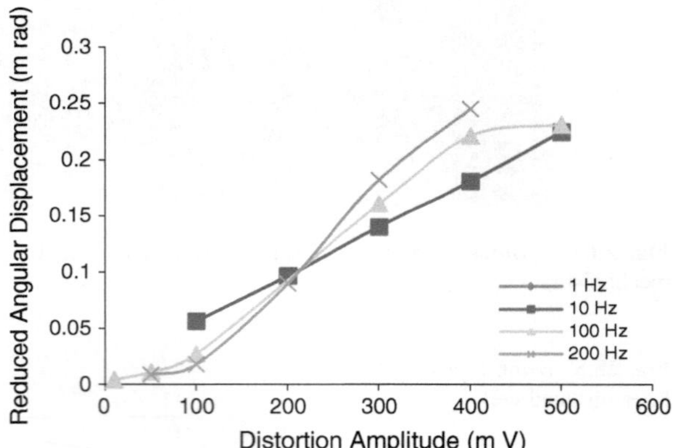

Fig. 28.9 Reduced angular displacement versus distortion amplitude

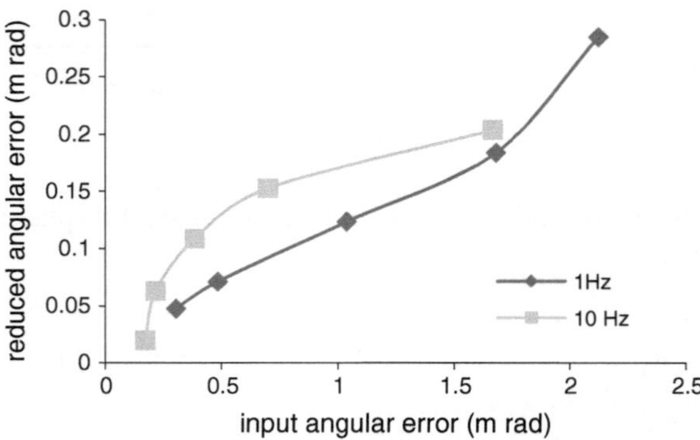

Fig. 28.10 Reduced angular displacement versus input angular error

28.6 Conclusion

Precision laser beam steering is crucial for numerous applications. In this paper, we have discussed the laboratory set up of fine pointing mechanism using quad cell. Experimental results and subsequent simulations are also performed. The random movement of FSM is produced by waveform generator with different frequency. In our experiment, we got about 10 times improvement in reduced angular error.

Acknowledgments The authors are thankful to Director, SAC and Deputy Director, SNPA for their support. The authors are also thankful to Optical Communication Division, SAC engineers for their support during alignment and testing to perform the experiment.

References

1. Zhou Q, Ben-Tzvi P, Fan D, Goldenberg AA (2008) Design of fast steering mirror system for precision laser beams steering. In: ROSE 2008-IEEE international workshop on robotic and sensors environment
2. Grigov C (2008) Evolution of coarse and fine pointing methods for optical free space communication. Master thesis, continuation courses, space science and technology, Department of Space Science, Kiruna
3. Wilkerson BL et al (2006) Concepts for fast acquisition in optical communication system. SPIE, vol 6304. International Society for Optics and Photonics, Bellingham
4. Pacific Silicon Sensor series 6 datasheet, Quad sum and difference amplifier. QP-506SD2
5. User Manual, FSM-300 fast steering mirror & FSM CD300B controller/driver. Newport Corporation
6. Portillo AA, Ortiz GG, Racho C (2001) Fine pointing control for optical communications (0-7803-6599-2/01/$10.00 Q 2001 IEEE)
7. Bailly M, Perez E (1991) The pointing, acquisition and tracking system of SILEX European program: a major technological step for intersatellite optical communication. SPIE, vol 1417. International Society for Optics and Photonics, Bellingham (Free-Space Laser Communication Technologies 111)
8. Hedding LR, Lewis RA (2005) Fast steering mirror design and performance for stabilization and single axis scanning. Proc SPIE 1304:14–24
9. Skormin VA, Busch TE, Givens MA (1995) Model reference control of a fast steering mirror of a pointing, acquisition and tracking system for laser communication. In: Proceedings of the IEEE NAECON, vol 2, pp 907–913
10. Application note on "Active beam stabilization between optical tables" from Newport Corporation

Part VIII
Micro-electronics and VLSI

Chapter 29
Power Effective Design of 10T D-FF Using MTCMOS Technique

Ankit Singh Kushwah and Shyam Akashe

Abstract In digital environment, circuit design is an important parameter to take under consideration for making a compact and power efficient design. A high speed low power compact design is a challenge to all digital circuit and from these parameters power consumption with high speed is always desirable. The main important point of this paper is to provide, power effective model of 10T D-FF using MTCMOS technique, as a new low power solution for VLSI design. MTCMOS (multi threshold CMOS) is more effective circuit level technique that gives a high performance and low power design by utility of both low and high threshold voltage transistor. After implementing MTCMOS technique in our desired circuit power consumption reduces from 602.9E-9 to 189.4E-12 in active mode and up to 22.28E-12 in standby mode, and the static leakage power reduces from 7.85E-12 to 4.14E-13 in standby mode. The overall process is carried out on cadence virtuoso tool at 45 nm technology.

29.1 Introduction

Power consumption is much essential parameter taken under investigation in the recent years. Digital integrated circuit (DIC) uses CMOS circuits as generally building blocks. The continuing scaling in feature size of CMOS circuits and analogous increase in chip density and operating frequency have made power consumption is a major anxiety in VLSI design. Due to more power consumption in any circuit it causes over heating of chip, reduces the life of chip and degraded the performance of the integrated circuit. Therefore power minimizing is an important to take under consideration for with increasing chip density, feasibility, reliability and cost [1]. Power dissipation of the circuit can be maintained by controlling the supply voltage. On other hand power scaling affects the speed of the circuit. That's

A.S. Kushwah (✉) · S. Akashe
Department of Electronics and Communication Engineering, ITM University, Gwalior, India
e-mail: akushwah13@gmail.com

V. Lakshminarayanan and I. Bhattacharya (eds.), *Advances in Optical Science and Engineering*, Springer Proceedings in Physics 166,
DOI 10.1007/978-81-322-2367-2_29

by industries put their efforts in designing low power high speed circuits based on FinFET technology [2]. MTCMOS is a promising alternative to use in logic gate operating at high speed with lowering of power dissipation as compared to CMOS circuits. MTCMOS technique is an effective circuit level technique with high performance and low power dissipation by using a pair of low threshold voltage and high threshold voltage transistors.

29.2 Power Consumption Using CMOS Technique

In above conventional D-FF using NAND Gate generally 18 transistors used and according to Moore's law and industry demand transistor minimizing of transistors are needed and minimum transistor also helpful to design a low power circuit design, a modified 10T D-FF shown in Fig. 29.1.

As we know in CMOS design the circuit is an addition of pull up and pull down network and both network shows their equal importance during their circuit operation. The supply voltage has been maintained such that power dissipation must be in control. So that threshold voltage of transistor has to maintain to maintain drive current and improve performance. But in other side due to lowering of threshold voltage, sub threshold leakage current increases. Total power of the circuit shows the combination of static (P_{static}) and dynamic P_{dyn} Power. And it can express as:

$$P_{total} = P_{dyn} + P_{static} \tag{29.1}$$

P_{dyn} represents dynamic power dissipation due to charging/discharging of capacitances when the output signal of a logic gate makes a transition. P_{static} represents the static power consumption due to the leakage current and their major components are the sub threshold leakage, gate direct tunnelling leakage, and junction band-to-band tunnelling leakage [3].

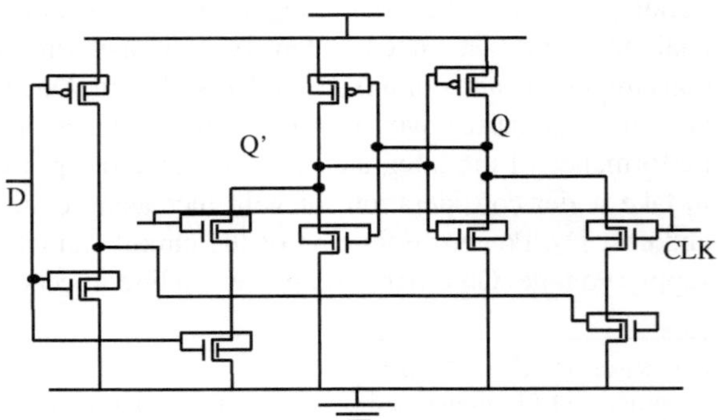

Fig. 29.1 Circuit of D-FF using FinFET

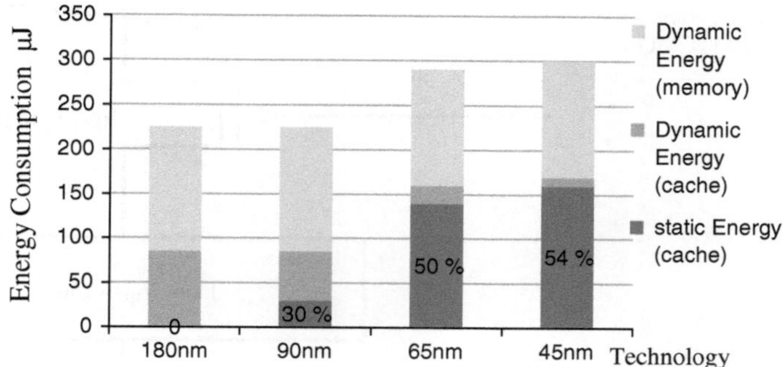

Fig. 29.2 Technology to energy consumptions

Power consumption by DIC is:

$$P_{\text{total}} = I_o \cdot V + \alpha C V_{\text{dd}}^2 f \tag{29.2}$$

where, I_o represents leakage current in any circuit and it defined by the diode equation I_s $(e^{qv/kT} - 1)$. In equation first term in the equation results the leakage power and the second term shows the dynamic switching power. As the features size decreases, V_{dd} has also decreased, forcing a reduction in the threshold voltage V_t of the transistor. Thus the leakage current I_o which depends on V_t, through the diode equation increases. A more detail expression for sub threshold leakage [4].

$$I_{\text{sub}} = A * \exp\left(\frac{q}{n'kT}(V_g - V_s - V_{\text{tho}} - \gamma V_s + \ldots \eta V_{\text{ds}})\right) * B \tag{29.3}$$

Static power consumption can be given as:

$$P_{\text{static}} = V_{\text{dd}} * I_{\text{static}} = V_{\text{dd}} * \left(I_{\text{gate}} + I_{\text{sub}}\right) \tag{29.4}$$

As we move from 180 nm technology to 45 nm technology there are different types of variations in Dynamic Energy and Static Energy occurs which shown in Fig. 29.2 using tabular form [5].

29.3 MTCMOS Technique

As per industries and market demand supply and threshold voltage reduces with the scaling of CMOS technology. As the threshold voltage lower downs different type of leakage comes out like sub threshold leakage, band to band tunnelling etc. In modern technologies 40 % of active power or energy dissipated as leakage current due to scaled down of transistors. To tackle with these possibilities, here we are

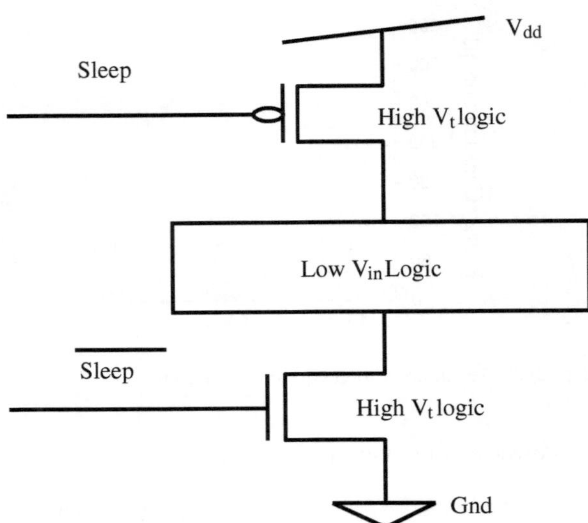

Fig. 29.3 Power gating technique using MTCMOS

going to use a low leakage technique namely the multi threshold voltage CMOS (MTCMOS) [6]. The multi threshold voltage technique has two main operational modes. Active mode and other is standby mode for power efficient model. MTCMOS technique attains high speed in active mode and lower leakage in standby mode [7] (Fig. 29.3).

In this technique high threshold voltage is apply to provide separate supply. And the lower V_{in} logic circuit or circuit under test (CUT) is separated by high threshold voltage in active mode and standby mode. The high threshold voltage will be turn on in active mode and it will be turns off in standby mode also known as sleep mode. In standby very low sub threshold passes from V_{dd} to ground. Using MTCMOS technique circuit attain high speed in active mode and low leakage current during standby mode [8, 9].

29.4 Comparison and Results

29.4.1 Total Power

Total power of the circuit results the power dissipation across the entire circuit. And it can state as:

$$P_{total} = P_{dyn} + P_{static} \tag{29.5}$$

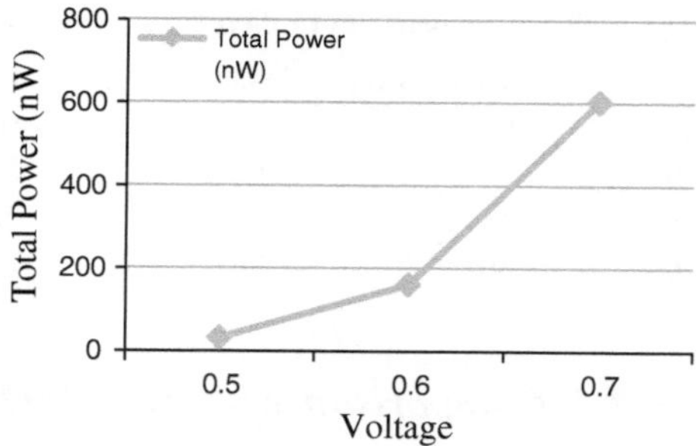

Fig. 29.4 Total Power consumption at different voltages

P_{dyn} represents dynamic power dissipation due to charging/discharging of capacitances when the output signal of a logic gate makes a transition. P_{static} represents the static power consumption because of the leakage currents (Fig. 29.4).

29.4.2 DC Power or Static Power of the Circuit

P_{static} is the static power consumption due to the leakage current whose major components are the sub threshold leakage, gate direct tunnelling leakage, and junction band-to-band tunnelling leakage.

Static power consumption can be given as (Fig. 29.5):

$$P_{static} = V_{dd} * I_{static} = V_{dd} * \left(I_{gate} + I_{sub}\right) \tag{29.6}$$

To compare all the result of total power, static power and leakage current with their respective rise time and fall time at different voltages shown in Table 29.1.

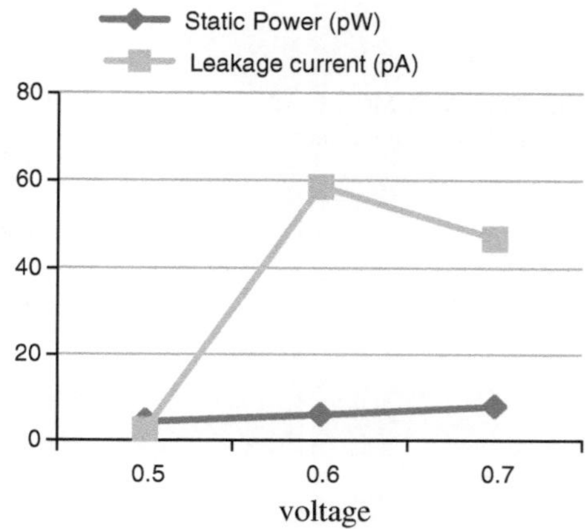

Fig. 29.5 Static power and leakage current at different voltage

Table 29.1 Comparison of different parameter at different voltage

Voltage	Total power (nw)	Static power (nw)	Leakage current (pA)	Rise time (ps)	Fall time (ps)
0.5	30.60	4.2	52.38	301.7	301.7
0.6	159.2	5.87	58.49	272.9	272.9
0.7	602.9	7.85	46.80	172.8	172.8

29.4.3 Comparison with MTCMOS Circuit

As we studied the MTCMOS technique works in two important modes, active and standby mode. In this part we compare the total power and static power of the CUT at different voltages in active and standby mode.

29.4.3.1 Total Power in Active and Standby Mode

This section shows the total power consumption due to different leakage and power dissipation across whole circuitry. In active mode, pull up and pull down high threshold voltage transistor connected in ON condition and vice versa for standby mode condition (Fig. 29.6).

MTCMOS active mode power is that power when both the high threshold voltage should be turn on. Pull up high threshold voltage should be connected to lower supply voltage and pull down high threshold voltage network connected to high supply voltage (Fig. 29.7).

29.4.3.2 Static Power in Active or Standby Mode

Active and Standby leakage current of low power design is calculated in this section. The basic equation of leakage current is

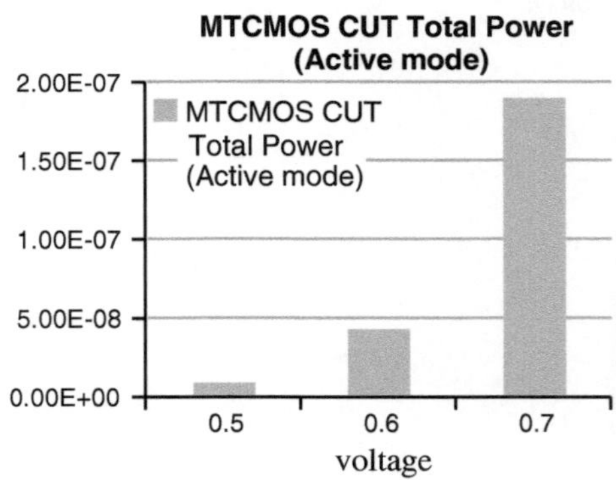

Fig. 29.6 MTCMOS CUT total power in active mode

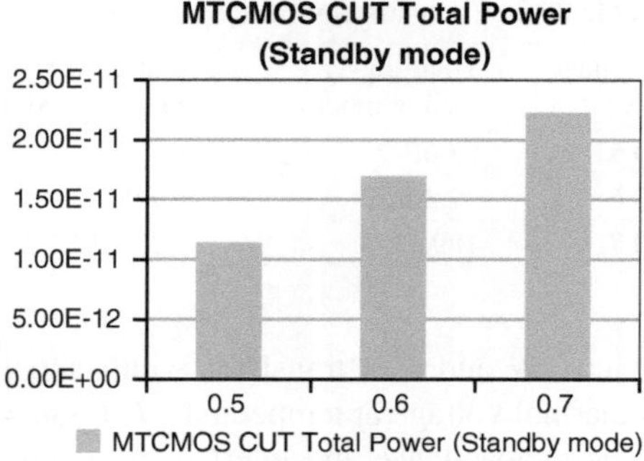

Fig. 29.7 MTCMOS CUT total power in standby mode

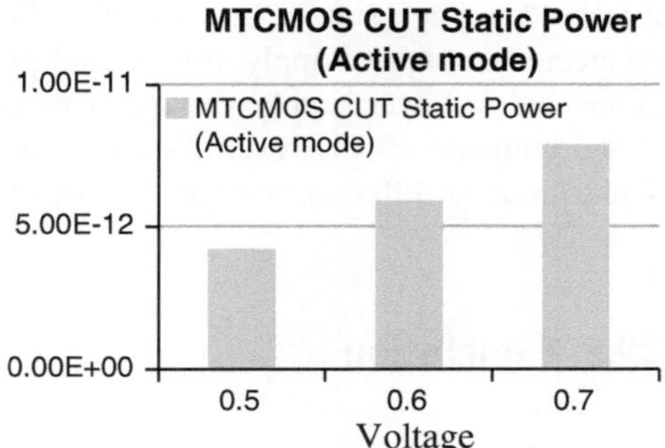

Fig. 29.8 MTCMOS CUT static power in active mode

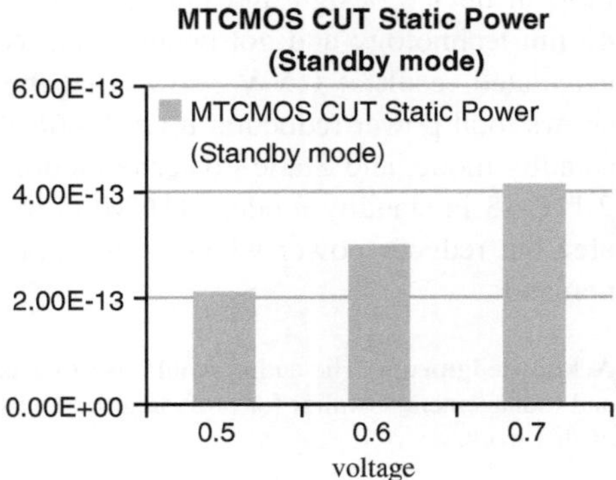

Fig. 29.9 MTCMOS CUT static power in standby mode

$$I_{\mathrm{sub}} = \frac{W}{L}\mu \cdot V_{\mathrm{th}}^{2}\, C_{\mathrm{STH}}\, e(\frac{V_{\mathrm{GS}} - V_{\mathrm{TH}} + \eta V_{\mathrm{DS}}}{\eta V_{\mathrm{TH}}}(1 - \mathrm{e}^{\frac{-V_{\mathrm{DS}}}{V_{\mathrm{TH}}}}) \tag{29.7}$$

Table 29.2 Comparison of different parameter at different voltage

Voltage (V)	Total power (active mode)	Total power (standby mode)	Static power (active mode)	Static power (standby mode)
0.5	8.66E-9	11.47E-12	4.2E-12	2.11E-13
0.6	42.80E-9	16.97E-12	5.87E-12	3.0E-13
0.7	189.4E-9	22.28E-12	7.85E-12	4.14E-13

where, W and L = transistor width and length, μ = carrier mobility, $V_{TH} = kT/q$ thermal voltage at temperature T, $C_{STH} = C_{DEP} + C_{IT}$ summation of the depletion region capacitance and interface trap capacitance both per unit area of MOS gate, η = Drain-induced barrier lowering (DIBL) coefficient (Figs. 29.8 and 29.9).

MTCMOS standby mode static power is that power when both the high threshold voltage should be turn OFF. Pull up high threshold voltage should be connected to higher supply voltage and pull down high threshold voltage network connected to lower supply voltage. This comes into existence due to leakages.

To compare all the result for total power and static power in both active and static mode at different voltages shown in Table 29.2.

29.5 Conclusion

In this report we introduced the characteristics results on power effective design of 10T D-FF using MTCMOS technique. The work carried out on cadence virtuoso tool, in device design and optimization at 0.5, 0.6 and 0.7 V supply voltage at 45 nm technology and got better result as compared to CMOS based circuit. The calculated result at 0.5 V shows that the circuit based on MTCMOS technique shows total power reduction up to 8.66E-9 in active mode and up to 11.47E-12 in standby mode, and Static Power reduction up to 4.2E-12 in active mode and up to 2.11E-13 in standby mode. MTCMOS technique based circuit generally increases area but reduces power which is an important factor in accordance of technology demand.

Acknowledgments The author would like to express his gratitude to the Institute of Technology and Management, Gwalior for providing the Tools and Technology for the successful completion of this work.

References

1. Nirmal U, Sharma G, Mishra Y (2010) Low Power Full Adder Using MTCMOS Technique. In: Proceeding of international conference on advances in information, communication technology and VLSI design. Coimbatore, India, Aug 2010

2. Bansal A, Mukhopadhyay S, Roy K (2007) Device optimization technology for robust and low-power FinFET SRAM design in nanoscale era. IEEE Trans Electron Devices 54(6):1409–1419
3. Mano MM, Ciletti MD (2008) Digital design, 4th edn. Prentice Hall, New Delhi
4. Gomathisankaran M, Tyagi A (2004) WARM SRAM: a novel scheme to reduce static leakage energy in SRAM arrays. In: Proceedings to IEEE computer society annual symposium on VLSI Emerging trends in VLSI systems design (ISVLSI'04), pp 105–112, Feb 2004
5. Alioto M, Consoli E, Palumbo G (2011) Analysis and comparison in the energy-delay-area domain of nanometre CMOS flip-flops: part 2, results and figures of merits. IEEE Trans VLSI 19(5):737–750
6. Jiao H, Kursun V (2011) Ground bouncing noise suppression techniques for data preserving sequential MTCMOS circuits. IEEE Trans Very Large Integr Syst 19(5):763–773
7. Dev MS, Akashe S, Sharma S (2011) Leakage power reduction techniques of 45 nm static random access memory (SRAM) cells. Int J Phys Sci 6(32):7341–7353
8. Calhoun BH, Honore FA, Chandrakasan AP (2004) A leakage reduction methodology for distributed MTCMOS. IEEE J Solid-State Circuits 39(5):818–826
9. Pakbaznia E, Pedram M (2012) Design of a tri-modal multi-threshold CMOS switch with application to data retentive power gating. IEEE Trans Very Large Integr (VLSI) Syst 22 (2):380–385

Chapter 30
High Performance FinFET Based D Flip Flop Including Parameter Variation

Pooja Joshi, Saurabh Khandelwal and Shyam Akashe

Abstract In present digital circuit's leakage current controlled delay flip flops are needed significantly for circuit stability and high performance. The use of delay flip flop in finite state machine is to store the circuit state which requires low power consumption for scaled down devices. In this paper we have employed delay flip flop using FinFET design with multiple threshold technique to reduce the leakage power which targets both combinational and sequential elements. Parameter variations are also done to minimize sub-threshold leakage current in FinFET based D flip flops. The proposed circuit shows a design with the aim to increase the throughput of D flip flop in comparison to the Conventional transmission gate based D flip flop by considering minimum transistor count with reduced circuit delay and leakage power. Short channel effect gets considerably reduced due to fin structure and geometry. Simulation result shows that the proposed MTCMOS based FinFET D flip flop has the least leakage power dissipation of nearly 9.20 pW at 0.7 V supply voltage.

30.1 Introduction

The scaling done on transistor sizes in CMOS circuits has lead to a spectacular enhancement of leakage currents [1]. The major contributors of total power consumption as well as leakage current dissipation are the sub-threshold and gate oxide leakages. One of an important design constraint in microprocessors is power consumption [2]. So the continuous scaling with power reduction in future technology will cause sub-threshold leakage current in becoming an increasingly outsized component of total power dissipation [3]. Therefore always it requires new techniques for providing low power circuit operation with high performance. A major component of digital circuits is flip flop which act as a bi-stable element [4].

P. Joshi (✉) · S. Khandelwal · S. Akashe
ITM Universe, Opposite of Sitholi Railway Station, NH-75, Jhansi Road, Gwalior, Madhya Pradesh, India
e-mail: poojajoshi1311@gmail.com

V. Lakshminarayanan and I. Bhattacharya (eds.), *Advances in Optical Science and Engineering*, Springer Proceedings in Physics 166,
DOI 10.1007/978-81-322-2367-2_30

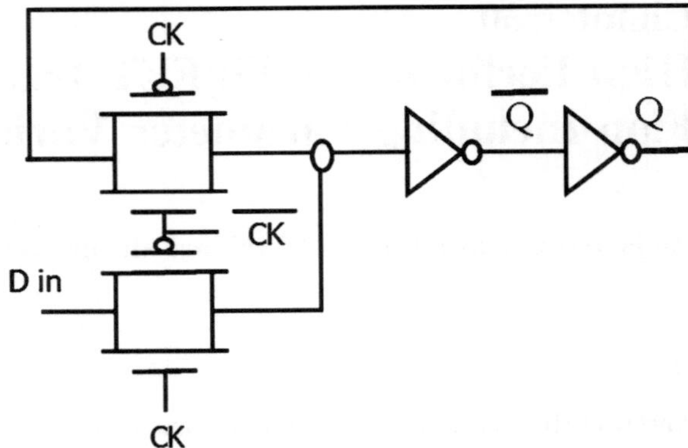

Fig. 30.1 Transmission gate based D flip flop implementation

The functioning of flip flop incorporates storing a logical state in response of an input clock pulse. A transmission gate is similar to a relay which can conduct or block in both directions by a control signal with almost any voltage potential [5].

The delay flip flop implemented by two transmission gate and redundant feedback path is shown in Fig. 30.1 which for **CK** = 1 causes $\mathbf{Q_{n+1}} = \mathbf{D}$, a bit is loaded. For **CK** = 0, $\mathbf{Q_{n+1}} = \mathbf{Q_n}$ thus a bit is stored through feedback. Feedback along the path causes an abrupt increase of leakage current.

30.2 Proposed Design

In the proposed flip flop as the memory cell of 1 bit is shown where five FinFET based transistors are used (two P-FETs and three N-FETs) through fin patterning's having quasi planar geometry in which a narrow fin structure get prepared which lead to a double gate design aligned with each other.

We have used the shorted gate FinFET mode as in (30.1) which reflects that the width of the fin introduced in the structure of FinFET depends mainly upon its two important parameters which are height (H_{fin}) and thickness (T_{fin}) of the fin.

$$W_{\text{fin}} = 2 * H_{\text{fin}} + T_{\text{fin}} \tag{30.1}$$

The introduced electric field in the FinFET structure at the drain appears to be completely screen from the source end due to channel nearness of the added gate which reduces the leakage current and provides short channel compatibility in devices by overcoming the known short channel effects. For having symmetrical rise and fall time of flip flop the channel width of P-channel FETs must be double to that of the channel width of N-channel FETs hence a channel width biasing of flip flop is also implemented as parameter variation. But the leakage is needed to be

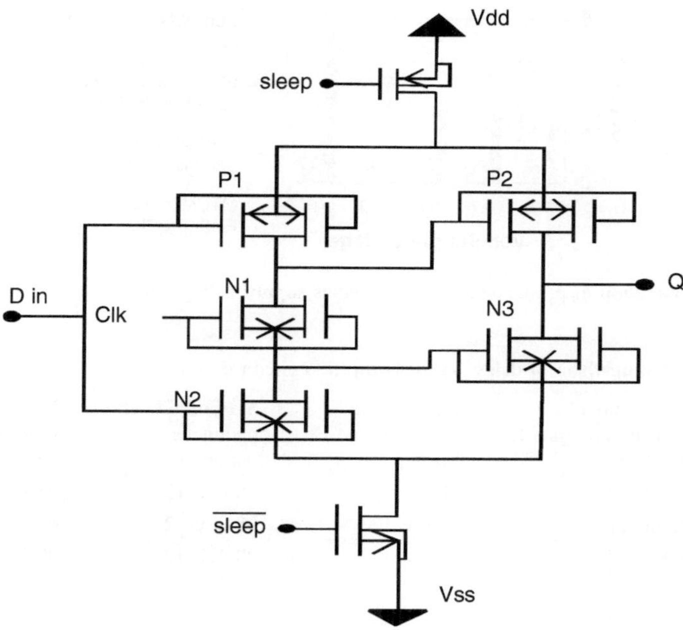

Fig. 30.2 Delay flip flop using shorted gate FinFET and MTCMOS technique

further reduced so another design has been projected called Multi–Threshold CMOS or MTCMOS. This is shown in Fig. 30.2 with the use of high V_t sleep transistor which is the gateway of power supply to the low V_t logic block results to decrease the power delay product along with the overall power consumption in the circuit.

30.3 Simulation Analysis

All the circuits for flip flops designs have been implemented using Cadence design tool in 45 nm technology. Thus for establishing an unbiased testing situation for each circuit we have tested all the circuits using the same input patterns. In the simulation, the transient responses of various D Flip Flops are compared with respect to the total power consumption in term of leakage power and delay known as power delay product shown in Fig. 30.3.

Delay is defined as the time required (in s) for the data to travel from input towards output. So the delay is encountered in various D Flip Flops design at different V_{DD} (supply voltage) as shown in Table 30.1.

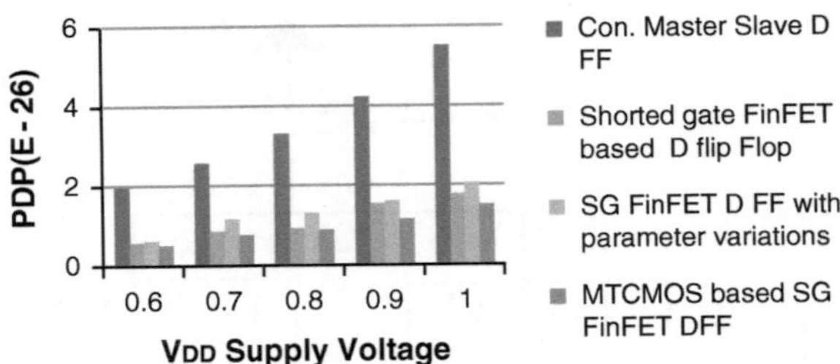

Fig. 30.3 Variation of power delay product versus supply voltage (V_{DD})

Table 30.1 Comparison of delay (s) for D flip flop circuit design

V_{DD} (V)	Conventional transmission gate D flip flop	Shorted gate FinFET based D flip flop	SG FinFET with parameter variation	MTCMOS SG FinFET D flip flop
1	9.33E-12	08.29E-12	08.23E-12	08.14E-12
0.9	9.36E-12	08.31E-12	08.28E-12	08.25E-12
0.8	9.41E-12	08.35E-12	08.31E-12	08.29E-12

30.4 Conclusion

The power consumption should be low as it is an essential requirement for high performance VLSI circuits with analog, digital and mixed signals. This paper concludes that MTCMOS based SG FinFET D flip flop design has minimum power consumption as it has the least power delay product which leads to best performance compared with other designs of D flip flop also the parameter variation done in SG FinFET shows a drastic change in certain characteristics of flip flop design as its power consumption get reduced to 13.93 pW from 24.86 pW at 0.7 V supply. The proposed flip flop shows a tremendous reduction in power consumption from 13.93 pW to 9.20 pW at 0.7 V supply showing nearly 34 % reduction in comparison to the SG FinFET D flip flop with parameter variation.

References

1. Mehta K, Arora N, Singh BP (2011) Low power efficient D flip flop circuit. In: Proceedings of international symposium on devices MEMS, intelligent systems and communication 1(8):16–19
2. Singh S, Singh SK, Singh Kumar M, Sagar Kumar R (2012) Design and analysis of clocked subsystem elements using leakage reduction technique. Int J Sci Technol Res 1(5):112–116
3. Sagar Daya CH, Moorthy Krishna T (2012) Design of a low power flip-flop using MTCMOS technique. Int J Comput Appl Inf Technol 1(1):19–20

4. Heo S, Shin Y (2007) Minimizing leakage of sequential circuits through flip-flop skewing and technology mapping. J Semicond Technol Sci 7(4):215–220
5. Baek D, Shin I, Shin Y (2013) Accurate gate delay extraction for timing analysis of body-biased circuits. J Circuits Syst Comput 22(8):1350072-1-16

Chapter 31
Memristive Power Optimization of Non-volatile Seven Transistors Static Random Access Memory Cell

Atibhi Jadon and Shyam Akashe

Abstract Starting from the consecutive properties of Memristor, researchers identify that using a non-linear element "Memristor" into the circuit, the power consumption of the design will be reduced to a great extent. Memristor acts as a conventional resistor, but it has a memory which stores data in the form of resistance. A new Seven Transistors (7T) based Static Random Access Memory (SRAM) cell using memristors is designed in this manuscript that provides low power as well as non-volatile functionality to the cell. Such hybrid CMOS-Memristor's combination gives robustness, consistency and higher functionality to the CMOS subsystem. With the proposed 7T SRAM circuit, we observed that the Average Power (0.234 μW), Total Power (10.08 μW), Static Power (20.83 pW) and Dynamic Power (0.01 μW) are reduced over the conventional 7T SRAM circuits. Different low-power optimization tools and device modeling were discovered to make more accurate power consumption calculations. There will be a need of designing such devices which are essential for the VLSI circuit designers. Memristor verifies itself to be a strong contender for the growing demands of Low power and Portable applications. Simulation of the SRAM design has been done in a 45 nm CMOS environment with the help of Cadence Virtuoso tool, and the recorded results depicted the advantages of proposed design over the CMOS based SRAM cell.

31.1 Introduction

Now-a-days, the electronic market is growing too fast. Semiconductor Electronics is performing an extensive role in our day to day life. Designers and researchers are working on new technologies due to the limited scaling of CMOS based technologies, which can also fulfill the portable and computing demands. A variety of

A. Jadon (✉) · S. Akashe
Department of ECE, ITM, NH-75, Jhansi Road, Gwalior 474001, Madhya Pradesh, India
e-mail: engg.atibhi27@gmail.com

V. Lakshminarayanan and I. Bhattacharya (eds.), *Advances in Optical Science and Engineering*, Springer Proceedings in Physics 166,
DOI 10.1007/978-81-322-2367-2_31

technologies are being studied and investigated by different researchers. In the literature, we studied different types of Static Random Access Memory (SRAM) cells like CMOS based SRAM cells [1], Field Effect Transistor (FinFET) based SRAM cells [2, 3], Carbon Nanotube FET (CNTFET) based SRAM cells [4], SRAM cell based on Transmission gates [5]. It also gave birth to a new fundamental element, called Memristor [6–8] other than the Resistor, Inductor and the Capacitor. Memristor based technology is a new technology that came into the market in order to create non-volatile and low power operated memories of smaller size. So, for the betterment of SRAM, we embed it in every integrated circuit (IC). The major target of the current researches is to design low power SRAM memory devices as they are the most crucial component in both handheld devices and high-performance processor. With the limited use of battery and expensive packaging of portable systems, power consumption is regarded as an important distinctive trait for electronic consumers. In case of memories, the main significance is given to low power designs. SRAM memories are extensively used in portable applications due to their low leakage (standby) and easy use [9, 10]. The SRAM cell finds its use mainly in embedded and portable devices [11].

The goal of this paper is to lower the power consumption of the SRAM memory device by incorporating memristor into the circuit and thus comparing the results of basic Seven Transistors (7T) SRAM cell and the proposed SRAM memory cell. Here, we designed 7T SRAM cell based on Memristors and investigated that the overall performance of Memristor based 7T SRAM is superior as compared to conventional SRAMs. The paper is systematized as follows: Firstly, the Introduction is given in Sect. 31.1. Section 31.2 discusses the overview of the conventional 7T SRAM cell. Memristor and its modeling are explained under Sect. 31.3. Implementation and analysis of Memristor based SRAMs are given in Sect. 31.4. Simulation results are included in Sect. 31.5. Section 31.6 showed the conclusion of the paper.

31.2 Overview of the Conventional 7T SRAM Cell

31.2.1 Device Parameters

Both Device and technology parameters are essential for the design of a circuit. Before designing any circuit, we assign the following values to the various parameters as shown in the Table 31.1.

31.2.2 CMOS Based Seven (7T) Transistors Static RAM Cell

Conventional 6T SRAM cell is the most standard SRAM cell that consists of two Bit-Lines (BL and BLB) and a Word Line (WL) for performing read and write operations. This 6T based SRAM cell uses an additional power and gives poor

Table 31.1 Device parameters for 7T SRAM cell

S. No.	Device parameters	Memory cell (7T SRAM cell)
1	Technology	45 nm
2	Supply voltage	0.7 V
3	Precharge voltage	1 V
4	Temperature	27 °C
5	S/D metal width	60 nm
6	Gate spacing	160 nm
7	Source drain length (left)	140 nm
8	Source drain (right)	140 nm
9	Source diffusion periphery	520 nm
10	Drain diffusion periphery	520 nm
11	Source diffusion area	16.8 fm^2
12	Drain diffusion area	16.8 fm^2

stability at low supply voltage [12]. To overcome these limitations, 7T SRAM is designed where one NMOS transistor (T7) is added in the conventional 6T SRAM, which is positioned between the two (I1 and I2) cross-coupled inverters. This 7T SRAM functions more efficiently than basic 6T cell at low voltages [13]. T7 controls the read and writes operation of the whole circuit. Figure 31.1 demonstrates schematic of basic 7T SRAM cell, which includes two cross-coupled inverters (I1 and I2), formed using four transistors (T1, T2, T3 and T4). Transistors (T5 and T6) are the write access transistors and transistor T7 is the read access one.

31.3 Memristior and Its Modelling

Nanoelectronics play an imperative role for the discovery of Memristor. A memristor is a two terminal non-volatile resistor (variable) whose Memristance is altered by applying either current or voltage across the device. The invention of memristor leads to new opportunities in the research market for the realization of non-volatile designs [14]. Chua introduced memristor based on symmetry gap between four circuit variables (Fig. 31.2a). Pinched Hysteresis Loop defines Memristor's fingerprint. The voltage across this charge controlled device is determined by:

$$V(t) = M(q(t) \cdot I(t)) \tag{31.1}$$

$$M(q) = \frac{\mathrm{d}\varphi(q)}{\mathrm{d}q} \tag{31.2}$$

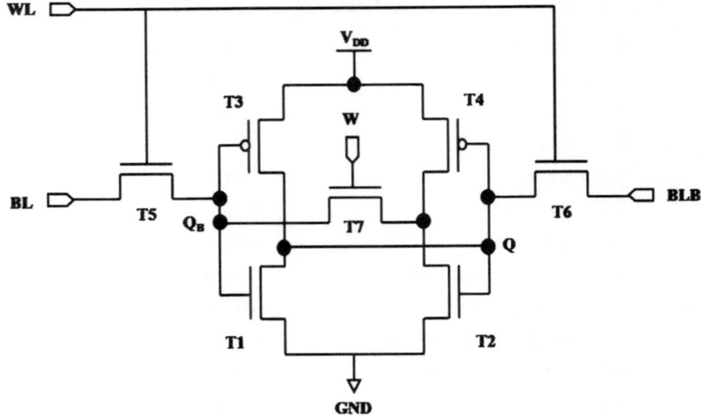

Fig. 31.1 Schematic of 7T SRAM memory cell

where M is the Memristance of this Memristor device. Equation 31.1 showed the resemblance with the Ohms' Law but Memristor's Memristance is different from resistor's resistance. Memristor behaved as a two resistors in series: the doped region resistance is R_{ON} and the undoped region resistance is R_{OFF} (Fig. 31.2b). The R_{init} and ΔR parameters are given by:

$$R_{init} = R_{ON}\left(\frac{w_0}{D}\right) + R_{OFF}\left(1 - \frac{w_0}{D}\right) \tag{31.3}$$

$$\Delta R = R_{OFF} - R_{ON} \tag{31.4}$$

where w_0 shows the initial position of the Memristors boundary region.

31.4 Memristior Based Non-volatile Seven (7T) Transistor SRAM Cell

There are enormous different paths to acquire low power, consistent memory cells of small sizes. Recently preferred low power reduction techniques are gated-V_{DD} [15], Multi-threshold (MT) CMOS [16], Self controllable voltage level (SVL) [17] etc. But all these techniques required an extra circuitry which increases the chip area, cost and affects performance. Instead of using such techniques, Memristor is used that gives low power consumption to the memory cell with superior performance. At small sizes, 6T SRAM cell gave poor stability [18, 19] and consumes high power. A new technology is used to a fully SRAM cell to replace the standard 6T. To do that, it has to be low powered while maintaining the high performance of

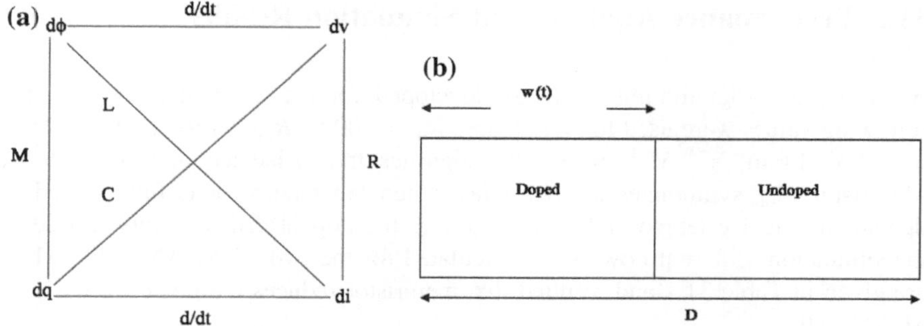

Fig. 31.2 **a** Symmetry diagram showing the missing link between the flux (φ) and charge (q); **b** Memristor's base structure

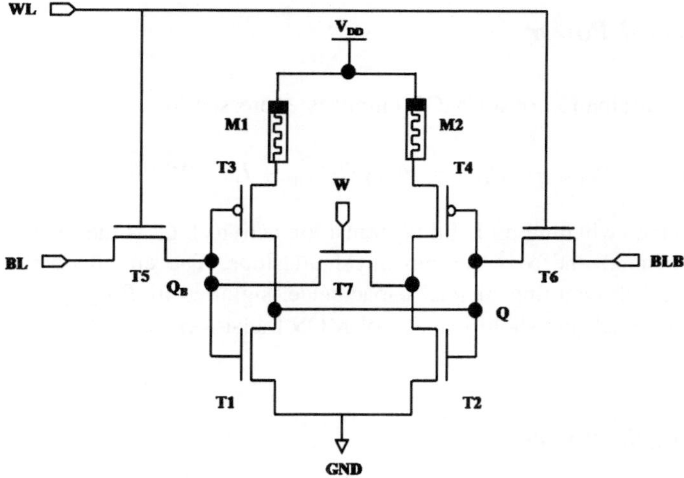

Fig. 31.3 Schematic of the memristor based NV 7T SRAM cell

the 6T SRAM. So, 7T SRAM cell is designed with memristors. Several other memristor based SRAMs were already designed and implemented as in [20, 21].

The schematic of Memristor based 7T SRAM cell is shown in Fig. 31.3, consists of two Memristors (M1 and M2) embedded with the design of basic 7T SRAM cell. BL are carrying the data from or to SRAM cell to a sense amplifier during read and write operations. Because of this last seventh transistor, 7T SRAM cell found to be more efficient and suitable for low power applications.

31.5 Performance Analysis and Simulation Results

A Verilog-A programming model is developed for the memristive device. The following values were used for simulation: $R_{on} = 100\ \Omega$, $R_{off} = 1600\ \Omega$, D = 10 nm, $\mu = 10e{-}14\ m^2\ s^{-1}\ V^{-1}$, where R_{off} signifies the resistance of fully undoped Memristor, R_{on} symbolizes the resistance when the memristor is fully doped; D denotes the device length and μ characterizes the dopant drift mobility. Based on the simulation, different powers are calculated for the two 7T SRAM cell's which are given in Table 31.2 and resulted that memristor reduces the power of basic 7T SRAM cell.

Following are the different powers, calculated and compared for the conventional and proposed 7T SRAM cell.

31.5.1 Total Power

The power dissipation of a CMOS circuit is expressed as:

$$P_{\text{Total}} = P_{\text{Change}} + P_{\text{Leak}} = \alpha f_{\text{CLK}} C_L V_{\text{DD}}^2 + I_{10} + 10^{-\left(\frac{V_{\text{th}}}{S}\right)} \cdot V_{\text{DD}} \tag{31.5}$$

where α is the switching probability (transition activity), C_L is the load capacitance, I_{10} is a constant, and S is the sub threshold slope. The circuit operation causes charging and discharging of load capacitance, signified as P_{Charge}. Static leakage current due to sub threshold leakage of MOS transistors is displayed as P_{leak}.

31.5.2 Static Power

Static dissipation is due to the static leakage current drawn continuously from the power supply. Hence, the static power consumption is given by the product of the leakage current and the supply voltage.

$$P_{\text{Static}} = I_{\text{Leakage}} \cdot V_{\text{Supply}} \tag{31.6}$$

Table 31.2 Different powers of 7T SRAM and memristor-based7T SRAM cell

S. No.	Parameters	7T SRAM	Memristor based 7T SRAM
1	Average power (P_{avg}) in watts	1.105e−6	0.234e−6
2	Total power (P_{total}) in watts	822.1e−6	10.08e−6
3	Static power in watts (P_{static})	21.03e−12	20.83e−12
4	Dynamic power in watts (P_{dyn})	0.823e−6	0.01e−6

The static power component is usually a small contributor to the overall power consumption.

31.5.3 Dynamic Power

Dynamic dissipation occurs due to switching transient current (referred to as short-circuit current) and charging and discharging of load (C_{load}) capacitances (referred to as capacitive switching current).

$$P_{\text{Dynamic}} = P_{\text{Total}} - P_{\text{Static}} = \frac{1}{2} \cdot C_{\text{load}} \cdot V_{\text{Supply}}^2 \tag{31.7}$$

31.5.4 Average Power

Depending on the design requirements, there are different power dissipation factors that need to be considered. The average power is one of them, which is related to the cooling or battery energy consumption necessities. The average power is defined by the following equation:

$$P_{\text{Average}} = \frac{1}{T} \int_0^T \left(I_{\text{Supply}}(t) \cdot V_{\text{Supply}} \right) \text{d}t \tag{31.8}$$

31.6 Conclusion

A novel memory device derived from the CMOS based Static RAM and Memristor device is proposed and studied in this work. The device combines the properties of both devices in terms of small size, non-volatility and low power memory. Instead of applying several power reduction techniques like Self-Voltage level, gated-V_{DD} technique and multi threshold voltage technique, etc., a small size, non-linear element Memristor is utilized in the SRAM cell in order to reduce its power consumption. Small size of memristor will provide better scalability to ICs. The Average Power (0.234 μW), Total Power (10.08 μW), Static Power (20.83 pW) and Dynamic Power (0.01 μW) of proposed SRAM cell get reduced due to memristor embed into it.

References

1. Singh J, Mathew J, Pradhan DK, Mohanty SP (2008) A subthreshold single ended I/O SRAM cell design for nanometer CMOS technologies. In: Proceedings of the IEEE international SOC conference. IEEE, Newport Beach, pp 243–246
2. Raj B, Saxena AK, Dasgupta S (2011) Nanoscale FinFET based SRAM cell design: analysis of performance metric, process variation, underlapped FinFET and temperature effect. IEEE Circuits Syst Mag 11(3):38–50
3. Jha NK, Chen D (2011) Nanoelectronic circuit design. Springer, New York, p 290
4. Zhang Z, Delgado-Frias JG (2012) Carbon nanotube SRAM design with metallic CNT or removed metallic CNT tolerant approaches. IEEE Trans Nanotechnol 11(4):788–798
5. Islam A, Hasan M (2012) A technique to mitigate impact of process, voltage and temperature variations on design metrics of SRAM Cell. Microelectron Reliab 52(2):405–411
6. Chua L (1971) Memristor-the missing circuit element. IEEE Trans Circuit Theory 18(5):507–519
7. Tetzlaff R, Schmidt T (2012) Memristors and memristive circuits—an overview. In: Proceedings of the IEEE international symposium on circuits and systems (ISCAS). IEEE, Seoul, pp 1590–1595
8. Mazady A, Anwar M (2014) Memristor: part II–DC, transient, and RF analysis. IEEE Trans Electron Devices 61(4):1062–1070
9. Osada K, Saitoh Y, Ibe E, Ishibashi K (2003) A 16.7-fA/cell tunnel leakage-suppressed 16-Mb SRAM for handling cosmic-ray-induced multi errors. IEEE J Solid-State Circuits 38 (11):1952–1957
10. Yamaoka FM, Shinozaki Y, Maeda N, Shimazaki Y, Kato K, Shimada S, Yanagisawa K, Osada K (2004) A 300-MHz, 25 μA/Mb leakage on-chip SRAM module featuring process-variation immunity and low-leakage-active mode for mobile-phone application processor. In: Proceedings of the IEEE international solid-state circuits conference digest of technical papers. 1, 494–542
11. Kuroda T, Sakurai T (1995) Overview of low-power ULSI circuit techniques. IEICE Trans Electron 78(4):334–344
12. Madiwalar B, Kariyappa BS (2013) Single bit-line 7T SRAM cell for low power and high SNM. In: Proceedings of the IEEE, international multi-conference on automation, computing, communication, control and compressed sensing (imac4 s). IEEE, Kottayam, pp 223–228
13. Grossar E, Stucchi M, Maex K, Dehaene W (2006) Read stability and write-ability analysis of SRAM cells for nanometer technologies. IEEE J Solid-State Circuits 41(11):2577–2588
14. Eshraghian K, Kavehei O, Rok Cho K, Chappell JM, Iqbal A, Al-Sarawi SF, Abbott D (2012) Memristive device fundamentals and modeling: applications to circuits and systems simulation. Proc IEEE 100(6):1997–2001
15. Powell M, Yang SH, Falsafi B, Roy K, Vijaykumar TN (2000) Gated-V_{dd}: a circuit technique to reduce leakage in deep-submicron cache memories. In: Proceedings of the international symposium on low power electronics and design (ISLPED). pp 90–95
16. Akashe S, Sharma S (2013) Leakage current reduction techniques for 7T SRAM cell in 45 nm technology. Wireless Pers Commun 71(1):123–136
17. Upadhyay P, Ghosh S, Mandal D, Ghoshal SP (2014) Low static and dynamic power MTCMOS based 12T SRAM cell for high speed memory system. In: Proceedings of the IEEE international joint conference on computer science and software engineering (JCSSE). Chon Buri, pp 212–217
18. Sakurai T, Newton AR (1990) Alpha-power law MOSFET model and its applications to CMOS inverter delay and other formulas. IEEE J Solid-State Circuits 25(2):584–594
19. Pedram M (1997) Low power design in deep submicron technology. Kluwer Academic Publishers, Netherlands, pp 179–201

20. Sarwar S, Saqueb SAN, Quaiyum F, Harun-Ur Rashid ABM (2013) Memristor-based non-volatile random access memory: hybrid architecture for low power compact memory design. IEEE Access 1:29–34
21. Chang MF, Chuang CH, Chen MP, Chen LF, Yamauchi H, Chiu PF, Sheu SS (2012) Endurance-aware circuit designs of nonvolatile logic and nonvolatile SRAM using resistive memory (memristor) device. In: Proceedings of the IEEE 17th Asia and South Pacific design automation conference (ASP-DAC). IEEE, Sydney, 329–334

Chapter 32
Modeling and Analysis of FinFET Based Schmitt Trigger with Stability Response and Gain-Bandwidth Product

Pawan Sharma, Saurabh Khandelwal and Shyam Akashe

Abstract Schmitt Trigger (ST) plays a vital role in analog circuit whose tremendous use in today's VLSI design has enlightened the implementation of ultra low power ST design. Power consumption and delay is needed to be minimized for better performance of CMOS analog circuits. FinFET structure greatly contributes to decreases leakage power due to fin configuration. In this paper FinFET based ST is represented which suggest good noise response, gain bandwidth product and slew rate. Our design has obtain considerable reduced leakage power of 0.9 pW with delay of 264 ps for a sine wave of frequency 1 GHz by implementing Adaptive voltage level technique and obtain power dissipation of 1.636 mW at 0.7 V supply. Various performance parameters of the circuit have been evaluated with cadence virtuoso design tool at 45 nm FinFET implemented in independent gate mode (IG). After simulation a comparative analysis is carried with previous published 4T ST at 45 nm along with the proposed design and evaluate results. This work reduces delay and leakage power considerably as FinFET based ST offers delay and leakage power reduction of 37.38 and 75.15 % respectively at 0.7 V power supply. The slew rate of this design is 6.492 V/ps at 229.6 ps is obtained at 0.7 V.

32.1 Introduction

The VLSI system design faces various challenges due to leakage power consumption when CMOS technology is scaled down [1]. In Standby mode by disconnecting the power rail leakage power can became negligible. However, under significant scaling circumstances planar device structures undergoes limitation of short-channel effects (SEC) [2]. This makes the designer to moves towards three-dimensional device structures likes FinFET and MugFET (Tri-Gate) [3]

P. Sharma (✉) · S. Khandelwal · S. Akashe
Department of ECE, ITM Universe, Opposite of Sitholi Railway Station,
NH-75 Jhansi Road, Gwalior, Madhya Pradesh, India
e-mail: sharmapawanec@gmail.com

V. Lakshminarayanan and I. Bhattacharya (eds.), *Advances in Optical Science and Engineering*, Springer Proceedings in Physics 166,
DOI 10.1007/978-81-322-2367-2_32

FinFET device technology has emerged as most promising substitutes for CMOS technology because of its better controlling of Gate, SCE and process variation. Moreover, FinFET still is facing lithographic and work-function engineering tasks [4]. Among the various types of Double Gate (DG) devices, quasi-planar Silicon on insulator (SOI) FinFET is easy to fabricate as compared to planar DG devices. DG devices can efficiently provide less power consumption and better performance because of proper controlling of front and back Gates. Independent Gate (IG) control can be used to combine parallel transistors in non-critical paths. As consequences of combining parallel transistors in non-critical paths there will be a reduction in effective switching characteristics and therefore power dissipation [5]. Static power dissipation have been becoming an restriction in overall reducing power of the design but due to technology scaling dynamic power can be reduced up to a greater extent [6]. This paper presents a FinFET-based Schmitt trigger (ST) circuits to contradict the unpleasant effects of perusing technology. FinFET based ST is described and for further reducing the gate leakage and sub-threshold leakage we have used AVL technique. The remaining paper is organized as follows. Section 32.2 gives the circuit description of the proposed design. Section 32.3 provides an overview of the AVL current reduction techniques. The simulation framework and results is formulated in Sect. 32.4 and finally we conclude the paper.

32.2 Circuit Description of FinFET Based Schmitt Trigger

In this paper we propose a FinFET based ST at 45 nm with finger width of 120 nm and V_{sine} pulse of amplitude 1 V and frequency 1 GHz. The FinFET based ST consists of four FinFET transistors as shown in Fig. 32.1 and the benefit of FinFET is less density chip area with respect to other available technology.

ST is an electronic circuit mainly helps to reduce noise by shaping the input pulses within two threshold voltage i.e. lower threshold voltage and upper threshold voltage. It actually consists of two inverter circuit both of which is driven by

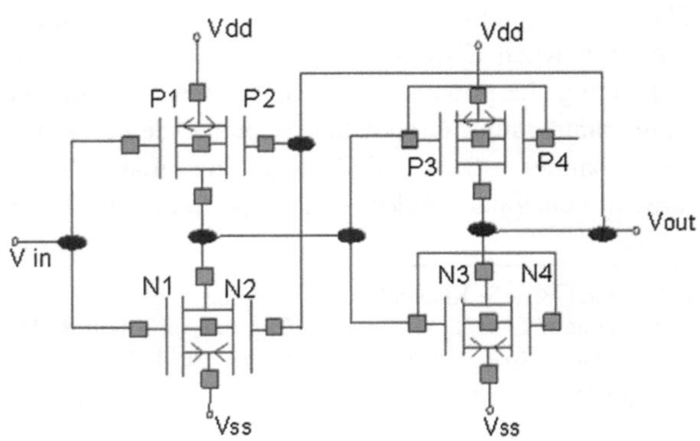

Fig. 32.1 Proposed FinFET based Schmitt trigger

regenerative positive feedback. The switching threshold of the design depends upon the pull-up to pull-down ratio of the transistor as this ratio increases switching threshold of the design will also increases and vice versa [7]. With the help of positive feedback steep slope are formed which in turn reduces leakage power by decreasing the direct current path. The calculation of leakage power is shown in (32.1) of leakage power (P_{leak}) which is a function of I_{Leak} (leakage current)

$$P_{\text{leak}} = V_{\text{dd}} \times I_{\text{Leak}} \tag{32.1}$$

32.3 Leakage Reduction of FinFET Based Schmitt Trigger Using AVL

Adaptive voltage level control circuit [8] can be apply in three ways: (1) To the upper end of the design to decrease supply voltage also known as AVLS Technique (2) To the lower end of the design to increase V_{ss} level voltage which is referred as AVLG Technique and (3) Applying AVLS and AVLG simultaneously in the design. To reduce the leakage current in this design we use both AVLS and AVLG simultaneously also shown in Fig. 32.2.

In AVLG technique switches provides low voltage i.e. V_{ss} during active mode whereas it offers high voltage during inactive mode. Figure 32.2 shows a schematic of FinFET based ST in which AVL technique is applied. The lower transistors provide LOW Voltage at the V_{ss} (ground) node during the active mode, whereas

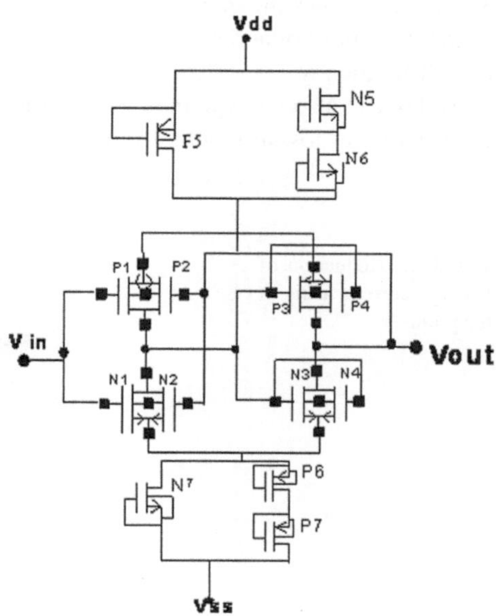

Fig. 32.2 FinFET based Schmitt trigger using AVL technique

increased voltage observed at ground node i.e. virtual ground when the design is in stand-by mode. An increase in the virtual ground voltage (V_{ss}) leads to reduce the Gate-source (V_{gs}) and the Gate-drain (V_{ds}) voltages of the proposed circuit. Hence due to decrease in V_{gs} of the circuit, results in sharp reduction in gate leakage currents from the lowers n-FinFETs.

32.4 Simulation Results

The simulation of all ST circuit was performed using Cadence Virtuoso tool at 45 nm technology. FinFET based ST design is compared with AVL technique and we observe that 52.3 % of leakage reduction can be obtained from AVL technique. FinFET based ST obtained by AVL technique shows leakage current of 1.309 pA at 0.7 V of supply voltage. FinFET has a better controlling of Gates as compared to CMOS technology because of two Gate terminals which provides higher driving current capability during the Active condition when it is connected to V_{dd} whereas it increases the threshold voltage in OFF condition (standby mode).

In the Fig. 32.3. Leakage current in FinFET based ST reduces about 52.1 % at 0.7 V supply after the application of AVL technique and evaluated from 0.6 to 1 V supply voltage range.

Stability response of the FinFET based ST shown in Fig. 32.4 gives the information that it is stable at 1 GHz frequency. The stability response of this design consists of Loop phase gain and loop Gain, at 0.7 V power supply we find that proposed design provides loop gain of −6.3 dB with a phase of −180°. Hence FinFET based ST gives more stable results as compared to conventional ST. Table 32.1 shows the gain-bandwidth product of the design. Gain is determined by output to input voltage ratio and bandwidth is the difference of high cut off to low cut off frequency.

Delay can also be explained as the time required for logic data to propagate from input terminal towards output terminal.

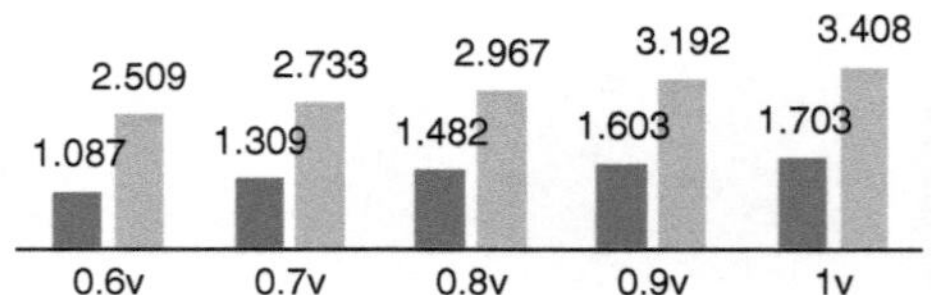

Fig. 32.3 Reduction of leakage current using AVL technique

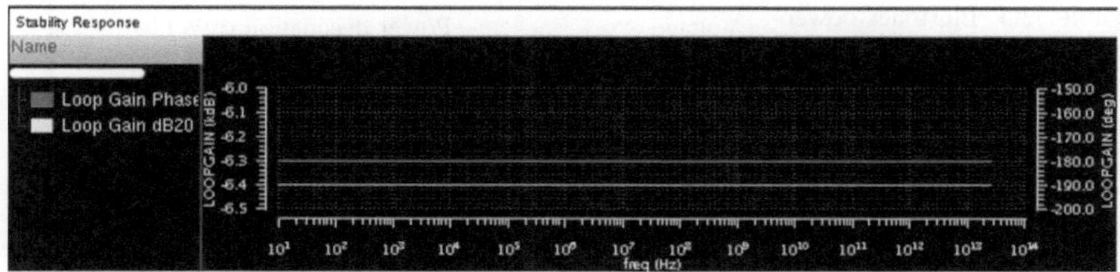

Fig. 32.4 Stability response of FinFET based Schmitt trigger

Table 32.1 Gain-bandwidth product of FinFET based Schmitt trigger

V_{dd} (V)	Gain	Bandwidth	Gain-bandwidth product
0.6	2.293×10^{-8}	371.5×10^{-12}	8.518×10^{-18}
0.7	6.397×10^{-9}	372.5×10^{-12}	2.383×10^{-18}
0.8	6.482×10^{-9}	361.6×10^{-12}	2.344×10^{-18}
0.9	1.862×10^{-8}	349.5×10^{-12}	6.508×10^{-18}

$$\text{Delay} = 0.69R_{\text{eq}} \times \text{CL} = \frac{\text{tphl} + \text{tplh}}{2} \tag{32.2}$$

where, tphl is propagation time for HIGH to LOW transition, tplh is propagation time for LOW to HIGH transition, R_{eq} is the equivalent resistance and CL is the load capacitance. However the power consumed by the circuit during the standby mode is nothing but leakage power consumption. Transistor's leakage current is also known as transistor's off-state current which is the current that is generated during the off-state of the design. As higher the value of load capacitance and resistance delay can reach up to few nanoseconds, therefore we obtained the distorted and delayed output waveform of the design [9].

Table 32.2 represents the variation of Delay and leakage power in three different types of design for ST and the total number of transistors used during single mode of operation. Total Power dissipation of the design is the summation of the static

Table 32.2 Simulation result in terms of delay, leakage power and their comparative analysis

Design name	Area (number of transistors)	Delay (ps)				Leakage power (pA)			
		Supply voltage (V_{dd})							
		0.6 V	0.7 V	0.8 V	0.9 V	0.6 V	0.7 V	0.8 V	0.9 V
4T Schmitt trigger [9]	4	430.7	405.5	379.7	353.2	1.67	7.7	17.28	15.97
FinFET based ST	4	258	264.0	261.1	297.4	2.509	2.733	2.967	3.192
AVL technique FinFET based ST	11	–	–	–	–	0.6522	0.9163	1.1856	1.4427

Table 32.3 Illustrates power dissipation of the FinFET based Schmitt trigger

Voltage	Power dissipation (mW)
0.6	0.989
0.7	1.636
0.8	2.599
0.9	4.065

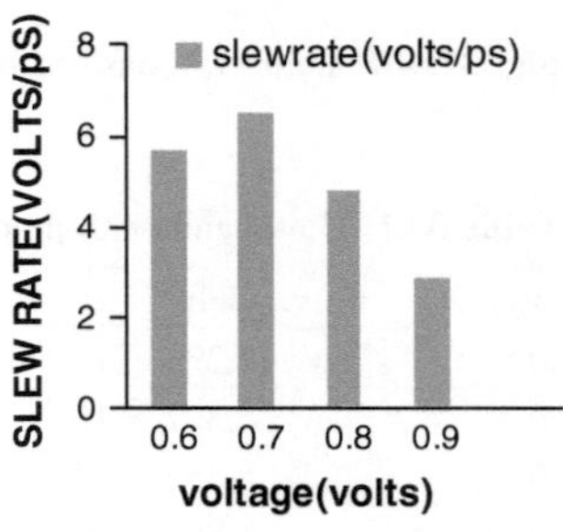

Fig. 32.5 Slew rate graph of FinFET based Schmitt trigger

and dynamic power dissipation as shown in 32.3. Table 32.3 gives information about total power dissipation and we observe that as the voltage of the design increase power dissipation increases.

$$\mathrm{PD} = I_{\mathrm{sc}} * V_{\mathrm{dd}} + \alpha C_{\mathrm{L}} \cdot V_{\mathrm{dd}}^{2} f \tag{32.3}$$

The Rate at which change in output voltage per unit time become maximum, we generally called it Slew rate. Equation (32.4) explains the mathematical expression of slew rate.

$$\text{Slew rate} = \frac{\mathrm{d}v_{\mathrm{o}}}{\mathrm{dt}}(\max) = \omega \cdot V_{\mathrm{m}} = 2\pi f_{\mathrm{m}} \cdot V_{\mathrm{m}} \tag{32.4}$$

where, $\frac{\mathrm{d}v_{\mathrm{o}}}{\mathrm{dt}}$ is change in output voltage per unit time, ω is the input frequency and it is equal to $2\pi f_{\mathrm{m}}$ and V_{m} is the maximum amplitude of the input sine pulse. The slew rate calculated is shown in Fig. 32.5.

32.5 Conclusion

The simulation is performed on Cadence Virtuoso tool using gpdk045 analog library and important parameter while configuring the FinFET based ST is calculated like delay (264 ps), leakage power (2.733 pW), power dissipation (1.636 mW), stability response, Slew rate (6.49 V/ps) and gain bandwidth product (2.383E-15) on nanometer range at 0.7 V supply. The AVL technique applied on FinFET based ST circuit decreases further the leakage current by AVL switches. ST widely finds its application in communication circuit like voltage controlled oscillators, memory elements like SRAM, Flip-flop etc. by the application of AVL technique to FinFET based ST offers minimum Leakage current of 1.087 pA. With the help of AVL technique there is 52.3 % leakage power reduction from FinFET based ST.

Acknowledgments We would like to thanks ITM University and cadence design system, Bangalore for their supports to our work.

References

1. Sakurai T (2004) Low power digital circuit design. In: Proceedings of 30th ESSCIRC 1356606, pp 11–18
2. Nowak EJ, Aller I, Ludwig T, Keunwoo K, Joshi RV, Ching-Te C, Bernstein K, Puri R (2004) Turning silicon on its edge. IEEE Circuits Devices Mag 20:20–31
3. Bohr M (2004) The new era of scaling in an SoC world. In: Proceedings of ISSCC 4977293, pp 23–28
4. Tang A, Yang Y, Lee C-Y, Jha NK (2014) McPAT-PVT: delay and power modeling framework for FinFET processor architectures under PVT variations. IEEE Trans 1
5. Roy K, Mahmoodi H, Mukhopadhyay S, Ananthan H, Bansal A, Cakici T (2006) Double-gate SOI devices for low-power and high-performance applications. In: Proceedings held jointly with 5th international conference on embedded systems and design, 19th international conference on VLSI design, p 1
6. Cotter M, Liu H, Datta S, Narayanan V (2013) Evaluation of tunnel FET-based flip-flop designs for low power, high performance applications. In: Proceedings of 14th international symposium on quality electronic design 6523647, pp 430–437
7. Rabaey JM (1996) Digital integrated circuits (a design perspective), Prentice-Hall, Englewood Cliffs, New Jersey
8. Saini S, Veeramachaneni S, Kumar AM, Srinivas MB (2009) Schmitt trigger as an alternative to buffer insertion for delay and power reduction in VLSI interconnects. In: TENCON IEEE region 10 conference, pp 1–5
9. Saxena A, Shrivastava A, Akashe S (2014) Design and performance evaluation of Schmitt trigger for nano scale CMOS. Quantum Matter 3:1–4

Chapter 33
Comparison of 6T and 8T SRAM Cell with Parameters at 45 nm Technology

Joshika Sharma, Saurabh Khandelwal and Shyam Akashe

Abstract Static random access memory (SRAM) plays a most significant role in the microprocessor world, but as the technology is scaled down in nanometers, leakage current, leakage power and delay are the most common problems for SRAM cell which is basically designed in low power application. In this paper we compares the performance, working and simulation results of two different SRAM cell methods i.e. Conventional six transistor SRAM cell and proposed eight transistor SRAM cell Designing of 8T SRAM cell is done due to high speed operation. By simulating and performing operations we confirmed that our proposed conventional 8T SRAM cell has amend their parameters. During write operation of 8T SRAM cell gives leakage current is 69 pA, leakage power is 7.581 nW and delay is 20.55 ns and for read operation of leakage current is 53.90 pA, leakage power is 1.709 μW and delay is 21.44 ns and SNM of 8T SRAM Cell has greater stability by 29 % as compared to 6T SRAM.

33.1 Introduction

Static random access memory (SRAM) has its own applications mainly in the various types of portable devices. As the size is reduced the effect of leakage current, leakage power is increased in the circuit [1]. As we know that the number of transistor is maximum so that leakage current of an SRAM cell is high as it dominates in stand-by mode which is in direct relation to the number of transistor [2]. Due to scaling perform on devices a different design challenge arises for the nanometer design of SRAM memory [3]. 6T SRAM cells results to a low memory density compared with the DRAM cells design. Hence in conventional SRAM cell which uses 6T SRAM cell faces problems to meet the demand of large memory area in mobile application [4].

J. Sharma (✉) · S. Khandelwal · S. Akashe
ITM, Opposite Sithouli Station, NH-75, Gwalior, Madhya Pradesh, India
e-mail: jsharma791@gmail.com

V. Lakshminarayanan and I. Bhattacharya (eds.), *Advances in Optical Science and Engineering*, Springer Proceedings in Physics 166,
DOI 10.1007/978-81-322-2367-2_33

33.1.1 6T SRAM Cell

In this case, we have minimizing the loading effect through PMOS transistor. In this SRAM cell design, we have used two inverter pair resulting two NMOS and 2 PMOS transistor are called active transistor and plus two additional NMOS transistor are connected in the row line. This architecture is known as 6T SRAM cell. This additional two transistor are used to retrieve data on bit lines i.e. BL (bit line) and BLB (bit line bar). Both bit line and word line (WL) are used for write and read operation. The main drawback of 6T SRAM cell is its huge size (Fig. 33.1).

33.1.1.1 Write Operation/Read Operation

When we performing a write operation, both the bit lines are at opposite voltages which represent if bit line BL is at high then BLB is at low and vice versa (BL = 1 and BLB = 0 or BL = 0 and BLB = 1). When WL enables transistors M5 and M6 then data writes on the output nodes Vout and Vout1 of back to back connected inverter. When we perform the read operation which is just opposite to the write operation, both the bit lines are at high voltages also behave as an output and WL is raised to high. Since one of the output nodes (Vout and Vout1) is at low then one of pre-charged bit lines start discharging and at that instant data is going to be read.

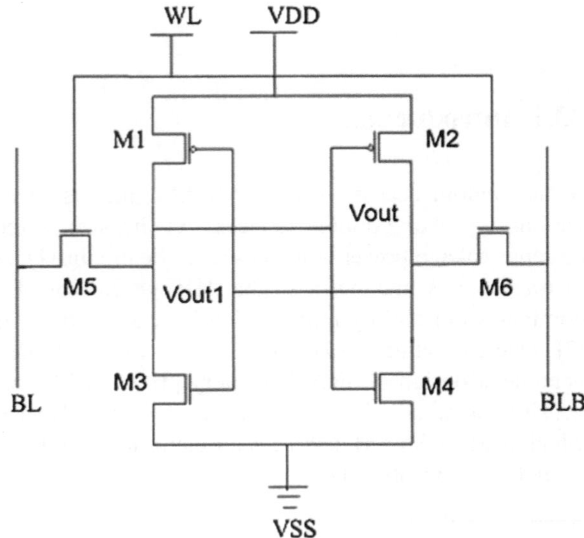

Fig. 33.1 Conventional 6T SRAM cell

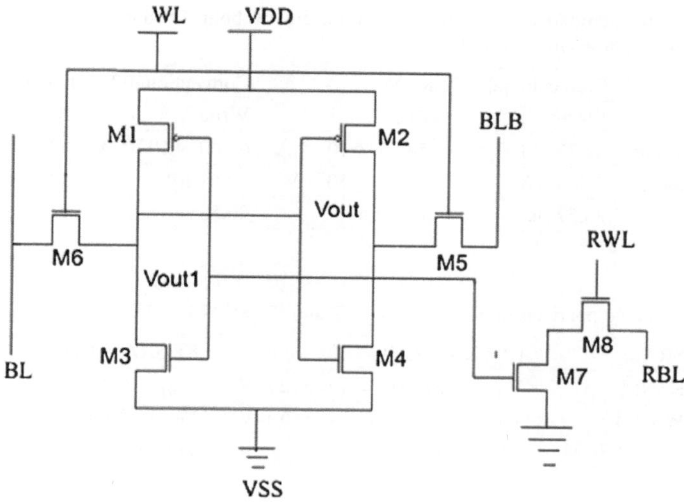

Fig. 33.2 Conventional 8T SRAM cell

33.1.2 8T SRAM Cell

When we designed the 6T SRAM cell so many problems are occurring due to continuous scaling of the technology. We add two more transistors for the low power design in 6T SRAM cell. In this cell we add two more transistor in 6T SRAM cell to access read bit line (RBL). Two transistors M7 and M8 are implemented to reduce leakage current.

Figure 33.2 shows the design of the 8T SRAM cell in which individual read word line (RWL) and RBL are used.

33.2 Perfomance Analysis and Simulation Result

For SRAM cell, the leakage current is the main basis of standby power consumption whose major components are the sub-threshold leakage, the reverse biased band-to-band tunneling junction leakage and the gate direct tunneling leakage in nano-scale devices. To define leakage power, whenever the CMOS inverter is in stable mode, it has its PMOS and NMOS transistor shot off (Table 33.1).

33.2.1 Static Noise Margin

To flip from one state to another in an SRAM cell Static Noise Margin may be defined as the minimum DC noise voltage by the cell. The Stability of the Cell is

Table 33.1 The comparison of different parameters of both 6T and 8T SRAM at 45 nm technology in cadence virtuoso tool

Parameters	Conventional 6T SRAM		Conventional 8T SRAM	
	Write	Read	Write	Read
Leakage current	69.22×10^{-12} A	54.88×10^{-12} A	69.00×10^{-12} A	53.90×10^{-12} A
Leakage power	7.346 nW	1.710×10^{-6} W	7.561 nW	1.709×10^{-6} W
Delay	20.57 ns	21.70 ns	20.55 ns	21.44 ns

Table 33.2 The comparison of SNM of both 6T and 8T SRAM at 45 nm technology

S. no.	SRAM	6T SRAM Cell	8T SRAM Cell
1.	RSNM	V_{OH} = 299 mV, V_{IH} = 392 mV	V_{OH} = 330 mV, V_{IH} = 375 mV
2.	WSNM	V_{IL} = 385 mV, V_{OL} = 315 mV	V_{IL} = 320 mV, V_{OL} = 390 mV
3.	Overall SNM	83.21 mV	117 mV

measured generally by the SNM. SNM of the SRAM cell depends on the supply voltage, pull up ratio (PR) and cell ratio (CR). CR is considered as the ratio of the driven transistor to the access transistor sizes. Pull-up ratio is defined as the ratio between sizes of the access transistor to the load transistor during write operation (Table 33.2).

This illustrates that SNM of 8T SRAM Cell is less than the 6T SRAM Cell. Therefore, 8T SRAM Cell is more stable than 6T SRAM cell.

33.3 Conclusion

For a high density and low leakage current, we propose a conventional 8T SRAM cell in which we perform both read and write operation and compare the result of both write and read operation with the 6T SRAM cell operation. By comparing we can say that, 8T SRAM cell possess low leakage current as compare to 6T SRAM cell and delay is also reduced in both read and write operation, but leakage power is increased in 8T SRAM cell than 6T SRAM cell. These designs also improve the read and write stability.

Refrences

1. Mishra K, Akashe S (2012) Estimation of leakage components and static noise margin for 7T static random-access memory cell. In: Proceedings of ACCT, pp 361–363
2. Jain SK, Agarwal P (2006) A low leakage and SNM free SRAM cell design in deep sub micron CMOS technology. In: Proceedings of 19th international conference on embedded systems and design

3. Athe P, Dasgupta S (2009) A comparative study of 6T, 8T and 9T SRAM cell. In: Proceedings of ISIEA, pp 889–894
4. Akashe S, Bhushan S, Sharma S (2011) High density and low leakage current based SRAM cell using 45 nm technology. In: Proceedings of ICONSET, pp 346–350, (2011)

Chapter 34
Estimation of High Performance 3T DRAM Cell at Nanometer Technology

Priyanka Kushwah, Nikhil Saxena, Saurabh Khandelwal and Shyam Akashe

Abstract In this paper the analysis of conventional DRAM logic compatible 3T gain cell has been shown. In this paper 3T dram with semantic design technique is presented. The read and write operation for single bit storage is useful in terms of leakage power, static power dissipation, signal to noise ratio and delay time. The simulation result shows that when a wide range of operating voltage is taken, which is from 0.7 to 1.3 V then it is observed that low voltage operation is suitable for low read access time but the leakage power dissipation increases as increase in the range of operating voltage. The design has been carried out at the 45 nm scale technology on cadence virtuoso simulating tool.

34.1 Introduction

In recent CMOS scaling has offered improvement in performance of memory as technology moves from one node to the next. In digital logic system design semiconductor memories are most vital microelectronics component such as microprocessor. To store a large amount of digital information semiconductor memory arrays are required. The amount of memory is always depends on the application for which it is used [1]. The continually increasing demands for huge memory capacity has obsessed the fabrication technology and design rules more compact and dense, this results in higher storage densities for data. On chip memory arrays are broadly used subsystem in modern VLSI based circuits, commercially used single chip memory has the capacity about 1 GB. This trend toward larger storage capacity and higher density will prolong to increase the leading edge of digital system design [1, 2]. We proposed conventional dynamic random access memory using 3 transistor cells with structural support to enable high performance designs, which is tolerant to process variations for future generation of high

P. Kushwah (✉) · N. Saxena · S. Khandelwal · S. Akashe
ITM, Opposite Sitholi station NH-75, Gwalior, Madhya Pradesh, India
e-mail: priyankakushwah03@gmail.com

V. Lakshminarayanan and I. Bhattacharya (eds.), *Advances in Optical Science and Engineering*, Springer Proceedings in Physics 166,
DOI 10.1007/978-81-322-2367-2_34

performance microprocessors. This paper describes the parameters variation in accordance with the operating voltage in nanotechnology [3].

34.1.1 3T DRAM Cell

In this section to aid the understanding of our proposed technique, we describe the basic operation of 3T conventional gain cell. Figure 34.1 shows the schematic diagram of conventional 3T DRAM cell. There are three transistors M_1, M_2, M_3 are connected, only one single transistor NM_2 utilise as the working transistor to store the information data. The other two transistor NM_1 and NM_3 are used as access switches, according to write and read operation respectively [4].

34.1.2 Read/Write Operation of 3T Dram Cell

In the proposed schematic of 3T DRAM the switching transistor is NM_1 and NM_3 connected with the bit line and bit bar line respectively, To perform the "write" operation, the write word line is enabled 1 or high therefore the transistor NM_1 is turned "on" then the voltage of the bit line is passed on to the storage transistor NM_2, data is stored in the junction capacitance of the transistor and the voltage of the bit line bar is discharged to the ground and hence write operation is performed [4, 5]. During the read operation read word line is enabled or high, the gate voltage of storage transistor is high and the transistor NM_2 and NM_3 both are turned on. A minor voltage difference generated in the bit line bar depends on the voltage stored in the cell [5, 6].

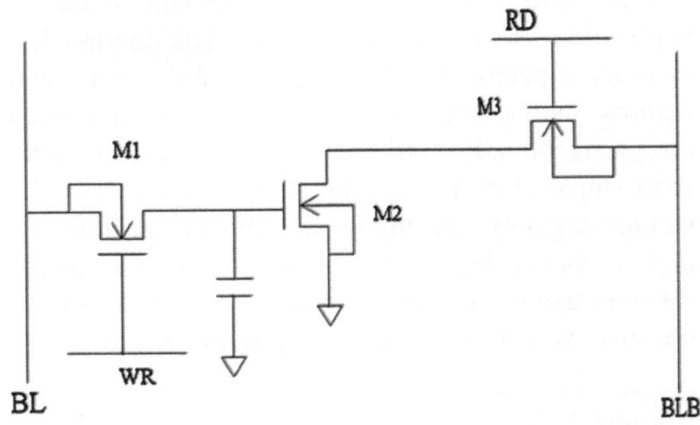

Fig. 34.1 Conventional 3T DRAM cell

34.2 Performance Analysis and Simulation Result

34.2.1 Leakage Power

See Fig. 34.2

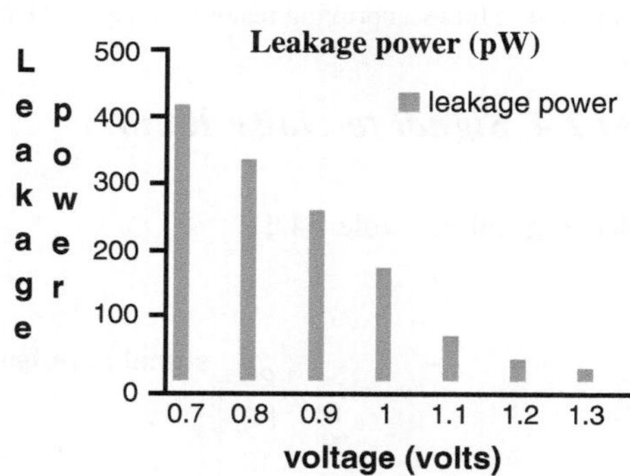

Fig. 34.2 Shows improving result of leakage power in 3T conventional DRAM cell during read operation

34.2.2 Static Power Consumption

See Fig. 34.3

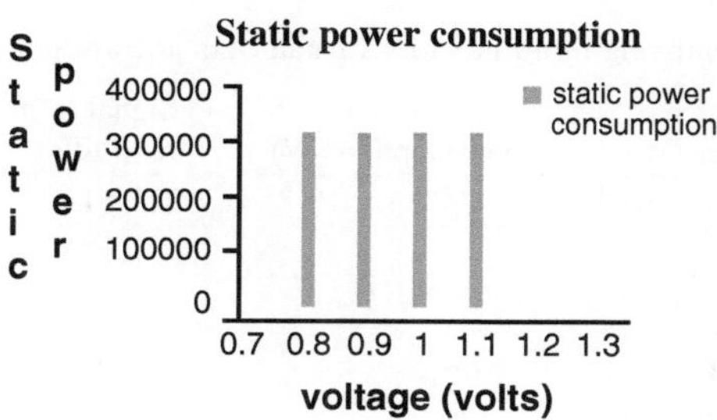

Fig. 34.3 Static power consumption in 3T conventional DRAM cell during read operation

34.2.3 Delay

See Fig. 34.4

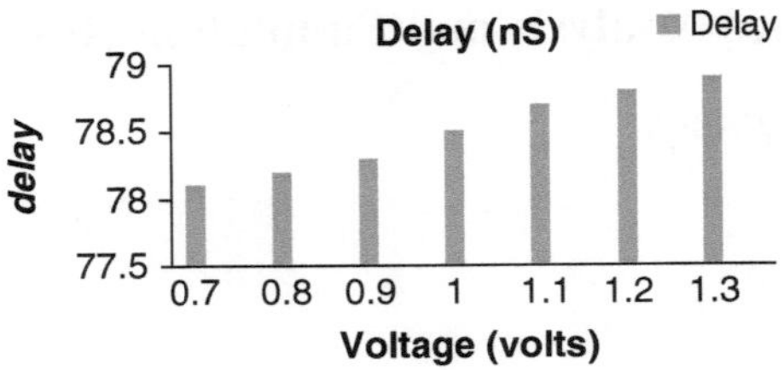

Fig. 34.4 Shows improving result of delay in 3T conventional DRAM cell during read operation

34.2.4 Signal to Noise Ratio:

See Fig. 34.5, Table 34.1

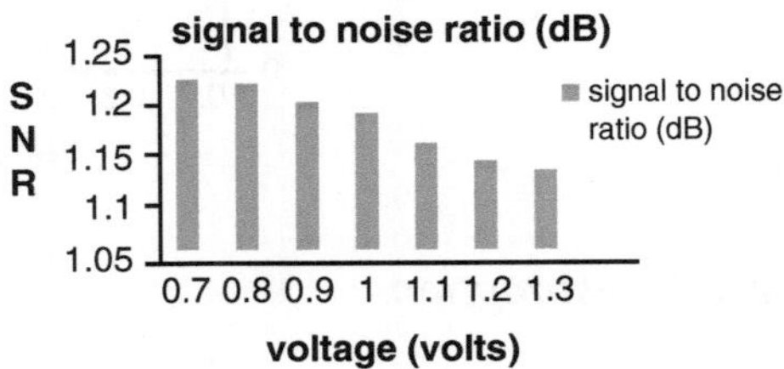

Fig. 34.5 Shows improving result of signal to noise ratio in 3T conventional DRAM cell during read operation

Table 34.1 Various improving result in various parameters at corresponding operating voltage

Operating voltage (V)	Leakage power (W)	Static power consumption (W)	Signal to noise ratio (dB)	Delay (s^{-1})
0.7	416.8×10^{-12}	0.004733×10^{-6}	1.2251	78.60×10^{-9}

34.3 Conclusion

We have proposed a conventional 3T DRAM cell in which we perform read/write operations at various operating supply voltage. In this paper it is observed that signal to noise ratio parameters are increased with increase in voltage, whereas delay, leakage current, leakage power is decreased with increase in voltage. It can be calculated that high performance and less power consumption as increase in operating voltage at 45 nm technology in Cadence virtuoso tool.

References

1. Gerik CM, Turi MA, Delgado-Frias JG (2012) FinFET 3T and 3T1D dynamic RAM cell. In: Proceedings of IEEE 55th International Midwest symposium, 454–457
2. Akashe S, Mudgal A, Singh SB (2012) Analysis of power in 3T DRAM and 4T DRAM cell design for different technology. Information and communication technology (WICT) world congress, 18–21
3. Singh L, Somkuwar A (2013) Dynamic random access memory with self-controllable voltage level to reduce low leakage current in VLSI. Int J Eng Res Appl (IJERA) 3:2248–9622
4. Singh L, Somkuwar A (2013) 4T DRAM based on self- controllable voltage level technique for low leakage power in VLSI. Int J Emerg Technol Comput Appl Sci (IJETCAS) 3(3):233–237
5. Borah M, Irwin MJ, Owens RM (1995) Minimizing power consumption of static CMOS circuits by transistor sizing and input reordering. In: Proceedings of the 8th international conference, 294–298
6. Kim NS, Austin T, Baauw D, Mudge T, Flautner K, Hu JS, Irwin MJ, Kandemir M, Narayanan V (2003) Leakage current: moore's law meets static power. IEEE Comput Soc 36(12):68–75 (Papers)

Chapter 35
Enactment of FinFET Based SRAM with Low Power, Noise and Data Retention at 45 nm Technology

Varun Sable and Shyam Akashe

Abstract In these days, FinFET is getting more preference than CMOS, as it is the most promising transistor known for their high short channel effects controllability and flexible threshold voltage (V_{th}) through double gate. The major challenge in this era is chip designing in Low Power with scaling of Integrated circuits (IC's). The thinner W_{fin} FinFET show less degradation in performance than thicker W_{fin}. In VLSI, the Area, Power and Delay are major key factors to improve circuit functionality, if any one of that can be reduced then the circuit performance can be enhanced. The types of memories offer Data Retention Voltage (V_{dr}) variables in FinFET due to different threshold voltage compare to CMOS based memories. This paper investigates implementation of SRAM cell using FinFET to focus on reducing any of the key factors of memory i.e. power dissipation, noise voltage and data retention voltage and also various simulations were carried out between conventional and FinFET based 6T, 7T and 8T SRAM cells. In our proposed FinFET based SRAM cell we get 15–20 % less power dissipation, 6–8 % reduction in data retention voltage and noise voltage reduced up to 30–40 % from conventional SRAM cell, this designing has been done using Cadence Virtuoso tool at 45 nm technology.

35.1 Introduction

In VLSI the on chip cell size are growing with increasing densities of transistors due to which thermal agitation generates power dissipation. This factor affects memory and day-by-day it also increases with generation, demands in various fields and speed with high performance is increasing [1]. Feature size (W/L) but wire size are also scaling down continuously enhancing process technology [2]. After Bi-polar junction transistor (BJT), Metal oxide semiconductor field effect transistor

V. Sable (✉) · S. Akashe
ITM University, NH-75 Jhansi Road, Gwalior 475001, Madhya Pradesh, India
e-mail: varunksable@gmail.com

V. Lakshminarayanan and I. Bhattacharya (eds.), *Advances in Optical Science and Engineering*, Springer Proceedings in Physics 166,
DOI 10.1007/978-81-322-2367-2_35

(MOSFET) turns the whole VLSI into sparkling path with scaling down the equipment. Multi-gate MOSFET (MuGFET) that was proposed by Sekigawa and Hayashi in 1984 [3]. Then the first DELTA (fully Depleted Lean channel TrAnsistor) was proposed by D Hisamoto in 89s [4]. That comes on chip in 1990s by UC Berkeley team coordinated by Dr. Chenming Hu who proposed FinFET, which reduces short channel effect, leakage current, power dissipation. A powerful concept of natural length:

$$\lambda = \sqrt{\frac{\varepsilon_{\mathrm{si}}}{2\varepsilon_{\mathrm{ox}}} t_{\mathrm{si}} t_{\mathrm{ox}}} \tag{35.1}$$

where λ the short-channel effect, t_{si} is device body thickness, t_{ox} is gate oxide thickness [5]. The side walls of the fins in FinFET known as Channel. The projected percentage of SRAM cell in Integrated Circuit (IC) by International Technology Roadmap for Semiconductor (ITRS) has increased from 84 to 94 % by the year 2014 [6, 7]. The SRAM cell is used to store single bit performance consisting of number of transistors in 45 nm technology. SRAM Cell is type of memory which uses cross coupled inverters (or bi-stable latching) circuitry [8]. The average power dissipation derive and it represented as:

$$P = CV^2 f \tag{35.2}$$

C is parasitic capacitance, *V* is supply voltage, and *f* is frequency. The FinFET has Fin shaped Source Drain in a single strip that is covered by the Gate, which works as either independent gate or with the same signal applied to it.

Another parameter that we examine in this paper is data retention voltage (V_{dr}) that is defined as the lowest supply voltage at which the data can be retained inside the SRAM Cell. And after that noise voltage level studied. Noise is unwanted signal it is generated by wire resistance and thermal agitations.

The RMS noise voltage is:

$$V_n = \sqrt{4K_B T \Delta f R} \tag{35.3}$$

K_B is Boltzmann constant i.e. $1.3806504 \times 10^{-23}$ J/K, Δf is bandwidth being considered (0.1 Hz to 100 GHz), *R* is resistance of circuit element.

35.2 Conventional SRAM Cells

6T SRAM have total of six transistors in which four are NMOS and two are PMOS. Out of six, four transistors i.e. NMOS and PMOS transistors combines to form a pair of inverters in bi-stable latched format [9] and the other two NMOS transistors (N3 and N4) are used to pass bit line data in bi-stable latched [10] that shown in Fig. 35.1.

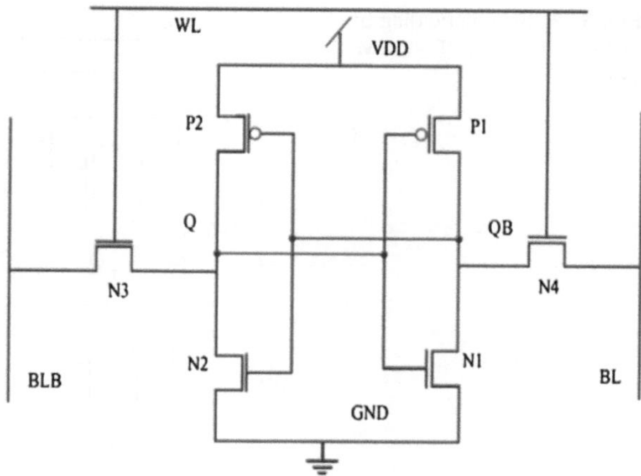

Fig. 35.1 Schematic diagram of conventional 6T SRAM cell

In 6T SRAM Cell there are two bit lines BL and BLB which are complement of each other but of same period and Q and QB are also complemented outputs and a single word line WL. There are three modes of operation in SRAM cell i.e. Hold mode, active mode, stand-by mode.

In stand-by mode WL goes high to low (WL = 0) and the bit line either 0 or 1 [11], it just holds the data inside the cross coupled inverters. In write mode, word line goes low to high (WL = 1) and new data is put into BL bit line, that data writes on Q and QB outputs using pass transistors. When word line goes high to low (WL = 0) and the data on bit line (BL) is precharged or left floating, the value stored during write operation at output Q goes through pass transistor to one of bit line that is discharging and another one line is precharging and this charging and discharging of data is sense by sense amplifier during read operation and amplifies the data which is used further at outputs Q and QB.

35.3 Proposed Finfet Based SRAM Cell

In previous era of VLSI, memory design was very simple just to hold single bit data but now in present scenario, the memory that we are using in our application purpose is the array of memory due to that noise immunity, data retention and power dissipation in density packed Integrated Circuits (IC's) are major issues, so future requires technology which has scaling capability itself to reduce size because of that FinFET becomes point of line for this issue [12]. For better reduction in short channel effects, power supply scaling capability, low power dissipation, FinFET is superior choice than conventional CMOS [13]. The MuGFET based emerging devices helps to reduce cell stability issues and leakage parameters in SRAM cells and E-memories. The proposed model for FinFET based SRAM cell [14, 15] schematic shown in Fig. 35.2 in this paper [16].

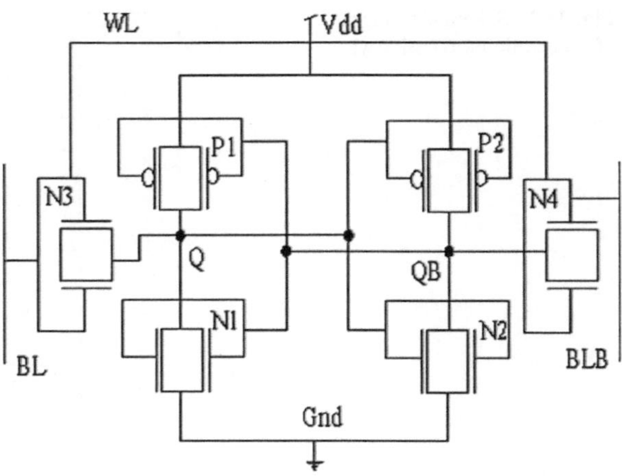

Fig. 35.2 Schematic diagram of FinFET based 6T SRAM cell

35.4 Simulation Results

35.4.1 Self-heating and Power Dissipation

The self-heating phenomena occur due to parasitic [17] present inside the circuit and this affects at local temperature (27 °C). Biasing, routing and loose interconnection affects threshold voltage V_{th} [18, 19] and this is controlled by directly voltage and technology variation. So FinFET based SRAM cell is used for low power consumption and results shown in tabular format as well as in graph Table 35.1 and Fig. 35.3a.

35.4.2 Average Noise

Noise is unwanted signal that distorts the data in SRAM, noise affects mainly when write operation is performed and the data in cross coupled inverter is in hold mode

Table 35.1 Representing average power, Noise Voltage and Data retention voltage of conventional and FinFET based SRAM cell

Mode	Conventional SRAM			FinFET based SRAM		
	6T SRAM	7T SRAM	8T SRAM	6T SRAM	7T SRAM	8T SRAM
Average power	52.32	41.44	77.31	21.687	30.90	70.15
Noise voltage	7.591	9.091	7.56	6.003	5.904	5.962
Transient	0.4	0.42	0.45	0.376	0.379	0.376
DC	0.5	0.5	0.5	0.161	0.269	0.215

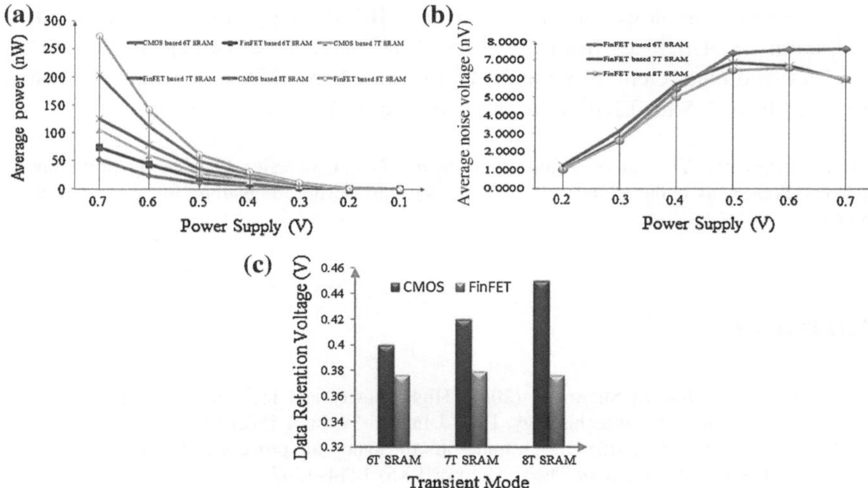

Fig. 35.3 Representation in simple and FinFET based SRAM cell. **a** Power; **b** Noise and **c** Data retention voltage

then the noise can vary the data at the output terminal at Q and QB. The comparative study between conventional and FinFET based SRAM cells are shown in Table 35.1 and in Fig. 35.3b.

35.4.3 Data Retention Voltage

Supply voltage takes down to achieve Data Retention Voltage (V_{dr}) to retain data at Q and QB node high in standby mode [20]. The level of power supply decreases from 0.7 V up till the state of SRAM cell remains same due to that Static Noise Margin (SNM) reduces to zero [21] in Table 35.1 and Fig. 35.3c.

35.5 Conclusion

In this paper, investigation of conventional and FinFET based SRAM cells at 45 nm technology were completed. The main purpose of this paper is to determine Data Retention Voltage (V_{dr}) because FinFET has variable threshold voltage and using that phenomena data retention voltage V_{dr} gives minimum value onto which SRAM Cell can work without any distortion in data that is stored in SRAM Cell. On behalf of this observation it concludes that proposed FinFET based 6T SRAM cell is better as compared to conventional 6T SRAM cell. Average power dissipation and noise voltage have better result in FinFET based compare to conventional based SRAM cell at 45 nm technology. FinFET based 6T SRAM Cell gives reduction in power dissipation i.e. it reduces from 76.65 to 21.687 nW, noise voltage from 7.591 to 6.003 nV and data retention voltage (V_{dr}) from 0.4 to 0.376 V for data in transient form and from 0.5 to 0.161 V in DC response at 0.7 V respectively.

Acknowledgments The author would like to thanks ITM University, Gwalior for providing the Cadence Tool with collaboration of Cadence System Design Bangalore for the work to be completed.

References

1. Akashe S, Bhushan S, Sharma S (2012) High density and low leakage current based 5T SRAM cell using 45 nm technology. Rom J Inf Sci Technol 15(2):155–168
2. Akashe S, Sharma S (2013) Design trade-offs for nanoscale process and material parameters on 7T SRAM cell. J Comput Theor Nanosci 10(5):1244–1247
3. Wu JJ (2010) A large σ V_{TH}/VDD tolerant zigzag 8T SRAM with area-efficient decoupled differential sensing and fast write-back scheme. IEEE symposium on VLSI circuits, 103–104
4. Khandelwal S, Akashe S (2012) Supply voltage minimization techniques for SRAM leakage reduction. J Comput Theor Nanosci 9(8):1044–1048
5. Yeoh YN, Wang B, Yu X, Kim TT (2013) A 0.4V 7T SRAM with write through virtual ground and ultra-fine grain power gating switches. IEEE international symposium on circuits and systems, pp 3030–3033
6. Khandelwal S, Raj B, Gupta RD (2013) Leakage current and dynamic power analysis of FinFET based 7T SRAM at 45 nm technology. Proceeding of the international arab conference on information technology, Khartoum Sudan (2013)
7. Akashe S,Shastri M, Sharma S, (2012) Multi Vt 7T SRAM cell for high speed application at 45 nm technology. International conference on nanoscience engineering and technology, 351–354 (2012)
8. Sikarwar V, Khandelwal S, Akashe S (2013) Optimization of leakage current in SRAM cell using shorted gate DG FinFET. In: IEEE 3rd International Conference On Advanced Computing And Communication Technologies, Rohtak India, 166–170 (2013)
9. Rahman N, Singh BP (2013) Static-noise-margin analysis of conventional 6T SRAM cell at 45 nm technology. Int J Comput Appl 66–20 (2013)
10. Akashe S, Sharma S (2010) Low power SRAM cell design based on 7T configuration. Int Electron Eng Math Soc IEEMS 4:11–18
11. Tawfik SA, Kursun V (2008) Low-power and compact sequential circuits with independent-gate FinFET's. IEEE Trans Electron Devices 55(1):60–70
12. Lin X, Wang Y, Pedram M (2013) Joint sizing and adaptive independent gate control for FinFET circuits operating in multiple voltage regimes using the logical effort method. IEEE international conference on computer aided design, San Jose CA, 444–449
13. Garg A, Kim TH (2013) SRAM array structures for energy efficiency enhancement. IEEE Trans Circ Syst II60(6):351–355
14. Cakici T, Kim K, Roy K FinFET based SRAM design for low standby power applications. IEEE 8th international symposium on quality electronic design, San Jose CA, 127–132
15. Muttreja A, Agarwal, N, Jha, NK CMOS logic design with independent gate FinFETs. 25th International conference on computer design, Lake Tahoe CA, 560–567
16. Ouyang J, Xie Y, Power optimization for FinFET-based circuits using genetic algorithms. IEEE international system-on-chip conference, Newport Beach CA, 211–214
17. Shrivastava AK, Akashe S (2013) Comparative analysis of low power 10T and 14T full adder using double gate MOSFET at 45nm technology. Int J Comput Appl 75(3):48–53

18. Fulde M, Schmitt-Landsiedel D, Knoblinger G (2007) Transient variations in emerging SOI technologies: modeling and impact on analog/mixed-signal circuits. IEEE international symposium on circuits and systems, New Orleans LA, 1249–1252
19. Molzer W, Schulz T, Xiong W, Cleavelin RC, Schruefer K, Marshall A, Matthews K, Sedlmeir J, Siprak D, Knoblinger G, Bertolissi L, Patruno P, Colinge J.-P (2006) Self-heating simulation of multi-gate FETs. IEEE 36th european solid-state device research conference, Montreux 311–314 (2006)
20. Pavlov A, Sachdev M CMOS SRAM circuit design and parametric test in nano-scaled technologies. Intel Corporation, University of Waterloo, Springer Science and Business Media B.V, 40, 1–202 (2008)
21. Khandelwal S, Akashe S (2013) Design of 10T SRAM with sleep transistor for leakage power reduction. J Comput Theor Nanosci 10(1):165–170

Chapter 36
Calculation of Power Delay Product and Energy Delay Product in 4-Bit FinFET Based Priority Encoder

Vishwas Mishra and Shyam Akashe

Abstract Priority encoder converts multiple binary inputs into binary representation of the index of active input bit with the highest priority. It is used where more than one device want to access the system, it decides the priority of the device to be serve by the system. Priority encoders are used when multiple devices have to share common resources. Several researches are made on these encoders but found no research work on FinFET based Priority encoder. The FinFET device has gained very much attention on recent VLSI designs and FinFET is the substitute for bulk CMOS at nano-scale because of its high short channel effect immunity, scalability and lower leakage power consumption. In this paper, a 4 input–3 output priority encoder is implemented using FinFET design.

36.1 Introduction

Nowadays, priority encoders have been widely utilized in high performance applications which persistently impose special design constraints in terms of low power consumption, high-frequency and minimum area. They provide considerable time savings in the area of On-chip testing, optical communications, and computer arithmetic applications [1].Priority Encoders (PE) are commonly used in computer systems. Priority Encoder algorithm are used in number of computing components, such ascomparators [2], fixed and floating point units [3], increments, decrements [4] and interconnection network routers [5], sequential address encoder of content addressable memories [6] are important sub-systems located on-chip or off-chip, which predominantly utilize the priority encoder function. As the computer systems data width gets longer and computer system become fasten, the speed of the Priority encoder becomes a key parameter in the performance of the computer system. At the same time, the overwhelming demand for compact electronics encourages the

V. Mishra (✉) · S. Akashe
ITM University, Gwalior, Madhya Pradesh, India
e-mail: vishwas9617692128@gmail.com

V. Lakshminarayanan and I. Bhattacharya (eds.), *Advances in Optical Science and Engineering*, Springer Proceedings in Physics 166,
DOI 10.1007/978-81-322-2367-2_36

development of a power optimized Priority Encoder. Priority encoder function is to select only one request out of number of request at is input port and then selected input is been served by the system. The proposed priority encoder accepts 4 input request lines and sets only one of the outputs that correspond to the request that has the highest priority. In this paper we simulate 4:2 Priority Encoder using FinFET circuit at different voltages and calculate different parameters of the circuit such as average power, leakage current, leakage power, delay and Power Delay product (PDP), Energy Delay Product (EDP). After comparing all simulation results at different voltage we found that on applying 0.3 V to the circuit we get the better parameter in result.

36.1.1 FinFET

The term FinFET was invented by the University of California, Berkeley researchers to describe a non-planner, double-gate transistor built on an SOI (silicon on insulator) substrate, based on the earlier single gate transistor structure. The distinctive property of the FinFET is that the conducting channel is enfolded by a thin silicon "fin", that is forms the body of the device. The effective channel length of the device is determined by the width of the fin. Due to multi gate structure FinFET has full control over the channel and thus helps in reducing the Shot-Channel effect and also helps to overcome the Sub Threshold Leakage Current problems of MOSFET devices.

The conventional Double Gate FinFET structure is shown in the Fig. 36.1 [7]. The tiny slab of undoped silicone perpendicular to the structure is the channel of the FinFET.

As the gate drain and source of the FinFET is doped with same type of dopant, so there is no p-n junction formation along the channel length and reduces the leakage current. The Higher mobility in the FinFET is due to the undoped channel, which

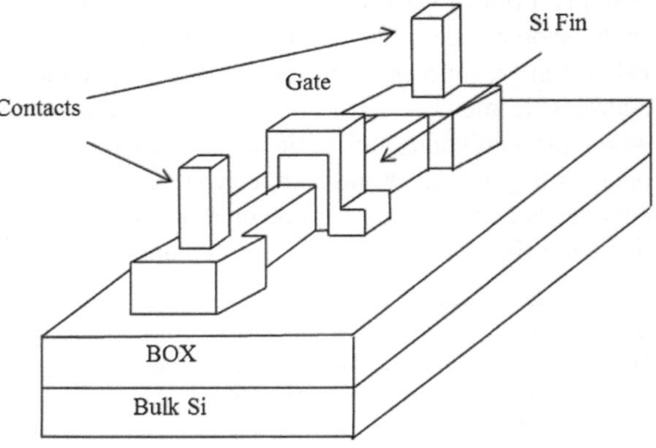

Fig. 36.1 Double gate FinFET

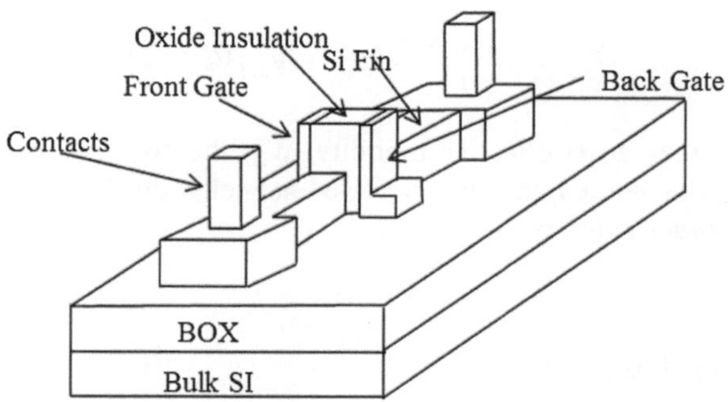

Fig. 36.2 Independent gate FinFET

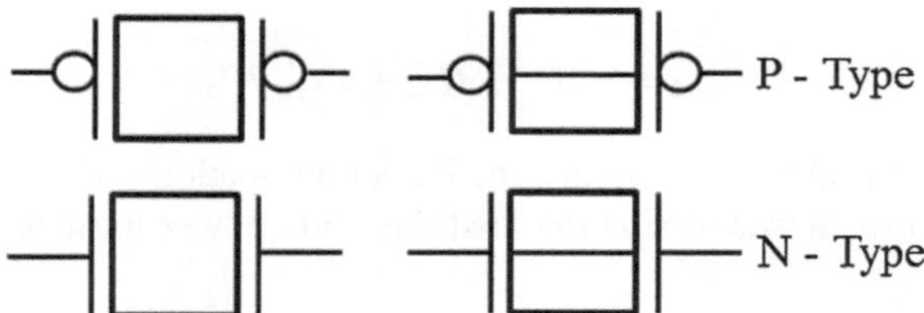

Fig. 36.3 Symbol of independent gate and double gate FinFET

eliminates the coulombs scattering [8]. The fraction of n-type to p-type mobility is higher in FinFET. In FinFET the source-body voltage variation did not affect the threshold voltage. This improvement in mobility and also covers the way for extended series stacked transistors in the pull-down or pull-up networks of logic gates.

Independent Gate FinFET is shown in Fig. 36.2 [9–13] is the double gate device in which the two Gate electrode are not connect together (the top portion of gate is etch out) and biased with different potential. MIGFET has many special features one of them is that the threshold voltage of one of the gates can be modified by the bias applied to the other gate. Here we present Priority Encoder circuit using double gate FinFET and the symbol for double gate FinFET is shown in Fig. 36.3 [14]. Here the priority encoder is simulated on virtuoso tool of cadence at 45 nm technology.

36.2 Parameters of FinFET

36.2.1 Sub Threshold Leakage Current

When the transistor is switches from cut-off region to saturation region ($V_{gs} > V_{th}$) a channel formation is occurs which allows current (I_{ds}) to flow between the drain and source. The current (I_{ds}) from drain to source is given in the below equation

$$I_{ds} = \frac{\mu_n C_{ox} W_{eff}}{L} \left(V_{gs} - V_{th} \right) V_{ds} - \frac{V_{ds}^2}{2} \tag{36.1}$$

with μ_n the charge-carrier effective mobility, W_{eff} the effective channel width and the L effective channel length. The effective channel width W_{eff} equals the channel width in First order approximation [12].

36.2.2 Power Dissipation

Power dissipation is an addition of static power dynamic power and leakage power. But in most of the cases power dissipation depends on dynamic power. Hence it is given by

$$P_{total} = \alpha_{in} I_{sc} V_{dd} f T_{sc} + \alpha V_{DD}^2 f C_{out} \tag{36.2}$$

Equation 36.2 is clearly shows that the Power dissipation is depends on dynamic power and for low power dissipation [16] the static power must be less.

36.2.3 Gate Delay

As the gate capacitance depends on time deviation after and before the change of gate state, gate delay is due to some, time deviation before or after the change of gate state.

$$T_g = \frac{C_g V}{I_{on}} \tag{36.3}$$

36.2.4 Threshold Voltage

Joining the gate work function, difference between the Gate and silicon fin, the increase in band gap and the increase in potential results in the following threshold voltage formula

$$V_{th} = V_{FB} + 2\alpha_B + \delta_w + \frac{q N_b X_{dep}}{C_{ox} \left(1 - \frac{x_n}{L} \right)} \tag{36.4}$$

where V_{FB} work function difference, Φ_B surface potential at threshold voltage, δ_w is a fitting parameter introduced to take into account the changes [15, 17].

36.3 Priority Encoder

Priority as the name suggests means something that is more important than other and should be dealt with first (high/low priority). Priority Encoder does the same thing, whose way of providing output is similar to simple encoder with one advantage of selecting the operation with the priority function. With this facility of priority function, the input having highest priority is served first whenever choice has to make among more than one output. In addition to the two output *x* and *y*, the circuit has the third output denoted as *V*; this is a valid bit indicator that is set to 1 when one or more input are equal to 1, if all the input are 0, there is no valid input and *V* is equal to 0 and other two output are not correct and are denoted as don't care condition, the highest priority is given to the most significant bit (MSB) that is highest priority is given to the highest subscript number of the input, here we have the input D1 D2 D3 and D3 has the highest priority next is D2 and in the last D1 has the lowest priority. When D3 is 1, the output of xy is 11 (binary 3), regardless of the value so the other three lower priority input. When D2 is 1 the output xy is 10, provided that D3 = 0, regardless of the value of the other two input. When D1 is 1 the output xy is 01, provided that D3 D2 = 00 and so on (Fig. 36.4).

As we construct SG FinFET by shorting two P-types MOSFET and two N-type MOSFET to form P-type and N-type FinFET respectively and its logic diagram is shown in the figure No 3. The Priority Encoder is realized using the virtuoso tool of cadence. The spectra simulator of cadence is used to simulate the output. The gate of two PMOS or NMOS transistors are connected together to formed a FinFET like structure. Then FinFET using Priority encoder circuit is simulated by applying different voltages such as 0.3, 0.4, 0.45, 0.5, 0.6 and 0.7 V respectively at 45 nm technology.

36.4 Results

In this section, we present the simulation results of Priority Encoder at 45 nm technology from virtuoso tool of cadence. Figure 36.5 shows the comparison among Energy Delay Product and Power Delay Product curves of DG FinFET at

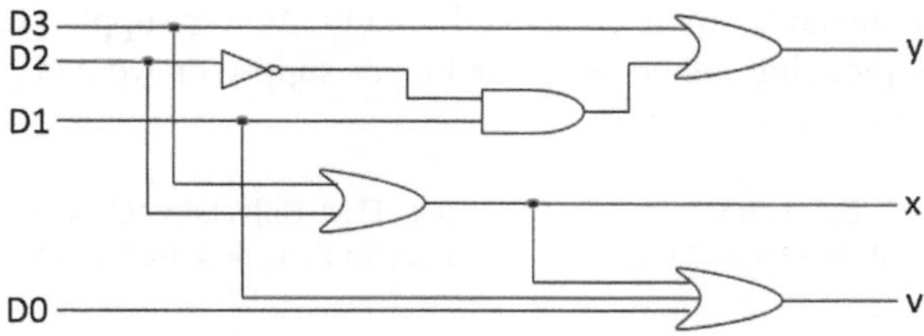

Fig. 36.4 Circuit diagram of priority encoder

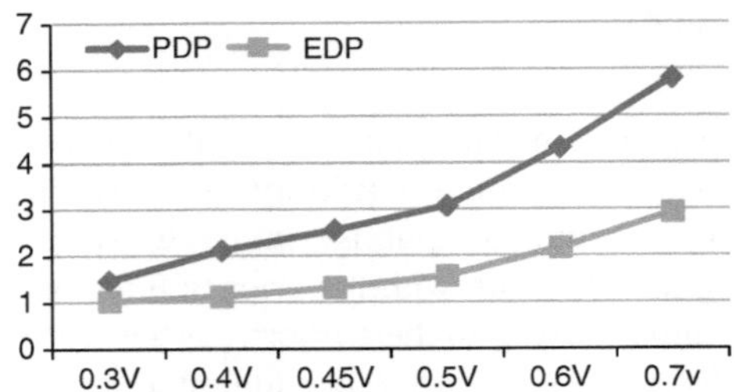

Fig. 36.5 Veriatin of PDP and EDP with respect to voltage

six different power supplies at 45 nm technology respectively. At 45 nm technology the 4:3 Priority encoder give better result at 0.3 V that is Average Power is 2.071 nW, Leakage Power is 54.39 pW rather than at 0.7 V that is Average Power is 11.6 nW, and Leakage Power is 131.1 pW. So, our experimental result gives minimum Delay and Average Power at 0.3 V supply.

36.5 Conclusion

In this paper, we work on Priority Encoder by using FinFET. FinFET is one of the emerging technology used in various applications, has replaced many conventional CMOS based devices. Low power consumption is one of the basic need in forming any electronic device and for that purpose scaling is done which leads to leakage of current but that problem is tackled by using FinFET. In this paper, Average Power, Delay, Leakage Current and Leakage Power of FinFET based Priority Encoder at different voltage was examined and better result at 0.3 V power supply is obtained. The Leakage Power, Average Power and Leakage Current were found to be 2.071 nW, 54.39 pW 13.74 pA respectively. Power Delay Product is 1.4641×10^{-16} J and Energy Delay Product is 1.0249×10^{-22} Js, than the 0.7 V power supply in which the results are Average Power is 11.56 nW, Leakage Power is 131.1 pW and Leakage Current is 26.19 pA, Power Delay Product is 5.8146×10^{-16} J and Energy Delay Product is 2.9247×10^{-22} Js. Here energy-delay product (EDP) is a useful metric for evaluating the quality of a design. As EDP is the product of PDP and Delay, PDP depends on Power supply. EDP has to be low, So EDP depends on the Delay (1/f) and the Power supply (V_{DD}), so we can reduce EDP by reducing either Delay or Power supply or we can get the tradeoff between delay and power supply.

Acknowledgments The author would like to thanks ITM University, Gwalior for providing the cadence tool with collaboration of Cadence system design Bangalore for the work to be completed.

References

1. Summerville DH, Delgado-Frias JG, Vassiliadis S (1996) A flexible bit-pattern associative router for interconnection networks. IEEE Trans Parallel Distrib Syst 7(5):447–485
2. Hennessy JL, Patterson (2002) Computer architecture: a quantitative approach, 3rd edn. Morgan Kaufmann, New York
3. Huang CH, Wang JS (2003) High-performance and power-efficient CMOS comparators. IEEE J Solid-State Circuits 38(2):254–262
4. Huang CH, Wang JS, Huang YC (2002) Design of high performance CMOS priority encoders and incremented/decrements using multilevel look ahead and multilevel folding techniques. IEEE J Solid-State Circuits 37(1):63–76
5. Delgado-Frias JG, Nyathi J, Summerville DH (1998) A programmable dynamic interconnection router with hidden refresh. IEEE Trans Circuits Syst Part I 45(11):1182–1190
6. Kadota H, Miyake J, Nishimichi Y, Kudoh H, Kagawa K (1985) An 8-Kbit content-addressable and reentrant memory. IEEE J Solid-State Circuits 20(5):951–957
7. Sekigawa T, Hayashi Y (1984) Calculated threshold-voltage characteristics of an XMOS transistor having an additional bottom gate. Solid-State Electron 27:827
8. Nuttinck S (2007) Double-gate FinFETs as a CMOS technology downscaling option: an RF perspective. IEEE Trans Electron Devices 54(2):279–283
9. Mathew L, Du Y, Thean AVY, Sadd M, Vandooren A, Parker C, Stephens T, Mora R, Rai R, Zavala M, Sing D, Kalpai S, Hughes J, Shimer R, Jallepalli S, Workman G, White BE, Nguyen BY, Mogab A (2004) Multi gated device architectures advances, advantages and challenges. In: International conference on integrated circuit design and technology, p 97
10. Zhang W, Fossum JG, Mathew L, Yang Du (2005) Physical insights regarding design and performance of independent-gate FinFETs. IEEE Trans Electron Devices 52(10):2198
11. Eminente S, Kyoung-Il N, Cristoloveanu S, Mathew L, Vandooren A (2005) Lateral and vertical coupling effects in MIGFETs. In: Proceedings of the IEEE international SOI conference, p 94
12. Endo K, Liu Y, Masahara M, Matsukawa T, O'uchi S, Suzuki E, Surdeanu A, Witters RL, Doornbos G, Nguyen VH, Van den bosch G, Vrancken C, Devriendt K, Neuilly F, Kunnen E, Suzuki E, Jurczak M, Biesmans S (2007) Independent double-gate FinFETs with asymmetric gate stacks. Microelectron Eng 84(9/10):2097
13. Masahara M, Surdeanu R, Witters L, Doornbos G, Nguyen VH, Van den bosch G, Vrancken C, Devriendt K, Neuilly F, Kunnen E, Suzuki E, Jurczak M, Biesmans S (2007) Independent double-gate FinFETs with asymmetric gate stacks. Microelectro Eng 84(9/10):2097
14. Dayal A, Akashe S (2013) A novel double gate finfet transistor: optimized power and performance analysis for emerging nanotechnologies. Comput Inf Syst Dev Inform Allied Res 4(4):75–80
15. van Rossem F (2009) Doping extraction in FinFET. University of Twente, thesis
16. Yu B, Chang L, Ahmed S, Wang H, Bell S, Yang CY, Tabery C, Ho C, Xiang Q, King T-J et al (2002) FinFET scaling to 10 nm gate length. In: Electron devices meeting. IEEE international conference on IEDM, pp 251–254
17. Poiroux T, Vinet M, Faynot O, Widiez J, Lolivier J, Ernst T, Previtali B, Deleonibus S (2005) Multiple gate devices: advantages and challenges. Microelectron Eng 80:378–385

References

1. Kornerville [illegible] (199[illegible]) [illegible] IEEE Trans Parallel Distrib Syst [illegible]
2. Lehmann [illegible] (19[illegible]) Compound [illegible] Hanser [illegible]
3. Hussain [illegible] High-performance and power-efficient [illegible] IEEE J Solid-State Circuits [illegible]
4. Huang [illegible], Yang [illegible], Huang [illegible] Exploring high-performance [illegible] IEEE J Solid-State Circuits [illegible]
5. Delgado-Frias [illegible] interconnection [illegible] IEEE Trans Circuits Syst [illegible]
6. Kuroda [illegible] IEEE J Solid-State Circuits [illegible]
7. Schegolev [illegible]
8. Nabhan [illegible] CMOS technology [illegible]
9. Rawlinson [illegible]
10. Zhang [illegible]
11. Dhanasekaran [illegible] Lateral and vertical [illegible] Proceedings [illegible] IEEE International SOI conference [illegible]
12. Ernst [illegible]
13. [illegible]
14. [illegible]
15. [illegible]
16. Yu B, Chang L, Ahmed S, Wang H, Bell S, Yang CY, Tabery C, Ho C, Xiang Q, King TJ et al (2002) FinFET scaling to 10 nm gate length. [illegible] International [illegible] IEDM, pp 251–254
17. [illegible]

Chapter 37
A Relative Investigation of TIQ Comparator and Dynamic Latched Comparator

Julia Soram and Shyam Akashe

Abstract This paper reports the use of threshold inverter quantization (TIQ) comparator as a replacement of the dynamic latch comparator in various Analog to Digital Converter (ADC) designs, memory sensing circuits and so on. The TIQ comparator offers low power, reduced delay and minimum area as compared to the dynamic latch comparator. The TIQ inverter comparator comprises of only four MOS transistors while there are nine MOS transistors in dynamic latch comparator. Thus using TIQ comparator results in better upgraded advanced electronic designs that earlier implemented dynamic latch comparator. Both the designs were simulated using Spectre simulator in Cadence.

37.1 Introduction

With the increase in speed, consumption of low power and small area the comparator has turn out to be a exceptionally significant circuit block for most electronic applications, such as fast analog to digital converters (ADC), memory sense amplifiers, data receivers and so on. The voltages that are seen at their inputs are measure and compared by the comparator then only it provides an output voltage that represents the net distinction between them. We know that the comparator is a circuit used to compare an analog signal with other analog signal and results in a binary signal depending on the comparison. If the positive input of the comparator is at a bigger prospective than the negative input, the output equals logic one else zero. Therefore, the comparators are significant elements in many high speed devices. The technology scaling of CMOS transistor minimizes power consumption and area that is occupied by the design. As we know that a clocked comparator usually consists of two parts. In that first part we see it interfaces the input signals. The second part consists of two cross coupled inverters, with each input connected to the output of the other. In a CMOS based latch design, the regenerative part

J. Soram (✉) · S. Akashe
Department of ECE, ITM University, Gwalior 474001, Madhya Pradesh, India
e-mail: soramjulia@gmail.com

V. Lakshminarayanan and I. Bhattacharya (eds.), *Advances in Optical Science and Engineering*, Springer Proceedings in Physics 166,
DOI 10.1007/978-81-322-2367-2_37

and its following stages take low static power because the power ground path is turned off either by a NMOS or PMOS transistor [1]. The conventional dynamic latched comparators implements a positive feedback mechanism with one pair of back-to-back cross coupled inverters to switch a small input-voltage difference to a full-scale digital value in a small time. Comparators can be differentiated into three types: Open-loop comparators, Pre-amplifier based latched comparators and fully Dynamic latched comparator.

37.2 Comparators

37.2.1 Comparators with Open Loop

The operational amplifiers without frequency compensation also known as Open-loop comparators are used to stimulate the biggest achievable bandwidth. It's immediate response as the precise gain and dimensionality are of no interest in comparator design and no-compensation does not produce a haul. And because of its restricted gain-bandwidth product, open-loop comparators are slow for many applications. One the opposite hand, a cascade of open-loop amplifiers typically includes a considerably larger gain bandwidth product than a single-stage amplifier with an analogous gain. However, since it prices extra space and power consumption, cascading does not offer wise benefits for several applications.

37.2.2 Pre-amplifier Based Latched Comparators

The main benefits of the pre-amplifier based latched comparators are their quick speed and low input referred latch offset voltage. Essentially a pre-amplifier consists of one or two stages of an open-loop comparator, features a gain of 4–10 V/V and with the help of this feature it can reduce the input inferred latch offset voltage by its gain. Latched comparators ordinarily use one or two clock signals (Clka and Clkb) to see the modes of operation: Track Mode (Reset): output is reset and input is tracked, Latch Mode (Evaluation): output is toggled by employing a positive feedback. Consequently it can be concluded that pre-amplifier latched comparators, that is a grouping of a pre-amplifier and a latch, provides a fast speed and little offset while they still consume static power [1, 2].

37.2.3 Dynamic Latched Comparators

The dynamic latched comparator consists of two stages as shown in Fig. 37.1. The primary stage is that the interface stage that consists of all the transistors except 2 cross coupled inverters.

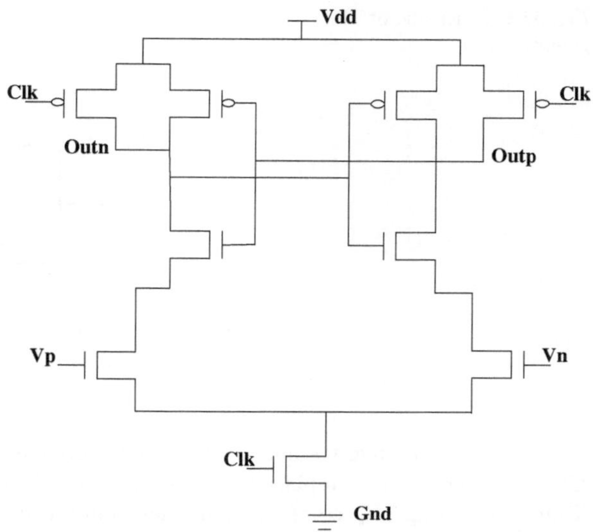

Fig. 37.1 Dynamic latch comparator

The next significant part of the dynamic latched comparator is the renewal stage which consists of two cascaded inverters with output feedbacks given to each other, whereby every input be linked to the output of the opposite. It operates in two phases. (1) Interface part and (2) Regeneration part. This half is made up of a single NMOS tail transistor coupled to the ground. As soon as the clock is zero, the tail transistor turns off and adds on to V_p and then V_n output reaches to VDD or the gnd. And once the clock becomes one, the tail transistor turns on and mutually the outputs are discharged to the ground [1, 2]. This significant part uses method of regeneration with lone pair of successive cross coupled inverters so as to decode a little input-voltage distinction to a complete digital level in a very small time. Therefore dynamic latched comparator is appropriate for both high speed and low power dissipation owing to decrease in transistor count that overcomes the matter of double tail latch and pre-amplifier based clocked comparators [3–7].

37.3 Threshold Inverter Quantization Comparator

In the design of ADC architecture, the comparator is mainly the principal component. The main objective of the comparator be to interpret an input voltage to logic one or else zero by means of making a comparison with the suggested voltage. If V_{in} is larger than V_{ref}, the output of the comparator is '1' else '0'. As the frequently used comparator styles in CMOS ADC design are the differential latched comparator and dynamic latched comparator. The earlier is also known as "clocked comparator", and the second is called a "auto-zero comparator" or "chopper comparator".

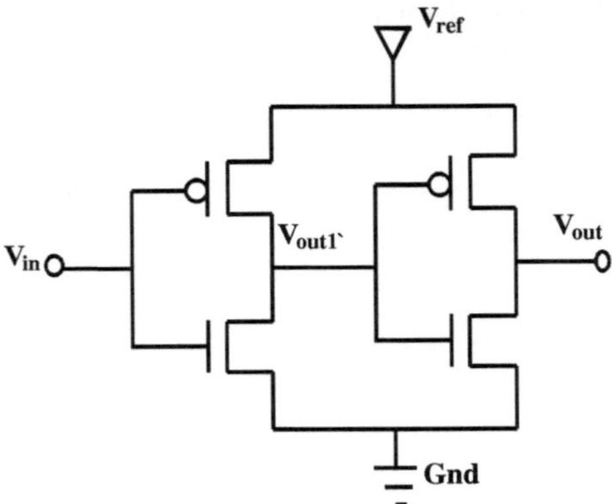

Fig. 37.2 Schematic of TIQ comparator

The threshold inverter quantization (TIQ) comparator uses two cross coupled CMOS inverters as a replacement for high speed and low power consumption. Tangel used this TIQ comparator for designing a high speed flash ADC [8]. The predicted TIQ comparator which has been briefed in this paper has been designed not just for top speed but additionally for elevated resolution also. As 2^n-1 comparators are required for n-bit ADC, Fig. 37.2 shows TIQ Comparator of a 3-bit Flash ADC involving 7 threshold inverter quantization (TIQ) comparators. The TIQ comparator comprises of 2 cross coupled CMOS inverters as shown in the Fig. 37.2 [2, 8].

The analog input signal quantization level is set in the first stage by changing the voltage transfer curve (VTC) by sizing the transistor [3]. As the channel length of the transistor is additionally potent than the channel thickness in calculating the performance, the channel length is reserved constant and only the width is varied throughout the design process. The second part of the two inverters is then applied for enhancing the gain and then inverting the logic level so that the behaviour of the circuit is same as a comparator that has been set internally. The necessary purpose with the next part is that the second inverter must be precisely the same as the first stage because the initial stage has to take care of a similar DC threshold and to stay the one-dimensionality in sense of stability for the voltage increasing and decreasing difference of high frequency signals [8].

The first part of the cross coupled inverters has the analog quantization signal level set by the voltage V_{m} at the entry point which represents the switching voltage of the comparator and depends on the width to length ratios of PMOS and NMOS. The second part of the two inverter in the TIQ comparator boosts the voltage gain and then to avoid an uneven propagation delay. The switching voltage for the TIQ comparators is calculated using the equation given below [9, 10]:

$$V_m = \left(\left(\sqrt{\mu pwp/\mu nwn}\right)\left(\mathrm{V_{dd}}|V_{\mathrm{tp}}|\right) + V_{\mathrm{tn}}/\left(1 + \sqrt{\mu pwp/\mu nwn}\right)\right) \tag{37.1}$$

here V_{tn} and V_{tp} are the electron and hole mobility, respectively. To derive the above equation, we presume that both transistors are in the active region, the gate oxide thickness (C_{ox}) for both transistors is the identical, and the lengths of both transistors (L_p and L_n) are also the same. From the above equation, we know that V_m is shifted depending the transistor width ratio (W_p/W_n). L_p and L_n denotes the channel lengths of the PMOS transistors and NMOS transistors respectively while W_p and W_n represents the width of PMOS transistors and NMOS transistors. This technique of varying of the breadths of the PMOS transistors and NMOS transistors devices while using a unchanging transistor length is the proposal of the TIQ comparator. The threshold voltage at the input of the inverter is here being employed as an in-house voltage reference to compare the input voltages applied [11–14].

37.4 Simulation and Results

Both the TIQ comparator and dynamic latched comparator has been designed in Virtuoso, simulated in Spectre in 45 nm technology and recorded the various power dissipation readings, delay and noise.

Table 37.1 shows the average, total and static power dissipation analysis of the TIQ comparator at various supply voltages. The least power dissipated is observed at 0.7 V. There is increase in dissipation of power with increase supply voltage. The dynamic latched comparator shows more power dissipation than the TIQ comparator at any supply voltage. Though the dynamic comparator shows minimum dissipation at 0.7 supply voltage, it is greater than the TIQ comparator (Table 37.2).

Table 37.1 Power analysis of TIQ comparator

Voltage (volts)	Average power (nW)	Total power (nW)	Static power (pW)
0.7	7.016	0.259	5.695
0.8	9.196	0.295	7.634
0.9	12.14	0.198	10.1
1	16.87	0.113	12.44

Table 37.2 Analysis of power in dynamic latched comparator

Voltage (volts)	Average power (nW)	Total power (nW)	Static power (pW)
0.7	8.977	6.18	7.13
0.8	11.89	10.0	9.96
0.9	15.33	0.827	13.74
1	19.69	2.388	18.88

Table 37.3 Delay and noise analysis

	TIQ comparator		DLC comparator	
Voltage (volts)	Delay (ns)	Noise (100 K)	Delay (ns)	Noise (100 K)
0.7	20.63	3.899 e^{-16}	20.70	8.577 e^{-16}
0.8	20.67	2.3795 e^{-16}	20.73	4.833 e^{-16}
0.9	20.70	1.778 e^{-16}	20.76	3.370 e^{-16}
1	20.71	1.47 e^{-16}	20.80	2.656 e^{-16}

Table 37.3 shows the delay and noise comparison of the TIQ comparator and the dynamic latched comparator. There is inverse relationship between the noise observed and the supply voltage in both the comparators, with the minimum delay seen in TIQ comparator at 0.7 V of 20.63 ns.

37.5 Conclusion

A comparison of the TIQ comparator and dynamic latched comparator has been made in this paper. The TIQ comparator with only 4 transistors offers better options while designing high speed ADCs than the dynamic latched comparator which comprises of 9 transistors. An average power of 7.016 nW and total power of 0.259 nW is observed in TIQ comparator whereas in dynamic latched comparator an average power of 8.977 nW and total power of 6.18 nW is recorded. In addition to power a reduced delay of 20.63 ns is seen in TIQ comparator as compared to 20.7 ns of dynamic latched comparator at 0.7 V supply voltage. An inverse relationship between the noise recorded and the supply voltage is also seen. All the simulation process has been carried out in Spectre Simulator of Cadence using 45 nm technology.

Acknowledgments The author would like to thank ITM University, Gwalior for providing us the tools and technology that were required for the work and help us in its completion.

References

1. Heung JJ, Yong BK (2012) A novel-low power, low offset, and high speed CMOS dynamic latched comparator. Analog Integr Circ Sig Process 70(3):337–346
2. Heung JJ, Yong BK (2010) A CMOS low-power, low-offset and high-speed fully dynamic latched comparator. In: Proceedings of IEEE ISOCC, pp 285–288
3. Mashhadi SB, Lotfi R (2014) Analysis and design of low voltage low power double tail comparator. IEEE Trans Very Large Scale Integr Syst 22(2):343–352
4. Rabaey JM, Chandrakasan A, Nikolic′ B (2003) Digital integrated circuits. A design perspective, 2nd edn. Englewood Cliffs, NJ
5. Kapadia DN, Gandhi PP (2013) Implementation of CMOS charge sharing dynamic latch comparator in 130 nm and 90 nm technologies. In: Proceedings of IEEE CICT, pp 16–20

6. Rajshekhar G, Bhatt MS (2008) Design of resolution adaptive TIQ flash ADC using AMS 0.35 μm technology. In: Proceedings of IEEE ICED, pp 1–6
7. Gurjar M, Akashe S (2013) Design low power encoder for threshold inverter quantization based flash ADC converter. Int J VLSI Des Comm Syst (VLSICS), 4(2)
8. Akashe S, Rajak V, Sharma G (2013) Optimization of fat tree encoder for ultra high speed analog to digital converter using 45 nanometer technology. Optik—Int J Light Electron Opt 124(20):4490–4492
9. Tangel A, Choi K (2004) The CMOS inverter as a comparator in ADC designs. Analog Integr Circ Sig Process 39:147–155
10. Segura J, Rossello JL, Morra J, Sigg H (1998) A variable threshold voltage inverter for CMOS programmable logic circuits. IEEE J Solid-State Circuits 33(8):1262–1265
11. Veeramachanen S, Kumar AM, Tummala V, Srinivas MB (2009) Design of a low power, variable-resolution flash ADC. In: Proceedings of IEEE ICVLSID, pp 117–122
12. Yoo J., Lee D., Choi K, Tangel A (2001) A 1GSPS CMOS flash A/D converter for SOC applications. Proceedings of IEEE computer society workshops on VLSI, pp 135–137
13. Mishra SN, Arif W,Mehedi J, Baishya J (2013) A power efficient 6-bit TIQ ADC design for portable applications. In: Proceedings of IEEE international conference on consumer electronics, pp 235–239
14. Khot SS, Wani PW, Sutaone MS, Tripathi S (2011) Design of a 45 nm TIQ comparator for high speed and low power 4-bit Flash ADC. ACEEE Int J Electr Power Eng 2(1)

Chapter 38
Design of Low Power Shift Register in Nano Scale Domain Using FinFET

Ankur Kumar Gupta and Shyam Akashe

Abstract In this paper, we have calculated and described the Shift Register Design using FinFET. Shift Register is used as to shift the value of any register, in microprocessor, microcontroller and shifted value is also equal to multiplication of any binary no. by 2. According to Moore's law the no. of transistor in a scrupulous chip area is doubled in every 18 months. This proclamation gives new epoch of VLSI meadow. If we want to increase the no. of component in chip area so we reduce the size of component. Applying this attribute in chip component, the size of transistor reduced. As we scale down the device parameter after a certain rule, the short channel effects like leakage power, surface scattering, velocity saturations, takes place. Fin-FET is a superior device to eradicate or decrease above mentioned problems. We analyze the various parameters like temperature effect to the total power, total power consumption, average DC power, calculation etc. For calculation these results we are using cadence tools. After simulating the circuit we get values of Average DC power is 777.6 nW, Average Transient Power Consumption is 67.20 nW, Delay is 20 nS.

38.1 Introduction

Shift Register is a fraction of sequential circuit. Sequential circuits are combination of a combinational circuit to which memory elements are linked form of a feedback path. Shift Register is used to store or shift or convert parallel data sequence to serial data or vice-versa the binary data of any given in binary sequence. Mostly shift of bit is used in microcontrollers and microprocessors. It may be used for right shift or left as per requirement. Shift Register also acts as a counter. Shift Register is cluster of Flip-Flops [1–3]. Shift Register is intended via D Flip-Flop. Four D Flip-Flop cascaded, make a Shift Register. Shift Register is also used to compact the

A.K. Gupta (✉) · S. Akashe
ECE Department, ITM University, Gwalior 474001, Madhya Pradesh, India
e-mail: erankurkumargupta@gmail.com

V. Lakshminarayanan and I. Bhattacharya (eds.), *Advances in Optical Science and Engineering*, Springer Proceedings in Physics 166,
DOI 10.1007/978-81-322-2367-2_38

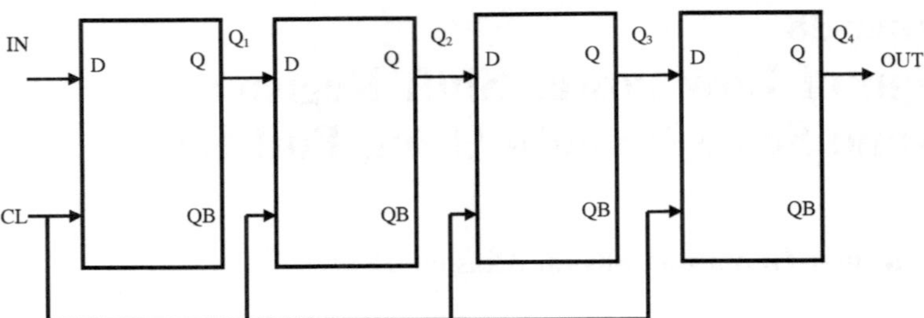

Fig. 38.1 Shift register

no. of wires. Shift Register provides interfacing between Serial to Parallel data translation. These all Flip-Flop are driven by same clock [4–7]. Block Diagram of Shift Register shown in Fig. 38.1 [1, 2].

Types of Shift Register:

- **Serial In Serial Out** (**SISO**) In this type of shift Register, data enters serially and at the output end data exits serially.
- **Serial In Parallel Out** (**SIPO**) In this type of Shift Register, data enters serially and outs as parallel. This is also known as serial to parallel converter.
- **Parallel In Serial Out** (**PISO**) In this type of Shift Register, data enters in parallel way and exits in serial way. This is also known as parallel to serial converter. When data transmits for a long distance then parallel data converts in serially this is economical.
- **Parallel In Parallel Out** (**PIPO**) In this type of Shift Register, data enters parallel and exits also in parallel.
- Ring Counter.

Data enters from Input end and exits from OUT end after shifting by four clock pulses. Shift Register is Designed using D Flip-Flop which is designed using NAND Gate. In this paper NAND Gate is designed using FinFET.

38.2 FinFET

A Field Effect Transistor which contains fins in its gate is called FinFET. FinFETs are multi-gate device. The two gates of a FinFET can be linked for higher performance and they can be independently controlled for lower leakage and reduced transistor count. CMOS designed NAND Gate in 45 nm regime has many drawbacks like short channel effect like surface scattering, velocity saturation, Impact ionization, Hot electron and DIBL. FinFET remove the short channel effect like surface scattering, velocity saturation, Impact ionization, Hot electron and DIBL. It is a promising candidate which overcomes to the short channel effects in this regime [8, 9] (Fig. 38.2).

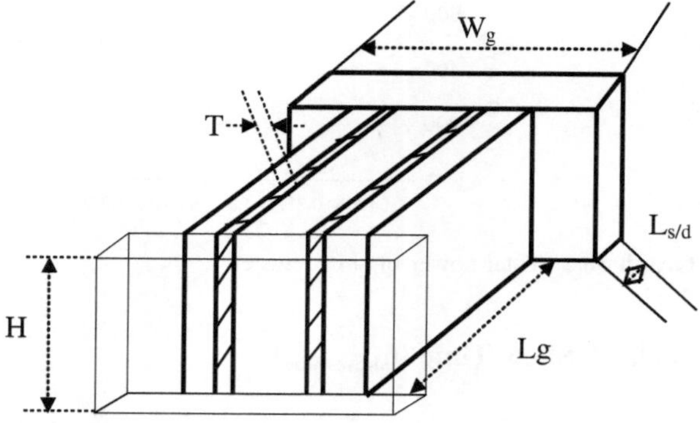

Fig. 38.2 FinFET device (*W* width of drain/source electrode, *T* Thickness of fin, *H* Height of fin, *L* physical gate length of fin, $L_{s/d}$ length of source/drain)

38.3 Analysis of Various Parameters

38.3.1 Delay

It is length of time, to response of any system or circuit to reach 50 % of the final value in first time after applying input. Minimal delay is characteristics of every circuit. The maximum Delay restriction for the flip-flop is

$$D = 1.05 * t_{\mathrm{CLK-Q}} + t_{\mathrm{setup}} \leq T - t_{\mathrm{logic}} - t_{\mathrm{skew}} \quad (38.1)$$

Calculated value of delay in FinFET based Shift Register is 20 nS, which is 35 % less than conventional Shift Register shown in Graph 38.1. FinFET based Shift register has low access time.

38.3.1.1 Set up Time ($t_{\text{set-up}}$)

It is the minimum time length in which the data must be stable before clock is transitioned.

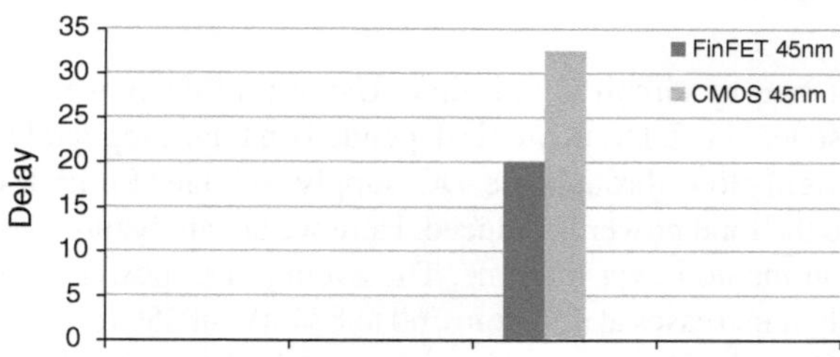

Graph 38.1 Delay comparison between CMOS and FinFET

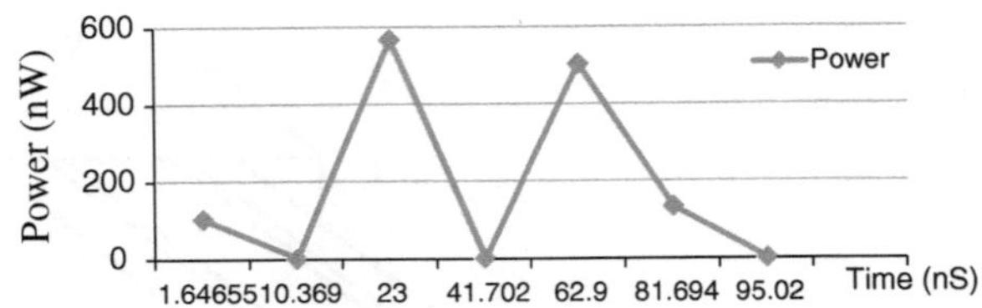

Graph 38.2 Total power of shift register

38.3.1.2 Skew Time ($t_{skew\ time}$)

It is the time difference of arrival of two adjacent signal.

38.3.2 Power Consumption

Power is the product of applied voltage to the circuit and current flow through the circuit due to applied voltage source. Power of the circuit which is designed using FinFET is less [10]. Total power consumption of circuit is product of total voltage supplied to circuit and current flow through the applied voltage sources. Results are shown below.

$$\text{Power}(P)(t) = \text{Voltage}(V_{\text{dd}}) * \text{Current}(I_{\text{dd}})(t) \tag{38.2}$$

The primary benefit of FinFET based Shift register is in its low power dissipation due to low short channel effect and low leakage current in FinFET based shift Register. Low leakage current reduces the power dissipation. FinFET can turned on at low voltage supply due to low threshold voltage so total supplied voltage is reduced to minimum level. When supply voltage is reduced then current flowing in circuit is reduced and leakage current also reduced so over all power dissipation is reduced (Graph 38.2).

38.3.3 Average DC Power

Energy is consumed by circuit in unit time. Using FinFET power is less so average DC power is also less [11]. DC power is depended on total supplied DC voltage to the system and current flow through the DC supply voltage. FinFET reduces supply voltage [11]. So the total power is reduced. Here we are analyzing average DC power using Integration method over the time. The average DC power Varies from 760 to 831 nW. Initially it increases at 0 V from 760 to 831 nW at 350 mV and then decreases till 760 nW at 600 mV. Power consumptions are shown in Graph 38.3 [12].

Graph 38.3 Average DC power of shift register

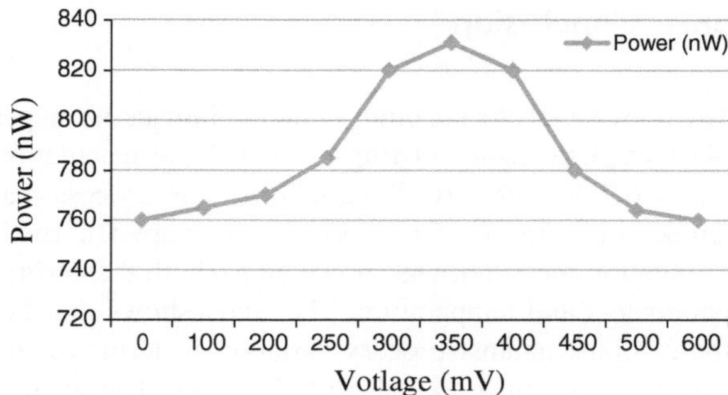

Graph 38.4 Power versus curve shift register

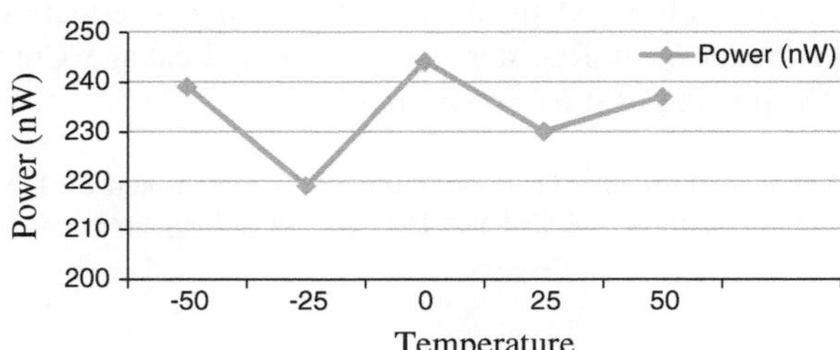

$$P_{\text{avg}} = E/T = 1/T \int_0^T (i_{\text{dd}(t)} * V_{\text{dd}(t)}) \tag{38.3}$$

T Time
E Energy in joule
i_{dd} Drain current
V_{dd} Drain voltage

38.3.4 Temperature V/S Power Curve

As vary the environmental Temperature of device the power varies because the resistance and mobility of electron and holes are also varying [13, 14] (Graph 38.4).

$$n_i^2 = A_0 T^3 e^{-E_{G0/kT}} \tag{38.4}$$

n_i Intrinsic concentration
A_0 Constant independent of temperature
T Temperature
E_{G0} Energy gap at 0 K

38.4 Conclusion

The above analysis concludes that the impact of temperature variations on power of Shift Register using D Flip-Flop. All the calculations are did at the voltage of V_{dd} = 0.7 V with ±10 % variations. The process parameters (±10 %) as well as temperature (from −50 to 50 °C) are varied normally using Gaussian function. Simulation measurements are taken for both the designs against variation of process parameters and temperature. The study shows that Shift Register using FinFET is more robust against process variation in terms of most of the design parameters compared to conventional Shift Register. The average DC power is found to be 1.410 nW while the average total power consumption is 777.20 nW. A delay of 20 ns is observed in the circuit showing a reduction of 35.38 % from the conventional Shift Register using CMOS. Leakage Current and Leakage Power are 200 pA and 200 pW respectively.

Acknowledgments This research work is ornamented by ITM University, Gwalior, MP India, with collaboration of Cadence Design system Bangalore.

References

1. Mano MM, Ciletti MD (2008) Digital design, 4th edn. New Delhi, India
2. Brown S, Vranesic Z (2012) Fundamentals of digital logic with verilog design, 2nd edn. New Delhi, pp 45, 382–405
3. Bhoj AN, Jha NK (2013) Design of logic gates and flip-flops in high performance FINFET technology. IEEE Trans VLSI Syst 21(11):1975–1988
4. Cheng LJ, Zhong YH (2003) A new low-power readout shift register for CMOS image sensors. In: IEEE proceedings on 5th international ASIC 2003, vol 2, pp 902–905
5. Geng D, Kang DH, Seok MJ, Mativenga M, Jang J (2012) High-speed and low-voltage-driven shift register with self-aligned Cplanar a-IGZO TFTs. IEEE Electron Device Lett 33 (7):1012–1014
6. Kartashev SI (1976) A microcomputer with a shift-register memory. IEEE Trans Comput 25 (5):470–484
7. Dubois M, Savaria Y, Haccoun D (2004) On low power shift register hardware realizations for convolution encoders and decoders. In: The 2nd annual IEEE northeast workshop on circuits and systems 2004, pp 213–216
8. Liu Z, Tawfik SA, Kursan V (2008) Statistical data stability and leakage evaluation of FinFET SRAM cells wit dynamic threshold voltage tuning under process parameter fluctuations. In: IEEE proceedings on 9th international symposium on quality electronic design (ISQED 2008), San Jose California, pp 305–310
9. Raj B, Saxena AK, Dasgupta S High performance double gate FinFET SRAM cell design for low power application. Int J VLSI Sig Process Appl 1(1):12–20. ISSN: 2231–3133
10. Saraswat R, Akashe S, Babu S (2013) Designing and simulation of full adder cell using FINFET technique. In: IEEE proceedings on 7th intelligent system and control (ISCO) 2013, Coimbatore, pp. 261–264
11. Shikarwar V, Khandelwal S, Akashe S (2013) Optimization of leakage current in SRAM cell using shorted gate DG FinFET. In: IEEE proceedings on advanced computing and communication technologies (ACCT 2013), Rohtak, pp 166–170

12. A Sayed, Al-Asaad H (2006) A new low power high performance flip-flop. In: IEEE proceedings on 49th IEEE international midwest symposium on circuit and systems, 2006 (MWSCAS'06), San Juan Puerto Rico, vol 1, pp 723–727
13. Markovic D, Nikolic B, Brodersen R (2008) Analysis and design of low-energy flip-flops. In: ACM proceedings on international symposium on low power electronics and design, pp 52–55
14. Rajaram A, Premlatha P, Sowmiya R, Saravanan S, Vijaysai R (2013) Design and analysis of high speed shift register using single clock pulse method. In: IEEE proceedings on international conference on computer communication and informatics (ICCI 2013), Coimbatore, pp 1–4

Chapter 39
Optimized Area and Low Power Consumption Braun Multiplier Based on GDI Technique at 45 nm Technology

Divya Billaiya and Shyam Akashe

Abstract In modern days, power dissipation is one of the biggest challenges in VLSI design. The numbers of transistors are reduced in the circuit and ultra-low power design. This chapter is based on full adders that are designed using EX-OR gates, and GDI technique is used for low power and delay in full adders. The main aim of this chapter was to reduce the power dissipation and area by reducing the number of transistors. Multipliers are the main sources of power dissipation in DSP. Braun Array is used to implement a multiplier, a relatively simple form of parallel adder. In our chapter, we designed a 4-bit Braun multiplier based on GDI and the simulations are performed by CADENCE VIRTUOSO based on 45-nm CMOS technology with the supply voltage of 0.7 V. The simulation results showed that proposed multiplier at 45 nm can reduced the average power from 291.4 to 133.9 nW, total power from 286.3 nW to 130.3 nW, static power from 1.15 nW to 4.10 pW, static current from 1.64 nW to 5.86 pW, and power consumption from 2.21 W to 1.009 μW.

39.1 Introduction

For computer arithmetic processes, the most common requirements are for addition and subtraction, but there is also a significant need for a multiplication capability. There are many applications especially in DSP field, which require the multiplication operation [1]. We implement the Braun multiplier based on GDI technique. Braun multiplier is generally known as carry-save multiplier and is an $n \times m$ bit parallel multiplier and constructed with m x $(n - 1)$ adders and $m \times n$ AND gates [2, 3]. The Braun multiplier due to the ripple-carry adder in the last stage of the multiplier has a glitch problem [4].

D. Billaiya (✉) · S. Akashe
ECE Department, ITM University, Gwalior 474001, MP, India
e-mail: divya.billaiya@gmail.com

V. Lakshminarayanan and I. Bhattacharya (eds.), *Advances in Optical Science and Engineering*, Springer Proceedings in Physics 166,
DOI 10.1007/978-81-322-2367-2_39

39.2 Architecture

A Braun's 4-bit multiplier is shown in Fig. 39.1. The details of the multiplication procedure are as follows. Let us assume that X is the multiplicand and Y is the multiplier.

$$X = X3X2X1X0; \quad Y = Y3Y2Y1Y0$$
$$X \times Y = X3X2X1X0 \times Y3Y2Y1Y0$$
$$= P7P6P5P4P3P2P1P0$$

$P7$ is the (MSB), and $P0$ is the (LSB).

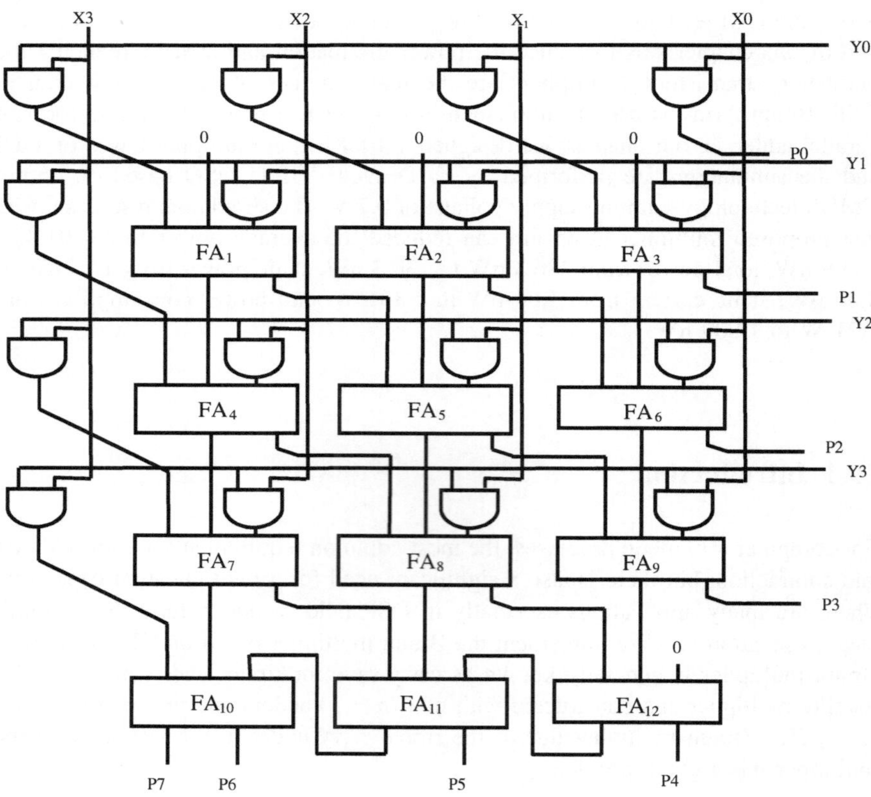

Fig. 39.1 Block diagram of Braun multiplier

39.3 Basic GDI Function

A new low power design technique that solves most of the problems is known as Gate Diffusion Input (GDI) technique [5, 6]. The GDI cell is shown in Fig. 39.2. The power usage, propagation delay, and area of digital circuits are reduced by this technique while maintaining low complexity of logic design [7].

The main difference between the CMOS- and GDI-based designs is that the source of the NMOS is not connected to GND and source of the PMOS in a GDI cell is not connected to VDD [8]. This is the main advantage that gives the GDI cell two extra input pins for use which provides the GDI design more flexibility as compared to CMOS design. GDI cell subsists of three inputs—G (common gate input of NMOS and PMOS), N (input to the source/drain of NMOS), and P (input to the source/drain of PMOS), and bulks of both NMOS and PMOS are associated with N and P, respectively [9]. Table 39.1 manifests the different logic functions implemented by GDI logic based on different input values. We can execute various logic functions with the help of GDI technique with less power and high speed in comparison with traditional CMOS design.

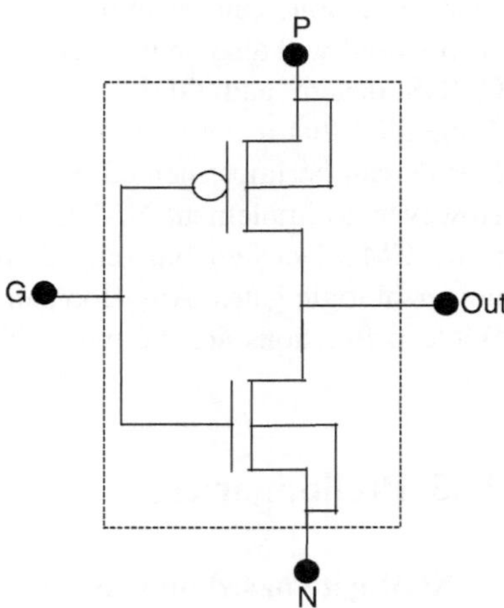

Fig. 39.2 Basic GDI cell

Table 39.1 Basic function using GDI cell

N	p	G	Out	Function
0	B	A	A'B	F1
B	1	A	A′ + B	F2
1	B	A	A + B	OR
B	0	A	AB	AND
C	B	A	A'B + AC	MUX
0	1	A	A′	NOT

Table 39.2 Comparison of transistor count of GDI and static CMOS

FUNCTION	GDI	CMOS
INVERTER	2	2
F1	2	6
F2	2	6
OR	2	6
AND	2	6
MUX	2	12
XOR	4	16
XNOR	4	16
NAND	4	4
NOR	4	4

39.4 Advantages of GDI Functions

The large number of functions can be resolved using the basic GDI cell. When we design the multiplexer with the help of traditional CMOS design, we have to use 8–12 transistors, but when we use GDI technique, we have to use only 2 transistors. Many functions can be implemented by GDI technique, and the number of transistors used will also be less. Table 39.2 displays the comparison between the static CMOS design and GDI in terms of transistor count. It can be noticed from Table 39.2 that using GDI technique, AND, OR, Functionl, Function2, XOR, and XNOR can be implemented more efficiently as compared to static CMOS design. However, to implement NAND and NOR gates, it requires 4 transistors as that in static CMOS design but only 2 in basic GDI design. NAND and NOR are the universal logic gates. Any Boolean function can be implemented using these gates. Boolean functions are the most efficient and popular with static design manner.

39.5 Preliminaries

1. **XOR gate based on Gate Diffusion Input cell**: Fig. 39.3 shows the execution of XOR gate using GDI technique. The main component of full adder circuit is XOR gate [10]. So, the optimization of XOR gate perks up the overall performance of the 1-bit full adder circuit. Less numbers of transistors are used in GDI technique in comparison with the conventional design of XOR gate using CMOS logic gates [11].
2. **GDI XOR full adder**: The full adder has three inputs (*A*, *B*, and *C*) and two outputs (Sum and Cout).The expression for the Sum and Cout of the full adder are described as in Eqs. 39.1 and 39.2.

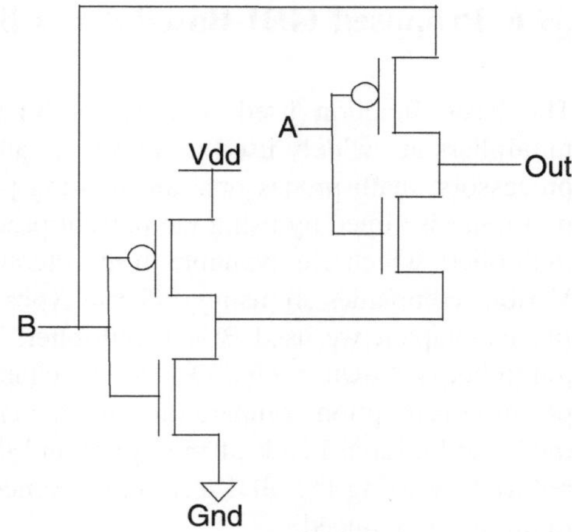

Fig. 39.3 GDI-based XOR gate

$$\text{Sum} = A(\text{XOR})B(\text{XOR})C \tag{39.1}$$

$$\text{Cout} = AB + BC + CA \tag{39.2}$$

The GDI XOR full adder consists of 10 transistors as shown in Fig. 39.4 in transistor level implementation [12]. One multiplexer and two GDI XOR gates are the three modules of the full adder circuit.

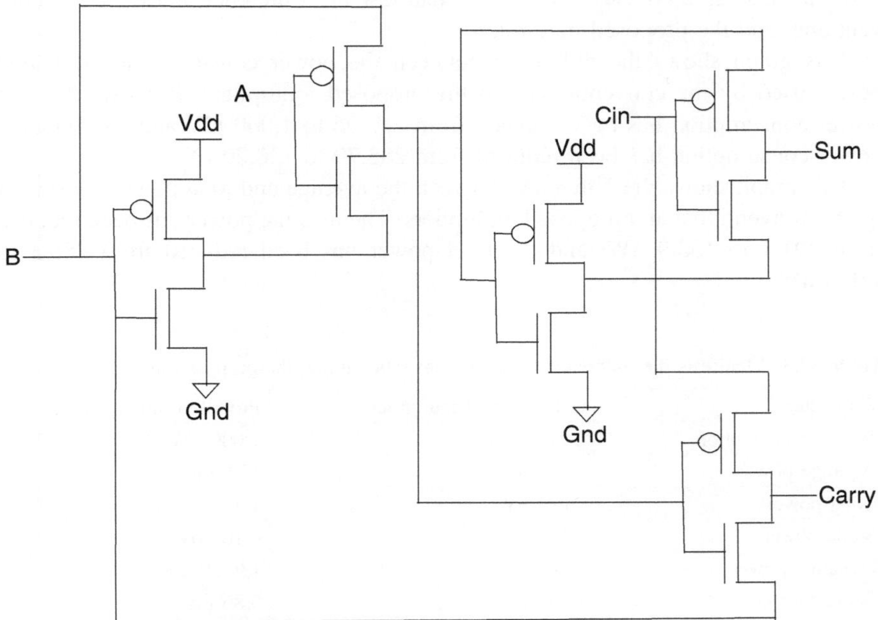

Fig. 39.4 GDI-based XOR full adder

39.6 Proposed GDI-Based 4 × 4 Braun Multiplier

The basic function used in different VLSI applications is multiplication. The multipliers are widely used in arithmetic and logical unit (ALU), FIR filters, DSP processors, math processors, and floating point units. The design of a low power multiplier is aimed by using the present processor [13]. There are different types of multipliers which are available in the literature, depending upon the requirements. Various companies are using different types of multipliers according to their needs. In this chapter, we used Braun multiplier. The block diagram of 4 × 4 bit Braun multiplier is shown in Fig. 39.1. In this chapter, the proposed multiplier reduces the power consumption compared to the conventional multiplier, because full adder is the basic building block of multiplier and the power consumption by full adder is reduced by using the GDI technique. Hence, the power consumption of proposed multiplier is reduced.

39.7 Simulation and Results

In this chapter, we determined the parameters such as static power, dynamic power, total power, and average power. Table 39.3 shows all the parameters of conventional multiplier and proposed multiplier which is designed in 45 nm technology with the help of CADENCE software.

Figures 39.5, 39.6, 39.7, and 39.8 manifest the difference between the conventional and the proposed multipliers.

This graph shows the difference between the power consumed and dynamic power used by the conventional and the proposed multipliers. It shows that the power consumption has been reduced from 2.2106 to 1.009 μW and the dynamic power consumption has been reduced from 285.79 to 130.29 nW.

This graph shows the difference between the average and total power consumed by the conventional and proposed multipliers. The average power has been reduced from 291.4 to 133.9 nW, and the total power has been reduced from 286.3 to 130.3 nW.

Table 39.3 Manifests the difference between conventional and the proposed multipliers

Parameter	Conventional multiplier	Proposed multiplier
Power consumption	2.2106 μW	1.009 μW
Average power	291.4 nW	133.9 nW
Total power	286.3 nW	130.3 nW
Static power	1.152 nW	4.107 pW
Dynamic power	285.79 nW	130.29 nW
Static current	1.64 nA	5.86 pA
Delay	686.1E-3	355.9E-3

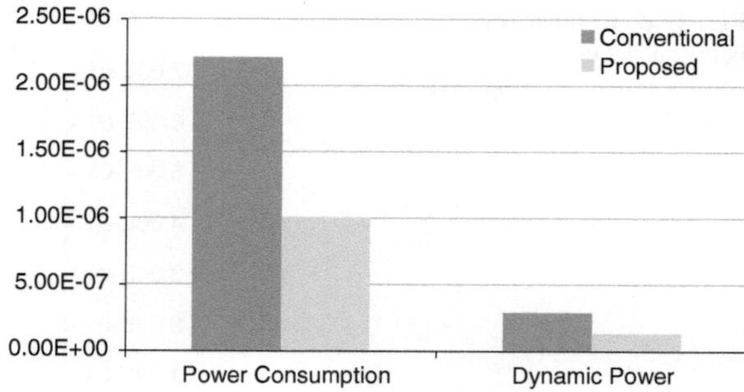

Fig. 39.5 Comparison between power consumption and dynamic power of conventional and proposed multipliers

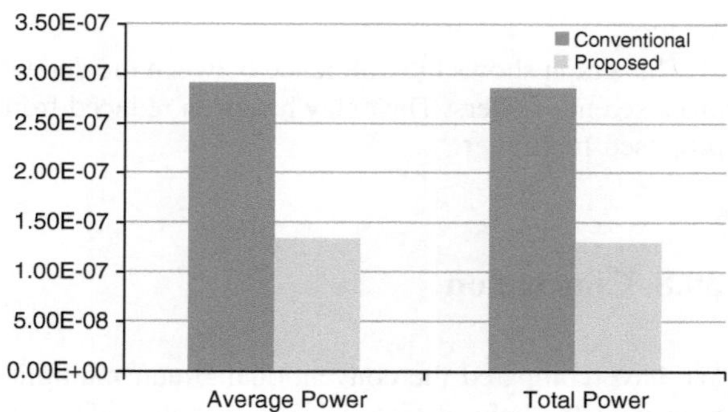

Fig. 39.6 Comparison between average power and total power of conventional and proposed multipliers

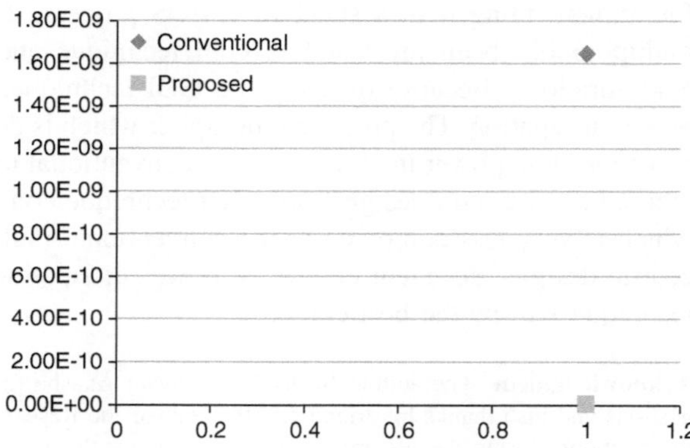

Fig. 39.7 Comparison between static current of conventional and proposed multipliers

This graph shows the difference between the static current between the conventional and proposed multipliers. The static current has been reduced from 1.64 nA to 5.86 pA in the proposed multiplier.

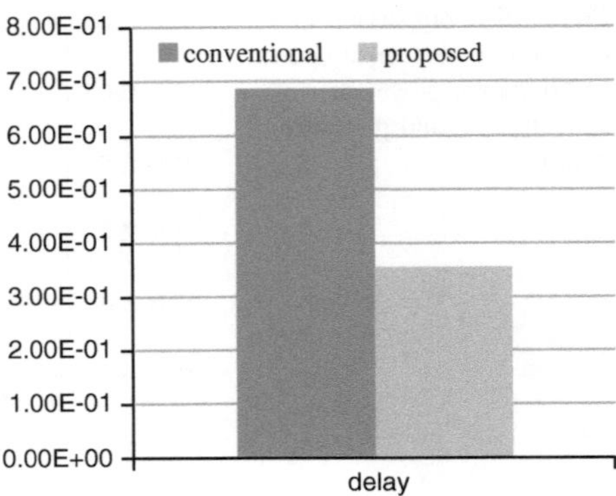

Fig. 39.8 Comparison between delay of conventional and proposed multipliers

This graph shows the difference between the delay between the conventional and proposed multipliers. The delay has been reduced from 686.1E-3 to 355.9E-3 in the proposed multiplier.

39.8 Conclusion

We have compared the conventional Braun multiplier and proposed Braun multiplier with the help of Gate Diffusion Input (GDI) technique for low power digital circuit design in this chapter. The proposed cell allows the implementation of a variety of logic functions, employing only two transistors, and it is fully compatible for implementing it in a standard CMOS process. The implementation of Braun multiplier has been presented in GDI technique and can be extended to higher configurations. Because of using the GDI technique, this design allows reducing power dissipation. The proposed multiplier which is designed by the GDI technique consumes less power in contrast to the conventional multiplier. The analysis shows that at 0.7 V, circuit design using GDI technique consumes static power 4.107 pW which is very less compared to the conventional multiplier. Therefore, for power-centric designs, stringent control of power dissipation is the essential use of GDI technique will be the best choice.

Acknowledgment The author thanks Prof. Shyam Akashe for his helpful annotations and suggestions and also thanks his friend and the staff of the Research Center of VLSI Laboratory for their support during the research. Finally, the author thanks the anonymous references for their thorough review and useful comments.

References

1. Abu-Khater IS, Bellaouar A, Elmastry MI (1996) Circuit techniques for CMOS low-power high-performance multipliers. IEEE J Solid-State Circ 31:1535–1546
2. Anitha R, Bagyaveereswaran V (2011) Comparative study of Braun's multiplier using FPGA devices. Int J Eng Sci Technol (IJEST) 3, ISSN: 0975-5462, June-2011
3. Prakash M, Kishor S, Kumar RN, Kiruthika S (2014) Design Of BRAUN multiplier using low power adder. In: Proceeding of 5th national conference on VLSI, Embedded, and communication and networks on 17 Apr 2014
4. Anitha R, Nelapati A, Lincy Jesima W, Bagyaveereswaran V (2012) Comparative study of high performance Braun's multiplier using FPGA. IOSR J Electron Commun Eng (IOSRJECE) 1:33–37
5. Morgenshtein Arkadiy, Fish Alexander, Wagner Israel A (2002) Gate-diffusion input (GDI): a power-efficnet method for digital combinatorial circuits. IEEE Trans VLSI Syst 10(5): 566–581
6. Morgenshtein A, Fish A, Wagner IA (2002) Gate-diffusion input (GDI)—a technique for low power design of digital circuits: analysis and characterization. In: IEEE international symposium on circuits and systems (ISCAS), vol 1, pp I-477–I-480
7. Morgenshtein A, Shwartz I, Fish A (2010) Gate diffusion input (GDI) logic in standard CMOS nanoscale process. In: 26-th convention of electrical and electronics engineers in Israel, 2010 IEEE
8. Dhar K, Chatterjee A, Chatterjee S (2014) Design of an energy efficient, high speed, low power full subtracter using gdi technique. In: Proceeding of the technology symposium, IEEE 2014
9. Balasubramanian P, John J (2006) Low power digital design using modified GDI method. In: Proceedings of international conference on design and test of integrated systems in nanoscale technology, IEEE 2006
10. Batta B, Choragudi M, Varma M (2012) Energy efficient full-adder using GDI technique. Int J Res Comput Commun Technol (IJRCCT) 1(6), ISSN 2278-5841, Nov 2012
11. Nishad AK, Chandel R (2011) Analysis of low power high performance XOR gate using GDI technique. In: Proceedings of the international conference on computational intelligence and communication systems, IEEE 2011
12. Shrivas J, Akashe S, Tiwari N (2012) Design and performance analysis of 1 bit full adder using GDI technique in nanometer era. In: Proceedings of the information and communication technologies, IEEE 2012
13. Rongali VK, Srinivas B (2013) Design of area efficient high speed parallel multiplier using low power technique on 0.18um technology. Int J Advan Res Comput Eng Technol (IJARCET) 2, July 2013
14. Moradi F, Wisland DT, Mahmoodi H, Aunet S, Caol TV, Peiravi A (2009) Ultra low power full adder topologies. In: Proceedings of the international symposium on circuits and systems, IEEE 2009
15. Kannan PM, Prathyusha K (2011) Implementation of low power RAM in GDI technique with full swing. In: Proceedings of international conference on signal processing, communication, computing and networking technologies (ICSCCN), IEEE 2011

Part IX
Nano-photonics, Bio-photonics and Bio-medical Optics

Chapter 40
A Method for Estimating the Wavefront Aberrations with Missing Spot Data in a Hartmann-Shack Aberrometer

R. Burman, A. Ommani, D. Thapa, K. Raahemifar, N. Hutchings and V. Lakshminarayanan

Abstract This paper reports a method of wavefront sensing based on Hartmann-Shack (HS) centroid displacements and predicts the number of HS spots that can be successfully deleted without hampering the prediction of the defocus of the computed wavefront, described via Zernike polynomials. The deletion of the HS spots was randomized. The experiment was performed on real data acquired through a custom made aberrometer tested on a model eye with various axial lengths to simulate refractive errors (defocus) between $\sim\pm1.50$D. Estimates of defocus were made from each of 1000 runs at each axial length. The paper presents the standard deviation of error and mean error for 1000 trials. The results indicate that as high as 50 % of the HS spots can be deleted without affecting the estimation of spherical defocus, within typical clinically acceptable limits of ±0.25D.

40.1 Introduction

Wavefront aberrations describe the optical imperfections of the eye by measuring the complete refractive elements of the eye [1]. One of the most popular instruments that measure the wavefront aberration is the Hartmann-Shack (HS) wavefront

R. Burman (✉)
Department of Electronics and Telecommunication Engineering, Jadavpur University, Kolkata, India
e-mail: ritambharkcdburman@gmail.com

A. Ommani · D. Thapa · N. Hutchings · V. Lakshminarayanan
School of Optometry and Vision Science, University of Waterloo, Waterloo, Canada

K. Raahemifar
Department of Electrical and Computer Engineering, Ryerson University, Toronto, Canada

V. Lakshminarayanan
Department of Physics, University of Waterloo, Waterloo, Canada

V. Lakshminarayanan and I. Bhattacharya (eds.), *Advances in Optical Science and Engineering*, Springer Proceedings in Physics 166,
DOI 10.1007/978-81-322-2367-2_40

sensor, where monochromatic light reflected back from the optical system is captured by a uniformly arranged lenslet array [2]. The lenslet array generates spots in the sensor according to the nature of the wavefront coming out of the eye. The pattern of the light imaged by the sensor is analysed to derive the Zernike aberration coefficients [3]. Zernike coefficients are the standard way for reporting the aberration of the eye as they provide a set of coefficients that are independent of each other. The technique uses several spots so that the wavefront is planar over each lenslet [4]. However, for some conditions, the spots can be extremely blurry and indistinct or non-existing [5]. In this study, we estimate wavefront aberrations from the missing spot data to test the reliability of the measurements. We randomly deleted a fixed percentage of spots and calculated the Zernike aberration coefficients from the reduced data set. The deletion of spots not only accounts for a possible reduction in the number of lenslets that drastically reduces the cost of the wavefront sensor but also helps in faster processing of the HS images.

The rest of the paper is divided as followed: Sect. 40.2 gives the proposed algorithm with elementary theory about Zernike polynomials. Section 40.3 gives the description of the apparatus along with the experimental results and discussion. Lastly, Sect. 40.4 concludes the paper outlining possible future works.

40.2 Proposed Algorithm

Let $I(x, y)$ be the subject image acquired using a model eye having 130 mm focal distance through the lenslet array by the CCD detector and R be the reference image acquired with a flat mirror. I and R are converted to binary images using a suitable threshold T. Next the images are serially subjected to spot analysis within the desired effective pupil radius. The brightest pixel in a particular spot represents the centroid of that spot. The brightest pixel of all spot centroids is treated as the overall centroid of the subject image, which is also taken as the reference point in the coordinate system. The geometrical center is forcefully made to realign with the centroid of the nearest spot in R, which results in shift of all the points in R. Each spot of I is corresponded with the nearest spot in R. Figure 40.1 shows the alignment of I with R along with the correspondence analysis.

The average slope for each lens i in a HS mirror is calculated from the geometry of the image by,

$$m_{x_i} = \frac{\Delta x_i}{f}; \quad m_{y_i} = \frac{\Delta y_i}{f} \tag{40.1}$$

where f denotes the focal length of the lenslet array, Δx_i and Δy_i are the respective shifts of each centroid along the x and y Cartesian coordinates. For each lens i, we arrive at (40.2).

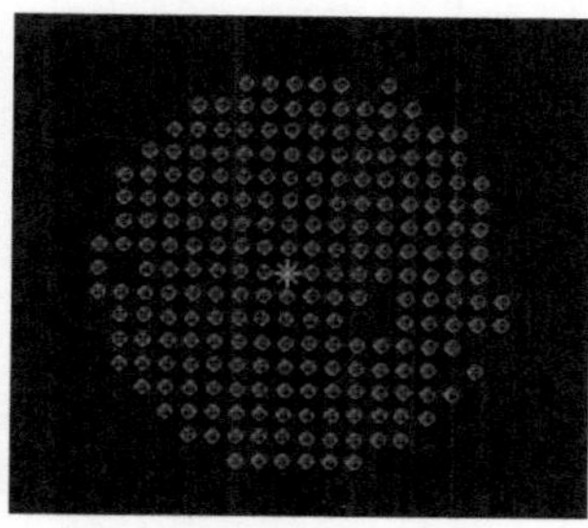

Fig. 40.1 It shows the subject image I centroids in *red pluses* within the effective pupil radius embedded on their corresponding reference image R centroids shown in *green*. The *green star* in the middle is the centroid of the entire image

$$m_{x_i} = \sum_{n=0}^{\infty} C_n \frac{\partial \overline{Z_{nx}(x,y)}}{\partial x}; \quad m_{y_i} = \sum_{n=0}^{\infty} C_n \frac{\partial \overline{Z_{ny}(x,y)}}{\partial y} \tag{40.2}$$

where $C_n = \oint W(x,y) Z_n(x,y)$, the integration occurs over the unit pupil circle. $\vec{c}$ denotes the matrix of all C_n. $Z_n(x,y)$ are the Zernike polynomials. $W(x,y)$ denotes the reconstructed wavefront. With q lenses, since the number of equations is finite and equal to $2q$, a finite number p of Zernikes can be calculated.$\vec{c}$ is unknown and thus needs to be estimated from 40.2. The number of equations is $2q$, which is generally greater than the number of unknowns p. A simple method to overcome the problem is to use Least Square Fit (LSF) method.

From the predicted values of $\vec{c}$, the spherical defocus, D in Diopters, of the model eye can be calculated using 40.3.

$$D = -\frac{4\pi\sqrt{3}}{r^2} c_5 - \frac{2\pi\sqrt{6}}{r^2} c_4 - \frac{2\pi\sqrt{6}}{r^2} c_6 \tag{40.3}$$

where r denotes the pupil radius. The aberrometer was calibrated using trial lenses where the power ranges from −1.50D to +1.50D.

The novelty of the paper lies in this section. Some HS spots may be missing if the optical components of the eye are compromised due to, for example, cataracts, dry eye, or if the shape of the eye is irregular. In such conditions, it can be impossible to get sufficient number of spots. The proposed technique deletes pre-fixed random number of spots from the image. The software then calculates the spherical power of the eye again by repeating equation numbers 1–3. By this deletion, it is possible to find the percentage of spots that can be absent from the HS image without hampering the prediction of the spherical power of the lens.

40.3 Experimental Results

40.3.1 Apparatus

The apparatus used for the experiment is a University of Waterloo custom made aberrometer. The schematic diagram of the aberrometer is given in Fig. 40.2. The focal length of each of the lenslet is 5.2 mm. Instead of a human eye, a model eye was used in order to simulate various refractive errors. For the reference, a flat plane mirror was used.

40.3.2 Experimental Result

The HS images were obtained from the CCD sensor and stored in Macintosh computer using i-Movie software. Table 40.1 gives the details of a simulation with a reference image and seven subject images having varied refractive errors from approximately −1.0D to +1.50D. Each of the subject images is captured 1000 times. The table shows how the algorithm performs in terms of average error and standard deviation of the error, with various proportions of HS spots deleted. It also averages the computational time for each run. The lens in the model eye has a focal point *f* of 130 mm and to simulate a refractive error of 0D, the axial length of the model eye was set to the focal length (130 mm). Subject images were also obtained

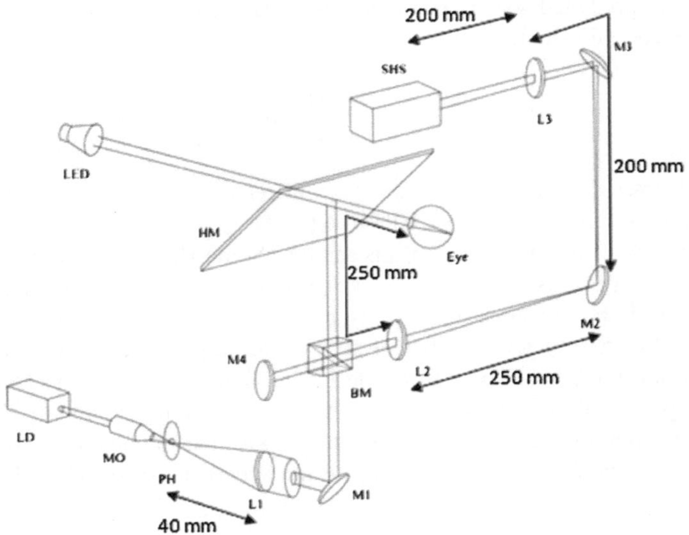

Fig. 40.2 Schematic illustration of the optical setup: *LD* laser diode; *MO* microscope objective; *PH* pinhole; *L1*, *L2*, and *L3* lenses; *M1*, *M2*, and *M3* mirrors; *M4* removable mirror used for calibration; *BM* beam splitter; *HM* viewing window (hot mirror); *HS* Hartmann-Shack sensor [6]

Table 40.1 Experimental outcomes for 1000 estimations of the spherical defocus at each of a series of axial lengths of a model eye, simulating different magnitudes of refractive error

% HS spots deleted	Axial length	116 mm	122 mm	130 mm (*f*)	138 mm	146 mm	161 mm
	Actual power (D)	0.928	0.504	0	−0.445	−0.843	−1.481
10	Avg. error (D)	0.046	0.101	0.081	0.153	0.118	0.100
	Std. dev	0.031	0.0572	0.038	0.107	0.033	0.114
20	Avg. error (D)	0.054	0.103	0.093	0.192	0.123	0.120
	Std. dev	0.041	0.073	0.051	0.140	0.049	0.118
30	Avg. error (D)	0.066	0.1203	0.096	0.222	0.134	0.134
	Std. dev	0.049	0.093	0.059	0.159	0.064	0.180
40	Avg. error (D)	0.085	0.151	0.09	0.223	0.158	0.230
	Std. dev	0.066	0.117	0.061	0.164	0.090	0.192
50	Avg. error (D)	0.094	0.156	0.109	0.133	0.175	0.231
	Std. dev	0.074	0.135	0.082	0.165	0.113	0.199
60	Avg. error (D)	0.111	0.169	0.126	0.446	0.20	0.603
	Std. dev	0.105	0.212	0.111	0.209	0.172	0.288
70	Avg. error (D)	0.181	0.406	0.178	0.775	0.299	0.880
	Std. dev	0.423	0.936	0.200	1.48	0.448	0.895
80	Avg. error (D)	3.371	2.482	8.10E + 7	5.072	1.979	2.87
	Std. dev	30.58	16.36	2.56E + 13	26.84	12.80	13.41

with the axial length of the model eye shifted from 130 mm to positions of 116, 122, 138, 146 and 161 mm. Axial lengths less than 130 mm simulated hyperopic defocus corresponding to +0.92D and +0.50D, and axial lengths greater than 130 mm simulated myopic defocus corresponding to −0.44D, −0.84D and −1.48D respectively. Figure 40.3 shows the contents of Table 40.1 for different defocuses in an error bar plot for 0–50 % deletion of HS spots.

40.3.3 Discussion

Spots in HS images were deleted in steps of 10 % up to 80 %. The software fails to run when 90 % of the spots were deleted, mainly because $2q$ becomes less than p in this condition and, as a result, the least squares fit is not applicable due to the number of equations being less than number of unknowns. From Table 40.1, it can

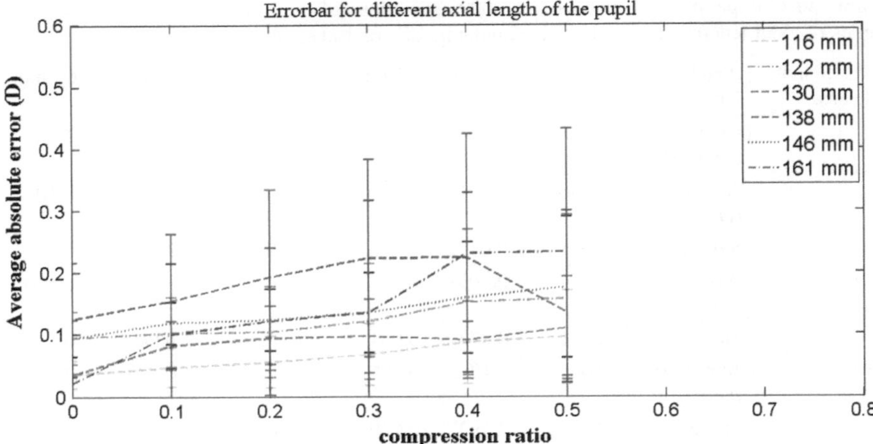

Fig. 40.3 It shows the errorbar plot with different axial lengths for compression ratio 0–0.5

be seen that up to 50 % of the spots can be safely deleted without hampering any of the parameters like average error and standard deviation of error appreciably. For these cases, the average error over 1000 runs is less than ±0.25D, which is the typical step size in a clinical estimation of the spherical component of refractive error. When it comes to 60 % deletion, the worst error combined with the standard deviation of the error in this condition is substantial, indicating that the method is unfeasible in practical terms for this condition.

Some errors in the estimations may arise from experimental conditions such as the precision of the measurement of the axial length and reflections on the background of the HS image due to room illumination. The latter, can pose a problem in accurate detection of the spots in the acquired image especially under clinical testing conditions.

40.4 Conclusion

The proposed method undoubtedly is a fast algorithm with minimal computation time. The novelty of the proposed method is that it statistically analyzes the effect of predicting the defocus term (in D) by randomly deleting some predetermined percentage of spots from the HS image. Even if 50 % of the spots are missing from the HS image acquired from a 'normal' model eye, it describes the defocus within the bounds of typical clinical limits (±0.25D). The error of prediction however increases as percentage of spots to be deleted exceeds 50 % and, indeed, is unable to estimate an outcome with 90 % or more of the spots deleted.

Other optimizations of the method may include investigation of procedures other than LSF to recover the unknown parameters and the impact on the error of the estimates when the deleted spots are spatially clustered versus spatially dispersed.

It should be noted, that this initial investigation only estimates the precision of the estimate of the second order defocus term in the wavefront aberration of a 'normal' eye. The second order defocus corresponds to the clinical measurement of myopia or hyperopia. Further investigation is needed to determine the precision of other Zernike coefficients (particularly the higher order terms) in 'normal' eyes and, in addition, the extension of the technique to eyes where the optical components depart from 'normal' due to the presence of disease. Nevertheless, this initial investigation provides evidence, in principle, that reasonable estimation of the ocular wavefront aberrations might be made in the absence a dense lenslet array.

References

1. Thapa D, Fleck A, Lakshminarayanan V, Bobier WR (2011) Ocular wavefront aberration and refractive error in pre-school children. J Mod Opt 58:1681–1689
2. Liang J, Grimm B, Goelz S, Bille JF (1994) Objective measurement of wave aberrations of the human eye with the use of a Hartmann-Shack wave-front sensor. J Opt Soc Am A 11:1949–1957
3. Lakshminarayanan V, Felck A (2011) Zernike Polynomials: a guide. J Mod Opt 58:545–561
4. Neal DR, Topa DM, Copland J (2001) Effect of lenslet resolution on the accuracy of ocular wavefront measurements. In: BIOS, the international symposium on biomedical optics. International Society for Optics and Photonics, Bellingham, pp 78–91
5. Thibos LN, Hong X (1999) Clinical applications of the Shack-Hartmann aberrometer. Optom Vis Sci 76:817–825
6. Diaz-Santana L, Guériaux V, Arden G, Gruppetta S (2007) New methodology to measure the dynamics of ocular wave front aberrations during small amplitude changes of accommodation. Opt Express 15(9):5649–5663

Chapter 41
Automated Detection of Optic Disc in Fundus Images

R. Burman, A. Almazroa, K. Raahemifar and V. Lakshminarayanan

Abstract Optic disc (OD) localization is an important preprocessing step in the automated image detection of fundus image infected with glaucoma. An Interval Type-II fuzzy entropy based thresholding scheme along with Differential Evolution (DE) is applied to determine the location of the OD in the right of left eye retinal fundus image. The algorithm, when applied to 460 fundus images from the MESSIDOR dataset, shows a success rate of 99.07 % for 217 normal images and 95.47 % for 243 pathological images. The mean computational time is 1.709 s for normal images and 1.753 s for pathological images. These results are important for automated detection of glaucoma and for telemedicine purposes.

41.1 Introduction

In color fundus images, the optic disc (OD) appears as a bright yellowish or white region inside the retina. It has a more or less circular shape, interrupted by blood vessels, but sometimes it may take the shape of an ellipse due to difference in nature of photographic projection. But reliable OD localization seems to be difficult due to the confounding bright pathologies and variable shapes of the OD. The most common methods for OD localization are based on intensity and shape detections. Methods based on intensity variance, principal component analysis, template

R. Burman (✉)
Department of Electronics and Telecommunication Engineering, Jadavpur University, Kolkata, India
e-mail: ritambharkcdburman@gmail.com

A. Almazroa · V. Lakshminarayanan
School of Optometry and Vision Science, University of Waterloo, Waterloo, Canada

K. Raahemifar
Department of Electrical and Computer Engineering, Ryerson University, Toronto, Canada

V. Lakshminarayanan
Department of Physics, University of Waterloo, Waterloo, Canada

V. Lakshminarayanan and I. Bhattacharya (eds.), *Advances in Optical Science and Engineering*, Springer Proceedings in Physics 166,
DOI 10.1007/978-81-322-2367-2_41

matching and pyramidal decomposition with Hausdorff-based template matching have been used in literature [1, 2] for OD detection. The detection deals with two complexities—accuracy and time. Any misclassification may degrade the performance, while the time complexity may play a large factor. So a trade-off is needed in terms of time complexity and success rate.

The paper is divided as followed: In Sect. 41.2, the various steps of the algorithm are discussed along with the mathematical formulations of multi-level image segmentation using interval type-II fuzzy sets. Section 41.3 gives a detailed statistical and computational analysis of the experimental results. Section 41.4 presents the conclusion and some of the algorithms that could be done to improve the accuracy.

41.2 Algorithm

41.2.1 Preprocessing

The first step in dealing with the retinal fundus image, the pre-processing stage, deals with the conversion of the RGB image to grayscale image, taking into account the weighted means of red, green and blue channels, that can distinguish the OD region from the rest of the retina more accurately than the RGB image or any other channel of the image.

41.2.2 Multi-level Segmentation

The proposed technique uses the intensity of the objects in the image to localize the OD. The brightest object of the segmented retinal image is the Optic Cup (OC) which lies inside the OD. By segmenting the grayscale image into various objects, we will be able to find the brightest object of the image, which happens to be the OC.

The multi-level image segmentation is a method to segment the image into various objects. Besides the various entropy based segmentation techniques (e.g., Shannon entropy [3], Tsalli's entropy [4] and Renyi's entropy [5]), Zhao et al. [6] thresholded the image by partitioning the histogram using fuzzy membership values and derived the optimal threshold condition, which was further modified by Tao et al. [7]. Their fuzzy entropy based technique partitioned the image histogram into various objects. Type II Fuzzy sets is a generalization of Type I Fuzzy Sets and can deal with more uncertainty. It uses a measure called ultrafuzziness to obtain the image thresholds. Burman et al. [8] introduced a new measure called Type II Fuzzy entropy which when maximized gives the best thresholds for a particular image. It uses Differential Evolution (DE) [9], a powerful meta-heuristic technique for faster convergence and less computational time complexity. The brightest region of the image is given by the upper threshold value in the multi-level segmented image. So the brightest object, being a part of OC, is automatically a part of OD.

41.2.3 Interval Type-II Fuzzy Sets

Type-I Fuzzy set [10] A, in a finite set $X = \{x_1, x_2, \ldots, x_n\}$ is represented as

$$A = \{x, \mu_A(x) | x \in X, 0 \leq \mu_A(x) \leq 1\} \tag{41.1}$$

A range of membership values were introduced instead of a single membership value. It is defined as:

$$A = \{x, \overline{\mu_A}(x), \underline{\mu_A}(x) | x \in X, 0 \leq \overline{\mu_A}(x), \underline{\mu_A}(x) \leq 1\} \tag{41.2}$$

where $\overline{\mu_A}(x)$ and $\underline{\mu_A}(x)$ are the upper and lower membership functions respectively.

Let I be a grayscale image having pixels $M * N$. L be the number of grey intensity values by which the image can be represented. Let the set of all gray levels be denoted by GL,

$$\begin{aligned} &GL = \{0, 1, 2, \ldots, L-1\} \\ &f(m, n) \in GL, \quad (m, n) \in I \\ &D_k = \{(m, n) : f(m, n) = k, k \in \{0, 1, 2, \ldots, L-1\}\}, \end{aligned}$$

where $f(m, n)$ denotes the intensity value of the pixel (m, n) inside the image I. $H = \{h_0, h_1, \ldots, h_{L-1}\}$ be the normalized histogram of the image, where $h_k = n_k/(M * N), n_k$ is the number of pixels in D_k. A measure called ultra-fuzziness, associated with fuzzysets, gives a zero value, when the membership values can be defined without any uncertainty, and increases to one when membership values can be indicated within an interval. The ultra-fuzziness for an image for the kth level can be defined mathematically as:

$$P_k = \sum_{i=0}^{L-1} h_i * \left(\overline{\mu_k}(i) - \underline{\mu_k}(i)\right) \tag{41.3}$$

where $k \in \{1, 2, \ldots, \text{Lev} + 1\}$, Lev is the number of levels of segmentation and μ_k is the trapezoidal fuzzy membership function for gray values to belong to kth level out of Lev + 1 levels of segmentation, defined as:

$$\mu_k(i) = \begin{cases} 0, & i \leq a_{k-1} \\ \frac{i - a_{k-1}}{c_{k-1} - a_{k-1}}, & a_{k-1} < i \leq c_{k-1} \\ \frac{i - c_k}{a_k - c_k}, & c_{k-1} < i \leq c_k \\ 0, & i > c_k \end{cases} \tag{41.4}$$

where a_k and c_k, $k \in \{1, 2, \ldots, \text{Lev} + 1\}$ are the fuzzy parameters and $a_0 = c_0 = 0, a_{N+1} = c_{N+1} = L - 1$. The fuzzy type-II entropy for kth level of segmentation is defined as:

$$H_k = -\sum_{i=0}^{L-1} \frac{h_i * \left(\overline{\mu_k}(i) - \underline{\mu_k}(i)\right)}{P_k} * \ln\left(\frac{h_i * \left(\overline{\mu_k}(i) - \underline{\mu_k}(i)\right)}{P_k}\right) \quad (41.5)$$

where $k \in \{1, 2, \ldots, \text{Lev} + 1\}$. The total entropy of I is given by the summation of entropies of all the levels.

$$H(a_1, c_1, \ldots, a_n, c_n) = \sum_{k=1}^{N+1} H_k \quad (41.6)$$

The optimal fuzzy parameters is obtained by making the total entropy maximum,

$$(a_1, c_1, \ldots, a_n, c_n)^* = \text{Argmax}(H) \quad (41.7)$$

The thresholds for the image is given by the equation

$$T_n = \frac{1}{2} * (a_n + c_n), \quad n \in \{1, 2, \ldots, \text{Lev}\} \quad (41.8)$$

DE is a novel heuristic optimization technique used to reduce the time complexity. The parameter set $(a_1, c_1, \ldots, a_n, c_n)$ acts as an individual of the population in DE.

41.2.4 Differential Evolution (DE)

Storn proposed DE, a global meta-heuristic optimizer to search for the global optimum point in a D-dimensional real parameter space R^D. Here a simple version of DE, DE/rand/1 scheme, is used. The ith individual, denoted by a vector of D dimension, can be represented as,

$$\overrightarrow{X_i}(t) = \left[\overrightarrow{X_{i,1}}(t), \overrightarrow{X_{i,2}}(t), \ldots, \overrightarrow{X_{i,D}}(t)\right] \quad (41.9)$$

In the search space, population is randomly initialized as,

$$\overrightarrow{X_{i,j}}(t) = X_{jmin} + \text{rand}(\text{NP}, D) * (X_{jmax} - X_{jmin}) \quad (41.10)$$

where NP is the total population members, X_{jmin} and X_{jmax} are the minimum and maximum bound of search space respectively and rand(NP, D) is a matrix having NP rows and D columns having random numbers between 0 and 1. In every iteration, one donor vector or mutant vector is obtained for each parent vector through differential mutation operation. Three parent vectors a_1, a_2 and a_3 are chosen randomly from the current population to create the ith donor vector $\overrightarrow{Y_i}(t)$ for the *i*th parent vector, so that they are not equal to one another and lie between 1 and NP. So the donor vector $\overrightarrow{Y_i}(t)$ is given by:

$$\overrightarrow{Y_i}(t) = \overrightarrow{X_{a_1}}(t) + F * \left(\overrightarrow{X_{a_2}}(t) - \overrightarrow{X_{a_3}}(t)\right) \tag{41.11}$$

where F is a scalar quantity called weighing factor. Trial vectors are created to enhance the potential diversity of the donor vectors to create the trial vectors. On each of the variables, a binomial crossover operation is performed as,

$$R_{i,j}(t) = \begin{cases} Y_{i,j}(t), & \text{rand}_j(0,1) \leq \text{Cr or } j = j_{\text{rand}} \\ X_{i,j}(t), & \text{otherwise} \end{cases} \tag{41.12}$$

where j = 1 to D and $\text{rand}_j \in [0,1]$ is a uniform random generator in the *j*th evaluation. $j_{\text{rand}} \in [1,2,\ldots,D]$ is randomly chosen to ensure $\overrightarrow{R_i}(t)$ getting atleast one component of $\overrightarrow{X_i}(t)$ and Cr is known as the crossover rate. If the fitness value of the trial vector in iteration t is more than the current parent vector for finding the maxima, the parent vector is replaced by the trial vector as the new parent vector in iteration t + 1.

$$\overrightarrow{X_i}(t+1) = \begin{cases} \overrightarrow{R_i}(t), & \text{if } f\left(\overrightarrow{R_i}(t)\right) \geq f\left(\overrightarrow{X_i}(t)\right) \\ \overrightarrow{X_i}(t), & \text{if } f\left(\overrightarrow{R_i}(t)\right) \leq f\left(\overrightarrow{X_i}(t)\right) \end{cases} \tag{41.13}$$

The above steps are repeated until the ending criterion is met. The termination criteria used here is the number of iterations.

41.2.5 Region of Interest (ROI)

After the detection of OD, the centroid of the OC can be found out by locating the geometrical centroid of the brightest object in the image. After the localization of the OC and hence the OD, the next thing is to find the Region of Interest (ROI), which deals with the cutting out a suitable portion of the retinal image by adding suitable number of pixels on either side of the centroid to show the OD for further processing applications.

41.3 Experimental Results

The simulation results were performed in Matlab 2013a environment in a workstation with Intel Core i-5 2.80 GHz processor. For testing, the algorithm was applied to 460 images, applied from the MESSIDOR dataset [11], which has 217

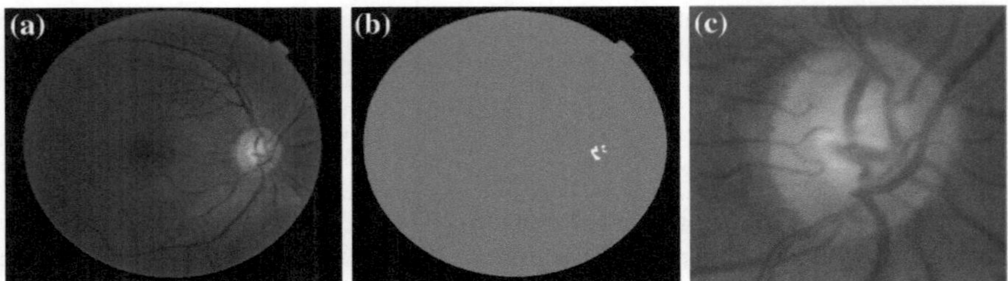

Fig. 41.1 **a** Is a sample fundus image in the grayscale, **b** segmented image, and **c** is the ROI. ROI is shown taking 350 × 350 pixel box

normal and 243 pathological fundus images. Pathological images contain micro-aneurysms, exudates and haemorrhages (retinal diseases), which were examined by different experts. Sample fundus image in grayscale, the segmented image and the localized OD are shown in Fig. 41.1.

An unsuccessful sample of pathological fundus images are shown in Fig. 41.2. The pathological image makes OD localization more difficult due to the illumination changes in the retina caused by the abnormality of the retina. Figure 41.2a is a sample pathological image where it has abnormal illumination in the leftmost part of the retina, while the OD is on the left most part. The proposed algorithm fails to run properly in this case as it detects the bright portion in the left part as well. Thus, the algorithm treats the entire bright region as the OD. Therefore, it miscalculates the centroid as the point that lies in the middle of the retina.

A constraint DE is applied here to achieve lower computational time. The parametric setup of DE is given in Table 41.1.

Table 41.2 shows that the MESSIDOR dataset has 217 normal images and 243 pathological images. The proposed algorithm was successful for 215 normal images achieving a success rate of 99.07 % and 232 pathological images achieving a success rate of 95.47 %. Overall the algorithm has a success rate of 97.17 %, being successful in 447 cases out of 460. The DE approach shows that the average computation time is more or less same for both normal and pathological images with

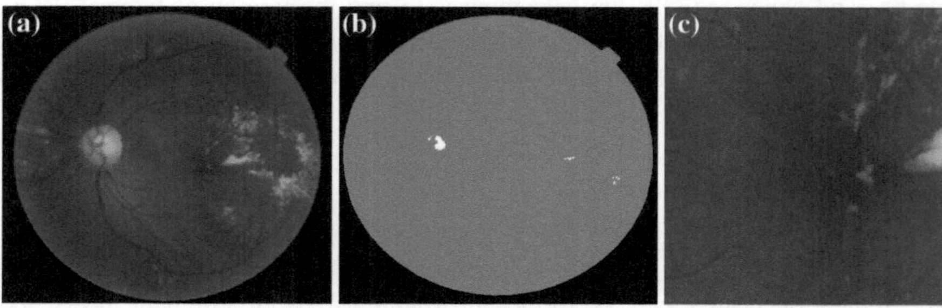

Fig. 41.2 **a** Is a sample of pathological image, **b** is the segmented image, and **c** is the ROI having 350 × 350 pixels

Table 41.1 Setup for DE

Parameter	Setup
Number of runs	5
Number of iterations per run	100
Dimension of the search space (*D*)	2 $*$ number of thresholds
X_{jmax}	Maximum gray level in image
X_{jmin}	Minimum gray level in image
Number of particles (NP)	10 $* D$
Weighing factor(*F*)	0.5
Crossover rate (Cr)	0.9

Table 41.2 Simulation result on the MESSIDOR dataset

	Normal images	Pathological images	Overall
No. of images	217	243	460
Success	215	232	447
% accuracy	99.07	95.47	97.17
Average time (s)	1.709	1.753	1.731

the latter is 0.05 s slower than the normal images. Overall the computation time for detecting the OD is 1.731 s per image. The implementation of the algorithm is unique and showed good results for both localization and computational time.

41.4 Conclusion

The algorithm undoubtedly manages to detect the centroid of the OD with near perfection for the normal fundus image and with very high accuracy for the pathological images. As result, Fuzzy Interval Type-II multi-level thresholding mechanism can undoubtedly be used in the case of OD localization. The use of the DE helps to maximize the ultrafuzziness measure and speeds up the process of detection. Better evolutionary computation variants may be used to further speed up the process. Also several other membership functions can be used to further enhance the accuracy.

References

1. Rotaru F, Bejinariu SI, Nit CD, Ramona R, Lazăr C (2014) Optic disc identification methods for retinal images. Comput Sci J Moldova, 22, 2(65) (invited article)
2. Kauppi T, Kälviäinen H (2008) Simple and robust optic disc localisation using colour decorrelated templates. In: Proceedings of 10th international conference on advanced concepts for intelligent vision system. Springer, Berlin, pp 719–729

3. Benzid R, Arar D, Bentoumi M (2008) A fast technique for gray level image thresholding and quantization based on the entropy maximization. In: 5th international multi-conference on systems, signals and devices, pp 1–4
4. Sarkar S, Das S, Chaudhuri SS (2012) Multilevel image thresholding based on Tsallis entropy and differential evolution. SEMCCO LNCS 7677:17–24
5. Sahoo PK, Arora G (2004) A thresholding method based on two dimensional Renyi's entropy. Pattern Recogn 37:1149–1161
6. Zhao MS, Fu AMN, Yan H (2001) A technique of three level thresholding based on probability partition and fuzzy 3-partition. IEEE Trans Fuzzy Syst 9(3):469–479
7. Tao WB, Tian JW, Liu J (2003) Image segmentation by three-level thresholding based on maximum fuzzy entropy and genetic algorithm. Pattern Recogn Lett 24:3069–3078
8. Burman R, Paul S, Das S (2013) A differential evolution approach to multi-level image thresholding using type II fuzzy sets. In: Swarm, evolutionary, and memetic computing. Springer, Switzerland, pp 274–285
9. Storn R, Price K (1997) Differential evolution-a simple and efficient heuristic for global optimization over continuous spaces. J Global Optim 11:341–359
10. Zadeh LA (1965) Fuzzy sets. Inf Control 8:338–353
11. MESSIDOR: Methods for evaluating segmentation and indexing technique dedicated to retinal ophthalmology. http://messidor.crihan.fr/index-en.php (2004)

Chapter 42
Diagnosing Heterogeneous Dynamics for CT Scan Images of Human Brain in Wavelet and MFDFA Domain

Sabyasachi Mukhopadhyay, Soham Mandal, Nandan K. Das, Subhadip Dey, Asish Mitra, Nirmalya Ghosh and Prasanta K. Panigrahi

Abstract CT scan images of human brain of a particular patient in different cross sections are taken, on which wavelet transform and multi-fractal analysis are applied. The vertical and horizontal unfolding of images are done before analyzing these images. Discrete wavelet transform (DWT) through Daubechies basis are done for identifying fluctuations over polynomial trends for clear characterization of CT scan images of human brain in different cross-sections. A systematic investigation of de-noised images are carried out through wavelet normalized energy and wavelet semi-log plots, which clearly points out the mismatch between results of vertical and horizontal unfolding. The mismatch of results confirms the heterogeneity in spatial domain. Using the multi-fractal de-trended fluctuation analysis (MFDFA), the mismatch between the values of Hurst exponent and width of singularity spectrum by vertical and horizontal unfolding confirms the same.

S. Mukhopadhyay · N.K. Das · N. Ghosh · P.K. Panigrahi
Indian Institute of Science Education and Research, Kolkata, India

S. Mandal (✉)
Institute of Engineering and Management, Kolkata, India
e-mail: somban637@gmail.com

S. Dey
Bidhan Chandra Krishi Viswa Vidyalaya, Kalyani, India

A. Mitra
College of Engineering and Management, Kolaghat, India

V. Lakshminarayanan and I. Bhattacharya (eds.), *Advances in Optical Science and Engineering*, Springer Proceedings in Physics 166,
DOI 10.1007/978-81-322-2367-2_42

42.1 Introduction

Significant research works are being undertaken for understanding of the dynamics of biomedical images over the last few decades. There are several conventional methods for early detection and diagnosis of abnormalities in cell tissues. For quantification of both the morphological and biochemical alterations associated with abnormalities development like cancer, the biomedical imaging based approaches have shown early promise [1–4]. Wavelets and MFDFA are very useful tool for their versatile applications from the field of environmental science to biomedical imaging [5, 9–12]. In previous works by Mukhopadhyay et al. the biomedical image analysis of DIC stromal as well as epithelium regions are done with wavelet and MFDFA [10, 11]. The significant variations are observed among different grades of cancer tissues for stromal and epithelium regions. In this paper heterogeneous dynamics for CT scan images of human brain are observed. The results are discussed in this paper elaborately.

42.2 Theory

42.2.1 Wavelet Transform

If the low pass and high pass coefficients are $c_k{}'s$ and $d_{j,k}{}'s$ respectively, the data set can be expressed for Discrete wavelet transform (DWT) as $f(t) = \sum_k c_k \phi_k + \sum_k \sum_{j=0}^{\infty} d_{j,k} \psi_{j,k}$. The trend components are extracted by father wavelet ϕ_k, located at k with level j and the deviations from the trend are picked up by mother wavelets $\psi_{j,k}$.

42.2.2 Multi-fractal De-trended Fluctuation Analysis (MFDFA)

Briefly, the profile $Y(i)$ (spatial series of length N, $i = 1\ldots N$) is first generated from the one dimensional spatial index fluctuations. The local trend of the series $[y_b(i)]$ is determined for each segment b by least square polynomial fitting, and then subtracted from the segmented profiles to yield the de-trended fluctuations. The resulting variance of the de-trended fluctuation is determined for each segment as

$$F^2(b, s) = \frac{1}{s} \sum_{i=1}^{s} [Y\{(b-1)s + i\} - y_b(i)]^2 \tag{42.1}$$

The moment (q) dependent fluctuation function is then extracted by averaging over all the segments as

$$F_q(s) = \left\{ \frac{1}{2N_s} \sum_{b=1}^{2N_s} \left[F^2(b,s) \right]^{\frac{q}{2}} \right\}^{1/q} \tag{42.2}$$

The scaling behaviour is subsequently determined by analysing the variations of $F_q(s)$ versus s for each values of q, assuming the general scaling function as

$$F_q(s) \sim s^{h(q)} \tag{42.3}$$

Here, the generalized Hurst exponent $h(q)$.

42.3 Results and Discussions

The CT scan images of human brain in different cross sections are taken for analysis purpose. The sample CT scan image is mentioned below as an example (Fig. 42.1).

We know that the biomedical images can be considered as irregular signals. Wavelets are ideal analysis tool for irregular data processing purpose. For any observed signal $x(t) = f(t) + e(t)$, where $f(t)$ is the signal and $e(t)$ is the noise, using wavelets $f(t)$ can easily be extracted out. In this paper, Discrete wavelet transform through Daubechies basis analysis has been done up to level-5 for identifying localized fluctuations (high pass coefficients) over polynomial trends for clear characterization and differentiation of CT scan images of human brain in different cross sections (Fig. 42.2).

The plot of Wavelet normalized energy versus level of decomposition (level 1–level 5) of high pass coefficients (fluctuations) have been performed here using db-4 for CT scan images of human brain.

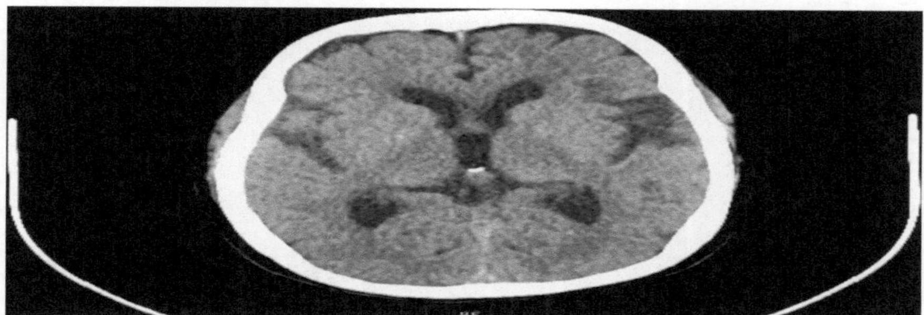

Fig. 42.1 CT scan image of human brain

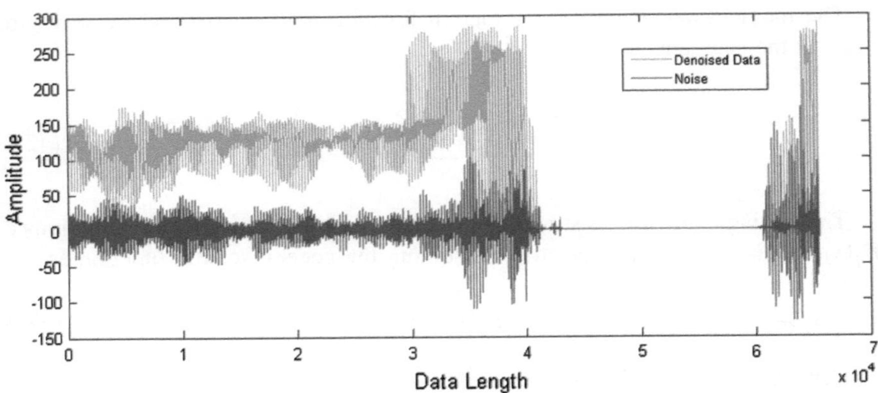

Fig. 42.2 Denoised data and noise (fluctuations) after applying Daubechies-4 (Db-4) on CT scan images

It is clearly observed that the normalized energy value differs from horizontal to vertical unfolding which is the evidence of medium heterogeneity.

Now the semi log plots for horizontal and vertical unfolding components are done to check the anisotropic nature of fluctuations using Morlet transform.

The mismatch between the horizontal and vertical unfolding from Fig. 42.3 result have confirmed the medium anisotropic nature. It is actually an evidence of medium heterogenity of CT scan images (Fig. 42.4).

We now turn to check the self-similar purpose. In this regard, MFDFA is helpful for subtle morphological changes which are otherwise hidden (Fig. 42.5).

The mismatch of hurst exponent and singularity spectrum width values for horizontal and vertical unfolding cases clearly have shown the heterogeneity nature of the human brain CT scan images.

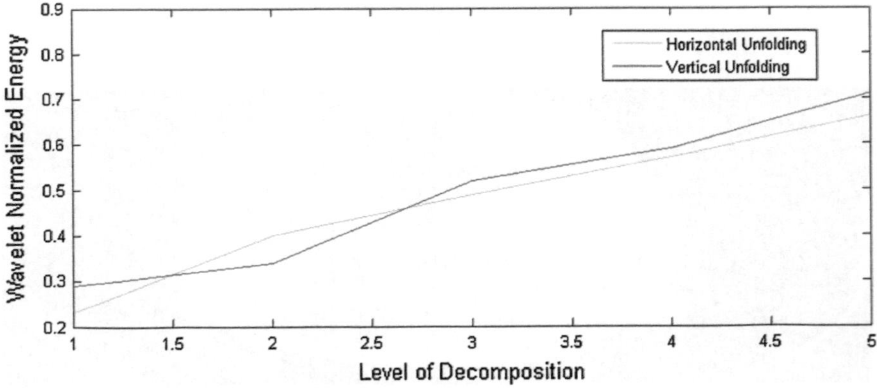

Fig. 42.3 Wavelet normalized energy for vertical and horizontal unfolding

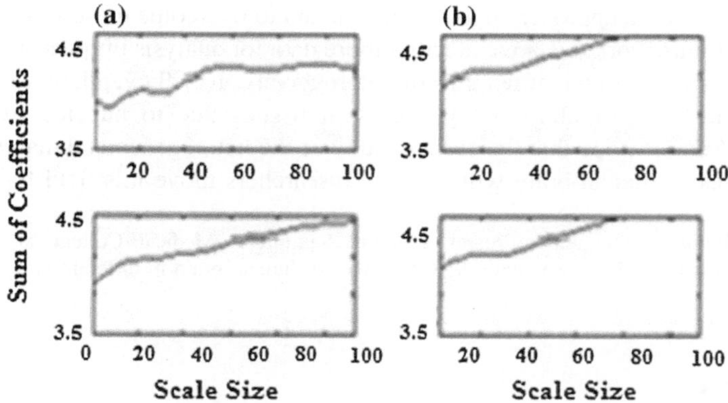

Fig. 42.4 Wavelet semi-log plot of brain CT scan images of human brain in different cross sections where column **a** stands for horizontal unfolding and column **b** stands for vertical unfolding. The mismatch among horizontal and vertical unfolding results clearly confirms the heterogeneity

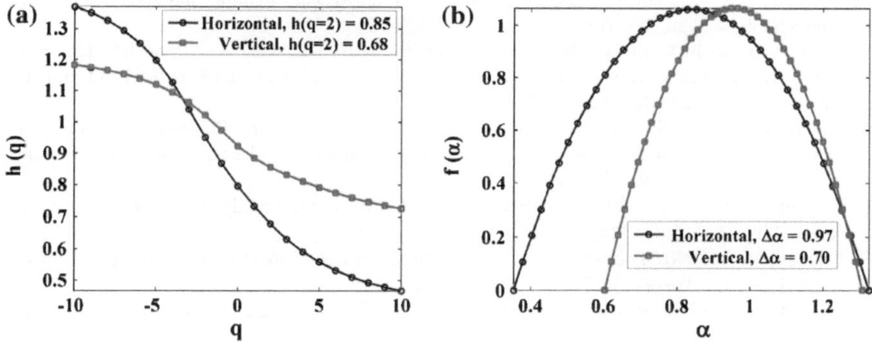

Fig. 42.5 **a**, **b** Plot of Hurst Exponent (Hq) and singularity spectrum [$f(\alpha)$] for CT scan images of vertical and horizontal unfolding respectively

42.4 Conclusion

We have described a robust method to automatically detect the heterogeneity in CT scan images of human brain. The current paper emphasizes upon the generalization of the method i.e., selection of wavelets, the scale of MFDFA analysis depend upon the judicious choices of the user. Thus the success of the methods depends upon that judicious choice. Wavelet normalized energy, Wavelet semi-log plots as well as MFDFA based hurst exponent and singularity spectrum based results further depict the heterogeneity of the CT scan images of human brain. In a nutshell, the proposed method is aimed to precisely extract out the substantial features which was unknown using traditional statistical operations over medical images. The wavelet and MFDFA based approach in this paper can aid to overcome those limitations. As a plan for future work, we are collecting more data for analysis purpose and hope to extract out more prominent features of heterogeneity. It will even help to predict if there is any kind of abnormality occurs in tissues due to affected by chronic diseases. Authors hope that the current study of the heterogeneous dynamics of CT scan images of human brain will help the researchers move this field forward.

Acknowledgments The authors thank Bankura Sammilani Medical College and Hospital, Bankura, West Bengal for providing the CT images of human brain in different cross-section.

References

1. Alfano RR, Das BB, Cleary J, Prudente R, Celmer E (1991) Light sheds light on cancer distinguishing malignant tumors from benign tissues and tumors. Bull NY Acad Med 67(2): 143–150
2. Ramanujam N (2000) Fluorescence spectroscopy of neoplastic and non-neoplastic tissues. Neoplasia 2(12):89
3. Boustany NN, Boppart SA, Backman V (2010) microscopic imaging and spectroscopy with scattered light. Ann Rev Biomed Eng 12(1):285–314
4. Schantz SP, Kolli V, Savage HE, Yu G, Shah JP, Harris DE, Katz A, Alfano RR, Huvos AG (1998) Invivo native cellular fluorescence and histological characteristics of head and neck cancer. Clin Cancer Res 4(5):1177–1182
5. Das N, Chatterjee S, Soni J, Jagtap J, Pradhan A, Sengupta TK, Panigrahi PK, Vitkin IA, Ghosh N (2013) Probing multifractality in tissue refractive index: prospects for precancer detection. Opt Lett 38(2):211–213
6. Kantelhardt JW et al (2002) Multifractal detrended fluctuation analysis of nonstationary time series. Phys A 316:87–114
7. Daubechies I (1990) The wavelet transform, time-frequency localization and signal analysis. IEEE Trans Inf Theory 36(5):961–1005
8. Modi JK, Nanavati SP, Phadke AS, Panigrahi PK (2004) Wavelet transforms: application to data analysis-II. Resonance 8–13
9. Mukhopadhyay S, Das N, Pradhan A, Ghosh N, Panigrahi PK (2014) Pre-cancer detection by wavelet transform and multi-fractality in various grades of DIC stromal images. In: SPIE BIOS, USA
10. Mukhopadhyay S, Panigrahi PK (2013) Wind speed data analysis for various seasons during a decade by wavelet and S transform. Int J Comput Sci Technol 3(4)
11. Mukhopadhyay S, Das NK, Kumar R, Dash D, Mitra A, Panigrahi PK (2014) Study of the dynamics of wind data fluctuations: a wavelet and MFDFA based novel method. In: Elsevier Science and Technology Proceeding, IEMCONGRESS, India
12. Mukhopadhyay S, Das NK, Pradhan A, Ghosh N, Panigrahi PK (2014) Wavelet and multi-fractal based analysis on DIC images in epithelium region to detect and diagnose the cancer progress among different grades of tissues. In: Proceedings of SPIE photonics Europe

Chapter 43
Growth of Blue Luminescent Cu Doped ZnO Nanowires by Modified Sol-Gel

U.P.S. Gahlaut, Vijay Kumar, R.K. Pandey and Y.C. Goswami

Abstract Copper doped ultra small highly luminescent ZnO nanowires were obtained by modified sol gel route. The zinc acetate mixed with ethanol was used as precursor for the synthesis. Copper doping was done by adding various concentrations of copper chloride. Gel was obtained by magnetically stirring the mixture at 60 °C for 2–3 h and then keeping it for another 24 h for aging. The particles obtained were characterized by XRD, AFM, Optical transmission and photoluminescence studies. X ray diffractogram peak is identified for ZnO. AFM micrograph shows particle alignment in linear direction and formation of nanowires on increasing Cu concentration. The optical band gap of copper doped ZnO shift from 3.2 to 4.76 eV. This strong shift confirms the ultra small size of particles. In PL studies, Undoped ZnO nanostructures exhibit a near-band-edge UV emission at 360 nm and a broad defect related blue emission at 440 nm. Addition of Copper improves the photoluminescence peak in UV region with an additional peak observed in middle UV region at 230 nm. Yellow green or blue makes the nanowires suitable for light emitting devices and biological sensing devices.

43.1 Introduction

Very high transmission in visible region (direct band gap 3.37 eV) [1], excellent conductivity with dopants like Al, Cu, special geometry and chemical physical properties make Zinc oxide promising candidate for transparent conducting electrodes and optoelectronic devices such as light emitting diodes, photo detectors, photo catalyst, window layer in solar cells, spintronics and sensing applications

U.P.S. Gahlaut (✉) · V. Kumar · Y.C. Goswami
School of Physical Sciences, ITM University, Turari, Gwalior 474001, Madhya Pradesh, India
e-mail: upsgahlaut@gmail.com; y_goswami@yahoo.com

R.K. Pandey
NIIT University Alwar, Rajasthan, India

V. Lakshminarayanan and I. Bhattacharya (eds.), *Advances in Optical Science and Engineering*, Springer Proceedings in Physics 166,
DOI 10.1007/978-81-322-2367-2_43

[2, 3]. Band gap engineering in ZnO based short wavelength devices is one of the most emerging area [4]. Luminescence properties of ZnO nanoparticles are also of great interest to analyze the surface defects and size. It has also been shown that transition metals are suitable candidate for luminescence property as well. The doping of transition metals in the ZnO to improve various physical qualities is already been reported [5]. Various transition metals are used as dopants. Out of which, Copper is one of the most promising dopant since it has comparable size with Zn ion, and hence it can easily substitute Zn ion. Synthesis of ZnO nanostructures has been reported using various deposition techniques like pulsed laser deposition, Sputtering, MOCVD, electrodeposition hydrothermal methods sol gel spin coating [6, 7]. Sol gel is a simple, low cost method known for high reproducibility, high controllability of composition and suitable for large scale production. In this paper we report the luminescence enhancement in Cu doped ZnO ultra small nanostructures grown by sol gel route.

43.2 Experimental Details

ZnO nanostructures were obtained by sol-gel chemical route. The Zinc acetate was (0.5 M) mixed with ethanol and stirred ultrasonically 30 min at room temperature. The doping was performed by adding Copper chloride in the ratio of 0.1 M (Sample A), 0.2 M (Sample B) and 0.3 M (Sample C). The homogeneous solution was obtained by magnetically stirring the mixture at 60 °C temperature for 2–3 h. The gel was obtained by sonicating the sol for another 3 h at 60 °C followed by aging of 24 h. The particles were obtained by centrifuge the gel.

The samples were characterized by XRD, AFM studies, UV-vis spectra and photoluminescence studies. The transmission spectra were obtained using Perkin Elmer Lembda 25 in the range 200–800 nm. The PL spectrum was obtained by Perkin Elmer LS55 with exciton at 200 nm of Xe lamp. The X-ray diffractograms were obtained in the 2θ range from 20 to 80° with Cu Kα radiation of wavelength 1.546 Å using Bruker D8 Advanced XRD. AFM micrographs were obtained by using digital Instrumentation—Inc, USA.

43.3 Results and Discussion

Figure 43.1 shows the X-ray diffractograms of undoped and Cu doped ZnO nanostructures grown by ultrasonic sol-gel method. Small peaks of (100) and (002) are observed and indentified as ZnO peaks [8]. Small intensity of the peaks indicates the formation of ultra small structures. The broad peak of at 22° may be due to glass substrate similar results also reported by others also for very thin layers [9].

Figure 43.2 shows the AFM micrographs of undoped and Cu doped ZnO nanostructures. Large numbers of islands are observed in for undoped ZnO. On

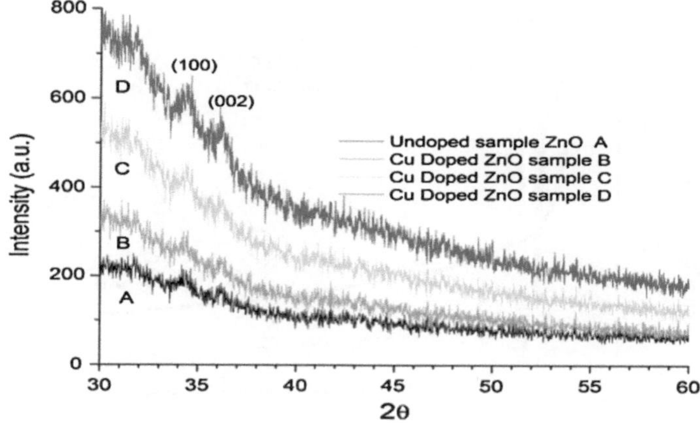

Fig. 43.1 X-ray diffractograms of undoped and Cu doped ZnO nanostructures

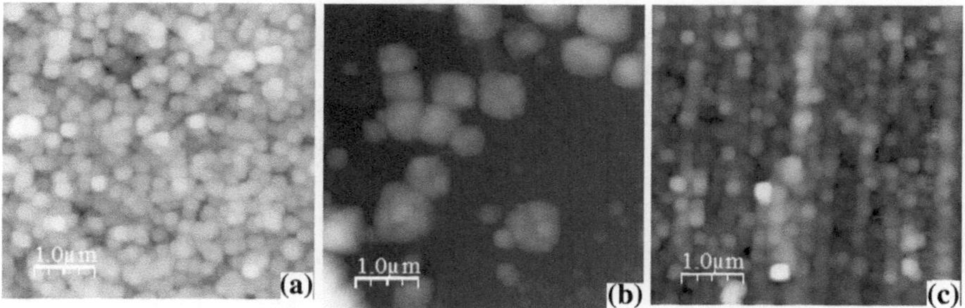

Fig. 43.2 AFM micrographs of **a** undoped ZnO, **b** and **c** Cu doped ZnO (Sample A and Sample C)

doping with copper these islands grow as more and more particles bind together. Further addition of Cu results in alignment of all islands in linear manner and formation of nanowires.

The Optical transmission spectra for undoped and doped ZnO particles (inset shown absorption spectra) are shown in Fig. 43.3. Constant absorption coefficient for ZnO for whole visible region indicates stable optical properties. Spectrum exhibit good transmission in the whole visible region and also exhibit strong blue shift in the absorption edge. The similar absorption is already reported with high Mg doping [10]. Cu provides free carriers that shift the Fermi level in the conduction band and widen the energy gap according to burston moss theory [11]. Cu doping affect the width of localized state outside the optical bandgap consequently optical band gap changed. Band gap increase from 5.2 to 5.5 eV with increasing in doping concentration [12].

The broad and sharp deep level emission observed around 409, 450 and 700 nm. With higher doping concentration intensity of peak shift towards lower energy.

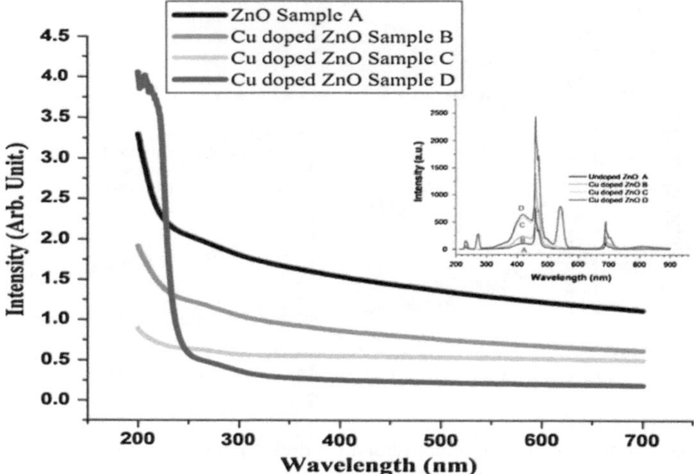

Fig. 43.3 Absorption and inset of PL spectra for undoped and Cu doped ZnO samples

Spectra show near band edge (NBE) emission at 230 nm generated by recombination of free electrons. DLE is usually attributed to structural defects such as at interstitials (red emission) and Vacancy (orange emission) in the crystal lattice.

43.4 Conclusions

Copper Doped ultra small ZnO nanoparticles were obtained by sol gel route. Ultra small size ZnO particles were confirmed by strong blue shift in the optical band gap. All the ZnO nanostructures also show strong photoluminescence. Addition of Copper improves the photoluminescence peak in UV region with an additional peak observed in middle UV Region at 270 nm. Other emissions such as yellow green or blue have also been observed. The UV emissions of the samples are centered at 3.42–4.90 eV, which is higher than the energy band gap of bulk ZnO. This confirms the size of particles comparable to the Bohr radius of ZnO.

Acknowledgments The authors are thankful for providing XRD and AFM facility to UGC-DAE consortium for scientific research Indore India and P C Ray research centre, ITM University Gwalior (MP) for providing optical and photoluminescence facilities.

References

1. Tang ZK, Wang GKL, Yu P, Kawasaki M, Ohtomo A, Koinuma H, Segawa Y (1998) Room-temperature ultraviolet laser emission from self-assembled ZnO microcrystalline thin films. Appl Phys Lett 72:3270–3272
2. Han WS, Kim YY, Kong BH, Cho HK (2009) Ultraviolet light emitting diode with n-ZnO:Ga/i-ZnO/p-GaN:Mg heterojunction. Thin Solid Films 517:5106–5109
3. Zhang L, Zhao J, Lu H, Li L, Zheng J, Li H, Zhu Z (2012) Highly sensitive and selective dimethylamine sensors based on hierarchical ZnO architectures composed of nanorods and nanosheet-assembled microspheres. Sens Actuators B Chem 171–172:1101–1109
4. Chow L, Lupan O, Chai G, Khallaf H, Ono LK, Cuenya BR, Tiginyanu IM, Ursaki VV, Sontea V, Schulte A (2013) Synthesis and characterization of Cu-doped ZnO one-dimensional structures for miniaturized sensor applications with faster response. Sens Actuators A 189:399–408
5. Ferhat M, Zaori A, Ahuja R (2009) Magnetism and band gap narrowing in Cu-doped ZnO. Appl Phys Lett 94:142502
6. Lu H, Wang Y, Lin X (2009) Structures, varistor properties, and electrical stability of ZnO thin films. Mater Lett 63:2321–2323
7. Dedova T, Acik IO, Krunks M, Mikli V, Volobujeva O, Mere A (2012) Effect of substrate morphology on the nucleation and growth of ZnO nanorods prepared by spray pyrolysis. Thin Solid Films 520:4650–4653
8. Thakur V, Verma UP, Rajaram P (2013) Solvent dependent growth of fibrous and non-fibrous nanocrystalline thin films of ZnO. J Sol-Gel Sci Technol 66(2):280–287
9. Kiyotaka M, Kazuyuki H, Shimotuma Y (2011) Nanotechnology and Nanomaterials. In: Abass H (ed) Nanowires—fundamental research, InTech, Rijeka
10. Ozgur U, Morkoç H (2006) Optical properties of ZnO and related alloys. In: Jagadish C, Pearton S (eds) Zinc oxide bulk, thin films and nanostructures, Elsevier, Amsterdam, pp 175–239
11. Burstein E (1954) Anomalous optical absorption limit in InSb. Phys Rev 93:632–633
12. Ilican S, Caglar Y, Claglar M (2008) Preparation and characterization of ZnO thin films deposited by sol-gel spin coating method. J Opto Electron Adv Mater 10:2578–2583

Chapter 44
Growth of Green and Blue Luminescent Cu Doped CdS Nanorods and Their Optical Structural Characterization

Nitin Kumar, Vijay Kumar, L.P. Purohit and Y.C. Goswami

Abstract Highly luminescent Copper doped CdS nanorods were prepared by modified sol-gel method. The sol was prepared by continuous stirring of mixture of ethylene glycol, ethanol and acetic acid with Cadmium chloride ($CdCl_2$) and thiourea as the precursors of Cd and S respectively at 50 °C temperature for 3 h. For Cu doping cuprous chloride was added slowly. Fine green and blue colored particles were obtained. The samples were analyzed by XRD, SEM, optical transmission and photoluminescence studies. Broad peak of (101) is observed in X-ray diffractograms for undoped CdS. Intensity of this peak increases with increase in copper (Cu) doping. In SEM, large number of islands is observed in undoped CdS. On doping with copper these islands grow as more and more particles bind together. Undoped structures are generally spherical with uniform size and start to reassemble in ordered rodlike structure on increase in doping.

44.1 Introduction

In recent years, semiconductor nanoparticles have gained a lot of attention due to their dramatic size dependent optical and electrical properties because of quantum confinement. The nanoparticles exhibit properties that are intermediate between those of bulk semiconductors and discrete molecules [1, 2]. Among various II–VI inorganic semiconductor nanocrystals, CdS nanoparticles are proved to be the most versatile materials as their wide applications in optoelectronic devices like thin-film light emitting devices, solar cells and biological labels. CdS quantum dots have been prepared by different researcher using various techniques such as chemical bath deposition, molecular beam epitaxy, spray pyrolysis, physical vapor deposition

N. Kumar (✉) · L.P. Purohit
Gurukula Kangari Vishwavidyalaya, Haridwar 249404, Uttrakand, India
e-mail: nitinmishra97@gmail.com; y_goswami@yahoo.com

V. Kumar · Y.C. Goswami
ITM University, Turari, Gwalior 474001, Madhya Pradesh, India

V. Lakshminarayanan and I. Bhattacharya (eds.), *Advances in Optical Science and Engineering*, Springer Proceedings in Physics 166,
DOI 10.1007/978-81-322-2367-2_44

and sol-gel method [3–5]. Sol-gel method includes atom-level mixing reported to be more suitable to synthesize quantum dots. It involves lower temperature processing of the raw materials and produces nano-structured powders and films. Recently lot of research has been initiated in doping of semiconductor nanocrystals. Thermal, photochemical stability and emission activity can be improved by modifying the properties of semiconductor [4]. However very few works have been reported on copper doped CdS [5, 6], even with small quantities of Cu have ability to tune the band gap of CdS nanocrystals which result in additional emission centers depending on their concentration and variations in the Photoluminescence (PL) and photoconductive properties make them suitable for wide range of applications such as light emitting displays and optical sensors [7–9]. In this paper we have reported high quality Cu:CdS nanocrystals by the sol-gel method. The doping dependent optical properties and self assembling behavior of the Cu doped CdS nanostructures have been investigated.

44.2 Experimental

CdS and Cu doped CdS low dimensional rods were grown by modified sol-gel method. Firstly ethylene glycol, ethanol and acetic acid were mixed together in a specific ratio and stirred for 30 min. Cadmium chloride (.1 M) and thiourea (.1 M) were added in the solution was kept at 50 °C for 3 h, yellow colour sol was obtained. Now cuprous chloride (.05 M) was added slowly in the prepared sol. The solution was kept for ultrasonic agitation for next 5 h and finally 24 h for aging process. Fine green and blue colored particles were obtained and filtered for characterization. Sample (a) is for undoped CdS, sample (b) contains low Cu doping and sample (c) contains Cu doping with high concentration. The spin coater with speed of 5000 rpm was used to prepare the thin film on the glass substrate. The sample was annealed at 300 °C for 2 h and analyzed by XRD, SEM, optical transmission and photoluminescence studies. The X-ray diffractograms were obtained in the 2θ range from 20° to 80° with Cu Kα radiation of wavelength 1.546 Å using XPERT-PRO XRD, SEM micrographs were obtained using SEM-Zeiss EVO-40 EP and transmission spectra were obtained using Perkin Elmer Lambda 25 in the range 200–800 nm. PL spectra was obtained using Perkin Elmer LS55 with excitation at 250 nm. All the chemicals were used as AR grade supplied by Ranbaxy.

44.3 Results and Discussion

X-ray diffractograms of undoped and Cu doped CdS nanostructures grown by ultrasonic sol-gel method dispersed on glass substrates are shown in Fig. 44.1. Broad peak of (101) is observed for undoped CdS of which intensity increases with

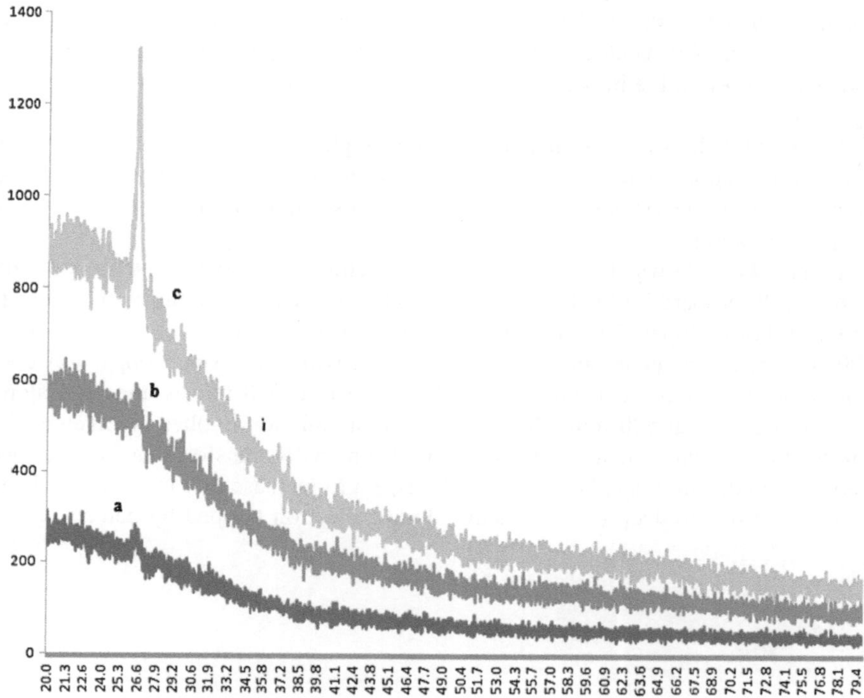

Fig. 44.1 X-ray diffractograms of undoped and Cu doped CdS nanostructures

increase in copper doping. Sample (c) gives the high intensity (101) peak. High background peak due to amorphous substrate might occur for very thin layers of CdS.

Figure 44.2 shows the SEM micrographs of undoped and doped CdS nanostructures. Large numbers of islands are observed in figure (a) for undoped CdS. On

Fig. 44.2 SEM micrographs of **a** undoped CdS, **b**, **c** Cu Doped CdS with increasing concentration

doping with copper these islands grow as more and more particles bind together as shown in figure (b). Undoped structures are generally spherical with uniform size and start to reassemble in ordered rodlike structure on increase in doping shown in figure (c).

Figure 44.3 shows the combined optical absorption spectra for undoped and Cu doped CdS nanoparticles with increase in copper doping. Spectrum exhibit good transmission in the visible region. Red shift in absorption occurs as increasing the Cu concentration.

Figure 44.4 shows the normalized PL spectra for undoped and doped CdS samples. PL properties of CdS nanoparticles are very sensitive to the surface defects and size. For undoped CdS very small luminescence is occurred in the range 380–480. Enhancement in luminescence is observed with increasing copper ion concentration. For higher concentration of Cu, a peak at 300 mm is appeared along with strong peak at 580 mm. Higher wavelength emission implies involvement of Cu in emission that indicates photoemission is gradually transforming from surface trap to deep dopant related emission with respect to increase in doping. Large Peak is observed due to deep traps on heavy doping electron trapped by defects.

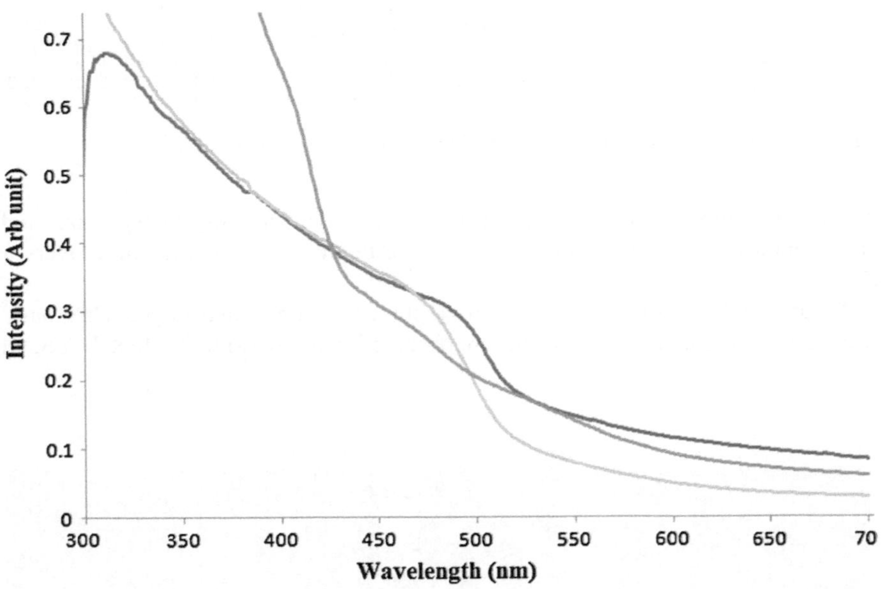

Fig. 44.3 Absorption spectra for undoped and Cu doped samples

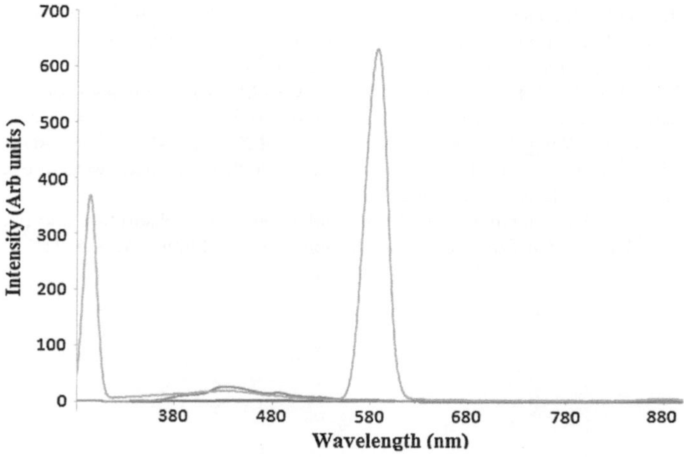

Fig. 44.4 Photoluminescence spectra for undoped and Cu doped CdS samples

44.4 Conclusions

Highly luminescence Cu doped CdS rodlike nanostructures were obtained by ultrasonic sol-gel method. Broad XRD (101) peak is observed which increases with increase in Cu doping. SEM micrographs show the assembling of nanostructures in rods on increasing the concentration of copper. Samples are transparent in visible region and gives good luminescence in visible and UV region, makes them suitable for optoelectronic devices.

Acknowledgments The authors are thankful to Wadia Institute of Himalyan Geology, Dehraudun India for providing XRD, SEM and PL, UV facility for P C Ray research centre, ITM University Gwalior for providing Optical and photoluminescence facilities.

References

1. Park JH, Kim JY, Chin BD, Kim YC, Park OO (2004) White emission from polymer/quantum dot ternary nanocomposite by incomplete energy transfer. Nanotechnology 15:1217
2. Pradhan N, Goorskey D, Thessing J, Peng X (2005) An alternative of cdse nanocrystal emitters: pure and tunable impurity emissions in ZnSe nanocrystals. J Am Soc 127:17586
3. Tang A, Yi L, Han W, Teng F, Wang Y, Hou Y, Gao M (2010) Synthesis, optical properties, and superlattice structure of Cu(I)-doped CdS nanocrystals. J App Phy 97:033112
4. Hasanzadeh J, Shayestes SF (2011) Luminescence of doped CdS nanocrystals: effect of doping and capping agent. Optica Applicata 4(41):4
5. Zhong H, Zhou Y, Ye M, He Y, Ye J, He C, Yang C (2008) Controlled synthesis and optical properties of colloidal ternary chalcogenide $CuInS_2$ Nanocrystals. Chem Mater 20:6434

6. Esakkraj E, Sheik SP, Henry J, Mohanraj K, Kannan S, Barathan S, Sivakumar G (2013) Optostructral and vibrational characteristics of Cu:CdS nanoparticles by precipitation method. Int J Light Electron Optics (OPTIK) 124(21):5229–5231
7. Tang A, Yi L, Han W, Teng F, Wang Y, Hou Y, Gao M (2010) Synthesis, optical properties, and superlattice structure of Cu(I)-doped CdS nanocrystals. Appl Phys Lett 97:033112
8. Tang W, Teng F, Wang Y, Hou YB, Han W, Yi LX, Gao MY (2008) Investigation on photovoltaic performance based on matchstick-Like Cu2S-In2S3 heterostructure nanocrystals and polymer. Nanoscale Res Lett 3:502
9. Unni C, Philip D, Gopchandran KG (2008) Studies on optical absorption and photoluminescence of thioglycerol-stabilized CdS quantum dots. Spectrochimica Acta A 71:1402

Chapter 45
Light Absorption in Nano-film of Wide Band Gap Semiconductor

Moumita Mukherjee and K.K. Ghosh

Abstract Optical absorption plays a significant role in measuring of direct bandgap of semiconductors. In the present paper absorption of light in nano-film of widegap semiconductor GaAs is studied. It is observed that such absorption is quantum in nature increasing stepwise with increase in photon energy. This is because of restructuring of energy levels in the quantum film. This interesting property of the light particle quantum absorption leads to the prospect of innovating ultrathin films to be used as novel optical absorber.

45.1 Introduction

Light absorption in thin nanofilm, in effect, is the absorption in quantum well. A thin layer of ($\sim$30–50 atomic layers) semiconductor (well material) is sandwiched between two outer relatively thick materials (barrier materials). Due to the confined motion in the direction of growth of the well while free motion in the other two directions the energy structure of electrons and holes are reorganized viz. energy in the growth direction is quantized with the conduction band splitting into subbands and energy in the planar direction remains free. The density of states in each subband of such 2D structure is a step function that starts at the appropriate confinement energy. By photon absorption electrons from the valence band(s) are raised preferably vertically to the conduction band obeying selection rule in addition to the momentum conservation principle also in the direction of the well growth. Optical absorption follow the 2D density of states. A high level of interest

M. Mukherjee (✉)
Centre for Millimeter-Wave Semiconductor Devices and Systems (CMSDS), DRDO, Kolkata, India
e-mail: mm_drdo@yahoo.com

K.K. Ghosh
Institute of Engineering and Management (IEM), Kolkata 700091, India
e-mail: kk_ghosh@rediffmail.com

V. Lakshminarayanan and I. Bhattacharya (eds.), *Advances in Optical Science and Engineering*, Springer Proceedings in Physics 166,
DOI 10.1007/978-81-322-2367-2_45

is put on understanding the physics of photon absorption in quantum well and exploring novel quantum light 2D absorbing devices. Realization of novel 2D absorbers mounted on virtually any support substrate has ushered a new era in nano-optoelectronic field [1–5]. In the present work, we have studied light absorption in ultra-thin film of wide bandgap material GaAs sandwiched between infinitely high barrier materials realized by outside vacuum (air). k.p perturbation formalism and single photon absorption is taken into account in this work. Theoretical modeling of the absorption process and simulation of results are presented in Sects. 45.2 and 45.3 respectively.

45.2 Theoretical Background

The associated electric and magnetic fields of an incident electromagnetic (e.m) wave interacts with the electrons in the material exerting both the electric and magnetic force as well. The total Hamiltonian describing a particle of charge q and free mass m_0 in presence of the e.m field is given by: $\hat{H} = \frac{\left[\hat{\vec{p}} - q\vec{A}(\hat{\vec{r}},t)\right]^2}{2m_0} + V(\hat{\vec{r}})$, $V(\vec{r})$ being the periodic crystal potential and $\vec{A}$ the vector potential. For electron q is $-e$, where e the magnitude of electron charge. The Hamiltonian in that case is

$$\hat{H} = \frac{(\hat{\vec{p}} + e\vec{A})^2}{2m_0} + V(\vec{r})$$

For weak electromagnetic radiation the Hamiltonian is simplified [6] into,

$$\hat{H} = \left[\frac{\hat{\vec{p}}^2}{2m_0} + V(\vec{r})\right] + \frac{e}{m_0}\vec{A} \cdot \hat{\vec{p}} \tag{45.1}$$

The last term in RHS of (45.1) is the optical perturbation to the electron system due to the light-matter interaction. Thus

$$H_{\text{int}} = \frac{e}{m_0}\vec{A} \cdot \hat{\vec{p}}$$

$\vec{A}$, the vector potential may be expressed by: $\vec{A} = \frac{\vec{A}_0}{2} e^{i(\vec{q}.\vec{r} - \omega t)} + \frac{A_0}{2} e^{-i(\vec{q}.\vec{r} - \omega t)}$, $\vec{A}_0$ being the amplitude ($=\hat{e}A_0$) of the vector potential, $\vec{q}$ the wave vector and ω the angular frequency of the e.m. wave, $\hat{e}$ being the unit vector in the direction of polarization of light wave.

The probability rate of electron transition from an occupied state $\vec{k}$ in valence band to an unoccupied state $\vec{k}'$ in conduction band by one photon absorption (OPA) follows from Fermi's Golden Rule [7] and is given by

$$W_{\vec{k}',k} = \frac{2\pi}{\hbar} \left| \langle \psi_{\vec{k}'} | H_{\text{int}} | \psi_{\vec{k}} \rangle \right|^2 \delta\left(E_{\vec{k}'} - E_{\vec{k}} - \hbar\omega\right) \tag{45.2}$$

The energy conservation is taken care of by the delta function. Transition probability thus depends on the interaction matrix element between the final and the initial electron states involved in interaction with photons. Considering the states are the Bloch states and the lattice periodic part is relatively rapid varying space function we can express the transition probability rate following some algebraic work as

$$W_{\vec{k}',\vec{k}} = \frac{2\pi}{\hbar} \left(\frac{eA_0}{2m^*}\right)^2 \left| \hat{e} \cdot \hat{\vec{p}}_{\text{cv}} \right|^2 \delta\left(E_{\vec{k}'} - E_{\vec{k}} - \hbar\omega\right) \tag{45.3}$$

$\hat{\vec{p}}_{\text{cv}}$ being the momentum matrix element between the conduction and valence bands involved in the photon transition due to the photon-electron interaction process. Considering all the valence band states to be filled and the all the conduction band states empty to take part in the photon absorption process we find the up transition rate as

$$Q_{\vec{k}',\vec{k}} = D(E) \times \frac{2\pi}{\hbar} \left(\frac{eA_0}{2m_0}\right)^2 \left\langle \left| \hat{e} \cdot \hat{\vec{p}}_{\text{cv}} \right|^2 \right\rangle \tag{45.4a}$$

$D(E)$, the joint density of states between the conduction and valence bands, is given by

$$D(E) = \frac{m_r^*}{\pi\hbar^2} \sum_{n,n'} \Theta\left[\left(E_{n,n'} + E_g\right) - \hbar\omega\right] \tag{45.4b}$$

taking into account the spin degeneracy. The sub-band levels in the valence band is indexed by n and those in the conduction band by n' while Θ denotes the Heaviside step function and $E_{nn'}$ being the sum of the electron energies in the n' conduction band, nth valence bands i.e. $E_{n,n'} = E_n + E_{n'}$.

The effective reduced mass m_r^* of the electron-hole system is given by $\frac{1}{m_r^*} = \frac{1}{m_n^*} + \frac{1}{m_p^*} m_n^*$ and m_p^* being the effective electron and hole masses on the interacting conduction sub-bands and valence sub-bands respectively.

It is thus clear that to calculate the transition rate, determination of the momentum matrix element between two states in the conduction and valence band states is important.

$\hat{\vec{p}}_{\text{cv}} = -i\hbar \int \psi_{\vec{k}'} \nabla \psi_{\vec{k}} d^3\vec{r}$ where the full wave function $\psi(\vec{r})$ is a product of the slowly varying modulating envelope function $\chi(\vec{r})$ and the appropriate Bloch function $u_{\vec{k}}(\vec{r})$ in the transverse direction $(\vec{r} = (x, y))$. Thus, $\psi(\vec{r}) \infty \chi(\vec{r})$ $u_{\vec{k}}(\vec{r}) = \frac{1}{\sqrt{AL_z}} e^{i\vec{k}_t.\vec{r}_t} \phi_{\vec{k}}(z) u_{\vec{k}'}(\vec{r}_t)$ $\phi_{\vec{k}'}(z)$ being the envelop function in the z-direction of the bound state of alignment in the conduction band of the well, $\vec{k}_t\left(=\vec{k}_x+\vec{k}_y\right)$ and $\vec{r}_t\left(=\hat{i}x+\hat{j}y\right)$ the in-plane (transverse) wave vector and position vector respectively, A the area of the well and L_z the well width which is the film thickness (t) in the z-direction.

Photon flux $\Phi=\rho_P n_r c$, ρ_P and n_r being the photon density and refractive index of the material. Therefore the light energy through unit area and in unit time is given by

$$I=\Phi\times\hbar\omega=\rho_P n_r c\hbar\omega$$

Absorption coefficient α for direct interband transition is defined as the power removed from the optic beam of unit flux density incident on unit volume of the material and may be expressed as the ratio of the rate of absorbing photons to the incident flux of photons. Thus

$$\alpha=\frac{Q_{\vec{k}',\vec{k}}\times\hbar\omega}{\Phi} \quad (45.5)$$

$$=\frac{e^2m_r^*}{\varepsilon_0\hbar^2 t\omega cn}\frac{1}{m_0^2}\left|\hat{e}\cdot\hat{\vec{p}}_{\mathrm{cv}}\right|^2 \quad (45.6)$$

where the symbols ε_0, t, c and n represent vacuum permittivity, film thickness, light velocity and refractive index of the material respectively while ω being the frequency of the incident radiation.

The momentum matrix element and the k.p effective mass is related by [6]

$$\left|\hat{\vec{p}}_{\mathrm{cv}}\right|^2=\frac{\hbar^2}{2m_r^*}\left|E_g\right| \quad (45.7)$$

Using (45.6) and (45.7) we finally get

$$\alpha=\frac{1}{2}\hbar\frac{e^2}{\varepsilon_0 t n_r c m_0^2\hbar\omega}E_g\sum_{n,n'}\Theta\left[\left(E_{n,n'}+E_g\right)-\hbar\omega\right] \quad (45.8)$$

have plotted the absorbance versus photon energy taking three different interacting subbands for three different film thicknesses.

45.3 Results and Discussions

In the simulation, GaAs is chosen as the material for its large band gap of 1.5 eV and to verify our results with results published elsewhere. The interacting valence bands with incident photons are the two heavy hole bands and one split off light

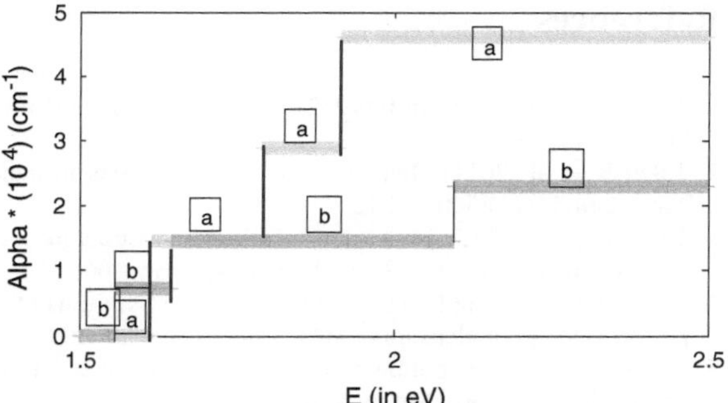

Fig. 45.1 Plot of absorption coefficient (alpha) versus photon energy E for well thickness **a** t = 5 nm and **b** t = 10 nm

hole band while the conduction band is taken as the first subband. The split off in the valence band is taken as 0.3 eV. In Fig. 45.1 we have plotted the absorption coefficient α versus incident photon energy for two different well thicknesses t.

It is observed that for smaller well thickness the material shows higher fundamental absorption edge compared to that in relatively higher well thickness. This is because the quantized energy levels in lower thickness goes higher up compared to those for higher well thickness. This shows that thinner material remains transparent to photons of higher energies which means for light of shorter wavelengths. Further, each vertical step signifies the electron transition from heavy hole (HH) to the first conduction subband (CB), light hole (LH) to (CB) serially. Variation of the absorption coefficient with film thickness for constant photon energy is shown in Fig. 45.2 for a given photon energy the absorption is a decreasing function of film thickness. The result confirms that the band gap increases effectively as a result of lowering the thickness and vice versa. At a given film thickness, the film is energetically more transparent to lower energy i.e. to higher wavelengths.

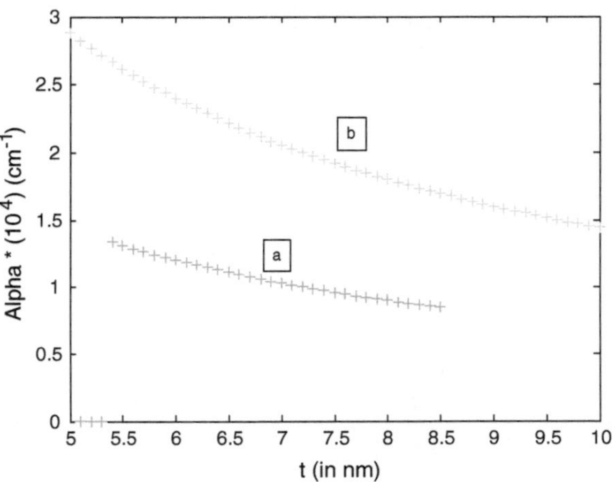

Fig. 45.2 Plot of absorption coefficient (alpha) versus well thickness t for photon energy **a** E = 1.6 eV and **b** E = 1.8 eV

References

1. Ando T, Fowler B, Stern F (1982) Electronic properties of two-dimensional systems. Rev Mod Phys 54:437–672
2. Takei K et al (2011) Quantum confinement effects in nanoscale-thickness InAs membranes. Nano Lett 11(11):5008–5012
3. Takei K et al (2012) Nanoscale InGaSb heterostructure membranes on Si substrates for high hole mobility transistors. Nano Lett 12(4):2060–2066
4. Hu P, Wen Z, Wang L, Tan P, Xiao K (2012) Synthesis of few-layer GaSe nanosheets for high performance photodetectors. ACS Nano 6(7):5988–5994
5. Ko H et al (2010) Ultrathin compound semiconductor on insulator layers for high performance nanoscale transistors. Nature 468(7321):286–289
6. Kireev PS (1978) Semiconductor physics, 2nd edn. Mir
7. Schawbl F (2007) Quantum mechanics, 4th edn. Springer

Chapter 46
SEM Imaging for Observation of Morphological Changes in Anaemic Human Blood Cell

Triparna Datta and Uttam Roychoudhury

Abstract Scanning Electron Microscopy (SEM) is utilized to elucidate the morphological changes in anaemic human red blood cells. Haemoglobin concentration in human blood is in the range of 11.5–13.5 g/dl in healthy adults. Haemoglobin concentration in anaemic red blood is below the lower limit of normal range. Sometimes, the nature of the abnormal shape of the blood cell determines the cause of anaemia. Normally, there occurs a variation in the diameter of the red blood cell (RBC) for different types of anaemia. Increased variation of size in blood cell is termed anisocytosis (a type of anaemia) (Mohan H, Text book of pathology, New Delhi). In case of anisocytosis, diameter of cells larger than normal cell is observed. The classification of anaemia by the size of blood cell is logical, i.e. common morphological abnormality of human blood cell (Davidson's principle and practice of medicine, Publisher Churchill Livingstone, London). Cells are studied under ZEISS SEM with different magnification and applied potential of kV range. Thus the diameters of RBCs in SEM have been compared with RBCs photographed with light microscope. Anaemic cells are observed overlapped with each other with increasing diameter.

46.1 Introduction

The detailed understanding and characterization of cells affecting the structure and function most requires analysis by electron microscopy. Light microscopy has certain limitation in resolution to discriminate between structures of different biomolecules. Scanning Electron Microscope (SEM) is often used to study the topological structural details of biomolecules as the resolving power of the SEM is

T. Datta (✉)
Institute of Engineering and Management (IEM), Saltlake, Kolkata 700091, India
e-mail: triparna.datta@iemcal.com

U. Roychoudhury
Department of Chemical Technology, University College of Science and Technology, University of Calcutta, 92, A.P.C. Road, Kolkata 700009, India
e-mail: urc@vsnl.net

V. Lakshminarayanan and I. Bhattacharya (eds.), *Advances in Optical Science and Engineering*, Springer Proceedings in Physics 166,
DOI 10.1007/978-81-322-2367-2_46

much larger than normal light microscope. As Scanning electron microscopy (SEM) studied the surface structure in high resolution so one can interpret the characteristic behaviour of normal and disease cells.

Blood that flows through the body in the vascular system is made up of plasma and three cellular components as follows:

- Red cells contain haemoglobin which transport oxygen from the lungs to the different body tissues,
- White cells, which protect against infection to help our immune system,
- Platelets regulate blood vessels and clotting factors to maintain vascular integrity.

Many light microscopy shows red blood cell structure which are known to scientist. But very few electron microscopy, specially through SEM structural study of blood cells are seen. When the level of haemoglobin in blood is get decreased from the normal range this symptom is known as Anaemia. That occurs due to the decrease in oxygen carrying capacity of the blood. Anaemia can be classified on the basis of morphology of red cell.

46.2 Materials and Methods

For morphological study of anaemic cells with comparison of normal blood sample were smeared on glass slides and then cut into pieces (1 cm × 1 cm) size. Then specimens were mounted on stubs and coated with gold (100 Å) thickness with vacuum coater at 10^{-6} torr for examination in a scanning electron microscope (ZEISS). Micrographs were taken at the accelerating voltage 15 kV with different magnification. The diameters were measured with the help of an inbuilt eyepiece fitted with micro-scale (Graticule Ltd. UK).

46.3 Result

This novel study concerns with the morphological changes which occur in anaemic blood with normal cells. At scanning electron micrograph as shown in Fig. 46.1 depicts the morphology of normal RBCs. The micrograph was taken at 20K magnification and average diameter of the cell was 6.3 μm. This value was fitted in the normal range [3]. But major morphological difference noted in Figs. 46.2, 46.3 and 46.4 at magnification 2K, 10K and 20K, respectively. Electron micrographs were shown variation in diameter of the anaemic cell with normal one. This difference indicates the type of anaemia. Figures 46.2, 46.3, and 46.4 show increased variation in the size of RBC which indicates anisocytosis, a common type of anaemia. The higher magnification shows the central hole [5] in blood cell as the diameter seen broader and broader.

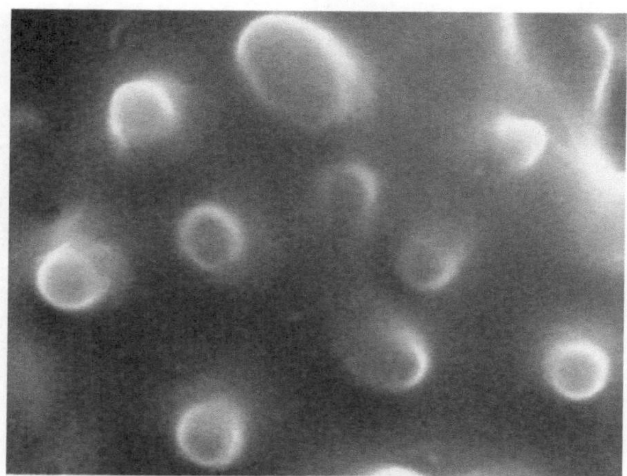

Fig. 46.1 SEM analysis of normal blood cells at 20K×

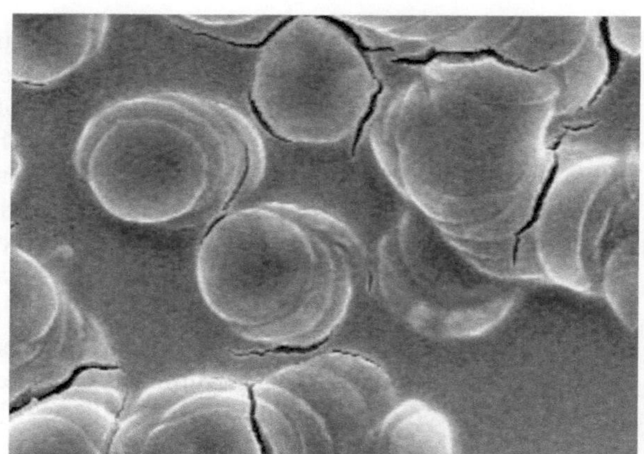

Fig. 46.2 SEM analysis of anaemic blood cells at 2K×

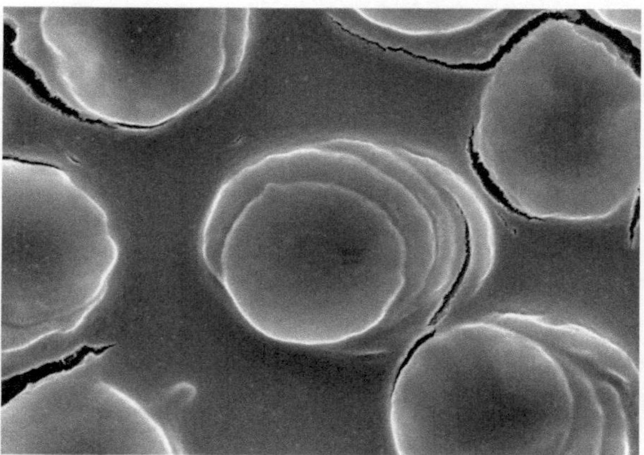

Fig. 46.3 SEM analysis of anaemic blood cells at 10K×

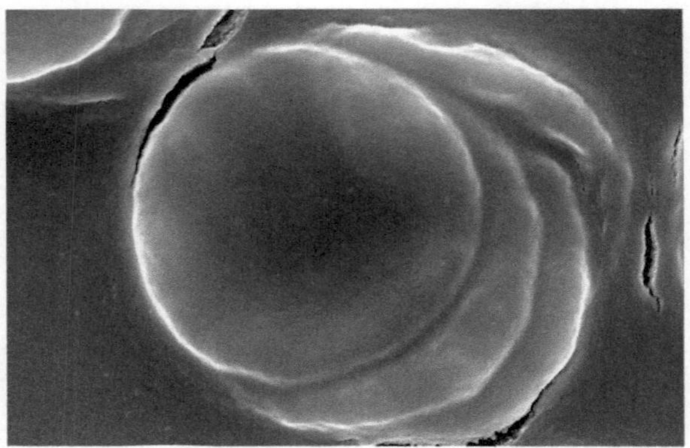

Fig. 46.4 SEM analysis of anaemic blood cells at 20K×

46.4 Conclusion

In the present study the characters of normal and anaemic blood samples were micrographed by SEM to specify the morphological changes. It is well known from the earlier study [4, 5] that the microstructural analysis of anaemic blood sample was performed by SEM and it reveals more information than that of light microscope. The electron micrographs clearly show the enhancement of diameter which signifies different kinds of anaemia. There are central holes which can be visualized at the high resolution magnifications [5]. In case of normal blood samples we have not seen any central holes as because the cells need more magnification. If we study the normal blood samples by the atomic force microscope (AFM) we will be able to see the central hole and the periphery of the cells in three-dimensional view. Though atomic force microscope depicts the surface structure of the cells like SEM, but the resolution achieved is more higher than the SEM.

Acknowledgments We are indebted to Dr. Anima Sen, Ex H.O.D of USIC for her constant encouragement and endeavour to support for this entire work. We thank Prof. G. Bhattacharya, Ex H.O.D, Dept. of Biochemistry, Calcutta Medical College & Hospital, Kolkata and Prof. T.K. Das, H.O.D, Dept. of Pathology, R.G. Kar Medical College, Kolkata. We also thank Centre for Research of Nano Science & Nano Technology (CRNN), University of Calcutta for their technical support.

References

1. Mohan H Text book of pathology. Jaypee Brothers, Medical publishers (P) Ltd, New Delhi
2. Davidson's principle and practice of medicine, Publisher Churchill Livingstone, London, p 894
3. Chakraborty P, Chakraborty G Practical pathology
4. Modak Dhar T et al (2004) Ind J Phys 78B(1):97–101
5. Guha T et al (2002) Curr Sci 83(6)

Chapter 47
Mueller Matrix Polarimeter with Diattenuation Error Calibration Approach

Kaustav Bhattacharyya, David Ignacio Serrano-García and Yukitoshi Otani

Abstract A calibration algorithm for the Mueller matrix dual-rotating retarder polarimeter is presented. The Mueller matrix dual-rotating retarder polarimeter is composed by two rotating retarders and two fixed polarizers with the final purpose of retrieving the polarization properties of a sample. Even algorithms to calibrate the retardation error and azimuthal error already exists, the accuracy obtained with the system is limited. Our proposal is to take into account the diattenuation error of the two retarders, which can provide more accuracy in the result. The analytical equations is described and supported with experimental results showing the accuracy incremental in our proposal.

47.1 Introduction

The polarization state of a light beam can be explained in terms of its Stokes vector. The Stokes vector consists of four parameters S_0, S_1, S_2 and S_3, associated with the polarization state of light separated in their basis polarization states. S_0 is the total intensity of light, S_1 is the difference between linear horizontal and vertical polarization states, S_2 is the difference between +45° and −45° linear polarization states and S_3 is the difference between right circular and left circular polarization states. A sample can be represented by a 4 × 4 matrix where all of its polarization properties are represented. By knowing the Mueller matrix of a sample, it is possible to know the output polarization state dependent of its input [1].

The Mueller matrix polarimeter is an optical device capable of retrieving the Mueller matrix of a sample by obtaining several intensity measurement for different polarizing states. The polarizing property of a sample can be determined by its

K. Bhattacharyya (✉) · D.I. Serrano-García · Y. Otani
Center for Optical Research and Education (CORE), Utsunomiya University, 7-1-2, Yoto, Utsunomiya 321-8585, Tochigi, Japan
e-mail: kaustav@opt.utsunomiya-u.ac.jp

V. Lakshminarayanan and I. Bhattacharya (eds.), *Advances in Optical Science and Engineering*, Springer Proceedings in Physics 166,
DOI 10.1007/978-81-322-2367-2_47

Mueller matrix. The Mueller matrix imaging has also many application in biomedical imaging, i.e. cancer diagnosis, tumor imaging, etc. [2, 3].

The dual rotating retarder polarimeter was proposed firstly by Azzam [4] and then errors occurred on the components, like non-ideal quarter-wave retarders and azimuthal angles of the retarders and analyzer, was proposed by Goldstein and Chipman [5, 7]. The approach presented here is to take into account, by the moment, only the errors due to the diattenuation of the retarders in order to calibrate them. The final purpose of this work is to pursue the enhancement of the accuracy of the dual rotating retarder polarimeter.

47.2 Dual-Rotating Retarder Polarimeter

The dual rotating retarder polarimeter, Fig. 47.1, consists of a polarization state generator (PSG) and a polarization state analyzer (PSA), each composed by a fixed polarizer (M_P, M_A) and a rotating retarder ($M_R(\theta)$, $M_R(5\theta)$ respectively. When the retarder of the PSA and the PSG rotates with an angular velocity 5:1, its output signal generates 12 harmonic frequencies in the Fourier spectrum. The Mueller matrix elements of the sample is retrieved from its Fourier complex coefficients [4, 5]. The output stokes vector will be:

$$S_{\mathrm{out}} = \begin{bmatrix} S_0 \\ S_1 \\ S_2 \\ S_3 \end{bmatrix}_{\mathrm{out}} = \left[M_A M_R(5\theta) M_S M_R(\theta) M_p\right] \cdot S_{\mathrm{in}} \tag{47.1}$$

where M_S is the Mueller matrix of the components, θ and 5θ is the rotation of the PSG and PSA retarder respectively. $M_R(\theta)$ is the Mueller matrix of a quarter-wave retarder oriented at an angle θ, and M_P and M_A is the Mueller matrix of a horizontally linear polarizer, given by Chipman [6]:

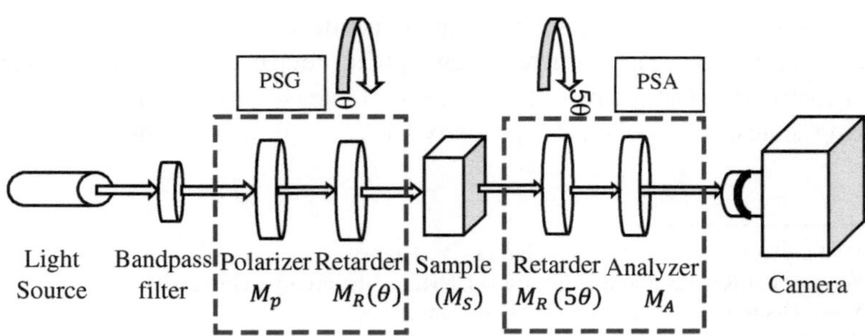

Fig. 47.1 Dual rotating retarder polarimeter

$$M_R(\theta) = \begin{bmatrix} 1 & 0 & 0 & 0 \\ 0 & \cos^2 2\theta & \sin 2\theta \cos 2\theta & -\sin 2\theta \\ 0 & \sin 2\theta \cos 2\theta & \sin^2 2\theta & \cos 2\theta \\ 0 & \sin 2\theta & -\cos 2\theta & 0 \end{bmatrix}$$
$$M_P = M_A = \begin{bmatrix} 1 & 1 & 0 & 0 \\ 1 & 1 & 0 & 0 \\ 0 & 0 & 0 & 0 \\ 0 & 0 & 0 & 0 \end{bmatrix} \quad (47.2)$$

By making the matrix multiplication with the Mueller matrix of the sample, (47.1), the output intensity captured by the detector can be represented as:

$$S_{0\text{out}} = \mu_{11}m_{11} + \mu_{12}m_{12} + \mu_{13}m_{13} + \cdots + \mu_{44}m_{44} \quad (47.3)$$

where m_{ij} are the elements of Mueller matrix of the sample, and μ_{ij} are the corresponding coefficients dependent of the angular position of the retarder θ. The output intensity can be expressed in the form of a Fourier series as:

$$S_{0\text{out}} = a_0 + \sum_{n=1}^{12} (a_n \cos 2n\theta + b_n \sin 2n\theta) \quad (47.4)$$

where a_n and b_n are the real and imaginary Fourier coefficients. From the inverse calculation, the elements of the Mueller matrix of the sample are successfully determined by [4, 5]:

$$\begin{aligned} m_{11} &= (a_0 - a_2 + a_8 - a_{10} + a_{12}) & m_{31} &= (b_{10} - b_8 - b_{12}) \\ m_{12} &= 2(a_2 - a_8 - a_{12}) & m_{32} &= 4(b_8 + b_{12}) \\ m_{13} &= 2(b_2 + b_8 - b_{12}) & m_{33} &= 4(a_8 - a_{12}) \\ m_{14} &= (b_1 + b_9 - b_{11}) & m_{34} &= 2(a_9 - a_{11}) \\ m_{21} &= 2(a_{10} - a_8 - a_{12}) & m_{41} &= (b_3 + b_7 - b_5) \\ m_{22} &= 4(a_8 + a_{12}) & m_{42} &= -2(b_3 + b_7) \\ m_{23} &= 4(b_{12} - b_8) & m_{43} &= 2(a_7 - a_3) \\ m_{24} &= 4(b_{11} - b_9) & m_{44} &= (a_6 - a_4) \end{aligned} \quad (47.5)$$

47.3 Diattenuation Error Calibration Model Approach

The errors due to non-ideal quarter-wave retarders and the azimuthal angles of the retarders and analyzer have been addressed before [5, 7]. Our approach is to take into account the errors due to the diattenuation of the retarders only, retardance values are obtained using Goldstein-Chipman-algorithm for complementation of the model (Fig. 47.2).

To get the Mueller matrix of the retarder with non-zero diattenuation characteristics, (T_{max}, T_{min}), and linear retardance ϕ, its Mueller matrix is constructed by

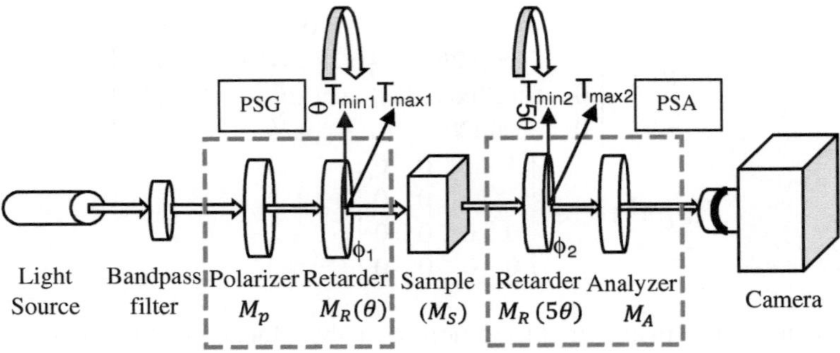

Fig. 47.2 Dual rotating retarder polarimeter with diattenuation and retardation error of the retarder

the matrix multiplication of a linear diattenuator, M'_{D} (θ, $T_{\max}$, $T_{\min}$) with a linear retarder M'_{R} (θ, ϕ) given by Chipman [6], Lu and Chipman [8]:

$$M'_{\mathrm{RD}}(\theta,\phi,T_{\max},T_{\min}) = M'_{\mathrm{R}}M'_{\mathrm{D}} = M'_{\mathrm{D}}M'_{\mathrm{R}}$$
$$= \begin{bmatrix} (T_{\max}+T_{\min}) & (T_{\max}-T_{\min})\cos 2\theta & (T_{\max}-T_{\min})\sin 2\theta & 0 \\ (T_{\max}-T_{\min})\cos 2\theta & (T_{\max}+T_{\min})\cos^2 2\theta + \sin^2 2\theta\cos\phi 2\sqrt{T_{\max}T_{\min}} & \left(T_{\max}+T_{\min}-\cos\phi 2\sqrt{T_{\max}T_{\min}}\right)\sin 2\theta\cos 2\theta & -\sin 2\theta\sin\phi 2\sqrt{T_{\max}T_{\min}} \\ (T_{\max}-T_{\min})\sin 2\theta & \left(T_{\max}+T_{\min}-\cos\phi 2\sqrt{T_{\max}T_{\min}}\right)\sin 2\theta\cos 2\theta & (T_{\max}+T_{\min})\sin^2 2\theta + \cos^2 2\theta\cos\phi 2\sqrt{T_{\max}T_{\min}} & \cos 2\theta\sin\phi 2\sqrt{T_{\max}T_{\min}} \\ 0 & \sin 2\theta\sin\phi 2\sqrt{T_{\max}T_{\min}} & -\cos 2\theta\sin\phi 2\sqrt{T_{\max}T_{\min}} & \cos\phi 2\sqrt{T_{\max}T_{\min}} \end{bmatrix} \quad (47.6)$$

where, $T_{\max}$ and $T_{\min}$ are the maximum and minimum intensity transmittance of diattenuation, ϕ is the retardance of the retarder and θ is the angular orientation of the fast axis. By using the same procedure in the first sections, (47.1) and (47.3), but with the non-ideal retarders, its corresponding output Stokes vector will be given by:

$$S'_{\mathrm{out}} = \left[M_{\mathrm{A}}M'_{\mathrm{RD}}(5\theta,\phi_2,T_{\max 2},T_{\min 2})M_{\mathrm{S}}M'_{\mathrm{RD}}(\theta,\phi_1,T_{\max 1},T_{\min 1})M_{\mathrm{p}}\right]\cdot S_{\mathrm{in}} \quad (47.7)$$

And

$$S'_{0\mathrm{out}} = \mu'_{11}m_{11} + \mu'_{12}m_{12} + \mu'_{13}m_{13} + \cdots + \mu'_{44}m_{44} \quad (47.8)$$

Just as previously explained, (47.4), m_{ij} are the elements of Mueller matrix of the sample, and μ'_{ij} are the corresponding coefficients dependent of the angular position of the retarder θ. The output intensity can be expressed in the form of a Fourier series as [4, 5]:

$$S'_{0\,out} = a_0 + \sum_{n=1}^{12}(a_n\cos 2n\theta + b_n\sin 2n\theta) \tag{47.9}$$

Following the same procedure explained at the introduction, where the Mueller matrix multiplication is done using (47.6–47.9), the Mueller matrix elements of the sample can be retrieve by the inverse calculation of the Fourier complex coefficients. In this case the Mueller matrix coefficients will be dependent of diattenuation and retardation parameters of the retarders.

$$\begin{aligned}
m_{11} &= \frac{1}{p_1p_2\alpha_2\beta_1}[\alpha_2\beta_1a_0 - \alpha_4\beta_1a_2 + (\alpha_4\beta_2 + \alpha_2\beta_3 - \alpha_2\beta_4)a_8 \\
&\quad -\alpha_2\beta_1a_{10} + (\alpha_4\beta_2 + \alpha_2\beta_3 - \alpha_2\beta_4)a_{12}] \\
m_{12} &= \frac{2}{\alpha_2\beta_1}(\beta_1a_2 - \beta_2a_8 - \beta_2a_{12}) \\
m_{13} &= \frac{2}{\alpha_1\alpha_2}(\alpha_1b_2 + \alpha_3b_8 - \alpha_3b_{12}) \\
m_{14} &= \frac{1}{\sin\phi_1p_2z_1}\left(b_1 + \frac{2\alpha_3}{\alpha_1}b_9\right) \\
&\quad - \frac{p_2q_1}{\sin\phi_1p_1p_2\alpha_1\alpha_2p_2z_1}[2p_1p_2\alpha_1b_2 + (4p_1p_2 - \alpha_4)\alpha_3b_8 + \alpha_2\alpha_3b_{10} - \alpha_3\alpha_4b_{12}] \\
m_{21} &= \frac{2}{\alpha_1\beta_1}(\beta_1a_{10} - \beta_3a_8 - \beta_3a_{12}) \\
m_{22} &= \frac{4}{\beta_1}(a_8 + a_{12}) \\
m_{23} &= \frac{4p_1p_2}{\alpha_1\alpha_2}(b_{12} - b_8) \\
m_{24} &= -\frac{4p_1p_2}{\sin\phi_1\alpha_1p_2z_1}b_9 + \frac{2p_2q_1}{\sin\phi_1\alpha_1\alpha_2p_2z_1}[(4p_1p_2 - \alpha_4)b_8 + \alpha_2b_{10} - \alpha_4b_{12}] \\
m_{31} &= \frac{2}{\alpha_1\alpha_2}(\alpha_2b_{10} - \alpha_4b_8 - \alpha_4b_{12}) \\
m_{32} &= \frac{4p_1p_2}{\alpha_1\alpha_2}(b_{12} + b_8) \\
m_{33} &= \frac{4p_1p_2}{\alpha_1\alpha_2}(a_8 - a_{12}) \\
m_{34} &= \frac{4p_1p_2}{\sin\phi_1\alpha_1p_2z_1}a_9 - \frac{2p_2q_1}{\sin\phi_1\alpha_1\alpha_2p_2z_1\beta_1}[(2\alpha_1\alpha_2 + 2p_1p_2\beta_1 - \alpha_2\beta_3)a_8 \\
&\quad +\alpha_2\beta_1a_{10} - (2p_1p_2\beta_1 + \alpha_2\beta_3 - 2\alpha_1\alpha_2)a_{12}] \\
m_{41} &= \frac{1}{\sin\phi_2p_1z_2}\left(\frac{2\alpha_4}{\alpha_2}b_7 - b_5\right) - \frac{p_1q_2}{\sin\phi_2p_1p_2\alpha_1\alpha_2p_1z_2}[\alpha_1\alpha_4b_2 + \alpha_3\alpha_4b_8 - 2p_1p_2\alpha_2b_{10} \\
&\quad +(4p_1p_2 - \alpha_3)\alpha_4b_{12}] \\
m_{42} &= -\frac{4p_1p_2}{\sin\phi_2\alpha_2p_1z_2}b_7 + \frac{2p_1q_2}{\sin\phi_2\alpha_1\alpha_2p_1z_2}[\alpha_1b_2 + \alpha_3b_8 + (4p_1p_2 - \alpha_3)b_{12}] \\
m_{43} &= \frac{4p_1p_2}{\sin\phi_2\alpha_2p_1z_2}a_7 - \frac{2p_1q_2}{\sin\phi_2\alpha_1\alpha_2p_1z_2\beta_1}[\alpha_1\beta_1a_2 - (\alpha_1\beta_2 + 2\alpha_1\alpha_2 - 2p_1p_2\beta_1)a_8 \\
&\quad +(2p_1p_2\beta_1 + 2\alpha_1\alpha_2 - \alpha_1\beta_2)a_{12}] \\
m_{44} &= \frac{1}{\sin\phi_1\sin\phi_2z_1z_2}\left[(a_6 - a_4) + \frac{4p_1q_2}{\alpha_1}a_9 - \frac{4q_1p_2}{\alpha_2}a_7 + \frac{4p_1p_2q_1q_2}{\alpha_1\alpha_2}(a_8 - a_{12}) + \frac{2q_1q_2}{\alpha_1\alpha_2\beta_1}\right. \\
&\quad \left.[\alpha_1\beta_1a_2 - (\alpha_1\beta_2 + 4\alpha_1\alpha_2 - \alpha_2\beta_3)a_8 - \alpha_2\beta_1a_{10} + (4p_1p_2\beta_1 + \alpha_2\beta_3 - \alpha_1\beta_2)a_{12}]\right]
\end{aligned} \tag{47.10}$$

For conveniences the parameters obtained were reduced by this relations:

$$\begin{aligned}
T_{\max 1}+T_{\min 1}&=p_1 \quad T_{\max 2}+T_{\min 2}=p_2\\
T_{\max 1}-T_{\min 1}&=q_1 \quad T_{\max 2}-T_{\min 2}=q_2\\
2\sqrt{T_{\max 1}T_{\min 1}}&=z_1 \quad 2\sqrt{T_{\max 2}T_{\min 2}}=z_2\\
\beta_1=p_1p_2-p_2\cos\delta_1z_1-p_1\cos\delta_2z_2+\cos\delta_1\cos\delta_2z_1z_2 &\quad \alpha_1=p_1(p_2-\cos\delta_2z_2)\\
\beta_2=p_1p_2-p_2\cos\delta_1z_1+p_1\cos\delta_2z_2-\cos\delta_1\cos\delta_2z_1z_2 &\quad \alpha_2=p_2(p_1-\cos\delta_1z_1)\\
\beta_3=p_1p_2+p_2\cos\delta_1z_1-p_1\cos\delta_2z_2-\cos\delta_1\cos\delta_2z_1z_2 &\quad \alpha_3=p_1(p_2+\cos\delta_2z_2)\\
\beta_4=p_1p_2+p_2\cos\delta_1z_1+p_1\cos\delta_2z_2+\cos\delta_1\cos\delta_2z_1z_2 &\quad \alpha_4=p_2(p_1+\cos\delta_1z_1)
\end{aligned} \tag{47.11}$$

where, $T_{\max 1}, T_{\min 1}, T_{\max 2}$ and $T_{\min 2}$ as the maximum and minimum intensity transmission, ϕ_1 and ϕ_2 as the retardance of polarizing state generator and polarizing state analyzer retarder respectively.

The diattenuation $\left(D=\frac{T_{\max}-T_{\min}}{T_{\max}+T_{\min}}\right)$ of the PSG and PSA retarders can be represented as: $\phi\delta$

$$D_1=\frac{q_1}{p_1} \quad D_2=\frac{q_2}{p_2} \tag{47.12}$$

In order to calibrate the Mueller matrix polarimeter, the error constants of the proposed model can be retrieved by taking into account a no-sample measurement. In this case the Mueller matrix can be taken ideally as identity matrix and its corresponding output stokes vector is given by:

$$S'_{0\,\mathrm{out}}=\mu'_{11}+\mu'_{22}+\mu'_{33}+\mu'_{44} \tag{47.13}$$

where, μ'_{11}, μ'_{22}, μ'_{33} and μ'_{44} are the same equation parameters obtained in (47.8). The error constants can be retrieved from the Fourier coefficients:

$$\begin{aligned}
p_1p_2&=0.5(a_0+a_2+a_8+a_{10}) & p_1z_2&=\frac{2(a_0+a_2)-3p_1p_2}{\cos\delta_2}\\
q_1q_2&=0.5(a_4+a_6) & p_2z_1&=\frac{2(a_0+a_2)-3p_1p_2}{\cos\delta_1}\\
q_1q_2&=0.5(a_1+a_9) & z_1z_2&=\frac{1}{2\sin\delta_1\sin\delta_2}(3a_6-a_4)\\
p_1p_2&=0.5(a_3+a_5) & &\\
D_1&=\frac{a_1+a_9}{a_0+a_2+a_8+a_{10}} & D_2&=\frac{a_3+a_5}{a_0+a_2+a_8+a_{10}}
\end{aligned} \tag{47.14}$$

D_1 and D_2 are the diattenuation of the PSG and PSA retarders respectively. The retardance of the retarder cannot be detected by this algorithm. To detect the retardance of the retarders we have used Goldstein-Chipman-algorithm [5, 7].

Table 47.1 Experimental result of the Mueller matrix for various objects. Result obtained by two algorithms has shown with the theoretical values and without calibration result

Sample	Theoretical values	Without calibration	Goldstein-Chipman algorithm	Proposed algorithm
No sample	$\begin{bmatrix} 1 & 0 & 0 & 0 \\ 0 & 1 & 0 & 0 \\ 0 & 0 & 1 & 0 \\ 0 & 0 & 0 & 1 \end{bmatrix}$	$\begin{bmatrix} 1.000 & 0.000 & -0.003 & 0.000 \\ 0.049 & 0.947 & 0.001 & 0.008 \\ -0.009 & -0.006 & 0.963 & 0.002 \\ -0.001 & 0.007 & 0.005 & 0.987 \end{bmatrix}$	$\begin{bmatrix} 1.000 & 0.004 & -0.008 & 0.000 \\ 0.004 & 0.988 & -0.015 & 0.008 \\ 0.001 & 0.010 & 1.005 & 0.002 \\ -0.001 & 0.007 & 0.005 & 0.988 \end{bmatrix}$	$\begin{bmatrix} 1.000 & 0.004 & -0.003 & -0.003 \\ 0.004 & 0.994 & 0.001 & 0.014 \\ -0.009 & -0.006 & 1.005 & 0.000 \\ 0.003 & 0.000 & 0.003 & 1.002 \end{bmatrix}$
Quarter wave plate	$\begin{bmatrix} 1 & 0 & 0 & 0 \\ 0 & 1 & 0 & 0 \\ 0 & 0 & 0 & -1 \\ 0 & 0 & 1 & 0 \end{bmatrix}$	$\begin{bmatrix} 1.000 & -0.043 & 0.011 & 0.006 \\ -0.031 & 0.997 & 0.014 & -0.012 \\ 0.005 & -0.008 & 0.007 & -0.982 \\ 0.003 & -0.010 & 0.930 & 0.008 \end{bmatrix}$	$\begin{bmatrix} 1.000 & -0.052 & 0.012 & 0.002 \\ -0.098 & 1.067 & 0.018 & -0.006 \\ 0.006 & -0.003 & 0.008 & -0.986 \\ 0.006 & -0.014 & 0.989 & 0.008 \end{bmatrix}$	$\begin{bmatrix} 1.000 & -0.052 & 0.011 & 0.002 \\ -0.094 & 1.063 & 0.015 & -0.004 \\ 0.006 & -0.008 & 0.008 & -0.996 \\ 0.000 & -0.004 & 0.983 & 0.015 \end{bmatrix}$
Linear polarizer (horizontal)	$\begin{bmatrix} 1 & 1 & 0 & 0 \\ 1 & 1 & 0 & 0 \\ 0 & 0 & 0 & 0 \\ 0 & 0 & 0 & 0 \end{bmatrix}$	$\begin{bmatrix} 1.000 & 0.943 & 0.036 & 0.004 \\ 1.032 & 0.922 & -0.004 & 0.013 \\ -0.017 & 0.092 & -0.091 & -0.068 \\ -0.005 & -0.026 & 0.025 & -0.006 \end{bmatrix}$	$\begin{bmatrix} 1.000 & 1.064 & 0.045 & 0.004 \\ 1.048 & 1.062 & -0.001 & 0.014 \\ -0.018 & 0.112 & -0.104 & -0.074 \\ -0.008 & -0.032 & 0.029 & -0.007 \end{bmatrix}$	$\begin{bmatrix} 1.000 & 1.056 & 0.040 & -0.019 \\ 1.040 & 1.055 & -0.005 & 0.061 \\ -0.024 & 0.105 & -0.103 & -0.063 \\ -0.028 & 0.037 & 0.032 & -0.007 \end{bmatrix}$
Linear polarizer (vertical)	$\begin{bmatrix} 1 & -1 & 0 & 0 \\ -1 & 1 & 0 & 0 \\ 0 & 0 & 0 & 0 \\ 0 & 0 & 0 & 0 \end{bmatrix}$	$\begin{bmatrix} 1.000 & -0.969 & -0.042 & 0.004 \\ -0.980 & 0.966 & 0.043 & -0.004 \\ 0.012 & -0.017 & 0.010 & 0.047 \\ -0.006 & 0.007 & 0.015 & 0.009 \end{bmatrix}$	$\begin{bmatrix} 1.000 & -0.971 & -0.044 & 0.004 \\ -0.981 & 0.968 & 0.044 & -0.005 \\ 0.006 & -0.016 & 0.013 & 0.045 \\ -0.007 & 0.008 & 0.014 & 0.008 \end{bmatrix}$	$\begin{bmatrix} 1.000 & -0.977 & -0.040 & 0.004 \\ -0.986 & 0.973 & 0.041 & -0.005 \\ 0.012 & -0.016 & 0.012 & 0.038 \\ -0.004 & 0.002 & 0.000 & 0.008 \end{bmatrix}$

47.4 Experimental Results

Experiments were done by using a tungsten halogen light source and a narrow band-pass filter of wavelength 632.5 ± 5 nm. By using a 14 bit CCD camera (Thorlab-8050 M-GE-TE), 36 images were taken covering a rotation of 180°, where each image was averaged by 20 frames for reducing noise. The obtained Mueller matrix by experiment is given below for various samples.

The average Mueller matrix value of 10 × 10 pixel at center for various samples were measured and presented in Table 47.1. The results are compared with both Goldstein-Chipman-algorithm [5, 7] and our proposed algorithm.

In order to show the imaging capability and accuracy of the implemented system, the spatial distribution of Mueller matrix Fig. 47.3a and its line profile along *Y* axis Fig. 47.3b for no sample measurement has been shown.

Taking a quarter wave plate as a sample, its corresponding Mueller matrix images are presented in Fig. 47.4a and its line profile along *Y* axis in Fig. 47.4b. All the line profiles are taken in the same position, which shows a noise in same position. This noise is due to the dust particles on the surface, improper polarizing property of the sample and defocus obtained by the imaging system. In this parts the authors are currently working to improve it.

By using the Lu-Chipman-algorithm [8, 9], retardance and diattenuation information were obtained and presented in Figs. 47.5a and 47.6a along with its

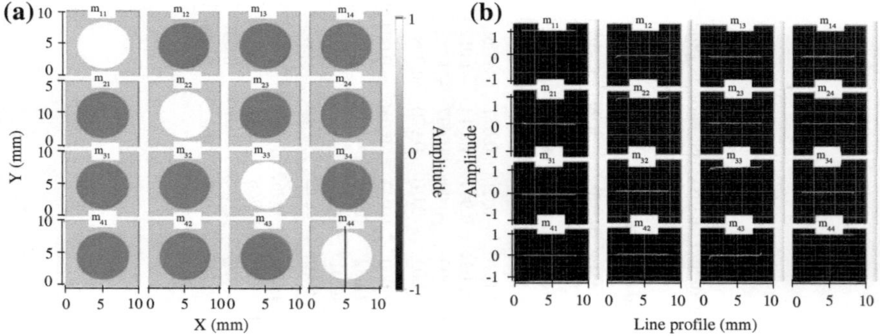

Fig. 47.3 Experimental images obtained for no sample. The Mueller matrix images and its one dimensional line profile along central *Y* axis has shown for the variation obtained. **a** The Mueller matrix image for no sample. **b** Line profile of Mueller matrix along the central Y axis (Marked in image m_{44})

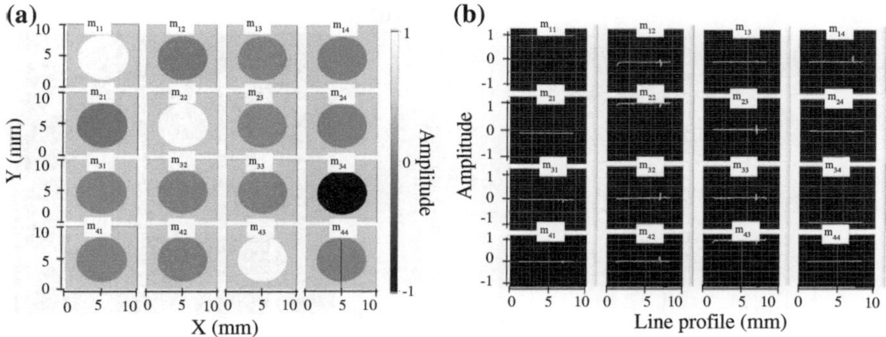

Fig. 47.4 Experimental images obtained for a quarter-wave plate (QWP) as a sample. The Mueller matrix imaging capability of the system, its line profile along the central *Y* axis has shown. **a** The Mueller matrix image for QWP. **b** Line profile of Mueller matrix along the central *Y* axis (Marked in image m_{44})

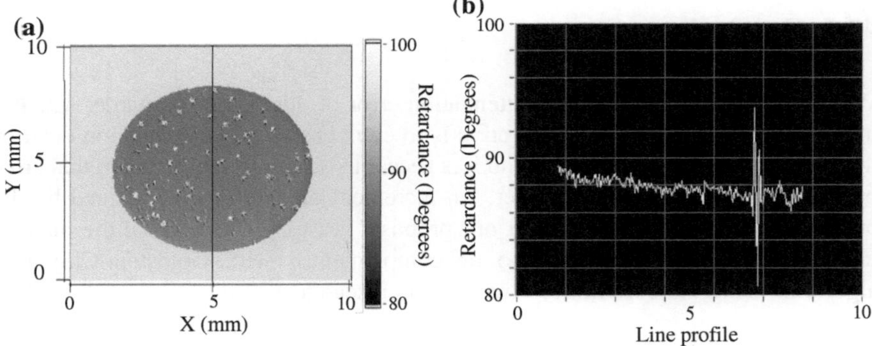

Fig. 47.5 By the decomposed images obtained for a quarter-wave plate (QWP), the spatial distribution of the retardance (**a**) and a line profile along central *Y* axis (**b**) is presented. The range was settled between 80° and 100° for better appreciation of the distribution. Higher and lower value points in the images corresponds to errors occurred on the implemented system. **a** Spatial variation of retardance of the QWP in degrees. **b** Line profile of retardance along the central Y axis (Marked in image)

corresponding line profile, Figs. 47.5b and 47.6b. The range was limited in both cases in order to show variations against the expected values. Retardance variations are shown in a range between 80° and 100°, where the ideal value is 90°. In the diattenuation the range was settled between 0 and 0.1 where the ideal value is 0. Values outside this range corresponds to error occurred on the system.

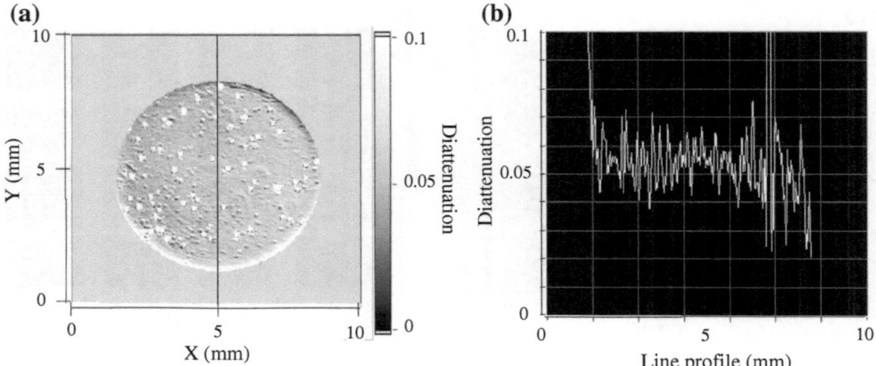

Fig. 47.6 Spatial distribution of the diattenuation of the QWP (**a**) and a line profile along central *Y* axis (**b**) is presented. The range was settled between 0 and 0.1 for better appreciation of the distribution. Higher and lower value points in the images corresponds to errors occurred on the implemented system. **c** Spatial variation of diattenuation of the QWP. **d** Line profile of diattenuation along the central *Y* axis (Marked in image)

47.5 Conclusions

We presented a calibration for diattenuation error of dual rotating retarder Mueller matrix polarimeter. The main objective is to compensate the diattenuation error of the retarders. To apply this algorithm it is necessary to minimize the azimuthal error of the retarders as well as polarizer. An increment of accuracy is obtained by the proposed method. By the moment our proposal is capable to retrieve the diattenuation parameters, but it need to be complemented with Goldstein-Chipman-algorithm being the current objectives of the research.

References

1. Collett, E., "Field Guide to Polarization", SPIE Press, (2005)
2. Anastasiadou M, De Martino A, Clement D, Liège F, Laude-Boulesteix B, Quang N, Dreyfuss J et al (2008) Polarimetric imaging for the diagnosis of cervical cancer. Physica Status Solidi (c) 5(5):1423–1426
3. Ghosh N, Vitkin IA (2011) Tissue polarimetry: concepts, challenges, applications, and outlook. J Biomed Opt 16(11):110801–11080129
4. Azzam RMA (1978) Photopolarimetric measurement of the Mueller matrix by Fourier analysis of a single detected signal. Opt Lett 2(6):148–150
5. Goldstein DH (1992) Mueller matrix dual-rotating retarder polarimeter. Appl Opt 31(31): 6676–6683
6. Chipman, R.A., "Polarimetry, Chapter 22, Handbook of Optics, Volume- II, Second Edition", McGraw-Hill Professional, (1995)

7. Goldstein DH, Chipman RA (1990) Error analysis of a Mueller matrix polarimeter. Opt Soc Am A 7(4):693–700
8. Lu SY, Chipman RA (1996) Interpretation of Mueller matrices based on polar decomposition. Opt Soc Am A 13(5):1106–1113
9. Goldstein D (2003) Polarized light, 2nd edn. Marcel Dekker, Inc., New York

Chapter 48
A Simple Configuration for Quantitative Phase Contrast Microscopy of Transmissible Samples

Chandan Sengupta, Koustav Dasgupta and K. Bhattacharya

Abstract Phase microscopy attempts to visualize and quantify the phase distribution of samples which are otherwise invisible under microscope without the use of stains. The two principal approaches to phase microscopy are essentially those of Fourier plane modulation and interferometric techniques. Although the former, first proposed by Zernike, had been the harbinger of phase microscopy, it was the latter that allowed for quantitative evaluation of phase samples. However interferometric techniques are fraught with associated problems such as complicated setup involving mirrors and beam-splitters, the need for a matched objective in the reference arm and also the need for vibration isolation. The present work proposes a single element cube beam-splitter (CBS) interferometer combined with a microscope objective (MO) for interference microscopy. Because of the monolithic nature of the interferometer, the system is almost insensitive to vibrations and relatively simple to align. It will be shown that phase shifting properties may also be introduced by suitable and proper use of polarizing devices. Initial results showing the quantitative three dimensional phase profiles of simulated and actual biological specimens are presented.

48.1 Introduction

Visualization of biological samples at cellular level with their 3D structure is of great scientific interest. Microscopic phase objects are those through which the intensity of the radiation transmitted will remain undetectable but minor phase variation of the sample will modulate the transmitted beam accordingly. Phase

C. Sengupta (✉)
SAMEER Kolkata Center, JC-30 Sector-III, Saltlake, Kolkata 700098, India
e-mail: mechandansengupta@gmail.com

C. Sengupta · K. Dasgupta · K. Bhattacharya
Department of Applied Optics and Photonics, University of Calcutta, JD-2 Sector-III, Saltlake, Kolkata 700098, India

V. Lakshminarayanan and I. Bhattacharya (eds.), *Advances in Optical Science and Engineering*, Springer Proceedings in Physics 166,
DOI 10.1007/978-81-322-2367-2_48

samples are usually visualized by staining which is by itself a destructive process. As a result, living cell dynamics cannot be studied.

Digital holography [1] is one of the techniques by which quantitative phase visualization is possible and has been extensively used for the purpose. However this process involves a complicated and precise optical alignment, which can make this procedure somewhat inadequate for in situ visualization. A standard, method for phase visualization is the use of Zernike phase microscopy [2] and normansky interference contrast microscopy [3] both of which are incapable of giving quantitative result.

Shearing interferometer [4] provides one of the solutions for microscopic 3D visualization of phase objects. Shear in differential interference contrast microscopy is achieved by using a birefringent prism, known as the normansky prism which generates two shifted versions of the wave front which interferes between each other. Standard shearing techniques also involve the constraints of complicated optical design and vibration isolation. In the present work, a cube beam splitter (CBS) accurately placed in front of the microscopic phase sample with a specific angle is used as a shear interferometers coupled to a microscopic system.

48.2 Experimental Setup

As shown in Fig. 48.1, the proposed setup consists of a CBS critically placed before a microscope objective (MO) with the sample (S) in between. The CBS is so oriented that it amplitude-splits the laser beam into two and makes them nearly converge on the sample. The output faces of the CBS are masked with orthogonally oriented linear polarizer and the two sheared and polarized imaging beams are made to pass through a quarter wave plate (QWP) and polarizer (P2) combination to allow for polarization phase shifting. If φ(x, y) is the phase distribution over the region of interest over the sample zone, two sets of four quarter-wave phase-shifted frames yielding δφ/δx and δφ/δy are recorded and utilized to reconstruct the sample phase. Between the two sets of measurements the CBS is rotated by 90°.

We propose simple setup having a 5 mW, 632.8 nm class 3 HE-Ne gas LASER as the source and two linear polarizer P1 and P2, an apochromatic MO of 50X optical magnification, a CBS masked two outputs by sheet polarizer, a QWP and CCD camera with computer control.

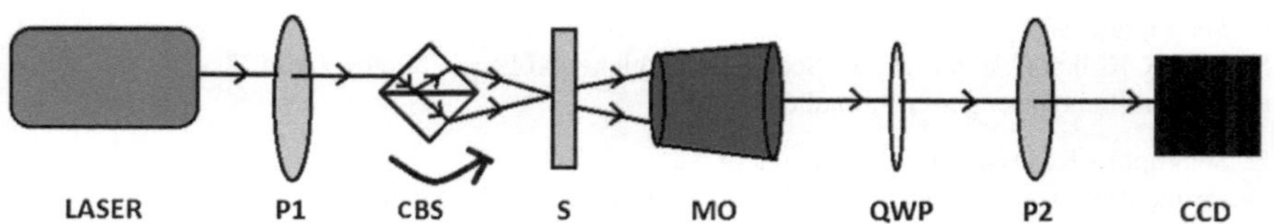

Fig. 48.1 Experimental setup

48.3 Theoretical Analysis

In term of Jone's vector the polarization state of the wave fronts ε_{s1} and ε_{s2} given as,

$$\varepsilon_{s1} = \begin{pmatrix} 0 \\ 1 \end{pmatrix}$$

and

$$\varepsilon_{s1} = \begin{pmatrix} 1 \\ 0 \end{pmatrix}$$

After passing through a QWP aligned with in fast axis 450 to the reference abscissa, both the beams are rendered circularly polarized, although in the opposite sense. If the beam energy from the QWP are given by ε'_{s1} and ε'_{s2}, then,

$$\varepsilon'_{s1} = \begin{pmatrix} 1 \\ i \end{pmatrix}$$

and

$$\varepsilon'_{s2} = \begin{pmatrix} 1 \\ -i \end{pmatrix}$$

Showing that the sheared wave fronts are circularly polarized in the opposite sense as expected. When a linear polarizer $P(\theta)$ intercepts the beam ε'_{s1}, the output is,

$$\begin{aligned}
\varepsilon''_{s1} &= \begin{pmatrix} \cos^2\theta & \sin\theta\cos\theta \\ \sin\theta\cos\theta & \sin^2\theta \end{pmatrix} \begin{pmatrix} 1 \\ i \end{pmatrix} \\
&= (\cos\theta + i\sin\theta) \begin{pmatrix} \cos\theta \\ \sin\theta \end{pmatrix} \\
&= \exp(i\Delta_1) \begin{pmatrix} \cos\theta \\ \sin\theta \end{pmatrix}
\end{aligned}$$

where $\Delta_1 = \theta$.

Similarly the beam ε'_{s2} may be represented as,

$$= \varepsilon''_{s2} = \exp(i\Delta_2)$$

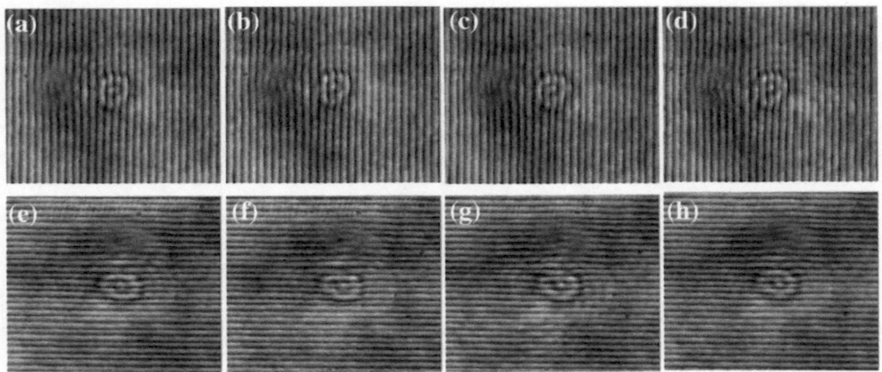

Fig. 48.2 Recorded fringes for horizontal shear (**a**)–(**d**) and vertical shear (**e**)–(**h**). Consecutive interferograms are polarization phase shifted by a uniform π/2 (from 0 to 3π/2) phase for both horizontal and vertical shear fringe sets

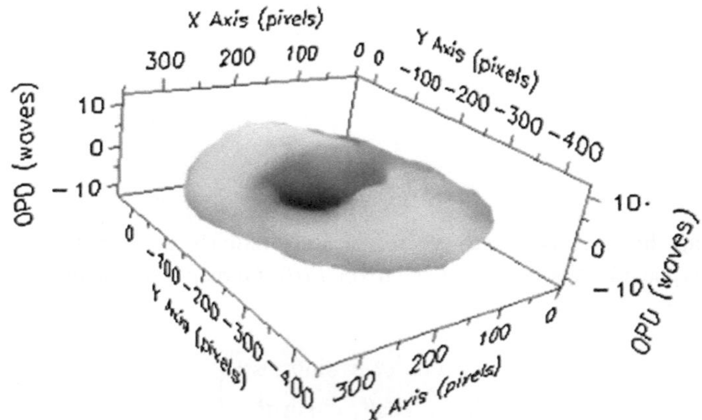

Fig. 48.3 3D phase profile of human erythrocyte

where

$$\Delta_2 = -\theta$$

This indicates that when the polarizer is rotated by an angle θ, a phase difference of 2θ is introduced between the sheared wavefronts.

Interferograms are recorded at θ = 0°, 45°, 90° and 135°, so as to introduce phase difference of 0, π/2, π and 3π/2 between the sheared wave fronts and the standard four-frame algorithm [5] is applied to reconstruct the differential phase waveform [δφ(x, y)/δx] and [δφ(x, y)/δy].

48.4 Experimental Results

The sample is an isolated erythrocyte from human blood. The sample was positioned between the CBS and the MO and the following phase shifted frames were captured on the CCD for shears along the x and y directions are shown in Figs. 48.2 and 48.3 respectively.

Integrating the data of x and y shear frames, preliminary results showing the three dimensional phase profile of human erythrocytes has been reconstructed as shown in Fig. 48.3. Although the doughnut structure is observed finer refinements are necessary and are being presently undertaken.

References

1. Tishco TV, Tishco DN, Titar VP (2012) Application of digital holographic interference microscopy for study of 3D morphology of human blood erythrocytes. In: Mecdez-Villas A (ed) Current microscopy contributions to advances in science and technology
2. Zernike F (1942) Phase contrast, a new method for the microscopic observation of transparent objects part I. Physica 9(7):686–698
3. Murphy D (2001) Differential interference contrast (DIC) microscopy and modulation contrast microscopy, in fundamentals of light microscopy and digital imaging. Wiley, New York, pp 153–168
4. Strjnik M, Paez G, Mantravadi M (2007) Lateral shear interferometers. In: Malacara D (ed) Optical shop testing, 3rd edn. Wiley, New York, pp 122–184
5. Schreiber H, John HB (2006) Phase shifting interferometry. In: Malacara D (ed) Optical shop testing, 3rd edn. Wiley, New York, pp 547–655

Part X
Non-linear Phenomena and Chaos

Chapter 49
Onset of Chaos for Different Non Linear Systems by Varying System Parameters

Mili Sarkar, Rajarshi Roy Chaudhuri, Subhajit Dutta Chowdhury, Nitish Kumar Thakur and Sabarno Chowdhury

Abstract Chaos is one of the most important behaviors exhibited by non linear dynamical system and refers to the predictable although seemingly random, motion of trajectories of a dynamical system. Chaotic systems display chaotic behaviors only for specific range of values of their parameters. In this paper, the generation of chaos in two chaotic systems, namely, Lorenz System and Rössler's model was shown using SIMULINK with few specific initial conditions and parameter values. The state space trajectories act as strange attractors and this shows the emergence of chaos for chosen set of control parameter values.

49.1 Introduction

Chaos is a new discipline in mathematics which describes a specific range of irregular behaviours in dynamical systems. Chaos is observed in systems with non-linear elements present within them [1–3]. Actually, in real world many common phenomena can be explained correctly in terms of non-linear models but for making our job of system analysis easier we generally tend to linearize them and then analyse. As a result of this approach we have not been able to account for various experimentally observed phenomena in these systems [4, 5].

A deterministic non-linear dynamical system is said to exhibit chaos when it exhibits non-periodic behaviour in the system state space with the state trajectories being highly sensitive to initial conditions. Though these systems are completely deterministic small differences in initial conditions would lead to widely diverging outcomes thus making long term prediction of the system behaviour difficult without having precise knowledge of the system's initial conditions [6–8].

R.R. Chaudhuri (✉)
Department of Electronics and Communication Engineering, Institute of Engineering and Management, Kolkata, India
e-mail: roychaudhurirajarshi@gmail.com

M. Sarkar · S.D. Chowdhury · N.K. Thakur · S. Chowdhury
Institute of Engineering and Management, Kolkata, India

V. Lakshminarayanan and I. Bhattacharya (eds.), *Advances in Optical Science and Engineering*, Springer Proceedings in Physics 166,
DOI 10.1007/978-81-322-2367-2_49

Continuous chaos can be observed in deterministic non-linear dynamical systems which have three or more degree of freedom [7, 9, 10].

In this paper we have dealt with two well known simple three-dimensional autonomous systems, namely, the Lorenz System, and, the Rössler's Model. We have modelled these two systems using SIMULINK and then simulated them for different values of the control parameter to observe and study the onset of chaos in this systems. This paper is divided into three broad sections. Sections 49.2 and 49.3 deal with the theoretical perspective of the Lorenz's system and Rössler's Model. Section 49.4 presents the simulation results which contain the phase portrait of the systems and the time series plots.

49.2 Lorenz Equations

49.2.1 The Basic Set of Differential Equations

Edward Lorenz modelled the two-dimensional thermal convection phenomenon [11] into a system of coupled ordinary differential equations (ODEs). The Lorenz equations are stated as follows:

$$\frac{\mathrm{d}x}{\mathrm{d}t} = -\sigma(x - y) \tag{49.1}$$

$$\frac{\mathrm{d}y}{\mathrm{d}t} = -xz + rx - y \tag{49.2}$$

$$\frac{\mathrm{d}z}{\mathrm{d}t} = xy - bz \tag{49.3}$$

The above set of equations is the model of a system whose state variables are x, y, and, z. The parameters σ, r, b are the control parameters of the modelled system. We have varied the control variable r keeping variables σ and b fixed.

49.2.2 Mathematical Analysis of the Equations

49.2.2.1 The Equilibrium Points

The equilibrium points of the system could be found out by letting $\dot{x} = 0, \dot{y} = 0$, $\dot{z} = 0$.

Whenever a system reaches the equilibrium point(s) its state variables remain there unchanged forever. It is so because the rate of change of the state variables x, y, z with respect to time at those points is zero. For the given set of differential equations in (49.1)–(49.3) the equilibrium points will be:

$$A(0,0,0);\ B(\sqrt{b(r-1)}, \sqrt{b(r-1)}, r-1);\ C(-\sqrt{b(r-1)}, -\sqrt{b(r-1)}, r-1)$$

It is evident that for $r < 1$ the equilibrium points B and C do not exist. The only equilibrium point which will exist is A (0, 0, 0). For $r > 1$ all the three equilibrium points A, B, C will exist.

49.2.2.2 System Dynamics

(i) **The Jacobian Matrix (J)**

$$J = \begin{pmatrix} -\sigma & \sigma & 0 \\ -z+r & -1 & -x \\ y & x & -b \end{pmatrix}$$

(ii) **The Eigenvalues and Eigenvectors**
At A (0, 0, 0) the eigenvalues would be:

$$\lambda = -b, \quad \lambda = \frac{-11 + \sqrt{81 + 40r}}{2}, \quad \lambda = \frac{-11 - \sqrt{81 + 40r}}{2}$$

We see that for all positive values of r the three eigenvalues are real. We also see that for all positive $r < 1$ the three eigenvalues are ***negative real numbers.*** This implies that for $r < 1$ the equilibrium point A (0, 0, 0) is stable [5].

Case 1 For $r < 1$
Let us see the system phase-space trajectory which shows how the system evolves with time given by (x, y, z) for $r < 1$.

We have taken $r = 0.9$ and simulated the model.

The time series response $x(t)$ versus t, $y(t)$ versus t and $z(t)$ versus t are given in Fig. 49.1.

The x versus y, y versus z and x versus z plots are shown in Fig. 49.2.

From the Fig. 49.2 we see that the values of x, y, z go to zero as time goes on increasing. It is due to the fact that the trajectory given by (x, y, z) spiral down towards the only existing stable equilibrium A (0, 0, 0).

It can be seen that as $r > 1$ occurs the equilibrium point A becomes unstable and the points B and C come into existence.

Case 2 For $r > 1$ taking $r = 15$;
We now have two stable equilibrium points B and C. We also see from the phase portrait that the trajectory starting from **the initial point** (15, 10, 40) is moving towards the equilibrium point $C\,(-6.11, -6.11, 14)$. It is equally probable for the phase trajectory of the system to evolve towards the equilibrium point B (6.11, 6.11, 14). The evolution of the trajectory of the system depends on the initial conditions of the

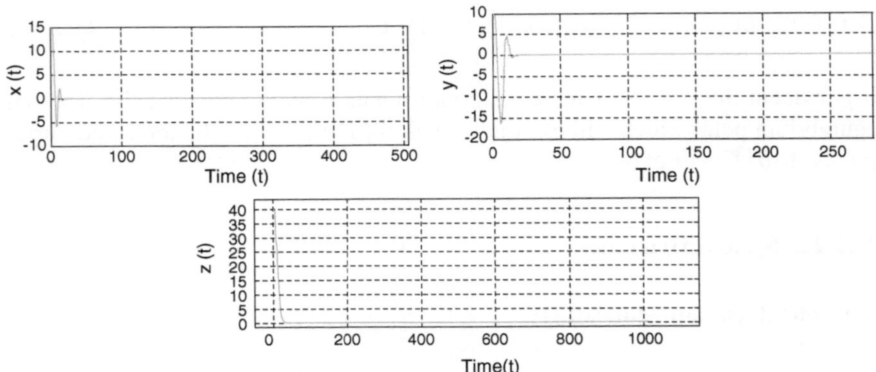

Fig. 49.1 The time series response $x(t)$ versus t, $y(t)$ versus t and $z(t)$ versus t

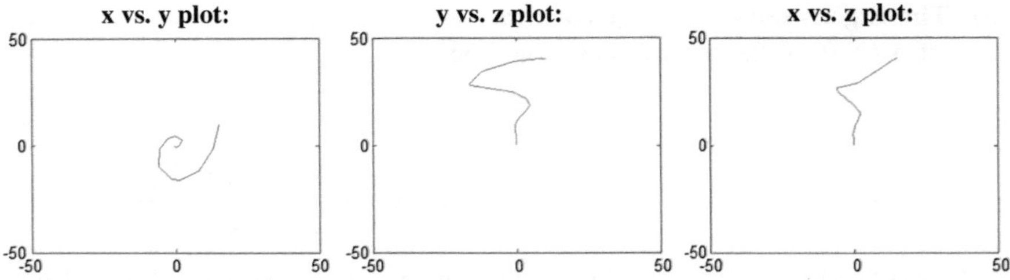

Fig. 49.2 The x versus y, y versus z and x versus z plots

state variables and hence for two different initial conditions the trajectory could evolve towards either the equilibrium point B or C [5].

Eigenvalues corresponding to the equilibrium point C are given by the following equation:

$$\det(J - \lambda I) = 0, \quad \text{where,} \quad J = \begin{pmatrix} -\sigma - \lambda & \sigma & 0 \\ 1 & -1 - \lambda & \sqrt{b(r-1)} \\ -\sqrt{b(r-1)} & -\sqrt{b(r-1)} & -b - \lambda \end{pmatrix}$$

$$\lambda^3 + (\sigma + b + 1)\lambda^2 + b(r + \sigma)\lambda + 2\sigma b(r - 1) = 0 \tag{49.4}$$

The stable equilibrium point C has three eigenvalues corresponding to it which gives the local nature of the phase-portrait about the equilibrium point. The three eigenvalues consist of one real negative numbered eigenvalue and two complex conjugate eigenvalues.

Thus one can guess the nature of the trajectory in the state space from the eigenvalues. Here, the Eigen direction corresponding to the negative real numbered eigenvalue would be the direction along which the trajectory would spiral down to the plane defined by the Eigen-vectors corresponding to the complex-conjugate eigenvalues. The time series response $x(t)$ versus t, $y(t)$ versus t and $z(t)$ versus t are given in Fig. 49.3. The variation of x, y and z with time has a response which looks like that of an oscillation dying down to a fixed value as time increases. This can be easily accounted by the presence of a negative real eigenvalue causing the exponential dying down of the oscillations which are caused by the presence of complex conjugate eigenvalues. The x versus y, y versus z and x versus z plots are shown in Fig. 49.4.

Case 3 For $r > 1$ taking $r = 23.5$;
We have seen that for r = *23.5* onwards the oscillations in the values of x, y, and, z with time perisits and it didn't die out as it was the case for the previous values of r.

This can be explained by the fact that the real part of the two complex conjugate eigen values obtained as the roots of the (49.4) are increasing, i.e., becoming less negative. As a result of this the rate of decay of the oscillations is also decreasing.

The time series response $x(t)$ versus t, $y(t)$ versus t and $z(t)$ versus t obtained from the computer simulation gives a clear picture of the decrease in rate of decay of oscillations in the values of the state variables.

The time series response $x(t)$ versus t, $y(t)$ versus t and z(t) versus t are shown in Fig. 49.5.

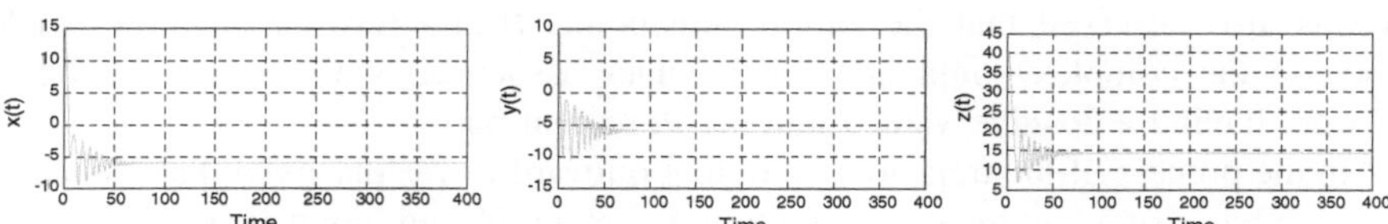

Fig. 49.3 The time series response $x(t)$ versus t, $y(t)$ versus t and $z(t)$ versus t

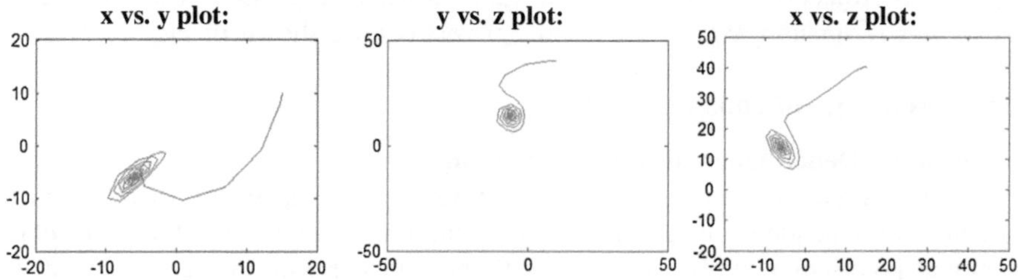

Fig. 49.4 The x versus y, y versus z and x versus z plots

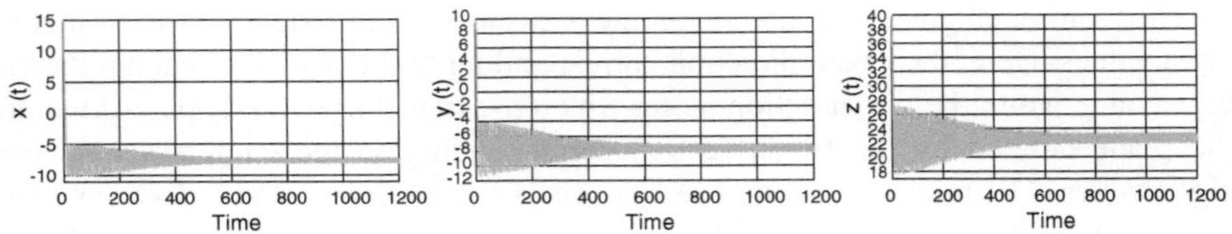

Fig. 49.5 The time series response *x*(*t*) versus *t*, *y*(*t*) versus *t* and *z*(*t*) versus *t*

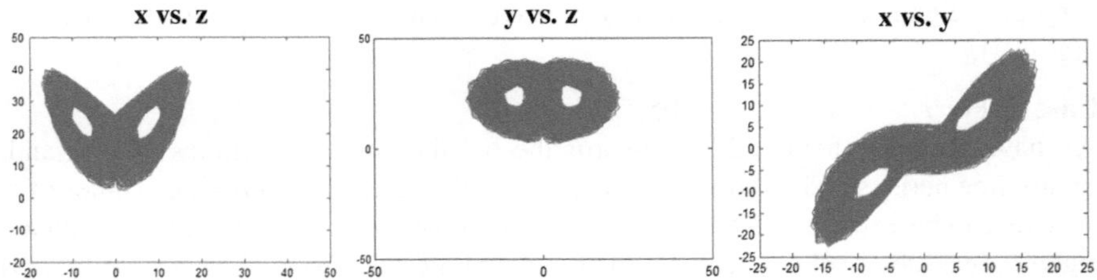

Fig. 49.6 The *x* versus *y*, *y* versus *z* and *x* versus *z* projections of the phase portrait

Case 4 For $r > 1$ taking $r = 24.74$;
When the model was simulated for $r = 24.74$ the system entered into the chaotic regime. It was seen that the amplitude of the oscillations of the state variables started to increase as the trajectory spiralled out in the eigen plane corresponding to the complex conjugate eigen values of the equilibrium point *C*. In the phase portrait it was also observed that the equilibrium points *B* and *C* were no more stable because the complex conjugate Eigen values associated with the points *B* and *C* were having a positive valued number in its real part.

In the process of the trajectory spiralling outward it was observed that it entered into the basin of attraction corresponding to the equilibrium point *B* and from there it again started spiralling outward to enter into the basin of attraction of the equilibrium point *C*. This is precisely the *stretching and folding* mechanism of the trajectory which leads to chaos. It can be clearly seen from *x* versus *y*, *y* versus *z* and *x* versus *z* projections of the phase portrait shown in Fig. 49.6. The time series response *x*(*t*) versus *t*, *y*(*t*) versus *t* and *z*(*t*) versus *t* are shown in Fig. 49.7.

Some properties of chaotic signals:

1. **Sensitive Dependence of initial conditions**
Regular stretching and folding of the trajectory in the phase space leads to sensitive dependence of the trajectory on the initial conditions. Two very close initial points may evolve in a very different manner with the separation of the trajectories increasing with time.

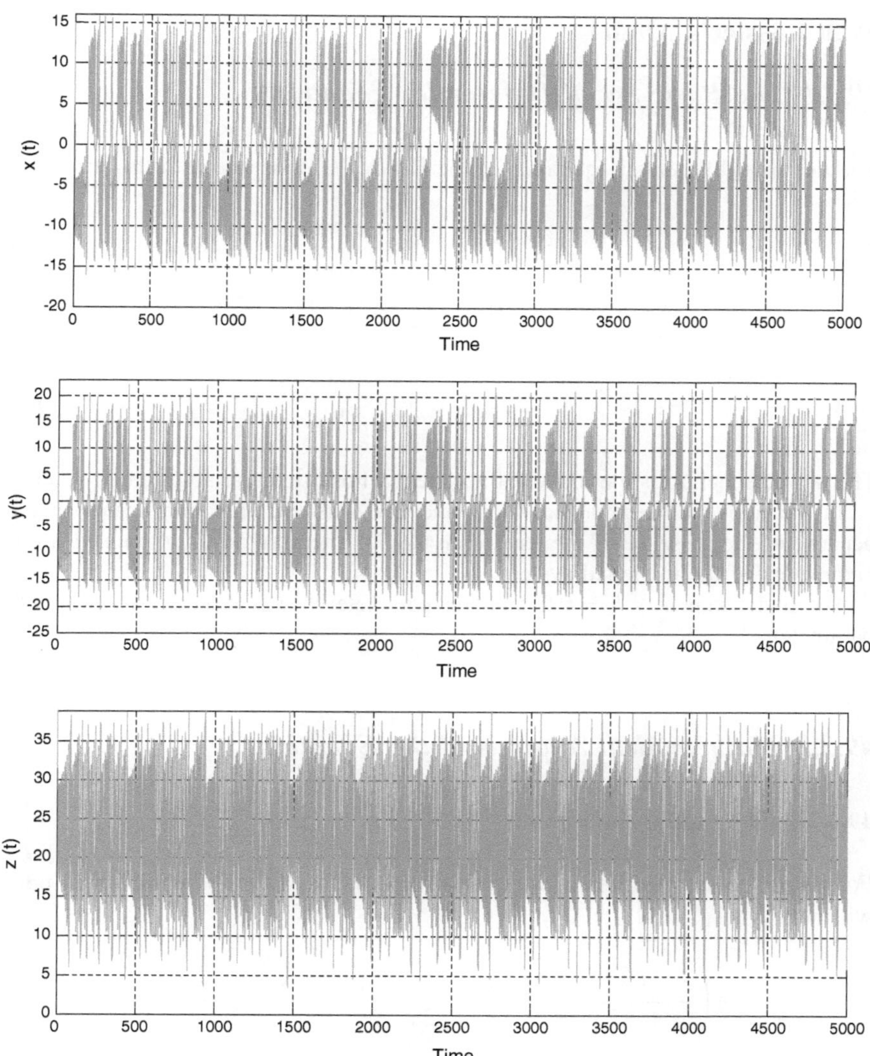

Fig. 49.7 The time series response $x(t)$ versus t, $y(t)$ versus t and $z(t)$ versus t

2. **Aperiodic Signal**
   In a chaotic system just as the one shown for $r = 24.74$ in a Lorenz system the state variables values when plotted with time do not show periodicity, i.e., there exists no pattern which repeats itself in the graphs of $x(t)$ versus t, $y(t)$ versus t and $z(t)$ versus t.
3. **Finite and Bounded Signal**
   The values of x, y, and, z do not become infinite but they remain bounded as seen from the graphs.

49.3 Rössler's Model

49.3.1 The Basic Set of Differential Equations

The defining differential equations of the Rössler's model are as follows:

$$\frac{dx}{dt} = -y - z \tag{49.5}$$

$$\frac{dy}{dt} = x + ay \tag{49.6}$$

$$\frac{dz}{dt} = b + z(x - c) \tag{49.7}$$

The above equation is the model of a system whose state variables are x, y, and, z.

In the above differential equations the (49.7) is the only non-linear equation because of the presence of the term xz in (49.7). The parameters a, b *and* c are the parameters of the modelled system.

We had taken $a = b = 0.2$ and the parameter c was selected as the control parameter.

49.3.2 Mathematical Analysis of the Equations

49.3.2.1 The Equilibrium Points

For the given set of differential equations in (49.5)–(49.7) the equilibrium points will be:

$$A\left(\frac{c+\sqrt{c^2-4ab}}{2}, \frac{-c-\sqrt{c^2-4ab}}{2a}, \frac{c+\sqrt{c^2-4ab}}{2a}\right)$$

$$B\left(\frac{c-\sqrt{c^2-4ab}}{2}, \frac{-c+\sqrt{c^2-4ab}}{2a}, \frac{c-\sqrt{c^2-4ab}}{2a}\right)$$

In our model $a = b = 0.2$, so, it is evident that for $\mathbf{c^2 < 4a^2}$ the equilibrium points A and B would exist.

49.3.2.2 Study of the Effect on System Dynamics Due to Variations of the Parameter '*C*'

(i) **Case 1: Taking $c^2 < 4a^2$, $c = 0.3$**
In this case, there would exist no equilibrium point. Hence, the state variables *x*, *y*, and, *z* would assume values that would be increasing with time implying an unstable system.

(ii) **Case 2: Taking $c^2 > 4a^2$, $c = 2$**
The system state space trajectory in this case exhibited the formation of a period-1 limit cycle. Trajectories from all possible initial conditions converge on this limit cycle as time increases. It can be clearly seen from *x* versus *y*, *y* versus *z* and *x* versus *z* projections of the phase portrait shown in Fig. 49.8.

(iii) **Case 3: Taking $c^2 > 4a^2$, $c = 3$**
In this case the state space trajectory exhibited the phenomenon of *period-doubling* and resulted in the formation of a period-2 limit cycle.
The time series plot *x*(*t*) versus *t* in Fig. 49.9 also reveals the presence of period-2 limit cycle.
The *x* versus *y*, *y* versus *z* and *x* versus *z* projections of the phase portrait is shown in Fig. 49.10.

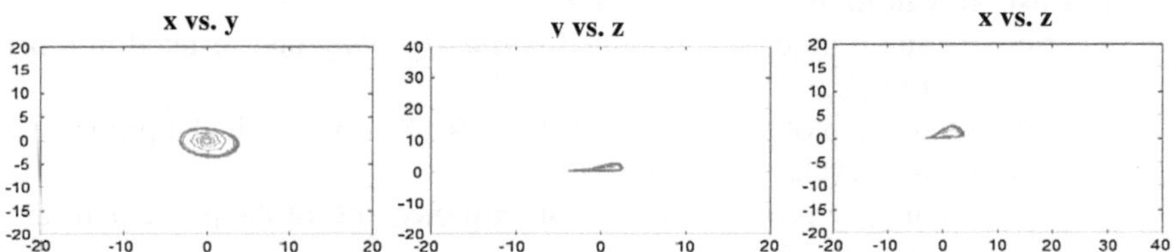

Fig. 49.8 The *x* versus *y*, *y* versus *z* and *x* versus *z* projections of the phase portrait

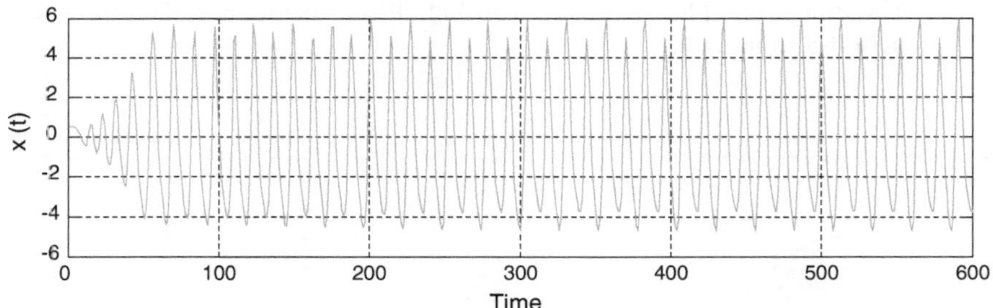

Fig. 49.9 The time series plot *x*(*t*) versus *t*

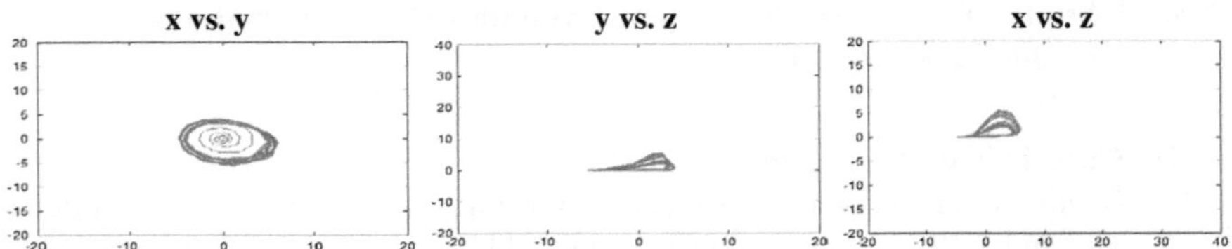

Fig. 49.10 The x versus y, y versus z and x versus z projections of the phase portrait

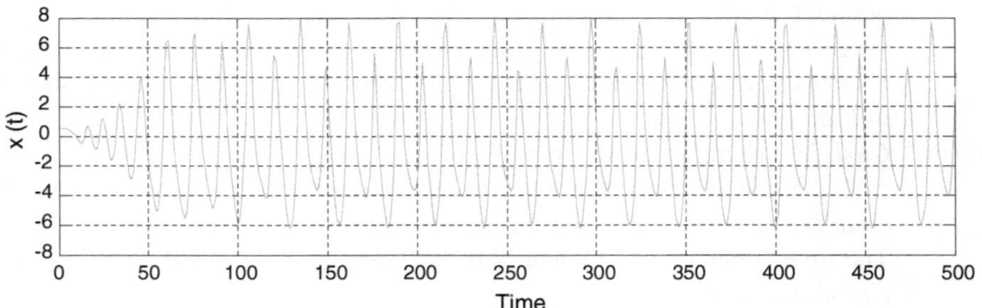

Fig. 49.11 The time series plot $x(t)$ versus t

(iv) **Case 4: Taking $c^2 > 4a^2$, $c = 3.9$**
The state space trajectory exhibited *period-doubling* and resulted in a period-4 limit cycle.
The time series plot $x(t)$ versus t in Fig. 49.11 also reveals the presence of period-4 limit cycle.
The x versus y, y versus z and x versus z projections of the phase portrait is shown in Fig. 49.12.

(v) **Case 5: Taking $c^2 > 4a^2$, $c = 4.15$**
The state space trajectory exhibited *period-doubling* and resulted in a period-8 limit cycle.
The time series plot $x(t)$ versus t in Fig. 49.13 also reveals the presence of period-8 limit cycle.

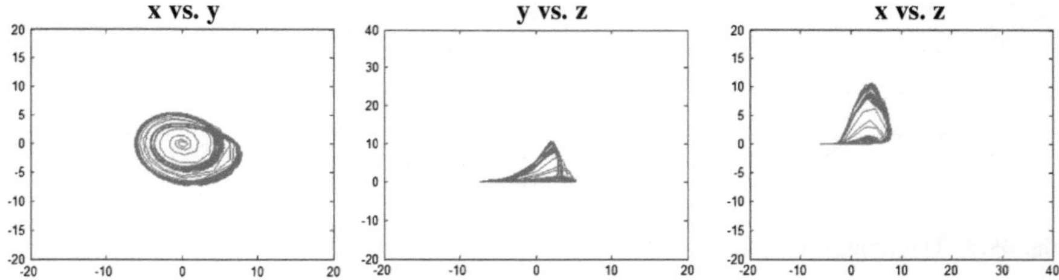

Fig. 49.12 The x versus y, y versus z and x versus z projections of the phase portrait

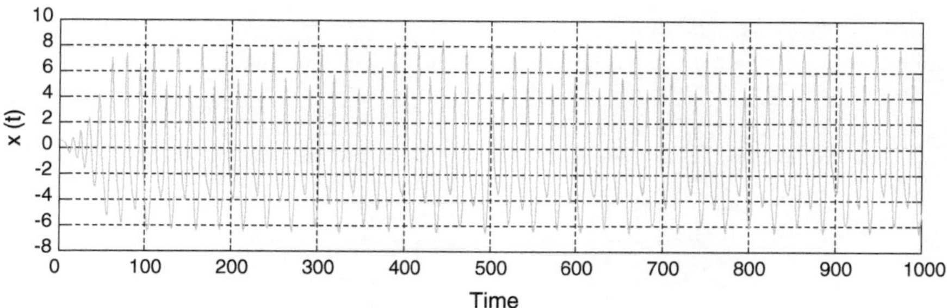

Fig. 49.13 The time series plot *x*(*t*) versus *t*

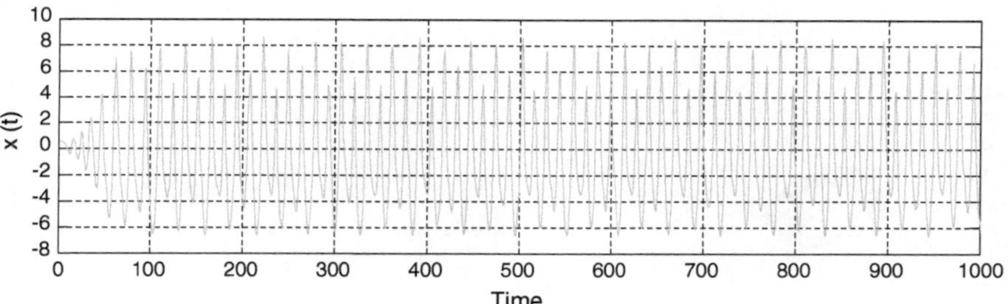

Fig. 49.14 The time series plot *x*(*t*) versus *t*

(vi) **Case 6: Taking $c^2 > 4a^2$, $c = 4.20$**
The state space trajectory exhibited *period-doubling* and resulted in a period-16 limit cycle.
The time series plot *x*(*t*) versus *t* in Fig. 49.14 also reveals the presence of period-16 limit cycle.
From the above variation of parameters we have observed that as the parameter c is increased the period-doubling phenomenon is happening for a smaller change in the parameter value. Thus the limit cycle period is changing very fast with changing parameter value.

(vii) **Case 7: Taking $c^2 > 4a^2$, $c = 4.60$**
In this case the system exhibited non-periodic chaotic behaviour. This can be explained from the period-doubling phenomenon which occurred with increment in value of the parameter *c*. The period of the orbit increased in

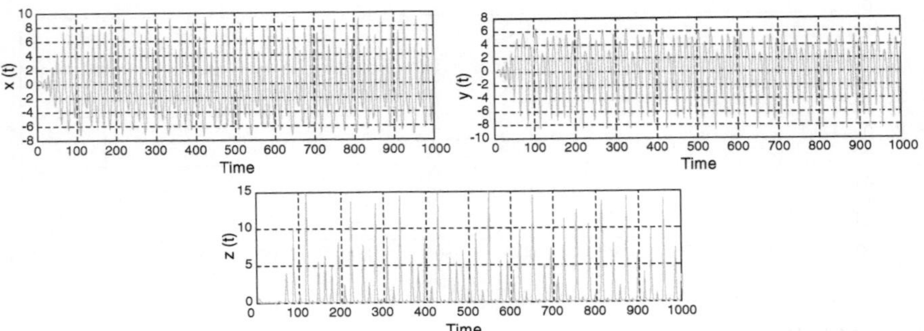

Fig. 49.15 The time series plot *x*(*t*) versus *t*, *y*(*t*) versus *t* and *z*(*t*) versus *t*

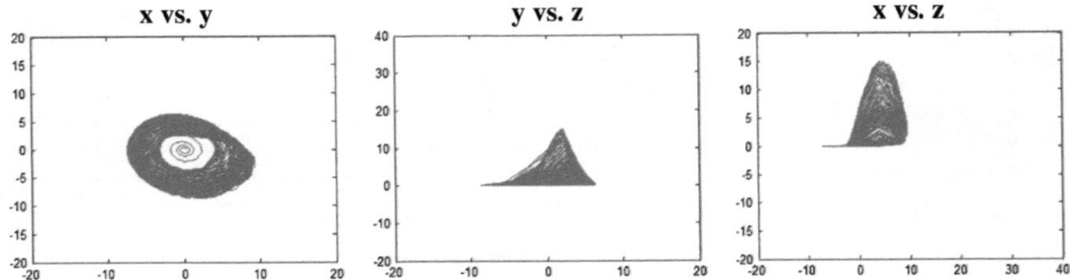

Fig. 49.16 The *x* versus *y*, *y* versus *z* and *x* versus *z* projections of the phase portrait

the following doubling sequence: 1, 2, 4, 8, 16, 32, 64 and so on with increase in the parameter *c*. Moreover the rate of change in the period of the limit cycle increased rapidly with the increase in parameter *c*. As the parameter value changed from 4.2 to 4.6 period doubling occurred for a large number of times making the limit cycle to be non-periodic (period infinity). The Rössler's model thus followed the period doubling route to chaos. The time series plot *x*(*t*) versus *t*, *y*(*t*) versus *t* and *z*(*t*) versus *t* in Fig. 49.15 also reveals the presence of the chaotic aperiodic signal.

The *x* versus *y*, *y* versus *z* and *x* versus *z* projections of the phase portrait is shown in Fig. 49.16.

49.4 SIMULINK Model Implementation and Experimental Results

49.4.1 SIMULINK Model Used for Simulating the Lorenz System

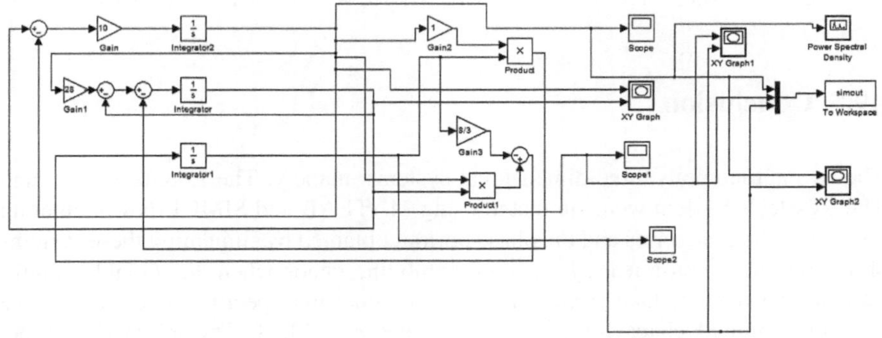

The block 'Gain 1' is the gain parameter '***r***'. The block 'Gain' is the parameter '**σ**'. And, the block 'Gain 3' is the parameter '***b***'.

49.4.1.1 Simulation Results and Diagrams for Lorenz System

See Figs. 49.1, 49.2, 49.3, 49.4, 49.5, 49.6 and 49.7.

49.4.2 The SIMULINK Model Used for Simulating the Rössler's System

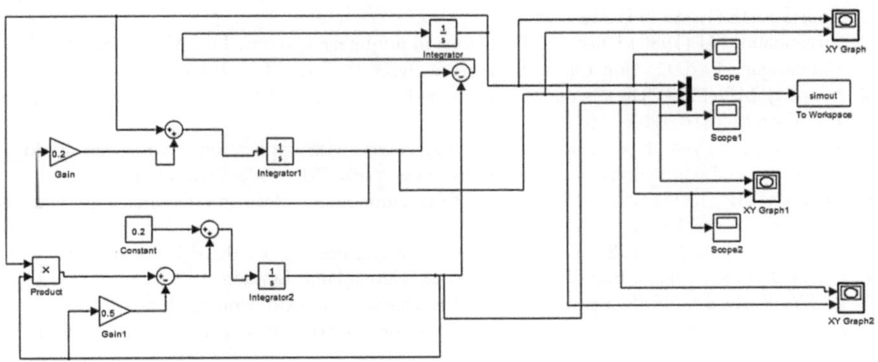

The block 'Gain 1' is the gain parameter '***c***'. The block 'Gain' is the parameter '*a*'. And, the block 'Constant' introduces the parameter '***b***'.

49.4.2.1 Simulation Results and Diagrams for Rössler's Model

See Figs. 49.8, 49.9, 49.10, 49.11, 49.12, 49.13, 49.14, 49.15 and 49.16.

49.5 Conclusion

The two autonomous three dimensional systems, namely, The Lorenz System and The Rössler's System were simulated using MATLAB and SIMULINK simulating software. The time series and the phase portrait obtained by simulating these systems show that these non-linear systems are exhibiting chaos when the variable control parameter is made to take particular values or values in a specific range. The Lorenz System exhibited chaos for control parameter $r > 24.74$. The Rössler's system exhibited the period doubling route to chaos with chaos occurring for $c = 4.60$. These chaotic signals were found to satisfy the identifying marks of chaos-sensitive dependence on initial conditions, non-periodic nature of the signals and finite and bounded signals. The state-space trajectories sketched by the systems as they evolved with time shows the chaotic nature of the systems. The continuous stretching and folding mechanism of the trajectories show the presence of chaos in these systems.

Acknowledgments The authors are grateful to Prof. P.K. Sinha Roy and Prof. G.S. Taki of Institute of Engineering and Management for their support and encouragement and their advice which has helped the authors to write this paper.

References

1. Kamil IA (2012) Demonstration of chaos in selected chaotic systems. J Emerg Trends Eng Appl Sci (JETEAS) 3(5):812–818
2. Feigenbaum MJ (1980) Universal behaviour in nonlinear systems. Los Alamos Sci I:4–27
3. Vidyasagar M (2002) Non Linear Systems analysis. Prentice-Hall, Engelwood Cliffs
4. Kennedy MP (1993) Three steps to chaos-part-I: evolution. IEEE Trans Circ Syst-I: Fundam Theor Appl 40(10):640–656
5. Strogatz SH (1994) Nonlinear dynamics and chaos: with applications to physics, biology, chemistry, and engineering. Perseus Books, New York. ISBN 9780201543445
6. Kennedy MP (1995) Experimental chaos from autonomous electronic circuits. Phil Trans R Soc Lond A 353:13–32
7. Parker TS, Chua LO (1987) Chaos: a tutorial for engineers. Proc IEEE 75(8):982–1008
8. Gleick J (1988) Chaos: making a new science. Heinemann, London
9. Li TY, Yorke JA (1975) Period three implies chaos. Am Math Mon 82:985–992
10. Lonngren KE (1991) Notes to accompany a student laboratory experiment on chaos. IEEE Trans Educ 34(1):123–128
11. Lorenz EN (1963) Deterministic nonperiodic flow. J Atmos Sci 20:130–141

Part XI
Optical and Digital Data and Image Processing

Chapter 50
Line Segmentation in Handwritten Assamese and Meetei Mayek Script Using Seam Carving Based Algorithm

Chandan Jyoti Kumar and Sanjib Kr. Kalita

Abstract Line segmentation is a key stage in an Optical Character Recognition system. This paper primarily concerns the problem of text line extraction on color and grayscale manuscript pages of two major North-east Indian regional Scripts, Assamese and Meetei Mayek. Line segmentation of handwritten text in Assamese and Meetei Mayek scripts is an uphill task primarily because of the structural features of both the scripts and varied writing styles. Line segmentation of a document image is been achieved by using the Seam carving technique, in this paper. Researchers from various regions used this approach for content aware resizing of an image. However currently many researchers are implementing Seam Carving for line segmentation phase of OCR. Although it is a language independent technique, mostly experiments are done over Arabic, Greek, German and Chinese scripts. Two types of seams are generated, medial seams approximate the orientation of each text line, and separating seams separated one line of text from another. Experiments are performed extensively over various types of documents and detailed analysis of the evaluations reflects that the algorithm performs well for even documents with multiple scripts. In this paper, we present a comparative study of accuracy of this method over different types of data.

50.1 Introduction

Pre-processing, Segmentation, Feature Extraction, Classification and Post-Processing are the main phases of any Optical Character Recognition System as shown in Fig. 50.1. The Segmentation plays a very crucial role in the overall performance of the

C.J. Kumar (✉) · S.Kr.Kalita
Department of Computer Science, Gauhati University, Gauhati, India
e-mail: chandan14944@gmail.com

S.Kr.Kalita
e-mail: sanjib959@rediffmail.com

V. Lakshminarayanan and I. Bhattacharya (eds.), *Advances in Optical Science and Engineering*, Springer Proceedings in Physics 166,
DOI 10.1007/978-81-322-2367-2_50

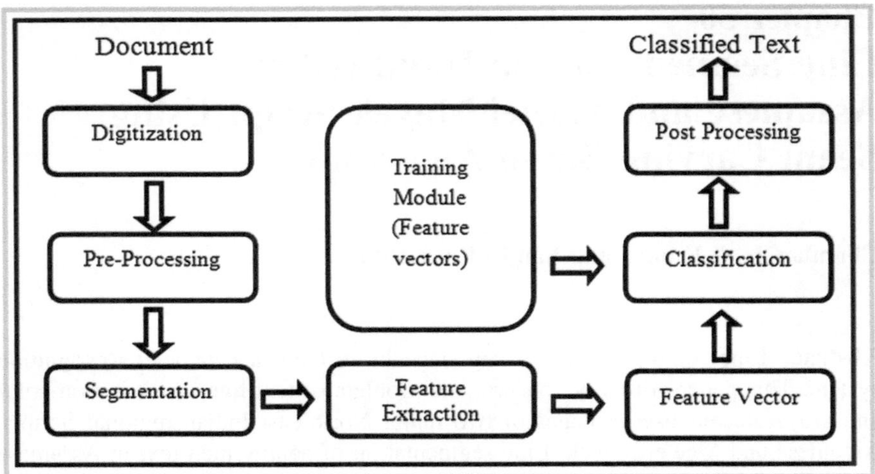

Fig. 50.1 Architecture of character recognition system

OCR System [1]. Segmentation can be further subdivided as Line, Word and Character Segmentation.

Text line extraction is an important step in the handwriting recognition system; it mainly focuses on extraction of individual text lines from the manuscript page. Various applications like keyword searching and word spotting also need to perform line and word segmentation. In his paper, we are presenting the implementation of a text line extraction method for recognition of Assamese and Meetei Mayek script using the algorithm based on seam carving [2]. Our basic objective is to compute seams that separate and reside between two consecutive lines of text without affecting the content of the document. For this purpose, Seam carving is an appropriate algorithm as it computes Seam with minimum energy in an image. In our notation, text components are extracted from high-energy regions and background are determined by low-energy regions. The seams computed are passing through the gaps between multiple texts lines, if these are the lowest energy regions of the neighbouring image space. By constraining the seam computation between two consecutive text lines, we are able to generate a separating seam that does not assign text parts to wrong lines. To address this problem, we use a modified version of the projection profile matching. This method can be applied to a grayscale manuscript page without doing the binarization of the image; this can be treated as an advantage. The generated seams can also be overlaid on the original colour page. The major advantage we achieve from this property, because even the best algorithms can also produce unreliable results when implemented over a binary image. The reason is that, binarization process involves a vital loss of information content based upon the type of document. In case of degraded manuscript, there is extensive loss of information in the binarization process, and any algorithm cannot produce reliable result based on a binary input. Algorithm we are implementing is a general one and can be applied to manuscripts of different languages, handwritings and for writing of various time periods, historical as well as current date document. We conduct experiments over Assamese, Meetei Mayek and multi-Script documents. For experimental purpose we are considering two types of Multi Script, Type I is of Assamese and English, Type II is of Meetei Mayek and English.

50.2 Literature Survey

Line Segmentation of Indian Regional Scripts is a challenging task. Most of the Line Segmentation algorithms are Script specific. If an algorithm is useful for one particular Script, it does not imply that it will also be useful for a different Script. The complicacy of the problem depends upon the separation between two lines or line gap between two adjacent line segments. This may lead to touching and overlapping between the lines, that makes significant complication to the problem. Difference in letter size or font in different lines is also an issue that makes line segmentation a difficult task [3]. Further, Skewness in lines is an inherent in handwritten documents. Although people are using various regional scripts in India, not much research work is done towards handwritten character recognition of all of these scripts. A lot of issues in most of the scripts yet to be explored. Among the nine regional scripts, Bangla and Devanagari are the two vastly explored scripts.

Starting with Bangla, A. Bishnu et al. in 1999, proposed a novel technique for extracting handwritten text lines from documents [2]. The technique was based on certain characteristics of Bangla writing methods, various zones across the height of the word are detected. The zones make available definite structural information about the component characters of the particular word. Based on the horizontal and run length histogram, cutting zone is detected and individual lines are extracted. In Bangla handwritten texts, generally there is overlap between rectangular hulls of successive characters. To watch out changeability concerned in the writing style of different persons, in 2003, Pal et al. [4] find out the valleys of the projection profile which are computed by sum of all black pixels in a row.

Roy et al. [5] in 2005, proposed a scheme for skew detection and then correcting the skewness in the text. The authors are explaining that, the most difficult case in character segmentation is that of the cursive script in any language. In this article first they estimated the angle of skewness, then the rotated the line to the angle estimated. Chaudhuri et al. [6] in 2009, proposed a interdependency between text-line and inter-line gap, new dual method based approach. The particular method draws various curves which simultaneously through inter-line gap points as well as the text, derived from strip-wise histogram peaks and inter-peak valleys. Some curves are generated after some iteration of the algorithm which pass through two consecutive lines. In doing this they always tried to keep the curves straight as much as possible. Once iterations are allowed occur for few time, then the curves stabilize and indicate the final text-lines and inter-line gaps. For them this particular technique was working fine for all types of documents with different geometric layouts. Saha et al. [7], in 2010, implemented a Hough transform based technique for line segmentation from digitized images. In Hough transform polar coordinate system is of more use than Cartesian coordinate. A line to point transformation is done from the Cartesian space to the polar coordinate space, in Hough transformation. Since a straight line in the Cartesian space is represented by the equation:

$$x \cos(\theta) + y \sin(\theta) = p \tag{50.1}$$

Obaida et al. [8], tackled the challenging issue of skew correction and explained a stroke-whitespace based technique [9]. The technique makes use of horizontal projection for correcting the skewness of writings, precisely for the languages they concerned. Saabni et al. [10] presented a new algorithm for extracting text-lines from handwritten document images, independent of language. The algorithm they used was based on the seam carving approach, which is mostly used for content aware image resizing. Adiguzel et al. [11] presented in 2012, an approach for text line segmentation which combines connected component based and projection based information to take advantage of aspects of both methods. Palakollu et al. proposed [12], a language specific property of header and base line for segmentation of Devanagari Text.

50.3 Proposed Methodology

Our approach for separating text blocks into lines of North East Indian Scripts is inspired and built upon the seam carving work [2], which resizes images in a content-aware fashion. Our objective of extracting lines starts with computing an energy map of the input image based on the signed distance transform. The process of extracting lines keeps on executing iteratively till extraction of all the lines complete. In our current implementation document image can be colored or grayscale. Two types of seams are generated, Type I seams approximate the orientation of each text line, and Type II seams separated one line of text from another. In the following sections we describe the stages of line segmentation in detail.

50.3.1 Computation of Type-I Seam

Our Type-I seam computation method is inspired by the projection profile matching approach of Arvanitopoulos et al. [13]. We split the page vertically into slices of width, $w = [m/r]$, where r is the number of slices. Sobel operator is used over the image I to compute its edge image $S \in R^{n \times m}$. We calculate smoothed horizontal projection profiles P_g^c of S in each slice independently smoothed horizontal projection profiles P_g^c of S in each slice independently:

$$P_i^c = \sum_{j=k}^{k+w-1} S_{i,j}, \quad P^c = \{P_i^c\} i = 1^n, \quad P_g{}^c = g(P^c)$$
$$c = 1, 2, \ldots r, \, k \in \{1, \, 1 + W, \, \ldots, \, 1 + (r-1)W\} \tag{50.2}$$

where, g is a cubic spline smoothing filter. For each maximum location of profile c, closest maximum location of profile $c + 1$ is find out and for each maximum location of profile $c + 1$, we find the closest maximum location of profile c. Then, they are connected with a line. This process is repeatedly executed for processing all the slices. This method creates piece-wise linear seams for approximating the medial axis of the text lines in the document page.

50.3.2 Computation of Type-II Seam

We adapt the seam carving algorithm projected by Avidan et al. [2] for computing the Type II seams i.e. separating seams. We include the regional constraints of the computed medial seams and modify the seam computation so that it can handle non-rectangular image regions [14]. The energy map is the derivative image of the grayscale manuscript page:

$$E_{ij} = \left| \left(I^{\sigma}_{i,j+1} - I^{\sigma}_{i,j-1} \right) /2 \right| + \left| \left(I^{\sigma}_{i+1,j} - I^{\sigma}_{i-1,j} \right) /2 \right| \tag{50.3}$$

where, I^{σ} is the original grayscale image smoothed with a Gaussian filter of standard deviation σ. On this map, high energy regions correspond to text components and low-energy regions correspond to parchment background. The first step is to traverse the image grid E_h from left to right and to compute the cumulative minimum energy for all possible connected seams for each pixel location $(y_h(j), j)$:

$$M_{yh(j);1} = E_{yh(j);1};$$
$$M_{yh(j);j} = E_{yh(j);j} + \min \begin{cases} M_{yh(j)-1\,,j-1} \\ M_{yh(j)\,,j-1} \\ M_{yh(j)+1\,,j-1} \end{cases} \tag{50.4}$$

The least value of the last column in M will designate the end of the minimal connected horizontal seam. In the next phase we traverse the cumulative energy M backwards to find the path of the optimal seam. The above procedure is repeated for each image grid E_h, until the whole manuscript page is processed.

50.3.3 Selection of Parameters

Our algorithm works on some parameters like, the number of slices 'r' for the medial seam computation, the smoothing parameter b of the cubic spline filter and the standard deviation of the Gaussian filter for the gradient image computation. Some selected values for the above parameters are being used on the applied datasets. The effective value of these parameters varies from document to document, so there is no automatic way to adjust these parameters. Different parameters were used inside the collections due to the different type of pages contained in them. The number of slices r depends on the image resolution and text layout. A value of $r = 8$ works relatively well for an average Assamese and Meetei Mayek manuscript page. The smoothing parameter b depends on the script and handwriting complexity.

50.4 Experiments and Results

50.4.1 Dataset Used

Since no benchmark dataset is available in Assamese and Meetei Mayek script, we have prepared a dataset for experimental purpose. Table 50.1 shows the details of the dataset used.

Each type of the dataset is prepared by people of different age group and working in different environment. Five different writers are used for obtaining each type of the dataset. Figure 50.2 and Fig. 50.3 shows some of the sample dataset.

Figure 50.2 is a good quality document of Assamese Handwritten Document, while Fig. 50.3 is a degraded document of Meetei Mayek Script.

Table 50.1 Details of the dataset used for the experiment

Scripts	No. of lines	Document type
Assamese	53	Good quality
	51	Degraded
Meetei Mayek	55	Good quality
	53	Degraded
Assamese and English	54	Good quality
	52	Degraded
English and Meetei Mayek	59	Good quality
	48	Degraded

Fig. 50.2 Handwritten Assamese good quality document

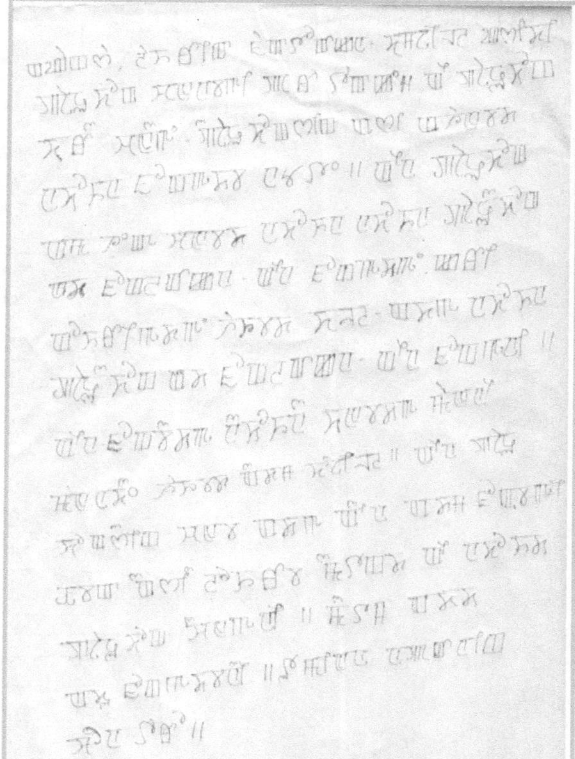

Fig. 50.3 Degraded Meetei Mayek document

50.4.2 Experimental Result

Experimental result shows that the algorithm implemented in this paper is useful for all types of documents in North East Indian scripts. Figure 50.4, shows the result of experiment over an image. We have used MATLAB platform to implement the proposed algorithms. MATLAB has predefined classes to perform operations on images. We tested our methodology over Assamese, Meetei Mayek and Multi Script documents as well. Number of lines per page varies from page to page as well as writer to writer. We have also considered historical documents which have several degradations and poor image quality. The algorithm performed successfully even in cases with characters of different size, or with text and non-text areas lying very near, or with warped text lines. Table 50.2 describes details of the experimental results.

The level of accuracy decreases with the skewness and because of the presence of header line, in North East Indian scripts. The line segmentation accuracy of degraded quality documents are less as compared to good quality documents.

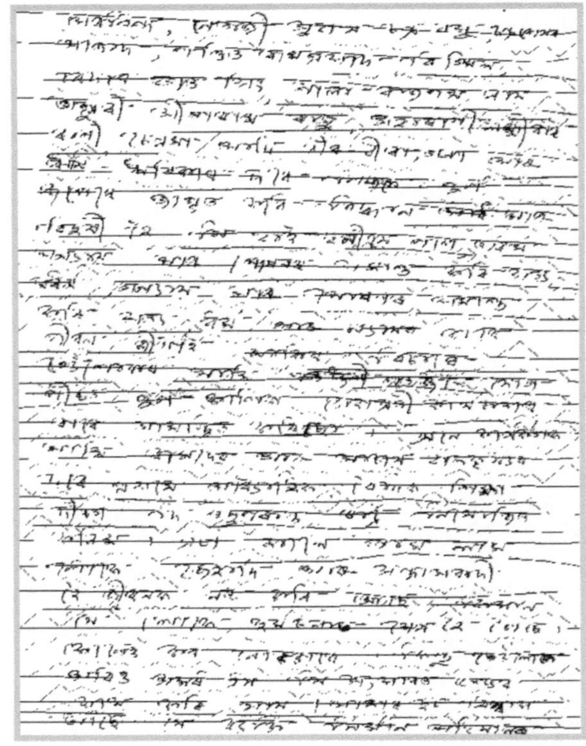

Fig. 50.4 Line Segmentation of Assamese document

Table 50.2 Details of the experimental result

Scripts	Document type	Segmentation accuracy (%)
Assamese	Good quality	91.74
	Degraded	65.32
Meetei Mayek	Good quality	92.13
	Degraded	64.48
Assamese and English	Good quality	92.60
	Degraded	67.37
English and Meetei Mayek	Good quality	93.25
	Degraded	68.14

50.5 Conclusion and Future Work

In this phase of our work we had processed up to the line segmentation phase. Skew related problem is resolved. Apart from these studies and implementations, the major contribution that is made is the use of Seam Carving for line segmentation of North East Indian Scripts. Till date no work has been reported on Seam Carving in North East Indian document analysis. But this can be a good segmentation model. We have concentrated on good quality handwritten documents as well as degraded or historical document analysis, which are still not explored extensively in major Indian Scripts. To some extent we have tried to consider degraded as well as, multi script documents in this report, but the area still needs to be developed, which can lead to a good research direction. Moreover, the issues regarding the overlapping lines are also interesting field of study. For English language documents these issues have been studied well, but have not studied in detail for North East Indian languages, apart from some works. Line segmentation Technique used here can be incorporated with development of OCR as well as Word Spotting Technique.

References

1. Chaudhuri BB, Pal U, Mitra M (2002) Automatic recognition of Printed Oriya Script. Sadhana 27(1):23–34
2. Avidan S, Shamir A (2007) Seam carving for content-aware image resizing. ACM Trans Graph (TOG) 26(3):9 (Proceedings of ACM SIGGRAPH 2007, article no 10)
3. Solanki P, Bhatt M (2013) Printed Gujarati script OCR using hopfield neural network. Int J Comput Appl 69(13):0975–8887
4. Pal U, Wakabayashi T, Kimura F (2007) Handwritten Bangla compound character recognition using gradient feature. In: 10th international conference on information technology
5. Roy A, Bhowmik TK, ParuiS K, Roy U (2005) A novel approach to skew detection and character segmentation for handwritten Bangla words. In: Proceedings of the digital imaging computing: techniques and applications, pp 125–132
6. Chaudhuri BB, Bera S (2009) Handwritten text line identification in indian scripts. In: Proceedings of 10th international conference on document analysis and recognition

7. Saha S, Basu S, Nasipuri M, Basu DK (2010) A hough transform based technique for text segmentation. J Comput 2(2):134–141
8. Obaida MA et al (2011) Skew correction function of OCR: stroke-whitespace based algorithmic approach. Int J Comput Appl 28(8):7–12
9. Priyanka N, Pal S, Mandal R (2010) Line and word segmentation approach for printed documents. IJCA Spec Issue Recent Trends Image Process Pattern Recogn RTIPPR 1:30–36
10. Saabni R, El-Sana J (2011) Language-independent text lines extraction using seam carving. In: International conference on document analysis and recognition (ICDAR 2011), pp 563–568
11. Adiguzel H, Sahin E, Duygulu P (2012) A hybrid approach for line segmentation in handwritten documents. In: Proceedings of international conference on frontiers in handwriting recognition, pp 501–506
12. Palakollu S, Dhir R, Rani R (2011) Segmentation of handwritten Devanagri Script. Int J Comput Sci Inf Technol 2(3):1244–1247
13. Arvanitopoulos N, Stisstrunk S (2014) Seam carving for text line extraction on color and grayscale historical manuscripts. In: ICHFR
14. Koppula VK, Negi A (2011) Fringe map based text line segmentation of printed Telugu document images. In: International conference on document analysis and recognition

Chapter 51
Information Retrieval Using Hadoop Big Data Analysis

Deepak Motwani and Madan Lal Madan

Abstract This paper concern on big data analysis which is the cognitive operation of probing huge amounts of information in an attempt to get uncovers unseen patterns. Through Big Data Analytics Applications such as public and private organization sectors have formed a strategic determination to turn big data into cut throat benefit. The primary occupation of extracting value from big data give rise to a process applied to pull information from multiple different sources; this process is known as extract transforms and lode. This paper approach extract information from log files and Research Paper, awareness reduces the efforts for blueprint finding and summarization of document from several positions. The work is able to understand better Hadoop basic concept and increase the user experience for research. In this paper, we propose an approach for analysis log files for finding concise information which is useful and time saving by using Hadoop. Our proposed approach will be applied on different research papers on a specific domain and applied for getting summarized content for further improvement and make the new content.

51.1 Introduction

Big data can be explained in terms of large amount of files as we say in terabytes. It can take place in terms of logs file as well as transaction records. Big data can be explained as processes and procedures tools that allow an organization to create and manipulate, and also manage huge data sets and storage facilities. By using

D. Motwani (✉)
Department of CSE, Mewar University, Chitordgarh, Rajasthan, India
e-mail: dmotwani20005@gmail.com

M.L. Madan
Faculty of Engineering and Technology, Mewar University, Chitordgarh, Rajasthan, India
e-mail: generalmadan@rediffmail.com

V. Lakshminarayanan and I. Bhattacharya (eds.), *Advances in Optical Science and Engineering*, Springer Proceedings in Physics 166,
DOI 10.1007/978-81-322-2367-2_51

pgadmin open source database for log file one can get the result on the basis of words, instead of character and runs the inverted index for two or more input files to get result in specific format [1].

In practice, we have run the word count example program using pgadmin open source database for log file and get the result on the basis of words, not character and also run inverted index for two input files and get result in specific format. A simple map reduces the job that makes an inverted index on the set of input doc. The location will not be based characters but on counting words [1–4].

Example: the input Documents.

$$\text{f1} = \{\text{Ram is good.}\}$$

$$\text{f2} = \{\text{Ram is not good?}\}$$

Output: Ram {(f1,1)(file2,1)} first parameter file name and second position (file1, position)

good {(f1,3)(f2,4)} is {(f1,2)(f2,2)} not {(f2,3)}

In this paper, we propose an approach for log files, big data files such as research papers for getting summarized information [5]. An example is given in Sect. 51.2. We present big data using Hadoop and file system, working on Hadoop, benefits of Hadoop in Sect. 51.3. We present pre solution and proposed approach in Sect. 51.4. We present result and analysis in Sect. 51.5. We present future scope in Sect. 51.6.

51.2 Motivating Example

As we take two input lines in which map and reduce functions are being used which can hold up the input data and get results. We can represent Map Reduce function as a heart of Hadoop.

Person who knows clustering can easily able to understand the concept of Map Reduce [6].

"Input: a set of key/value pairs, User supplies two functions

$$\text{Map}(\text{k}, \text{v}) \rightarrow \text{list } (\text{k1}, \text{v1}) \tag{51.1}$$

$$\text{Reduce}(\text{k1}, \text{ list } (\text{v1})) \rightarrow \text{v2} \tag{51.2}$$

(k1, v1) is an intermediate key/value pair Output is the set of (k1, v2) pairs [6, 7]. For our example, assume that system breaks up files into lines, and calls map function with value of each line, Key is the line number". Hadoop supports Multi-Node Cluster which is similar to Google File system (GFS) is a framework written in Java and provide high throughput access and fault tolerance [8, 9] (Fig. 51.1).

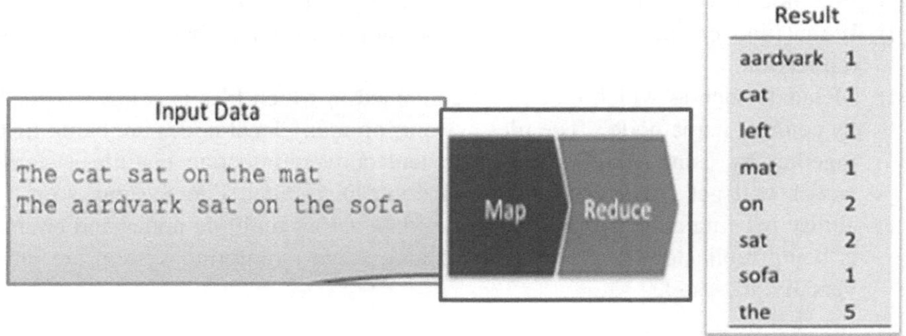

Fig. 51.1 Hadoop map reduce-input data and result

51.3 Big Data Using Hadoop

Map Reduce has one open source implementation named as Hadoop, it is a powerful tool designed for transformation of large data sets, deep analysis and explores complex data, which is being distributed across the cluster servers having high degree of fault tolerance and data availability [7] (Fig. 51.2).

51.3.1 Hadoop File System

Hadoop Distributed File System (HDFS) are being used by Hadoop. It provide parallel processing, high throughput, and solve problem the problems where you have lots of data.

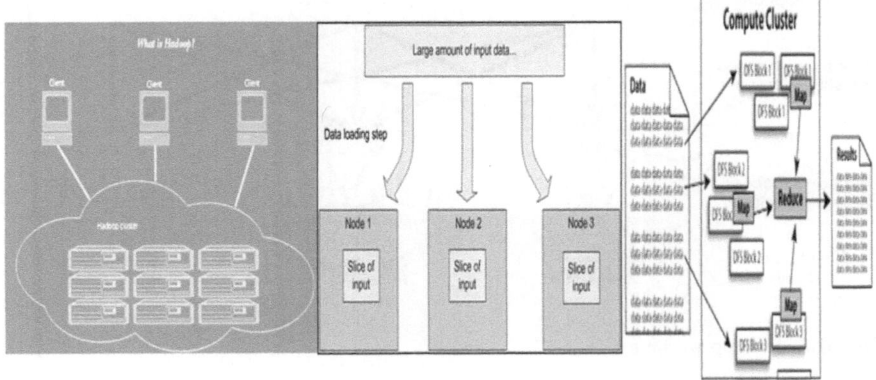

Fig. 51.2 Hadoop map reduce process

(i) File blocks are being partitioned across many machines.
(ii) If machine crashes data is neither lost nor unavailable because of blocks replication.
(iii) Metadata such as which blocks are contained in which files is being provided by central "name node". The files systems on many local nodes are being link together by using HDFS file system and convert into one big file system. Nodes will get fail is assumed by Hadoop Distributed File System, so reliability is being achieved by replicating data across multiple nodes and chunk is being replicated, single namespace and active monitoring system are universally accessible.

51.3.2 Working of Hadoop

Hadoop can not only execute any program. It can also dispense data across a cluster. Hadoop is lonely map reduce process. It restrict the total number of node communication, which can be done by the unusual process, each and every individual record is processed by a task in isolated from one to another and a major limitation at initially, it creates the whole framework much more faith full as compared to others. Records are executed and divided tasks identified by Mappers and outcome from the Mappers is brought together into other part of tasks known Reducers, where results from unlike Mappers can be combined together (Fig. 51.3).

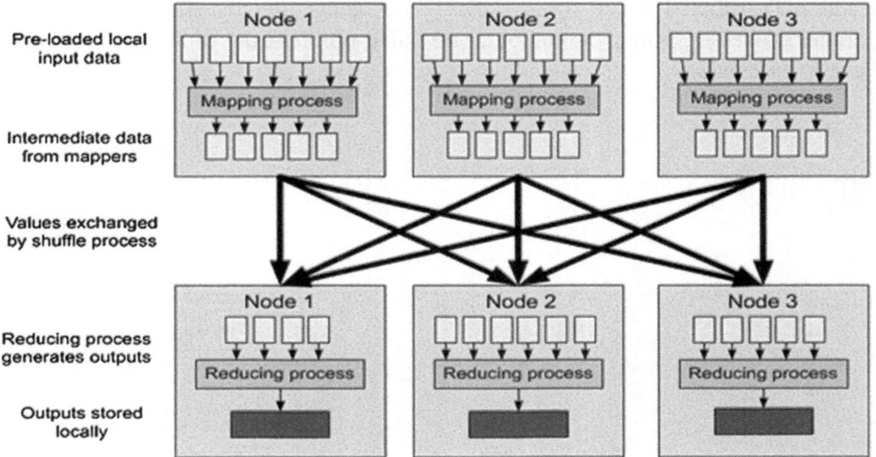

Fig. 51.3 Hadoop working

51.3.3 Benefits of Hadoop

Hadoop brings massively parallel computing and without needing New nodes can be added as essential, and added without needing to alter date formats, schema less and absorb any type of data structured, multiple sources can be joined and aggregated, avoid costly communication, redundancy of data allows, batch based computing framework, allows parallel work.

51.4 Pre Solution and Proposed Solution

We have data in several no of GB's. So manually it takes a lot of time, but by using Hadoop it can be possible in seconds. We can use this approach for searching a particular string in multiple file which is useful and time saving. In this paper, We run the number of word count program and run inverted file program using three files as a input for getting word summarized information like "which particular word find in which file with filename", line number, number of times. Proposed approach will be applied on different research papers on a specific domain and applied for getting summarized information such as Author name, year, objective of paper, Algorithm or method name, strength of paper, limitation of method for further improvement and to make easy literature review [5, 10] (Table 51.1; Figs. 51.4 and 51.5).

Table 51.1 Prcedure for input and output files

1. wordcount—Input (log file pgadmin) 2013-05-29 12:27:51 ERROR: ERROR: permission denied for relation student 2013-05-29 12:28:36 ERROR: ERROR: permission denied for relation classroom 2013-05-29 12:36:32 ERROR: ERROR: duplicate key value violates unique constraint "department_pkey" DETAIL: Key (dept_name) = (Biology) already exists 1. Output-occurrence—wordcount 56 3 BIO-101 2 BIO-301 2 Biology 2 CS-319 1 Comp 3 DETAIL 22 ERROR 134 Elec 2	2. Inverted Index (03 Files-Input) 1 name 2 file1 3 pgadminlog Output-Inverted Index given{(pgadmin.log,1039)} hello {(check,89)} host{(pgadmin.log,148)} iit{(file1,26)(file1,17)(file1,8)} iitb{(check,91)} in{(check,39)(check,34)} (check,5)(check,51)(check,75) (check,14)} Deepak{(name,2)} Motwani{(name,3)}

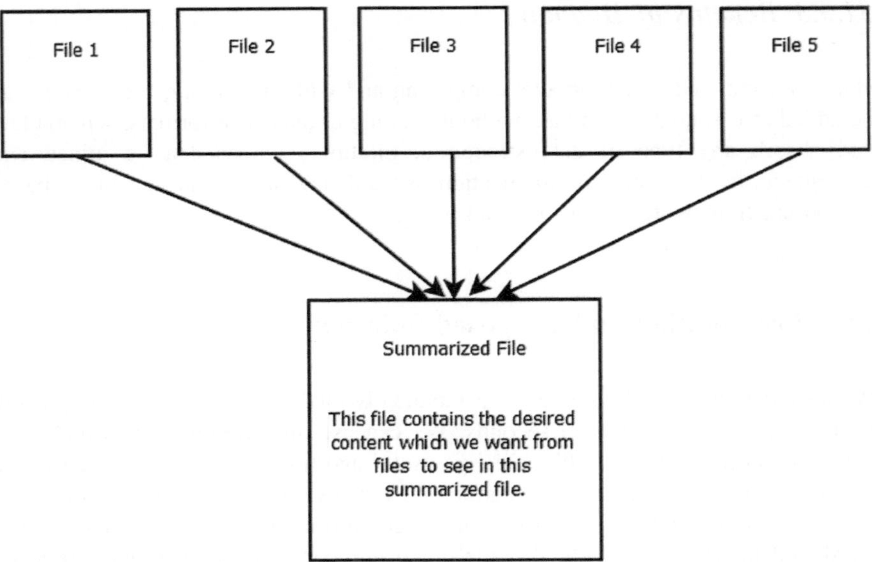

Fig. 51.4 Inverted index for input files summarize

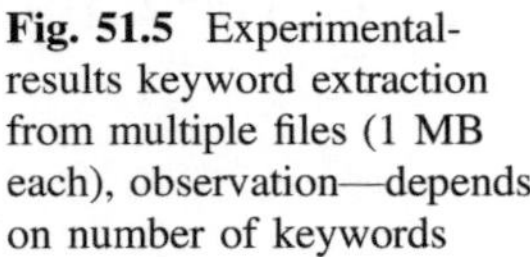
Fig. 51.5 Experimental-results keyword extraction from multiple files (1 MB each), observation—depends on number of keywords

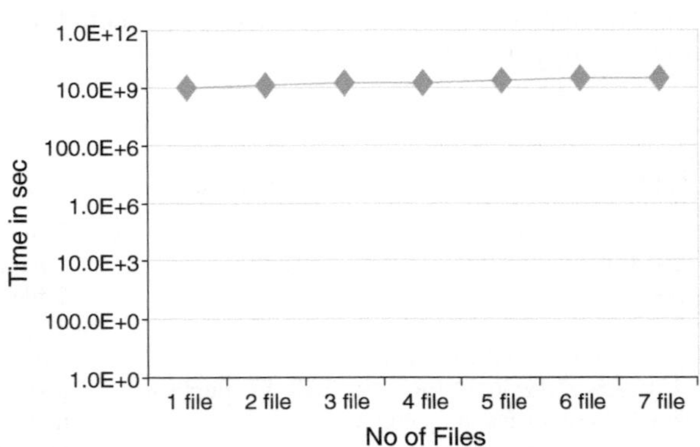

51.5 Conclusion

In this paper, we discuss approaches of Hadoop with small testing data on log file to count words and the use of inverted index in a single environment further; we can apply this technique for research papers summarization and finding latest research topic and summarized literature review. Hadoop avoids the expensive transmission when working with very bulky data sets. Hadoop is a flexible tool that allows new users to access the benefits of distributed computing, storage and transferring code instead of data.

51.6 Future Scope

Further research will be take place taking words with more accuracy so that experts will not find any difficulty using the same word based on same concept. Additionally, information technology can hopefully be applied so that new topics can also be discovered, but without wasting time and to make literature review easy, this idea can be converted into a model. Hopefully, model which is known by BODH model. This problem can be resolved out in future using data mining algorithms based on Hadoop features to identify the most frequently used words in an article become the feature in text mining.

References

1. Banerjee E, Patil A, Kakade P, Nandimath J (2013) Big data analysis using Apache Hadoop. In: Proceedings of the 14th international conference on information reuse and integration (IRI). IEEE, San Francisco, CA, pp 700–703
2. Jiang ZM, Hemmati H, Adams B (2013) Assisting developers of big data analytics applications when deploying on Hadoop clouds. In: Proceedings of the 35th international conference on software engineering (ICSE). IEEE, San Francisco, CA, pp 402–411
3. http://en.wikipedia.org/wiki/Inverted_index
4. De Capite D (2014) Techniques in processing data on Hadoop. Paper SAS033-2014
5. Tu Y-N, Seng J-L (2009) Research intelligence involving information retrieval—an example of conferences and journals. Int J Expert Syst Appl 36:12151–12166 (Elsevier)
6. http://labs.google.com/papers/mapreduce.html
7. Hadoop in action book. Manning, pp 10–20 (2011)
8. Silberschatz A, Korth HF, Sudarshan S (2010) Data base system concept (Chap 20), 6th edn. McGraw-Hill, New York
9. Managing Gigabytes for Java a free full-text search engine for large document collections written in Java
10. http://hpc.uva.nl/uploaded_files/inlineitem/Lecture_2_Mapreduce.pdf

Chapter 52
FANET Based Flights Monitoring Simulation System Over Cloud

Vipul Tiwari, Kapil Sharma and Brijesh Kumar Chaurasia

Abstract As mounting population and need of advance commutes systems, available vehicular ad hoc networks (VANET) systems will not suffice the communication complex issues between vehicles on land, water and air. In order to meet ever growing challenges, VANET cloud computing techniques start contributing with its supple state line less storage, high end virtual techniques and logical schema connectivity to make commute rich, safe and pleasure experience to users. In this paper, we proposed an architectural model for flight monitoring simulation, which optimizes the scheduling of flights in the most prominent manner by centralizing all resources at single cloud which optimizes the time management of effective communication among resources. This model will be an immense resource not only for commercial flights but also for air force navigation systems. VANET existing systems were designed for vehicular networks while our proposed solution flying ad hoc networks (FANET) or Flights Air Networks adapts VANET approach on high speed air traffic, FANET may use electromagnetic waves for data transfer between flights as electromagnetic waves speed is much higher than flights speeds, it may ensure timely communication between flights and other communicating sources.

V. Tiwari (✉) · K. Sharma · B.K. Chaurasia
Department of CSE, ITM University Gwalior, Gwalior, India
e-mail: getvipultiwari@gmail.com

K. Sharma
e-mail: kapil.rjit@gmail.com

B.K. Chaurasia
e-mail: bkchaurasia.itm@gmail.com

V. Lakshminarayanan and I. Bhattacharya (eds.), *Advances in Optical Science and Engineering*, Springer Proceedings in Physics 166,
DOI 10.1007/978-81-322-2367-2_52

52.1 Introduction

Cloud Computing is a computing for enabling convenient, dynamic scalable, on demand network access to a shared of pool of configurable computing recourses over the internet [1]. The cloud computing provides many advantages such as on-demand self service, ubiquitous network access, storage utility, software utilization, rapid elasticity, platform and infrastructure utilization, managed distributed computing power etc. [2]. It basically is a collection of storage systems, servers and devices to amalgamate data, software and computing power distributed across the network at various points. Cloud computing proposes a model that gives an on-demand, economy of scale and pay for use IT services over internet. The infra structure setup by the cloud computing architecture is known as cloud. Infrastructure as a Service (IaaS), Software as a Service (SaaS) and Platform as a Service (PaaS) [3], are prime services offered by cloud computing techniques.

In this paper, we are proposing cloud computing in flying ad hoc networks (FANETs). It will work as similar works as in vehicular ad hoc networks (VANETs) on the roads [4]. VANET is a collection of vehicles moving on the road. In VANET, cars use as mobile nodes and every participating vehicle plays role of wireless router with 100–300 m transmission range. Communication in VANET may be uses infrastructure as RSU and infrastructure less as MANET with 1000 and 300 m transmission ranges respectively [5]. Similarly, FANETs [6] are ad hoc networks connecting the unmanned air vehicle (UAVs). FANET can be understood as special form of MANET and VANET specially designed for flying objects at very high speeds. FANET collects data from the FCLOUD environment and relays to the wireless sensor networks, FANET support VANET approach of peer to peer communication. In this work, we are proposing cloud computing techniques to design "FCloud" which updates flights with latest information on routes, traffic and other important alerts. However, secure communication is an issue of cloud computing in FANET.

52.2 Literature Survey

Scalability of multi-UAV applications is proposed in limited FANET models and still requires further research. In [7], a FANET design was proposed for the range extension of multi-UAV systems. In this paper, forming a link chain of UAVs by utilizing multi-hop communication can extend the operation area is presented. However, the existing systems used by airports deploy open architecture conforming with ISO/OSI. It uses commercial Intel processors from HP for servers and workstations and LINUX Red-Hat operating system. High resolution displays of order: 2048 × 2048: SDD screen for radar controllers, 1600 × 1280: FDD screen, 1600 × 1280: SDD screens for tower are used. The data base requirements for DBM used are postgre SQL. The existing system uses high-level languages like ADA,

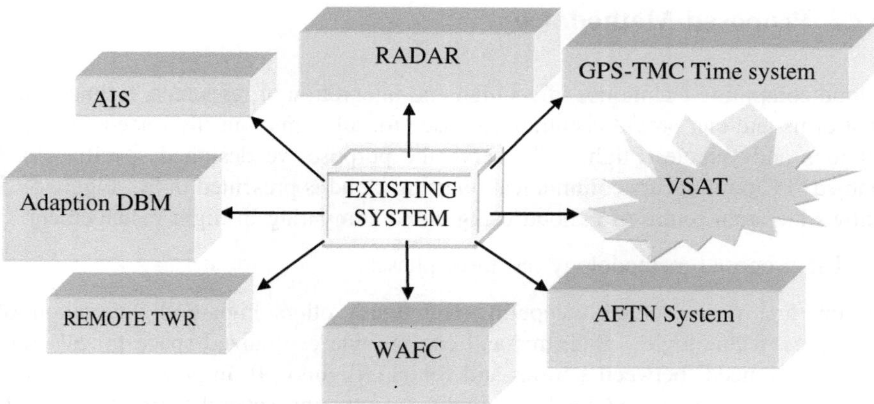

Fig. 52.1 The architecture of existing flight communication system

C and C++. The system uses standard graphics like X-Windows and Motif and the existing network is based in the Ethernet standards. All servers and workstations have connection to the double operational LAN. Figure 52.1 shows the architecture of existing flight communication system. There is collection of so many modules as radar. One for surveillance radar and other for communication meteo radar, GPS-TMC time system synchronizes location of flights at particular span, VSAT is remote positions system, AFTN system is ATS messages, WAFC is for meteo messages, remote TWR is for radar tracks, FP messages, adaption DBM is environment and adaption data, AIL FPLs, NOTAMs etc. Data-centric routing solutions may be used in FANETs for different types of application on the same multi-UAV system. Data-centric solution are needed to perform in network data aggregation. Space decoupling: communicating parties can be anywhere. Time decoupling: data can be dispatched to the sub-scribers immediately or latter. Flow decoupling: delivery can be performed reliably. FANET cross-layer architecture is introduced in 1961, where the interaction between the first three layer OSI reference model is facilitated.

52.3 Problem Statement

The existing system deploys many servers at each station under air traffic controllers (ATC) to communicate between different stations and between pilots, the more number of servers increases the probability of human errors, communication delays, signal losses and weak time management system on higher traffic air spaces. There is a need of synchronization of resources like communication devices, flights and communication networks to develop vital importance and emerged as a problem with existing systems. Cloud may be better solution for FANET.

52.4 Proposed Methodology

Cloud computing techniques offers high end integration of resources, technologies, platforms and can provide centralized space for all communication needs between stations and vehicle (Flights). To serve the purpose we designed specific cloud named "FCloud" secure communication using cloud is presented in [8]. Figure 52.2 shows the architecture of FCloud using cloud computing in flight management.

The proposed methodology has three phases.

In the first phase, cloud computing techniques offers high end integration of resources, technologies, platforms and can provide centralized space for all communication needs between stations and vehicle (flights) [9]. In phase two, network is established between FCloud and flights by utilizing several network protocols [10], it enable network with essential features as communications front-end for radars, Interpretation and broadcasting of radar messages, validates several signals through link definition algorithms as and when required (Fig. 52.3).

In the third phase, different models are connected to other layers like actuator models, which define various speed monitoring tools and gets information from network and cloud simultaneously to fix on parameters in run time environment.

Non linear aircraft models which changes directions as per need based or instructions provided are also connected by FANET cloud and updated as soon as system alerts them hence reduces risk situations. Several sensor models are also deployed at this stage which are connected to FCloud and upload information sensed from the surrounding and other sources. Figure 52.4 shows UAV system, which work in between phase 2 and phase 3.

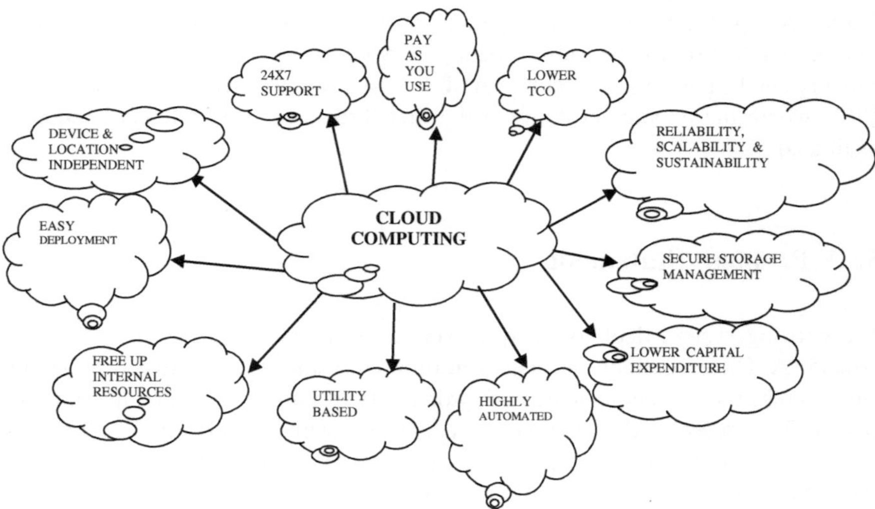

Fig. 52.2 The architecture of FCloud

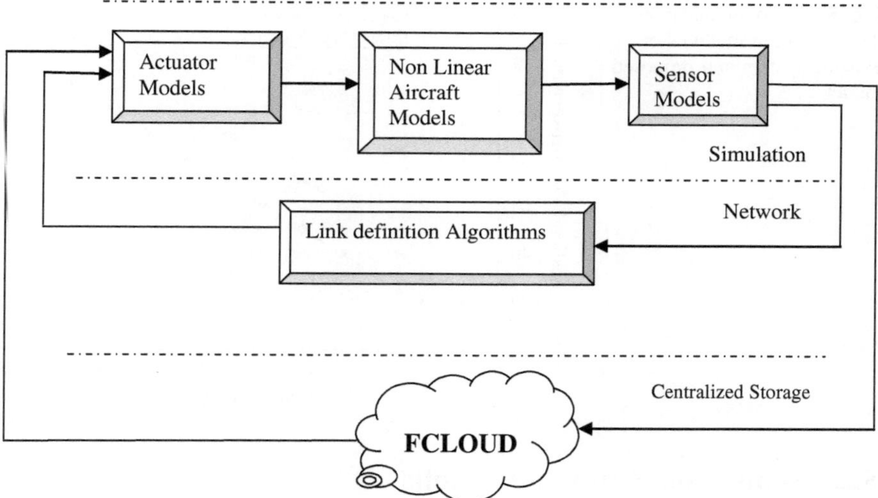

Fig. 52.3 The architecture of proposed methodology

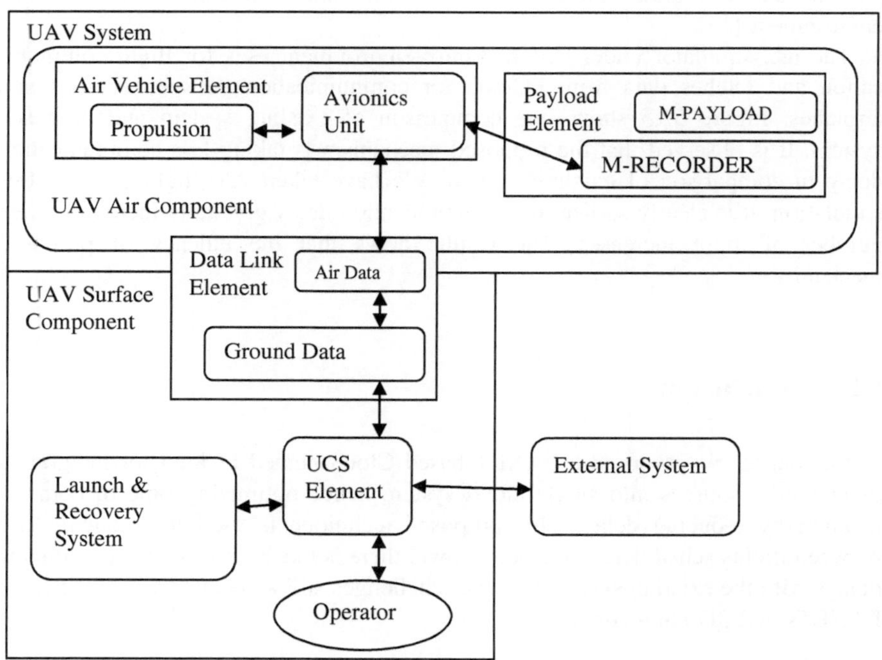

Fig. 52.4 UAV system interoperability architecture [11]

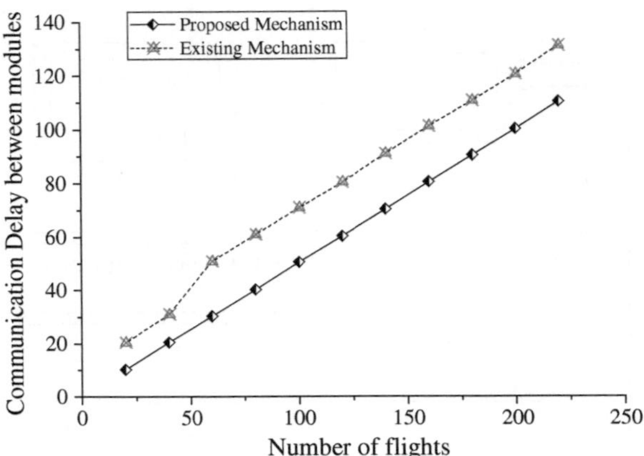

Fig. 52.5 Communication delay between modules of existing system and proposed model

52.5 Simulation Setup and Results

To illustrate some results of the proposed FCloud, we have taken 'Eucalyptus (software)' [12] which stores all data and recourses required for communication purpose and run under CentOS 6.5 (64 bits) under VMware virtual machine environment [13].

The ns2 simulator under Fedora Linux environment uses for flight communication and fetches data from FCloud for communication between stations and airplanes. Figure 52.5 shows the comparison of existing system and proposed system. It is observed that the proposed mechanism is taking less communication delay in comparison of existing system. We have taken 220 flight data for this simulation. It is clearly shown that communication delay gradually increases when number of flight increases. The result shows that the efficacy of proposed mechanism.

52.6 Conclusion

In this paper, we proposed a FANET based Cloud named FCloud for integrating diversified resources into single cloud system hence optimizing time of communication by reducing delay. This proposed technique is used for reducing risk between flights schedules. The result shown, there is fair improvements in reducing delays with the existing system however challenges still associated with security of FANETs and 3D simulations.

References

1. Takabi H, James BD Joshi, Ahn GJ (2010) Security and privacy challenges in cloud computing environments. IEEE Secur Soc 24–31
2. Kim W (2009) Cloud computing: today and tomorrow. J Object Technol 8(1):65–72
3. Robert LG (2009) The case of cloud computing. IT Prof 11(2):23–27
4. Al Mamun Md Ali, Khairul Anam Md Fakhrul Alam Onik, Esfar-E-Alam AM (2012) Deployment of cloud computing into VANET to create ad hoc cloud network architecture. WCECS-2012 pp 210–214
5. Chaurasia BK, Shekhar Verma (2010) Maximising anonymity of a vehicle. Int J Auton Adapt Commun Syst (IJAACS) 3(2)198–216 (Special Issue on: Security, trust, and privacy in DTN and vehicular communications, Inderscience)
6. Bekmezci I, Sahingoz OK, Temel S (2013) Flying ad-hoc networks (FANETs): a survey. Ad Hoc Netw 11:1254–1270
7. Olsson P, Kvarnström J, Doherty P, Burdakov O, Holmberg K (2010) Generating UAV communication networks for monitoring and surveillance. In: Proceeding of the 11th international conference on control, automation, robotics and vision (ICARCV), pp 1069–1087
8. Chaurasia BK, Shahi A, Verma S (2013) Authentication in cloud computing environment using two factor authentication. In: 3rd international conference on soft computing for problem solving (SocProS2013), organized by Springer, Greater Noida Extension Center of IIT Roorkee, pp 779–786
9. Cozzetti HA, Caragnano G, Goga K, Brevi D, Terzo O, Scopigno R (2012) Design and quantitative assessment of a novel hybrid cloud architecture for VANET simulations. In: IEEE vehicular technology conference (VTC Fall), pp 1–5
10. AIRCON 2100 system, supply, installation, testing and commissioning of tower ATS automation system AIRCON 2100 technical introduction. Online available at: http://www.icao.int/MID/Documents/2014/AIDC-OLDI%20Seminar/4%20Seminario_ElCairo_ AIDC % 20and%20OLDI.pdf
11. Alshbatat AI, Dong L (2010) Cross-layer design for mobile ad-hoc unmanned aerial vehicle communication networks. In: ICNSC, pp 331–336
12. https://www.eucalyptus.com/eucalyptus-cloud/reference-architectures
13. http://www.vmware.com/in

Chapter 53
Trust Based Scheme for Location Finding in VANETs

Sonam Soni, Kapil Sharma and Brijesh Kumar Chaurasia

Abstract VANET aims to improve road safety and reducing traffic accidents. However, false messages can result in serious conditions like collision. In this paper, trust based scheme for location finding in VANET is proposed. In the proposed trust based scheme has two phases computation of direct and indirect trust. However, evaluation of direct trust system using infrastructure (RSU) and trusted authority (TA) and evaluation of indirect trust uses watchdog approach. Results show that the proposed scheme suitable for the actual situation of VANET.

53.1 Introduction

VANET is used to give the safety and traffic report about congestion, earthquake, floods etc. to its user for reducing the road accidents, fuel wastage and provides secure driving environment [1]. In VANET, security applications will provide secure information to the users and introduce mechanisms to overcome the problem of road accidents, traffic jams etc. Trust management is one well known scheme and basic need to maintain a reliable, secure faithful communication in VANET by which vehicles can believe through their close eye and could drive for a safe journey [2]. Reputation and trust are two main components of security that are used to perform decision-making ability in VANETs. In general, reputation is the opinion of one entity as vehicles in VANETs about another [3]. Normally, reputation works as attribute of vehicles to VANETs and helps to recognize the reputation of the node. If it significance trustworthiness of a vehicle of node in VANETs. Trust is the level of confidence on messages those are passed to establish in VANET [4].

S. Soni (✉) · K. Sharma · B.K. Chaurasia
Department of CSE, ITM University Gwalior, Gwalior, India
e-mail: shona1813@gmail.com

K. Sharma
e-mail: kapil.rjit@gmail.com

B.K. Chaurasia
e-mail: bkchaurasia.itm@gmail.com

V. Lakshminarayanan and I. Bhattacharya (eds.), *Advances in Optical Science and Engineering*, Springer Proceedings in Physics 166,
DOI 10.1007/978-81-322-2367-2_53

53.2 Related Work

A significant amount of work have been developed so far in the field of trust management, concerning point to point networks, wireless sensor networks, vehicular ad hoc networks etc. There are very few work about trust establishment in VANETs [5–7]. Chaurasia et al. [5] proposed a trust computation method is based on Perron–Frobenius theorem in the VANET environment. The trust management based on AHP [6] provides a secure alternative to replace the current security mechanism and putted effort to give the better security environment. TRIP, a trust and reputation infrastructure aided approach for VANETs is presented in Marmol and Perez [7]. In this work, trust is calculated using reputation score assigned by other vehicles. Obimbo et al. [8] represent the intrusion detection system by using watchdog scheme for evaluating the misbehave nodes. Marti et al. [9] gives two techniques that improve throughput in an ad hoc network. Hasswa et al. [10] introduces intrusion detection scheme produces a more actual system, which is more reliable of dealing with malicious nodes.

53.3 Proposed Methodology

The proposed methodology is categorized into two phases. In the first phase, computation of direct trust based location finding is presented by using infrastructure such as RSU or trusted authority (TA). In the second phase, evaluation of indirect trust is presented by using watchdog. However, indirect trust system is used able to work in the absence of RSU.

53.3.1 Phase I: Direct Trust Computation

In this phase, vehicle wants to find some location information about its nearby hospital, hotels etc. then it broadcasts its request (REQ) message nearby vehicles in the network. Then various vehicles in its transmission range reply the source query with the help of (REP) reply messages. When the source vehicle gets REP from the other vehicles within transmission range. Source vehicle calculates the trust value on the behalf of REP. In this proposed work, there are three cases for calculating the trustworthy location.

Case 1 (*Calculate the Trusted Location using Vehicles*)
Source vehicle receives the (REP) messages from the various vehicles then vehicle computes ratio of trusted information about location.

If source vehicle getting REP messages from 20 vehicles and 18 vehicles say that in the range of 100 km forward direction located a good hospital and 2 vehicles tell there is no any hospital within the range of 250 km. There is a situation of the ratio is

[(18/20) * 100] = 90 % of the trusted information about hospital location. Source vehicle may trust of broadcast location until ratio is greater than 50 %. Below 50 % ratio, vehicle can compute location by case 2. Proposed scheme may also report to RSU or TA about malicious vehicles regarding given wrong message as in [11].

Case 2 (***Calculate the Trusted Location using RSU***)
If ratio below 50 % then the source vehicle asks to RSU about location. Source vehicle will receive location information from RSU and that information will be fully trusted received by RSU. If RSU is not present in the range of source vehicle then compute the location by using case 3.

Case 3 (***Calculate the Trusted Location using TA***)
If RSU is absent in transmission range of source vehicle. Then source vehicle will receive location information from TA (as police vans, ambulance and post office vehicles).

```
Algorithm- TRUSTED_LOCATION_DECETION
CASE 1:
Source vehicle broadcast REQ
Receive REQ by vehicles within communication entity range
vehicles Reply REP
For each (REP)
{
if (Location_present)
{
α ++ ;                    // α is used to count of vehicle who are saying the location are present.
}
else
{
β ++                      // β is used to count of vehicle who are denying the location are present.
}
}
Compute Ratio for α = [(α/γ)*100]          // Total number of reply by vehicles.
Compute Ratio for β = [(β/γ)*100]
If (α > β)
{
Path is trusted
}
else if (α ≤ β )
Go to second case

CASE 2:
    If (α ≤ β)
    {
  Ask to RSU about location
    }
    If (RSU is absent) then Go to third case
CASE 3:
  Ask to TA about location
```

53.3.2 Phase II: Indirect Trust Computation

In the second phase, indirect trust calculates in the absence of RSU and TA by using of watchdog technique. Watchdog is used to detect the malicious vehicles. Watchdog is set in the each vehicle and maintains a buffer. Buffer maintains all the information about the vehicle. On the basis of buffer information, we calculate the misbehaving vehicles. To reduce the weakness of watchdog, we implement this proposed scheme. The proposed working of watchdog technique is divided into three steps (Fig. 53.1):

1. **Initialization**—In the first step, source vehicle broadcasts the packet in the communication range and watchdog set in the each vehicle stores the sending packet information in own buffer. Many vehicles in the range receive the packet and forward to the next vehicle and put information in buffer.
2. **Sharing**—Second step perform sharing, each vehicle in the range of transmission sharing the buffer/table information among them.
3. **Update**—In the third step, after sharing all the information, vehicles maintains and updates the buffer. On the basis of buffer information we find out some vehicles receive the packet but do not forward to next vehicle. This type of misbehaving performance of vehicle, that vehicle is tagged as malicious vehicle.

Figure 53.2 shows, watchdog resides on all vehicles and maintains the buffer. Source vehicle A sends packet in the transmission range and watchdog stores the sending packet information in the buffer. Vehicles B, C, M receive the packet and maintain information and vehicle C forwards the packet to next vehicle and maintains it, and the vehicles sharing sending and receiving packet information among them. Then we evaluate vehicle M is a malicious vehicle it receives the packet from vehicle A and C, but does not forward to next vehicle D. Vehicle M performs misbehaving, Watchdog detect the vehicle M is a malicious vehicle and puts the information in the record (Table 53.1).

Algorithm- TRUSTED_LOCATION_DECETION

```
If (location_found)
{
α = α + trust value;
}
If (location_not_found)
{
α = α - trust value;
}
```

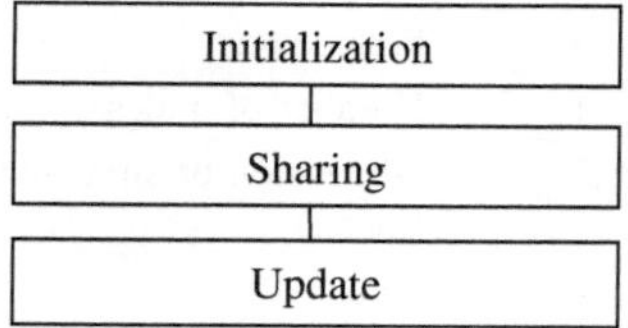

Fig. 53.1 Steps of watchdog technique

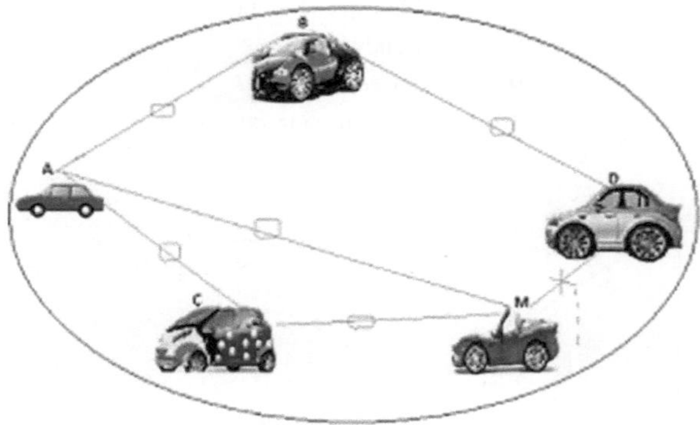

Fig. 53.2 Watchdog scenario. *Source* Vehicle receive reply from 100 vehicles and all 100 vehicles say there is a good hospital then we assign trust value 6. Then the ratio is 100:100

Table 53.1 Trust value calculate

Total no. of reply by vehicles (γ)	Location are present (α)	Trust value
100	100	6
100	80	4
100	60	2
100	50	1
100	Below 50	Malicious

53.4 Simulation Setup and Results

In this paper, we implement our experiment on a network simulator tool (NS-2.35). All the simulation parameters that are in Table 53.2 lists, we have used. VANET scenario composed of 10–50 vehicles distributed in a 2 km^2 area. Vehicles considered with variable speeds between 5–55 km/hr, inclusive of accelerations and decelerations and a maximum pause time of 2 s. Simulation time was setup up to 1000 s and, during this time, vehicle moves forward direction along with road.

Figure 53.3 shows that if up to 50 vehicles simultaneously send information, then packet delivery rate varies from 10 to 50 vehicles. It is observed that when density is high then average packet delivery ratio (PDR) is very much low around

Table 53.2 Simulation parameters

S. No.	Parameters	Values
1	Number of nodes	10, 35, 50
2	Dimension of simulated area (meter)	2000 × 2000
3	Antenna	Omni directional
4	Routing Protocol	AODV
5	Simulation time in seconds	1000
6	Transport Layer	UDP
7	Max queue length	50
8	Bandwidth	6 Mb/s
9	Speed range	5–55 km/h
10	Transmission rate	4 pckt/s

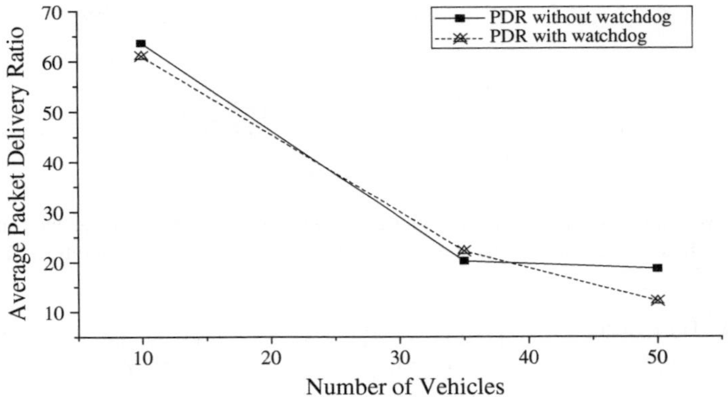

Fig. 53.3 Packet delivery ratio for direct trust computation without using watchdog technique and with watchdog scheme

19 %. However, when vehicle density in VANET is less than PDR is high around 63 %. PDR is decreases due to implement of watchdog over VANET scenario. The packet size is 512 bytes is considered. The average PDR is 60 % at low density and 12 % at high density, but 20 % at medium density environment.

Similarly, Fig. 53.4 shows that the average end to end delay is high to compute indirect trust using watchdog approach. However, to compute direct trust end to end delay is less around 162 ms at high density and 35 ms at low density. After calculating PDR we have evaluated trust computation of location finding information by two ways. Direct trust and indirect trust are two ways using infrastructure and without using infrastructure respectively Figs. 53.3 and 53.4 shows the packet delay ratio and end to end delay simulation and watchdog technique is shown by dotted line after simulation we know that watchdog technique is better one to find out the malicious vehicle.

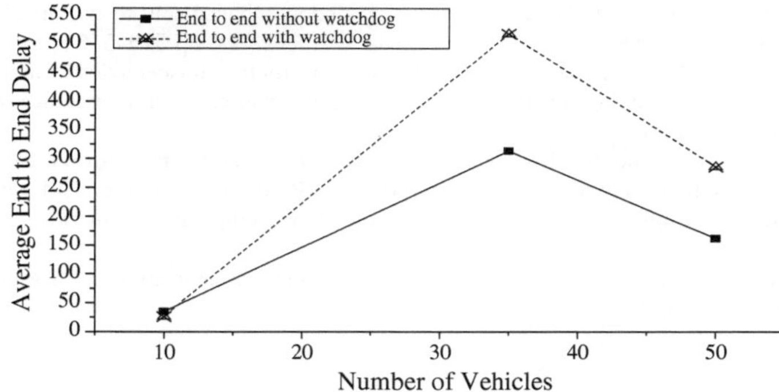

Fig. 53.4 End to end delay for direct trust computation without using watchdog technique and with watchdog scheme

53.5 Conclusion

This paper concern about a trust based scheme for location finding in VANETs. This proposed technique is used for identifying the trusted location. To evaluate the trusted location we introduce the trusted based mechanism using RSU, TA for computing direct trust and watchdog scheme for indirect scheme. In this scheme, the simulation results show that our proposed work can effectively find out the trustworthy location and this schemed performs satisfactorily in the realistic environment scheme.

References

1. Toor Y, Muhlethaler P, Laouti A (2008) Vehicle ad hoc networks: applications and related technical issues. IEEE Commun Surv Tutorials 10(3):74–88
2. Sharma K, Soni S, Chaurasia BK (2014) Reputation and trust computation in VANETs. In: Proceedings of the IEMCON-14 conference on electronics engineering and computer science, pp 118–122
3. Ding Q, Jiang M, Li X, Zhou XH (2010) Reputation based trust model in vehicular ad hoc networks. International Conference on Wireless Communications and Signal Processing (WCSP), pp 1–6
4. Sumra IA, Hasbullah H, Lail J, Rehman M (2011) Trust and trusted computing in VANET. Comput Sci J 1(1):29–51
5. Chaurasia BK, Verma S, Tomar GS (2013) Trust computation in VANETs. In: Proceedings of the international conference on communication systems and network technologies (CSNT), pp 468–471
6. Saraswat D, Chaurasia BK (2013) AHP based trust model in VANETs. In: Proceedings of the IEEE 5th international conference on computational intelligence and communication networks (CICN), pp 391–393
7. Marmol FG, Perez GM (2012) TRIP, a trust and reputation infrastructure based proposal for vehicular ad hoc networks. J Netw Comp Appl 35(9):34–941

8. Obimbo C, Arboleda LM, Chen Y (2006) A watchdog enhancement to IDS in MANET. In: Proceedings of the IASTED conference on wireless networks, pp 510–515
9. Marti S, Giuli TJ, Lai K, Baker M (2000) Mitigating routing misbehavior in mobile ad hoc networks. In: Proceeding of the 6th ACM annual international conference on mobile computing and networking, pp 255–265
10. Hasswa A, Zulkernine M, Hassanein H (2005) Routeguard: an intrusion detection and response system for mobile ad hoc networks. In: Proceedings of the IEEE international conference on wireless and mobile computing, networking and communications, vol 3. pp 336–343
11. Chaurasia BK, Verma S (2011) Infrastructure based authentication in VANETs. Int J Multimedia Ubiquit Eng 6(2):41–54

Chapter 54
Point Spread Function of Apertures Masked by Two-Dimensional Polar Walsh Filters

I. Bhattacharya, A. Saha and L.N. Hazra

Abstract Binary polar Walsh filters derived from two dimensional polar Walsh functions provides an effective tool for tailoring the three dimensional intensity distributions of a point spread function (PSF) in the pupil plane in a diffraction limited imaging system. This paper reports the study of radial as well as azimuthally variant polar Walsh filters and observation on their imaging characteristics near the focal plane of a rotationally symmetric imaging system.

54.1 Introduction

Walsh functions [1] are characterized by a closed set of functions satisfying orthogonality conditions over a finite interval [2, 3]. The functions take values +1 or −1 within the pre-specified domain except at a finite number of points of discontinuity where the value is zero. The number of occurrences of zero crossings or phase transitions within the specified domain defines the order of the Walsh function.

Walsh functions find many applications in communication, signal coding and transmission, allied problems of information processing, and in digital image processing [4–7], optical micromanipulation [8]. For tackling problems of optical

I. Bhattacharya (✉)
Department of ECE, Institute of Engineering and Management, Y-12, Sector-V, Salt Lake, Kolkata 700091, India
e-mail: indrani.bhattacharya@iemcal.com

A. Saha
Department of ECE, B.P. Poddar Institute of Technology, VIP Road, Poddar Vihar, Kolkata 700052, India
e-mail: arijit_sh@yahoo.com

L.N. Hazra
Department of Applied Optics and Photonics, University of Calcutta, JD-2,Sector III, Salt Lake, Kolkata 700098, India
e-mail: lnaphy@caluniv.ac.in

V. Lakshminarayanan and I. Bhattacharya (eds.), *Advances in Optical Science and Engineering*, Springer Proceedings in Physics 166,
DOI 10.1007/978-81-322-2367-2_54

image formation and radial Walsh functions were developed for the azimuth invariant case for tackling problems of apodization and adaptive optics [9–12]. Azimuthal Walsh filters for radial invariant cases may be considered as generalized zone plates and their scope of practical implementations in the field of micromanipulation has also been studied [13, 14]. Diffraction properties of Walsh filters on the transverse far-field plane have also been studied in the context of tailoring of resolution in microscopic imaging and optical encoding [15, 16].

54.2 Polar Walsh Functions

54.2.1 Two-Dimensional Polar Walsh Functions

Polar Walsh functions in two dimensions are defined as:

$$\begin{aligned} W_{np}(r,\theta) &= W_n(r)W_p(\theta) \Rightarrow W_{np}(r,\theta) \\ &= \prod_{m=0}^{\tilde{n}-1} \mathrm{sgn}\{\cos[k_m 2^m \pi r^2]\} \prod_{q=0}^{\tilde{p}-1} \mathrm{sgn}\left\{\cos\left[k_q 2^q \frac{\theta}{2}\right]\right\} \end{aligned} \tag{54.1}$$

Where the symbols have their usual meanings

The orthogonality conditions satisfied by the set of Polar Walsh functions are given by:

$$\frac{1}{\pi}\int_{r=0}^{1}\int_{\theta=0}^{2\pi} W_n(r)W_m(r)W_p(\theta)W_q(\theta)r\mathrm{d}r\mathrm{d}\theta = \delta_{mn}\delta_{pq} \tag{54.2}$$

where δ_{mn} and δ_{pq} are defined as Kronecker delta:

$$\begin{aligned} \delta_{mn} &= 1, \quad m = n \\ &= 0, \quad m \neq n; \end{aligned} \tag{54.3}$$

and

$$\begin{aligned} \delta_{pq} &= 1, \quad p = q \\ &= 0, \quad p \neq q \end{aligned} \tag{54.4}$$

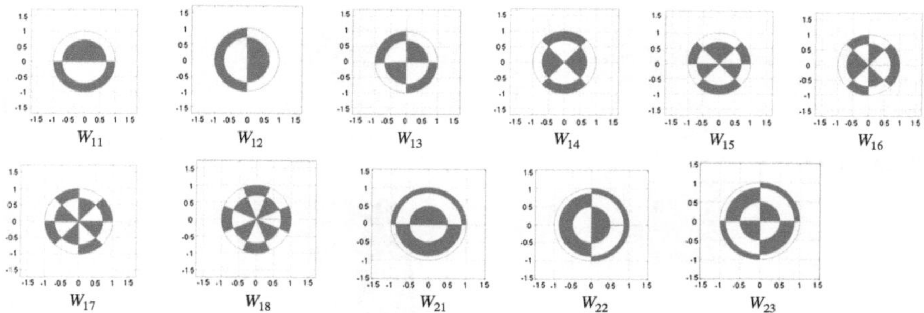

Fig. 54.1 The different orders of two-dimensional polar Walsh functions generated. The x- and y-axis are ranging from −1.5 to +1.5 in the graduated scale

By proper evaluation of k_m and k_q and applying orthogonality conditions, different radially and azimuthally varying two dimensional polar Walsh functions have been generated (Fig. 54.1).

54.3 Far-Field Diffraction Pattern of Two-Dimensional Polar Walsh Filters

The far-field diffraction pattern of a circular aperture masked by two dimensional radially and azimuthally varied Walsh filter can evaluated by determining the integral [17]:

$$U_p(\bar{r},\delta) = \frac{E_0}{\lambda_0}\int\limits_{R_1}^{R_2} \rho \mathrm{d}\rho \int\limits_{\varphi_1-\delta}^{\varphi_2-\delta} f(\psi)\mathrm{e}^{-j(kr\rho/f)\cos\psi}\mathrm{d}\psi \tag{54.5}$$

where

$$\begin{aligned} f(\psi) &= +1, \quad 0 \le \psi \le \delta \\ &= -1, \quad \delta \le \psi \le 2\pi \end{aligned} \tag{54.6}$$

The radiant flux density at any point $(\bar{r},\delta)$ in the plane of diffraction is given by:

$$\begin{aligned} &I_p(\bar{r},\delta) = U_p(\bar{r},\delta)U_p^*(\bar{r},\delta) \\ &\Rightarrow I_p(\bar{r},\delta) = \left(\frac{E_0^2}{\lambda_0^2}\right) \times \left[c^2(\bar{r},\delta) + s^2(\bar{r},\delta)\right] \end{aligned} \tag{54.7}$$

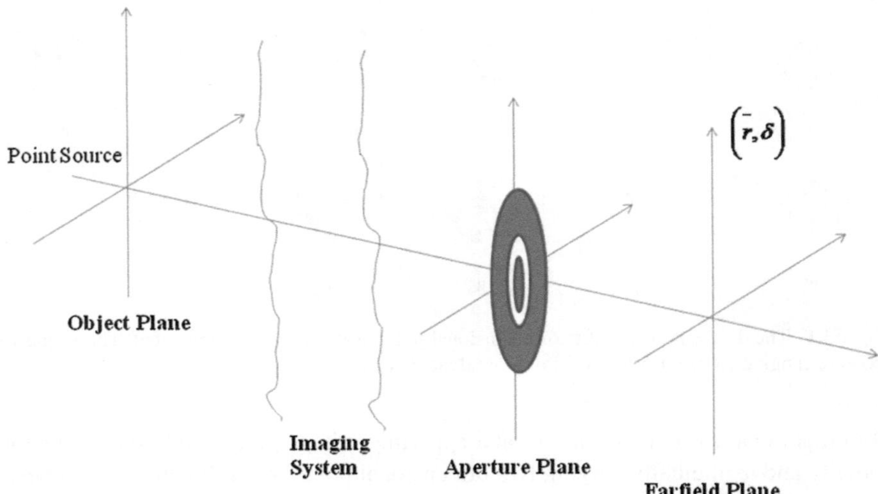

Fig. 54.2 For the centrally symmetric aperture, the superposition of the co-ordinate systems of the aperture plane on the diffraction plane occurs when they are projected into the plane of the aperture

where $c(\bar{r},\delta)$ and $s(\bar{r},\delta)$ are the real and imaginary parts of the intensity of the diffraction pattern due to the polar Walsh filters placed in the pupil plane referring to Fig. 54.2 and are given by as per Mahan's formula [18]:

$$c(\bar{r},\delta)amp; = R_2^2\left\{(\phi_2-\phi_1)\left[\frac{J_1(z_2)}{z_2}-\frac{R_1^2}{R_2^2}\frac{J_1(z_1)}{z_1}\right]+\sum_{n=1}^{\infty}\frac{2n+1}{n(n+1)}\left[\frac{J_{2n+1}(z_2)}{z_2}-\frac{R_1^2}{R_2^2}\frac{J_{2n+1}(z_1)}{z_1}\right]\right.$$
$$\left.\times\sum_{m=1}^{n}(-1)^m[\mathrm{Sin}2m(\phi_2-\delta)-\mathrm{Sin}2m(\phi_1-\delta)]\right\} \quad (54.8)$$

and

$$s(\bar{r},\delta) = 8R_2^2\left\{\sum_{n=1}^{\infty}\frac{n}{(2n-1)2n+1}\times\left[\frac{J_{2n}(z_2)}{z_2}-\frac{R_1^2}{R_2^2}\frac{J_{2n}(z_1)}{z_1}\right]\right.$$
$$\left.\times\sum_{m=1}^{n}(-1)^{m-1}\times[\mathrm{Sin}(2m-1)(\phi_2-\delta)-\mathrm{Sin}(2m-1)(\phi_1-\delta)]\right\} \quad (54.9)$$

where $1 < n < \infty$ and m, from each value of n, is summed from 1 to n, and z_1 and z_2 are defined as:

$$z_1 = \frac{k\bar{r}R_1}{f} \quad (54.10)$$

$$z_2 = \frac{k\bar{r}R_2}{f} \tag{54.11}$$

54.4 Illustrative Results

The far-field diffraction pattern for two dimensional polar Walsh filters have studied, where the maximum and minimum values of radii and azimuth ranges from 0 to 1 and 0 to 2π respectively and the results are reported.

For two dimensional polar Walsh filter for order (1,1) i.e., W_{11} (refer to Fig. 54.1), the value of radius varies from $r_1 = 0, r_2 = 0.7075, r_3 = 1$ and the values of azimuths are changing from $\theta_1 = 0, \theta_2 = 3.15738, \theta_3 = 2\pi$ within which the intensity changes from $E_0 = +1$ (Blue = Dark) to $E_0 = -1$ (White = Bright). The intensity of the diffraction pattern reduces to:

$$I_p(\hat{r}, \delta) = \left(\frac{2}{\lambda_0^2}\right) \times \left(\pi^2 \left[\frac{J_1(z_2)}{z_2}\right]^2\right) \tag{54.12}$$

Expression (54.12) shows that the intensity of the far-field diffraction pattern due to W_{11} at the aperture plane is symmetric and is independent of the azimuth dependent term δ. Equation (54.12) when plotted against r gives the well-known Airy pattern, independent of azimuth, δ, Fig. 54.3a and symmetrical contour plot, Fig. 54.3b.

The results for the Intensity of the far-field diffraction pattern calculated and plotted for azimuthal Walsh filters of order W_{12} and it is clear from Fig. 54.4a, b that the intensity distribution and corresponding contour plot is asymmetric showing azimuth dependent characteristics.

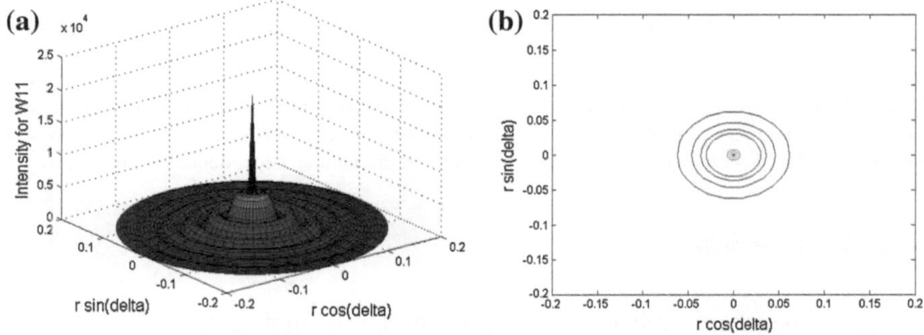

Fig. 54.3 **a** shows the symmetrical intensity distribution of the far-field diffraction pattern due to W_{11} and **b** shows the corresponding contour plot

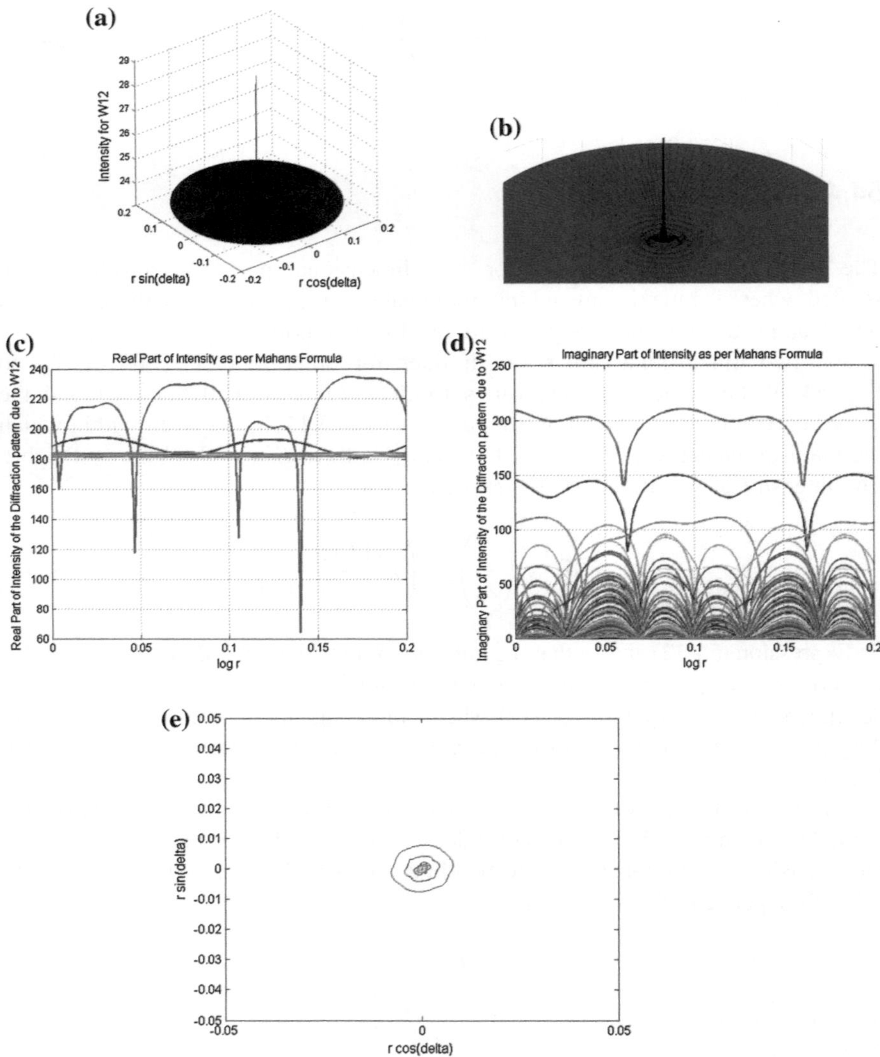

Fig. 54.4 **a** and **b** show total intensity of the far-field diffraction pattern due to W_{12}; **c** and **d** show the real and imaginary plot of intensity against log r for different values of azimuths; **e** shows the contour plot for W_{12} which shows asymmetry showing azimuthal dependence of the intensity

54.5 Conclusions

Further scope of research includes the possibility of finding a more generalized computational procedure to predict the nature of the intensity distribution of the far-field diffraction pattern for the apertures masked by higher orders of two dimensional polar Walsh filters and to explore the self-similarity existing between different orders [19–21]. The observation may provide valuable tool to tackle the problems of focusing and imaging of extreme UV and X-ray radiation, where the wavelengths vary approximately from 0.1 to 100 nm, to produce 3D light distribution in nano-lithography, optical super-resolution, optical micro-manipulation, optical tomography, microstructure fabrication and so on.

References

1. Walsh JL (1923) A closed set of normal orthogonal functions. Am J Math 45:5–24
2. Harmuth HF (1972) Transmission of information by orthogonal functions. Springer, Berlin. p 31
3. Beauchamp KG (1985) Walsh functions and their applications. Academic, London
4. Fu M, Wade G, Ning J, Jakobs R (2004) On Walsh filtering method of decoding CPM signals. IEEE Commun Lett 8:345–347
5. Andrews HC (1970) Computer techniques in image processing. Academic, London
6. Hazra LN, Banerjee A (1976) Application of Walsh function in generation of optimum apodizers. J Opt 5:19–26
7. De M, Hazra LN (1977) Walsh functions in problems of optical imagery. Opt Acta 24:221–234
8. Jahns J, Helfert S (2012) Introduction to Micro- and Nanooptics. VCH Wiley
9. De M, Hazra LN (1977) Real-time image restoration through Walsh filtering. Opt Acta 24:211–220
10. Hazra LN (1977) A new class of optimum amplitude filters. Opt Commun 21:232–236
11. Hazra LN, Guha A (1986) Farfield diffraction properties of radial Walsh filters. J Opt Soc Am A: 3:843–846
12. Hazra LN (2007) Walsh filters in tailoring of resolution in microscopic imaging. Micron 38:129–135
13. Vierke T, Jahns J (2014) Diffraction theory for azimuthally structured Fresnel zone plate. J Opt Soc Am A 31(2) (Micro- and Nano-photonik University), Germany
14. Lessard RA, Som SC (1972) Imaging properties of sector-shaped apertures. Appl Optics II(4)
15. Mukherjee P, Hazra LN (2013) Farfield diffraction properties of annular Walsh filters. Adv Opt Technol 2013:360450
16. Mukherjee P, Hazra LN (2014) Self-similarity in radial Walsh filters and axial intensity distributions in the far-field diffraction pattern. J Opt Soc Am A 31(2)
17. Born M (1933) Optik. Julius Springer-Verlag, Berlin, p 159
18. Mahan AI, Bitterli CV, Cannon SM (1964) Far-field diffraction patterns of single and multiple apertures bounded by arcs and radii of concentric circles. J Opt Soc Am 54(6)
19. Bhattacharya I, Hazra LN (2014) Self-similarity in polar Walsh filters. In International conference on optical electronics, ICOL2014, Dehradun, India
20. Bhattacharya I, Hazra LN (2014) Far-field diffraction characteristics of azimuthal Walsh Filters. In: Proceedings of 1st international science and technology congress 2014. Elsevier Publications, Amsterdam (ISBN 9789351072485)
21. Bhattacharya I, Hazra LN (2014) Study of imaging characteristics of azimuthal Walsh Filters. Am J Electron Commun I(2)

Chapter 55
Pan-Sharpened Image Optical Encryption

Isha Mehra and Naveen K. Nishchal

Abstract Satellites often capture multispectral and panchromatic images. The fusion of these two images results into a high resolution pan-sharpened image. In this paper, we apply image encryption scheme on the pan-sharpened four color band image. We fuse the multispectral data with panchromatic data in order to form the high resolution pan-sharpened image. Then this pan-sharpened image is separated into four spectral bands. These bands are encrypted using amplitude-, and phase-truncation approach with different fractional Fourier transform orders.

55.1 Introduction

The multispectral data received from satellites are being increasingly available for further processing and analysis for various remote sensing applications [1, 2]. The efficient fusion schemes are often needed for the high-resolution satellites which are being used for various remote sensing applications [3, 4]. These satellites results into a panchromatic image which is of high resolution and a low-resolution multispectral image. The panchromatic sensor provides high spatial resolution of grayscale images but gives improper color data, while the multispectral sensor provides the complete color information, but low spatial resolution. The fusion of the multispectral and panchromatic images, or pan-sharpening scheme, is suitable for the application where both high spatial and spectral resolutions are in demand. The fused information of various sensors is the source of comprehensive data. Discrete wavelet transform (DWT) based fusion scheme is one of the solution for merging the low and high frequency data and enhancing the quality of image [4]. Since many of such data are significant from the security point of view, therefore, it becomes necessary to develop an optical encryption technique for securing the multispectral data containing several bands [5].

I. Mehra · N.K. Nishchal (✉)
Department of Physics, Indian Institute of Technology Patna, Patna 800013, Bihar, India
e-mail: nkn@iitp.ac.in

V. Lakshminarayanan and I. Bhattacharya (eds.), *Advances in Optical Science and Engineering*, Springer Proceedings in Physics 166,
DOI 10.1007/978-81-322-2367-2_55

Most of the multispectral data consists of more than three spectral bands. In this paper, we fused the multispectral data with panchromatic data in order to form the high resolution pan-sharpened image. Then this pan-sharpened image has been separated into four spectral bands. These bands are encrypted using amplitude-, and phase-truncation approach with different fractional Fourier transform (FRT) orders [6–8].

55.2 Pan-Sharpened Image Optical Encryption

We have processed an image which has been captured by IKONOS satellite [2]. It's multispectral data consist of four bands: red, green, blue, and IR bands and the panchromatic image is a gray-scale image of high resolution. The two data has been fused using DWT method as shown in Fig. 55.1a. The encryption scheme in order to encrypt the pan-sharpened image has been shown in Fig. 55.1b. After obtaining the high quality pan-sharpened image it's four bands are separated into individual

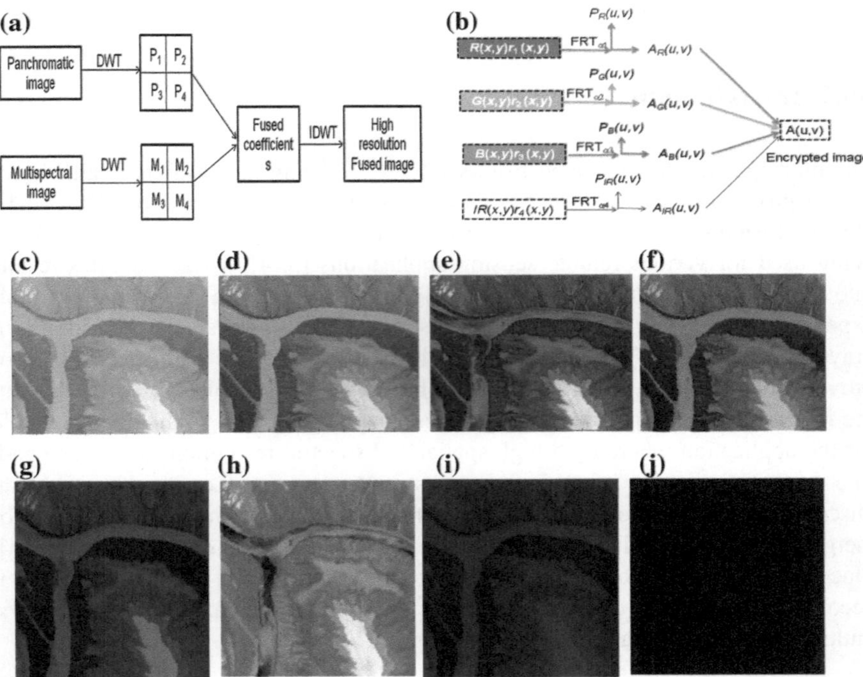

Fig. 55.1 **a** Fusion of multispectral and panchromatic image, **b** pan-sharpened image encryption scheme, **c** multispectral IKONOS image of poor resolution, **d** panchromatic *gray-scale* image of high resolution, **e** pan-sharpened image using DWT operation, **f–i** all four bands of the pan-sharpened image, and **j** encrypted image

functions given as $R(x,y), G(x,y), B(x,y)$, and $\mathrm{IR}(x,y)$, respectively. The red component is bonded with a random phase mask (RPM), $r_1(x,y)$ which is distributed uniformly in the range [0, 2π]. This function is FRT operated resulting in

$$F_R(u,v) = K\iint \{R(x,y) \times r_1(x,y)\} \times \exp\left[j\pi\frac{x^2+y^2+u^2+v^2}{\tan\alpha_1} - 2j\pi\frac{xyuv}{\sin\alpha_1}\right]\mathrm{d}x\mathrm{d}y \tag{55.1}$$

where K is a constant. The function $F_R(u,v)$ is phase-truncated (PT) and amplitude-truncated (AT) resulting in $A_R(u,v)$ and $P_R(u,v)$, respectively. Similarly, other three bands are encrypted resulting in $A_\mathrm{G}(u,v), P_\mathrm{G}(u,v), A_\mathrm{B}(u,v)$, $P_\mathrm{B}(u,v)$, $A_\mathrm{IR}(u,v)$, and $P_\mathrm{IR}(u,v)$, respectively. Now all the AT operated functions are combined together to form a single function as,

$$A(u,v) = A_R(u,v) \times A_G(u,v) \times A_B(u,v) \times A_\mathrm{IR}(u,v) \tag{55.2}$$

The decryption key in order to retrieve red band,

$$k_R = \frac{\mathrm{AT}[F_R(u,v)]}{A_G(u,v) \times A_B(u,v) \times A_\mathrm{IR}(u,v)} \tag{55.3}$$

Similarly, other three bands are retrieved. On applying the generated keys on the encrypted image, $A(u, v)$ and inverse FRT operation will yield the input image.

55.3 Simulation Results and Discussions

Figure 55.1c shows the multispectral image containing four bands. Figure 55.1d shows the corresponding panchromatic image. After application of DWT operation the two images are fused together in order to result into a high resolution pan-sharpened image as shown in Fig. 55.1e. Figure 55.1f–i show the red, green, blue, and IR bands of the pan-sharpened image, which are encrypted using FRT operation and amplitude-, and phase-truncation approach. The arbitrarily chosen FRT orders are $\alpha_1 = 0.50$, $\alpha_2 = 0.53$, $\alpha_3 = 0.55$, and $\alpha_4 = 0.58$, respectively. The encrypted image has been shown in Fig. 55.1j. After applying the decryption keys, all the four bands are retrieved successfully with correlation coefficient (CC) = 1.0.

55.4 Conclusion

The multispectral sensors provide multi-band images with accurate color data with low spatial resolution. On the other hand, panchromatic sensors yield gray-scale images with high spatial resolution but with imprecise color data. Remote sensing applications often require images with both high spectral and spatial resolutions. In this study, we have pan sharpened these images in order to provide high resolution. Then this pan-sharpened image has been encrypted using asymmetric cryptosystem.

Acknowledgments The authors acknowledge the funding from CSIR, Government of India, under Grant No. 03/(1183)/10/EMR-II.

References

1. Pohl C, Van Genderen JL (1998) Multisensor image fusion in remote sensing; concepts, methods, and applications. Int J Remote Sens 19:823–854
2. http://www.math.ucla.edu/ ~ wittman/pansharpening/
3. Javidi B, Do CM, Hong SH, Nomura T (2006) Multi-spectral holographic three-dimensional image fusion using discrete wavelet transform. J Disp Tech 2:411–417
4. Mehra I, Nishchal NK (2014) Image fusion using wavelet transform and its application to asymmetric cryptosystem and hiding. Opt Express 22:5474–5482
5. Muniraj I, Kim B, Lee BG (2014) Encryption and volumetric 3D object reconstruction using multispectral computational integral imaging. Appl Opt 53:G25–G32
6. Rajput SK, Nishchal NK (2012) Asymmetric color cryptosystem that uses polarization selective diffractive optical element and structured phase mask. Appl Opt 51:5377–5386
7. Rajput SK, Nishchal NK (2013) Known-plaintext attack on encryption domain independent optical asymmetric cryptosystem. Opt Commun 309:231–235
8. Mehra I, Nishchal NK (2015) Wavelet-based image fusion for securing multiple images through asymmetric keys. Opt Commun 335:153–160

Part XII
Optical Communications and Networks

Chapter 56
Comparison Between Three Different Types of Routing Algorithms of Network on Chip

Neetu Soni and Khemraj Deshmukh

Abstract Network on Chip (NoC) is an on-chip communication technology in which a large number of processing elements and storage blocks are integrated on a single chip. Due to scalability, adaptive nature, well resource utilization NoCs have become popular in and has efficiently replaced SoCs. NoCs performance depends mainly on the type of routing algorithm chosen. In this paper three different types of routing algorithms are being compared firstly one is deterministic routing (XY routing algorithm), secondly three partially adaptive routing (West-first, North-last and Negative-first) and two adaptive routing (DyAD, OE) are being compared with respect to Packet Injection Rate (PIR) of load for random traffic pattern for 4 × 4 mesh topology. All these comparison and simulation is done in NOXIM 2.3.1 simulator which is a cycle accurate systemC based simulator. The distribution of packets is Poisson type with Buffer depth (number of buffers) of input channel FIFO is 8. Packet size is taken as 8 bytes. The simulation time is taken 50,000 cycles. We found that XY routing is better when the PIR is low. The partially adaptive routing is good when the PIR is moderate. DyAD routing is suited when the load i.e. PIR is high.

56.1 Introduction

With the growing demand for the small size electron IC device, chip integration has become more and more complex. System on Chip (SoC) has been a better solution in this context but with the rapid increase in the number of transistors every year,its performance has degraded. Network On Chip (NoC) has solved the problems of SoC very efficiently. NoC ensures scalability, adaptive nature, well resource

N. Soni (✉) · K. Deshmukh
SSTC Bhilai, CSVTU, Bhilai, Chhattisgarh, India
e-mail: neetu.soni29@gmail.com

K. Deshmukh
e-mail: khemraj.deshmukh@gmail.com

V. Lakshminarayanan and I. Bhattacharya (eds.), *Advances in Optical Science and Engineering*, Springer Proceedings in Physics 166,
DOI 10.1007/978-81-322-2367-2_56

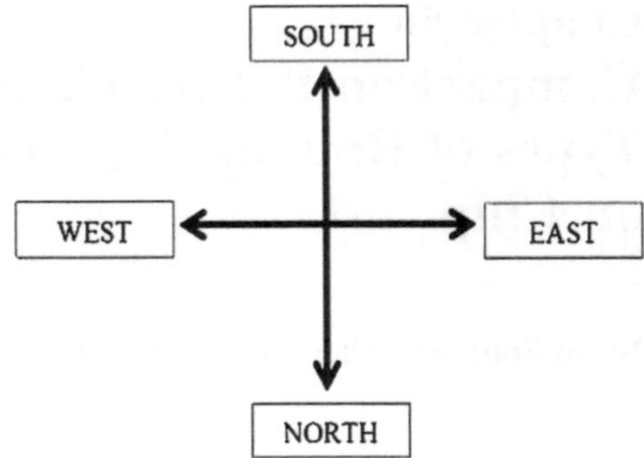

Fig. 56.1 Direction reference in NOXIM simulator

utilization, less power and energy consumption, complexity and cost reduction [1–5]. The NoC performance majorly depends upon its routing algorithms [6]. The routing algorithm determines an efficient path or route for the data or packets to share or transfer from its source to the destination. It essentially consist of a router which has a switching action. Router has five ports that is East, West, North, South and PE. In NOXIM simulator the directions taken for reference is as shown in Fig. 56.1, here the south is above and north is at the bottom.

In the next section we will discuss about the various routing algorithms. In third section comparison of these routing algorithms is done. In Sect. 56.4, we have the simulation results and analysis based on mesh topology. Conclusion is done at the last.

56.2 Various Routing Algorithms

In this section we will discuss about three types of routing algorithms such as deterministic (XY), partially adaptive (West-first, North-last and Negative-first) and adaptive routing (OE, DyAD).

56.2.1 XY Routing

In XY routing [2], we route the packets first in the horizontal (X) direction and then in the vertical (Y) direction. It is a deterministic routing algorithm which uses a fixed routing route throughout the process and is deadlock free as well as livelock free. We can implement this algorithm for both for regular and irregular network topology. This routing is also called dimension order routing (DOR). Here each node or router of NoC is identified by the (x, y) co-ordinates of that node for a 2D mesh. According to this algorithm the data packets will move in X-direction first and then in Y-direction towards the destination column. The packets cannot move in Y-direction first. It has turn restrictions shown in Fig. 56.2.

Fig. 56.2 Allowed turns in XY routing algorithm

As per travelling direction there are eight kind of turns, but algorithm XY allows only 4 turns out of them i.e. West to North, East to South, West to South and East to North. Due to these restrictions it becomes deadlock free. According to this algorithm, Current router is represented as (Cx, Cy) and Destination router as (Dx, Dy). The XY algorithm is as follows:

```
Algorithm XY(Cx,Cy,Dx,Dy)
{           Get Initial directions;
            if(Dx==Cx && Dy==Cy)
                        Return (Cx,Cy);// current router is destination
            else if (Dx>Cx)
                        Route towards East;
            else if (Dx<Cx)
                        Route towards West;
            else if (Dy>Cy)
                        Route towards South;
            else
                        Route towards North;
}// end
```

In the above algorithm the current router is compared with the destination router address if both are equal then destination and current router are the same. If destination router x-coordinate address is greater than the current router x-coordinate address then packets are routed towards east otherwise towards west. If the y-coordinate of destination router address is greater than the current router y-coordinate address then packets are routed towards south otherwise towards north [7]. However, this algorithm suffers from the disadvantage of mid network congestion. Example for XY Routing: Consider a 4 × 4 mesh and here we want to source (0, 0) to the destination (3, 1) as shown in Fig. 56.3 (Fig. 56.4).

The current router is at (0, 0) and the destination router is at (3, 1). Comparing the x-coordinates we found that Dx is greater than Cx, so path is proceeded towards east i.e. x-direction at (1, 0) then this coordinate becomes next current router becomes, like this the search goes on till the destinated x-coordinate is reached i.e. till (3, 0). Now at (3, 0), the router proceeds its path search in the y-direction i.e. towards south as Dy is greater than Cy.

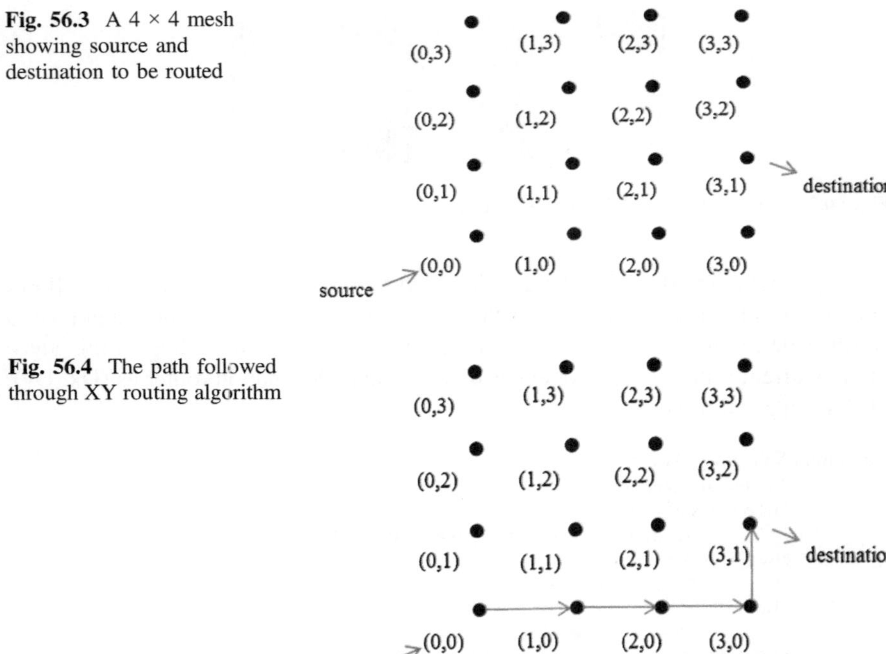

Fig. 56.3 A 4 × 4 mesh showing source and destination to be routed

Fig. 56.4 The path followed through XY routing algorithm

56.2.2 West-First Routing

The west-first routing is one of the partially adaptive routing algorithm in which 90° turns are allowed. In this routing [3], the packets have to be moved first in the west direction i.e. first movement should be in the west direction and then it can take turns in north, south and east direction. In this algorithm the path travelled can be minimal or non-minimal. This algorithm allows more turns then XY routing algorithm, it allows six turns out of eight turns shown in Fig. 56.5.

This algorithm puts two turn restrictions at any node in a mesh network: South to West Turn and North to West turn. In other words, the packets cannot take South to West Turn and North to West. The West-first algorithm is as follows:

```
Algorithm Westfirst (Cx,Cy,Dx,Dy)
{           Get initial directions;
            if ((Dx <= Cx) || (Dy ==Cy))
                        return (XY routing(Cx,Cy));//follow the XY routing
            else if (Dy<Cy)
                        Route towards North;  Route towards East;
            else
                        Route towards South;  Route towards East;
            Return directions;
} // end
```

Fig. 56.5 Allowed turns in West-first routing algorithm

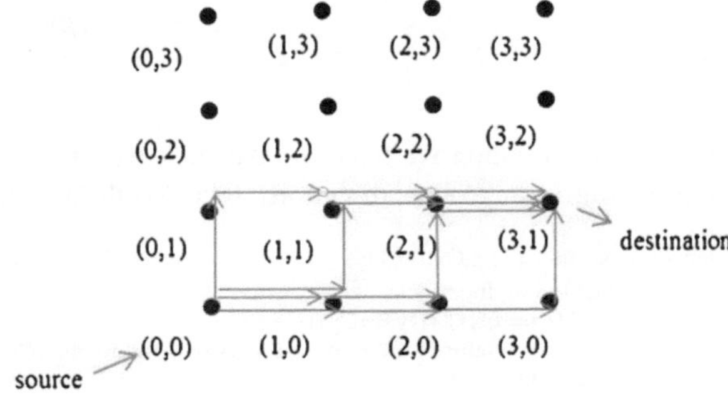

Fig. 56.6 The possible paths followed through West-first routing algorithm

Suppose in a 4 × 4 mesh source is (0, 0) and destination is (3, 1), then Fig. 56.6 shows the possible routing paths in West-first routing algorithm.

In this routing the packets must move in West direction and then it can adaptively choose any of the directions: North, South or East.

56.2.3 North-Last

It is another partially adaptive routing algorithm in which 90° turns are allowed. In this algorithm the packets will be routed in North direction at the last. This algorithm allows more turns then XY routing algorithm, it allows six turns out of available eight turns shown in Fig. 56.7.

This algorithm allows more turns then XY routing algorithm, it allows six turns out of available eight turns shown in Fig. 56.7. This algorithm puts two turn restrictions at any node in a mesh network: North port to West port or North port to East port. In other words, the packets cannot take North to West Turn and North to

Fig. 56.7 The allowed paths in North-Last routing algorithm

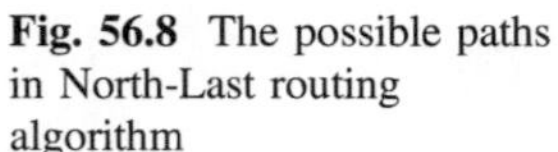

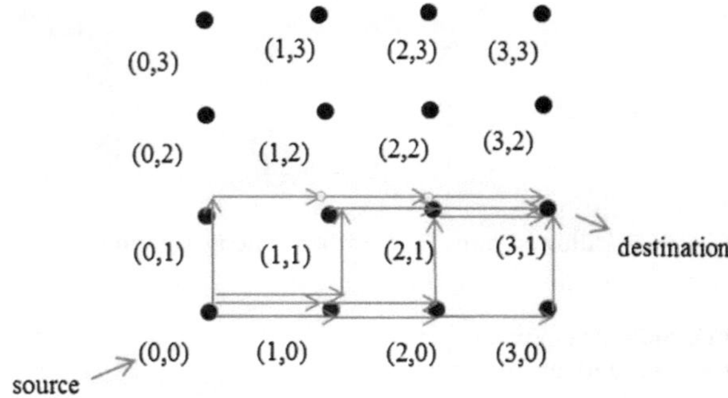

Fig. 56.8 The possible paths in North-Last routing algorithm

East. In any situation we have to take such path that ends in North directions after transversing in any other directions. The North-last algorithm is as follows

```
Algorithm Northlast (Cx,Cy,Dx,Dy)
{           Get initial directions;
            if ((Dx == Cx) || (Dy <=Cy))
                        return (XY routing(Cx,Cy)); // follow the XY routing
            else if (Dx<Cx)
                        Route towards South; Route towards West;
            else
                        Route towards South; Route towards East;
            Return directions;
} // end
```

Suppose in a 4 × 4 mesh source is (0, 0) and destination is (3, 1), then Fig. 56.8 shows the possible routing paths in North-last routing algorithm.

56.2.4 Negative-First Routing

Negative-First Routing is a partially adaptive routing algorithm in which two 90° turn can be avoided [3]. In this algorithm the turns towards west (negative X) and south (negative Y) are taken as negative direction first and the negative path is taken first and then east or north turns are taken according to the destination.

This algorithm allows more turns then XY routing algorithm, it allows six turns out of available eight turns shown in Fig. 56.9. This algorithm puts two turn restrictions at any node in a mesh network: East port to South port or North port to West port. In other words, the packets cannot take East port to South port or North port to West port. In any situation we have to take such path that starts with the negative directions then it can follow in any other directions. The Negative-first routing algorithm is as follows:

```
Algorithm Negativefirst (Cx,Cy,Dx,Dy)
{          Get initial directions;
           if (((Dx <= Cx) && (Dy <=Cy) ) || ((Dx >= Cx) &&(Dy >=Cy)))
           return (XY routing(Cx,Cy)); // follow the XY routing
           else if ((Dx>Cx) && (Dy <Cy))
           Route towards North; Route towards East;
           else
           Route towards South; Route towards West;
           Return directions;
} // end
```

Fig. 56.9 The allowed turns in Negative-First routing

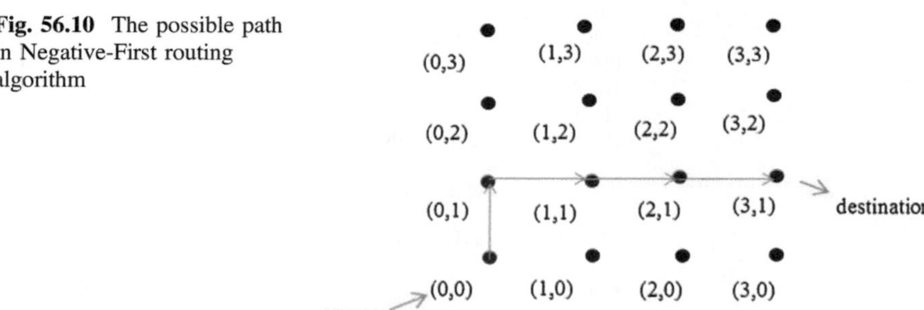

Fig. 56.10 The possible path in Negative-First routing algorithm

Suppose in a 4 × 4 mesh source is (0, 0) and destination is (3, 1), then Fig. 56.10. shows the possible routing paths in Negative-First routing algorithm.

Negative-First routing algorithm can be minimal or non-minimal. Being non-minimal version it will be more adaptive and fault tolerant.

56.2.5 Odd-Even (OE) Routing

It is a turn model based adaptive routing algorithms in in which more than one routing path can exist between source and destination node and one path will be selected for routing according network congestion conditions. It is basically a distributed adaptive routing algorithm which is based on odd-even turn model. This model do not need any virtual channel in 2D mesh topology [8]. According to the turn model, if its x dimension element is even then that column is considered as even column and if column's x co-ordinate is an odd number then that column is considered as odd column. According to this routing algorithm the following theorems are necessary to avoid deadlocks.

Theorem 1 *If a node is present on an even column the packets can't take East to North turns shown in* Fig. 56.11 *and if a node is present on an odd column the packets can't take North to West turns shown in* Fig. 56.12.

Theorem 2 *If a node is present on an even column the packets can't take East to South turns shown in* Fig. 56.11 *and if a node is present on an odd column the packets can't take South to West turns shown in* Fig. 56.12.

```
Algorithm Oddeven(Cx,Cy,Sx,Sy,Dx,Dy)
{           Get initial directions; int Ex=Dx-Cx; int Ey=-(Dy-Cy);
            if(Ex==0)
            {           If(Ey>0)
                                    Route towards North;
                        Else
                                    Route towards South;
            }
            else
            {           If(Ex>0)
                        {           If(Ey==0)
                                      Route towards East;
                                    Else
                                    {if((Cx%2==1) || (Cx==Sx))
                                    {If(Ey>0)
                                                Route to North;
                                                Else
                                                Route to South;
                                    }
                                    if((Dx%2==1) || (Ex!=1))
                                    Route toEast;
                                    Else
                                    Route to West;
                                    If(Cx%2==0)
                                    {If(Ey>0)
                                                Route to North;
                                                Else
                                                Route toSouth;
                                    }
                                    }
                        }
}
            Return directions;
} // end
```

Suppose in a 4 × 4 mesh source is (0, 0) and destination is (3, 1), then Figs. 56.11 and 56.12 show the possible routing paths in Odd-even routing algorithm. Odd-Even routing avoids the usage of virtual channels which can consume a lot of energy (Fig. 56.13).

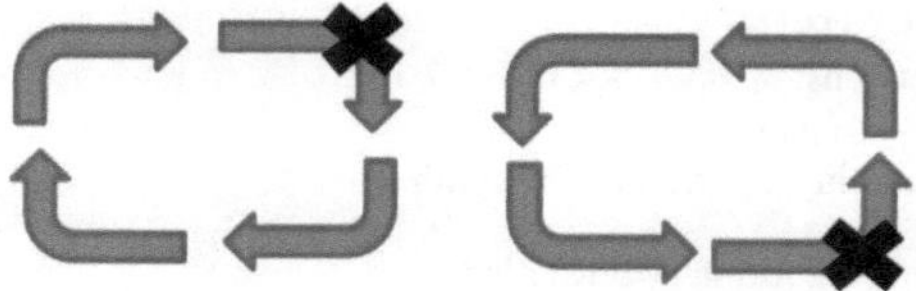

Fig. 56.11 The allowed turns for Even columns in Odd-even routing

Fig. 56.12 The allowed turns for Odd columns in Odd-even routing

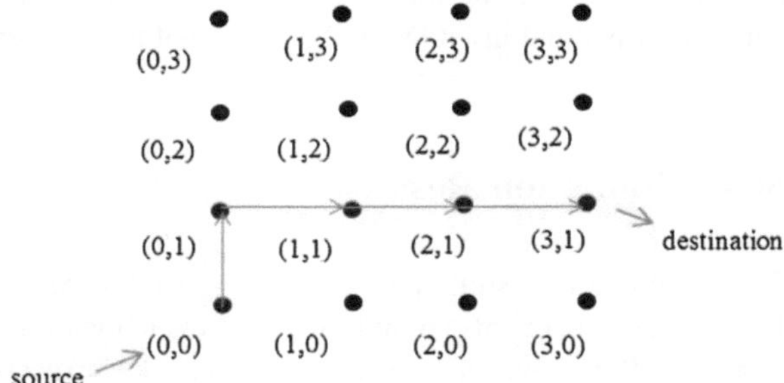

Fig. 56.13 The possible path in Negative-First routing algorithm

56.2.6 DyAD Routing

DyAD stands for Dynamically Adaptive and Deterministic routing [9]. This routing algorithm is dynamic in nature depending on the network traffic congestion. The traffic congestion threshold value depends upon the user. This algorithm uses two types of routing techniques Adaptive and Deterministic due to their advantages at different situations. Deterministic routers works well when packet injection rate (PIR) is low whereas if PIR increases, adaptive routing works well and gives good throughput. This is one kind of smart or intelligent routing algorithm [9]. The DyAD router will operate in adaptive mode when the network traffic congestion is more. It gives the advantage of avoiding the congested links by following other routing paths resulting in higher network throughput desired for NoC applications. It is free from deadlock and livelock due to mixed deterministic and adaptive routing modes into the same NoC. The algorithm for DyAD is as follows:

```
Algorithm Northlast (Cx,Cy,Dx,Dy)
{           Get initial directions;
            If (Congestion = 0) then
            Deterministic routing algorithm (XY Routing)
            Else
            Adaptive routing algorithm (OE Routing)
            Return directions;
} // end
```

56.3 Experimental Setup

The experimental setups for the evaluation of the routing algorithms are 4 × 4 mesh topology where each node is connected with a random traffic generator. The distribution of packets is Poisson type Buffer depth (number of buffers) of input channel FIFO is 8. Number of virtual channels per physical channel is 1. Packet size is taken as 8 bytes. The simulation time is taken 50,000 cycles. All these comparison and simulation is done in NOXIM 2.3.1 simulator, a systemC based software.

56.4 Simulation Results

The simulation results are shown in Tables 56.1, 56.2, 56.3 and 56.4 and Figs. 56.14, 56.15, 56.16 and 56.17. Packet Injection Rate (PIR) was varied from 0.001 to 0.01 and the corresponding packets received, Global average delay, throughput and energy readings were observed and plotted. With increasing PIR in all the parameters increases for all algorithms increases. It can observed from Fig. 56.14 that the packets received in XY, West-first and Negative-first routing algorithms are lower than in North-last and DyAD algorithms. The Fig. 56.15 shows the global average delay in receiving packets for the various algorithms. From Fig. 56.16 it is seen the throughput for all algorithms. DyAD routing has higher average throughput compared to the other algorithms. The Fig. 56.17 shows the Total energy consumption for all the algorithms. XY routing has the least energy consumption whereas DyAD routing has the highest energy consumption.

Table 56.1 Readings for received packets when PIR is varied from 0.001 to 0.01

PIR	XY	West-first	North-last	Negative-first	Odd-even	DyAD
0.001	583	679	620	654	624	654
0.003	2005	1890	1919	1859	1909	1932
0.005	3214	3216	3126	3183	3221	3106
0.007	4509	4470	4519	4530	4376	4421
0.009	5833	5590	5711	5684	5757	5944
0.01	6332	6426	6516	6403	6379	6594

Table 56.2 Readings for global average delay when PIR is varied from 0.001 to 0.01

PIR	XY	West-first	North-last	Negative-first	Odd-even	DyAD
0.001	7.44597	7.57732	7.55645	7.42202	7.62081	7.48012
0.003	7.77756	7.84868	7.91767	7.89242	7.74699	7.66874
0.005	8.07841	8.16511	8.36468	8.29218	8.20118	8.24372
0.007	8.45332	8.51454	8.64461	8.50773	8.39237	8.67948
0.009	8.82788	8.87692	8.77867	8.80155	8.92687	9.17749
0.01	9.05654	9.13243	9.19208	9.18257	9.10205	9.21174

Table 56.3 Readings for throughput when PIR is varied from 0.001 to 0.01

PIR	XY	West-first	North-last	Negative-first	Odd-even	DyAD
0.001	0.0047109	0.00540625	0.00477344	0.00512187	0.0049313	0.005094
0.003	0.0157094	0.0145906	0.0148844	0.0144187	0.0146969	0.015083
0.005	0.0255172	0.0252203	0.024525	0.0248531	0.0250562	0.024652
0.007	0.035025	0.0346547	0.035425	0.0351069	0.0339328	0.034408
0.009	0.0457391	0.0437437	0.0446734	0.0446391	0.0447641	0.046639
0.01	0.0492406	0.0507047	0.0513328	0.0501078	0.0497812	0.051533

Table 56.4 Readings for total energy consumption when PIR is varied from 0.001 to 0.01

PIR	XY	West-first	North-last	Negative-first	Odd-even	DyAD
0.001	3.17926	3.21279	3.18606	3.20135	3.19497	3.1883
0.003	3.88895	3.84013	3.87286	3.82502	3.8381	3.84968
0.005	4.51547	4.5321	4.49692	4.54668	4.5399	4.49075
0.007	5.18535	5.193	5.23945	5.19913	5.11605	5.1555
0.009	5.8796	5.78456	5.81295	5.78056	5.82707	5.93982
0.01	6.10106	6.1753	6.20172	6.19817	6.15934	6.26125

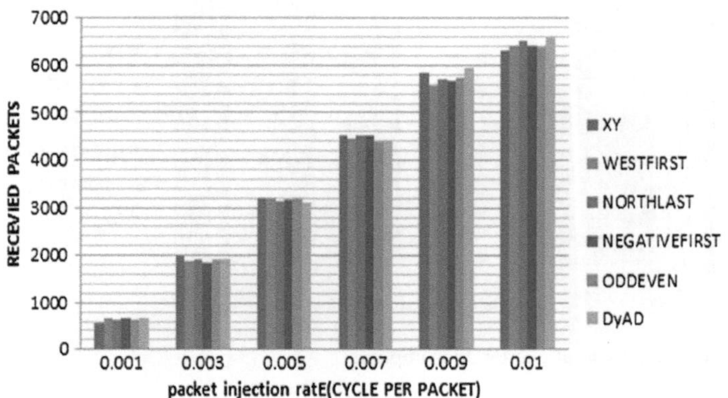

Fig. 56.14 PIR versus received packets

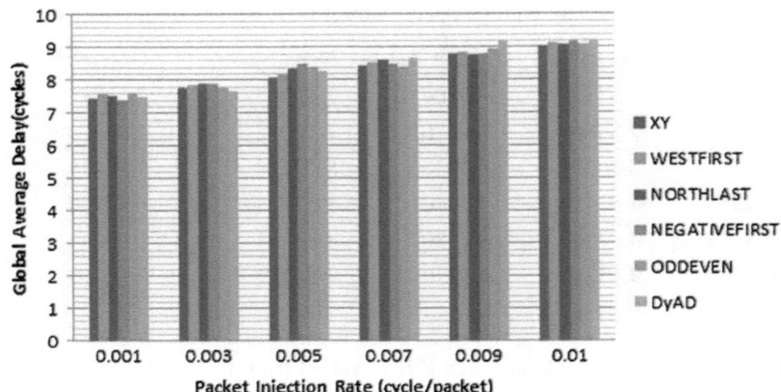

Fig. 56.15 PIR versus global average delay

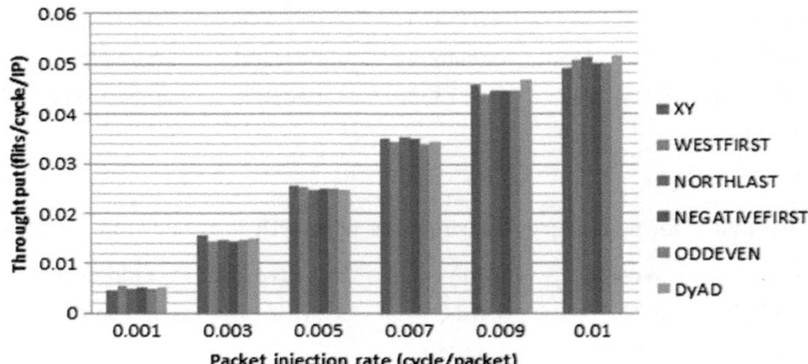

Fig. 56.16 Packet injection rate versus throughput

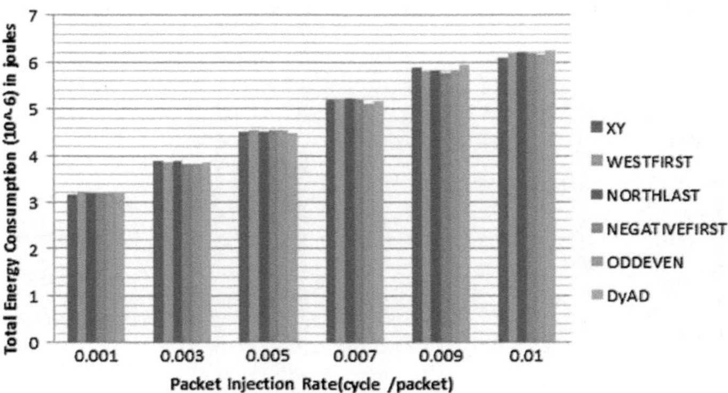

Fig. 56.17 PIR versus total energy consumption

56.5 Conclusion

In this paper we have brought comparison between different parameters of the various algorithms such as XY, West-first, North-last, Negative-first, Odd-even, and DyAD. We found that the packets received in XY, West-first and Negative-first routing algorithms are lower than in North-last and DyAD algorithms and DyAD routing has higher average throughput compared to the other algorithms. XY routing has the least energy consumption whereas DyAD routing has the highest energy consumption. The XY algorithm shows better results when the PIR is low and DyAD works better than all other algorithms. Hence we conclude that XY routing is better when the load i.e. PIR is low. The partially adaptive routing is good when the PIR is moderate i.e. when load is medium. DyAD routing is suited when the load i.e. PIR is high.

References

1. Guerrier P, Greiner A (2000) A generic architecture for on-chip packet-switched interconnections. In: Proceeding of the design, automation and test in europe conference and exhibition, pp 250–256. doi:10.1109/DATE.2000.840047
2. Hemani A, Jantsch A, Kumar S, Postula A, Oberg J, Millberg M et al (2000) Network on a chip: an architecture for billion transistor era. In: Proceedings of the IEEE NorChip, p 332
3. Dally W, Towles B (2001) Route packets, not wires: on-chip interconnection networks. In: Proceedings of the design automation conference, pp 684–689
4. Sqroi M, Sheets M, Mihal A, Keutzer k, Malik S, Rabaey J et al (2001) Addressing the system-on-a-chip interconnect woes through communication-based design. In: Proceedings of the design automation conference, pp 667–672
5. Kumar S, Jantsch A, Soininen J, Forsell M, Millberg M, Oberg J et al (2002) A network on chip architecture and design methodology. In: Proceedings of the IEEE computer society annual symposium on VLSI, pp 105–112. doi:10.1109/ISVLSI.2002.1016885
6. Marculescu R, Jingcao H, Ogras UY (2005) Key research problems in NoC design: a holistic perspective. In: Proceedings of the 3rd IEEE/ACM/IFIP international conference on hardware/software codesign and system synthesis (CODES + ISSS'05), pp 69–74. doi:10.1145/1084834.1084856
7. Dally WJ (2004) Principles and practices of interconnection networks. Morgan Kaufmann, San Fransisco
8. Cheng PYK, Mak T (2011) Adaptive routing in network-on-chips using a dynamic programming network. IEEE Trans Ind Electron 58(8):3701–3716
9. Hu J, Marculescu R (2004) DyAD: smart routing for networks-on-chip. In: Proceedings of the 41st design automation conference, pp 260–263
10. Zhang W, Hou L, Wang J, Geng S, Wu W (2009) Comparison research between XY and odd-even routing algorithm of a2-Dimension 3X3 mesh topology network-on-chip. Global Congress on Intelligent Systems, pp 329–333

Chapter 57
Effect of Phase-Shifter Domains in Quasi-Phase Matching Devices

Toijam Sunder Meetei, Sundararaman Hari Hara Subramani, Shanmugam Boomadevi and Krishnamoorthy Pandiyan

Abstract With the inception of Quasi-Phase Matching (QPM) technique, there has been a wide interest by the researchers to utilize the concept of domain engineered superlattices structures in optical networking. We present a theoretical approach to study the effect of phase-shifter (PS) domains in engineered optical superlattice structures and analyze the second harmonic spectral response. QPM device with PSs is the easiest approach to generate multiple and tunable second harmonic spectrum to utilize the same in optical communication system.

57.1 Introduction

The development of Quasi-Phase Matching (QPM) technique is significantly important due to the advantages of the process non-linear optical frequency conversion or mixing. This approach has been successfully acquired in ferroelectric crystal such as lithium niobate ($LiNbO_3$), lithium tantalate ($LiTaO_3$) etc., by various methods of domain reversals like electric field poling, caliographic poling and UV irradiation. Using the electric field poling method, various domain structures of QPM has been fabricated for desired applications. The engineered domain structures of QPM are designed to compensate the mismatch among the wave vectors of the interactive waves which is concluded in generating higher harmonics. In this operation, the non-linear optical coefficient (d_{eff}) of the ferroelectric crystal is polarized periodically or aperiodically by applying high electric field resulting into a grating structure of period $\Lambda = 2l_c$, where l_c is the coherence length and L is the

T.S. Meetei · S. Hari Hara Subramani · K. Pandiyan (✉)
Centre for Nonlinear Science and Engineering (CeNSE), School of Electrical and Electronics Engineering, SASTRA University, Thanjavur 613401, Tamil Nadu, India
e-mail: krishpandiyan@ece.sastra.edu

S. Boomadevi
Department of Physics, National Institute of Technology, Tiruchirappalli 620005, Tamil Nadu, India

V. Lakshminarayanan and I. Bhattacharya (eds.), *Advances in Optical Science and Engineering*, Springer Proceedings in Physics 166,
DOI 10.1007/978-81-322-2367-2_57

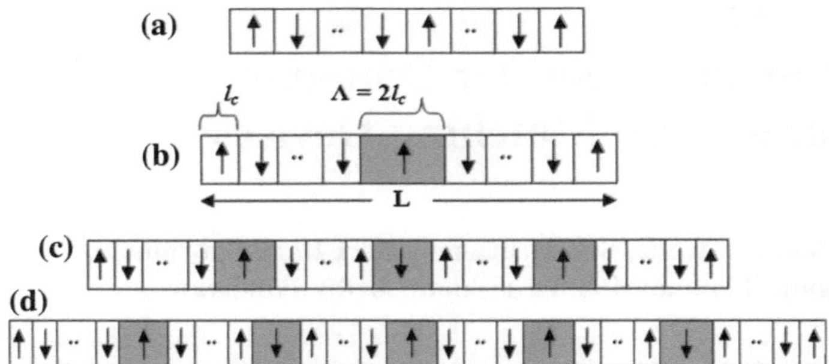

Fig. 57.1 Schematic diagram of QPM device **a** with no phase-shifter domain or ideal periodic structure, **b** with one phase-shifter domain at the centre of the device, **c** with three phase-shifter domains, and **d** with five phase-shifter domains

length of the device. Several domain engineered grating structures like periodic, aperiodic and periodic phase-reversal optical superlattices have been proposed to achieve multiple QPM frequency. In accordance with the standard optical frequency grid (ITU-T G.691) whose spacing should be in multiple numbers of 50 GHz, it is necessary to tailor the separation between the QPM peaks in order to accommodate the same in the optical grid. This could be easily attained by introducing phase-shifter (PS) domains in periodic grating structure which is shown in Fig. 57.1b–d. Hence, in this paper, we present a novel design to generate tunable multiple QPM devices to employ same in the optical communication systems.

57.2 Theory

We can derive the expression for frequency doubling of QPM device by taking the electric field components of the fundamental A_1 and the second harmonic wave A_2. The wave equation can be represented as,

$$\nabla^2\vec{A} + \frac{n^2}{c^2}\frac{\partial^2\vec{A}}{\partial t^2} = -\mu_0\frac{\partial^2\vec{P}_{NL}}{\partial t^2} \tag{57.1}$$

The fundamental form of slowly varying amplitude equation for the growth of second harmonic generation is given by one of the coupled modal equations [1, 2], which is given as

$$\frac{\mathrm{d}A_2}{\mathrm{d}z} = i\kappa A_1^2 \mathrm{e}^{-i\Delta kz} \tag{57.2}$$

where, $A_{1(2)}$, is the electric field amplitude of the fundamental (second-harmonic) wave, κ is proportional to the effective nonlinear optical coefficient and the wave vector mismatch. The wave vector mismatch for the QPM interaction, Δk is given by

$$\Delta k = k_2 - 2k_1 \pm G = 0 \tag{57.3}$$

where, G is the grating vector factor that provides phase mismatch compensation.

By assuming negligible depletion of the fundamental wave, we can directly integrate (57.2) to obtain a sin*c* spectrum, if κ is constant in the medium. In general, κ is not a constant, it is a function of z, i.e., $\kappa(z)$. Let $\Delta k = 2\pi q$, then SHG amplitude can be expressed as Fourier transform of $\kappa(z)$,

$$A_2(q) = iA_1^2 \int \kappa(z) e^{-i2\pi qz} dz = iA_1^2 \Im[\kappa(z)] \tag{57.4}$$

For our studies, we consider a grating structure which includes π-phase shifting domain(s) or aperiodic domain(s) of width Λ which is distributed within the periodically poled device or grating structure of device length L (Fig. 57.1). This type of structure is called phase reversal grating structure. The second harmonic (SH) amplitude of the phase reversal grating structure can be written as [3, 4],

$$A_2(q) = i\frac{1}{2} aLA_1^2 e^{ip\chi'} \frac{\sin\chi' \sin p\chi'}{\chi' \cos\chi'} \tag{57.5}$$

57.3 Results and Discussion

We studied the SH spectral analysis of a phase reversal QPM devices by increasing number of PS in the structure. The increase in number of phase shifter domain plays major role in changing the intensity of the dual peak and engineering separation between the peaks. Further, we studied how the peak separation between peaks changes with respect to the length of the device.

Figure 57.1a shows the periodic grating structure of QPM device without any PS domain(s). The second harmonic response of periodically poled QPM device without PS domains is shown in the Fig. 57.2a, which is resulted to a high intensity single peak (Table 57.1).

Further, we introduce PS domains to the periodic grating structure of QPM device which is shown in the Fig. 57.1b–d. With the presence of PS domains, the phase of the second harmonics is reversed by π giving two peaks near the phase matching point [5–9]. We introduced one, three and five numbers of PS domains in the periodic grating structure and theoretically analyzed the dependence of PS domains to the intensity of SH dual peaks and separation between the peaks. Figure 57.3a revealed that the intensity of the dual peaks has been exponentially decreased with the augmentation of PS domains in the grating structure. Also from Fig. 57.2b, it is clear that the separation between the peaks becomes wider with

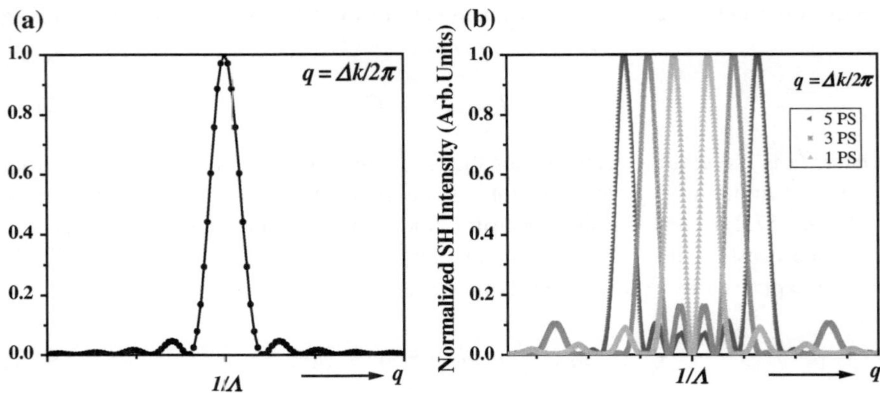

Fig. 57.2 Normalized second harmonic intensity spectrum of QPM **a** with no phase-shifter domains and **b** with phase-shifter domains (one, three and five)

Table 57.1 Calculation of spacing between SH dual peaks, intensity w.r.t variation of device length and PS domains

No. of PS domains	Normalized SH intensity (arb. units)	Device length (L) μm						
		100	200	300	400	500	600	700
		Spacing between SH dual peaks (Δk) μm^{-1}						
1	1	0.1170	0.0058	0.0038	0.0028	0.0024	0.0020	0.0016
3	0.46289	0.0295	0.0142	0.0096	0.0072	0.0058	0.0048	0.0042
5	0.24665	0.0451	0.0225	0.0154	0.0113	0.0086	0.0076	0.0068
7	0.16419	0.0592	0.0306	0.0202	0.0154	0.0123	0.0105	0.0088
9	0.12101	0.0777	0.0387	0.0259	0.0194	0.0156	0.0126	0.0111

more numbers of PS domains in the periodic grating structure and vice versa. Moreover, device length (L) also plays a vital role in tuning the spacing between the peaks. We have theoretically calculated the spacing between the peaks (i.e., $\Delta k = 2\pi/L$) of various QPM devices of length ranged from 100 to 700 μm by introducing various numbers of PS domains (*one, three, five, seven and nine*) in the grating structure of grating period $\Lambda = 8$ μm. Figure 57.3b revealed the spacing between the peaks is decreased with increased in device length. This concept gives a major advantage in desirable fine tuning of separation between the peaks which is useful in optical frequency grid for frequency conversion. We observed the side-bands or noises in the phase-matching which are due to the occurrence of deviation from the ideal periodicity in the nonlinear crystal are effectively suppressed with the augmentation of PS domains in the grating structure (Fig. 57.2b), resulting less cross-talk in the optical communication systems. Furthermore, we are analyzing the approach to avoid cross-talk for better applications in the optical communication system.

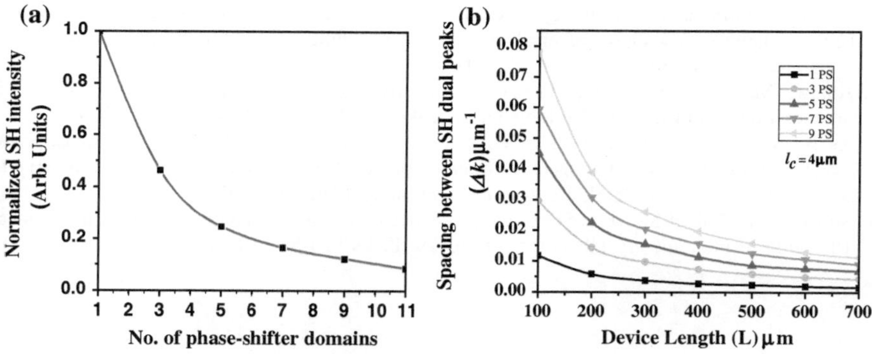

Fig. 57.3 **a** SH dual peaks intensity variation with phase-shifter domains, and **b** variation of SH dual peaks spacing with PS domains

57.4 Conclusion

We have successfully analyzed the theoretical concept of second harmonic spectral behaviour of domain engineered phase reversal QPM devices. We observed the tunability of the spacing between second harmonic peaks is a function of phase shifter domain(s) and the device length. With introduction of more number of PS domains or varying the device length in the periodic QPM structure could easily accommodate in the optical communication systems. Hence, by adopting these functions, one can evolve to a new QPM grating structure for better performance in the optical communication system.

Acknowledgments The authors wish to express their sincere gratitude to the Department of Science and Technology (DST), New Delhi, India for their financial support (Ref. No.: SB/FTP/PS-089/2013).

References

1. Fejer MM, Magel GA, Jundt DH, Byer RL (1992) Quasi-phase-matched second harmonic generation tuning and tolerances. IEEE J Quantum Electron 28(11):2631–2654
2. Armstrong JA, Bloembergen N, Ducuing J, Pershan PS (1962) Interactions between light waves in a nonlinear dielectric. Phys Rev 127:1918–1939
3. Pandiyan K, Kang YS, Lim HH, Kim BJ, Cha M (2009) Nondestructive quality evaluation of periodically poled lithium niobate crystals by diffraction. Opt Express 17(20):17862–17872
4. Chou MH, Parameswaran KR, Fejer MM (1999) Multiple-channel wavelength conversion by use of engineered quasi-phase-matching structures in $LiNbO_3$ waveguides. Opt Lett 24(16):1157–1159
5. Antonosyan DA, Kryuchkyan GYu (2012) Phase-reversed structures in superlattice of nonlinear materials. Opt Commun 285:795–799

6. Hari Hara Subramani S, Karthikeyan K, Mirunalini A, Prasath RK, Boomadevi S, Pandiyan K (2013) Analysis of a phase reversal quasi-phase matching device for the dual peak second harmonic response. J Opt 15(5):055205–055212
7. Ahlawat M, Tehranchi A, Pandiyan K, Cha M, Kashyap R (2012) Tunable all-optical wavelength broadcasting in a PPLN with Multiple QPM peaks. Opt Express 20(24):27425–27433
8. Asobe M, Tadanaga O, Miyazawa H, Nishida Y, Suzuki H (2005) Multiple quasi-phase matched device using continuous phase modulation of $\chi^{(2)}$ grating and its application to variable wavelength conversion. IEEE J Quantum Electron 41(12):1540–1547
9. Simeonov D, Saltiel S (2004) Statistical properties of second harmonic generation in nonideal 1d nonlinear photonic crystals, vol 5, pp 26–29. ISBN 954-580-120-4

Chapter 58
Hybrid Radio Frequency/Free-Space Optics (RF/FSO) Wireless Sensor Network: Security Concerns and Protective Measures

Koushik Banerjee, Hemant Sharma and Anasuya Sengupta

Abstract Wireless sensor networks (WSNs) are ad hoc wireless networks that are written off as spread out structure and ad hoc deployment. Sensor networks have all the rudimentary features of ad hoc networks but to altered points—for instance, considerably lesser movement and far more energy necessities. Commonly used technology for communication is radio frequency (RF) communications. Free-space optics (FSO) is relatively new technology which has the prospective to deliver remarkable increases in network lifetime of WSN. Hybrid RF/FSO communications has been suggested to decrease power consumption by a single sensor node. It is observed that security plays a very important role for either RF WSN or hybrid RF/FSO WSN as those are vulnerable to numerous threats. In this paper, various possible attacks in RF/FSO WSN are discussed and aimed to propose some way out from those attacks.

58.1 Introduction

From the beginning of our civilization human beings have placed great importance on the sharing and gathering of knowledge. Whether being utilized to research, inventing things or discovering philosophies, knowledge has been a vital part of human society. Before technology made the sharing of knowledge, knowledge was predominantly distributed via word of mouth. After that it was being shared in an inscription. In an ever evolving society, technology has a very central role in daily life. As technology upgraded over epochs, the speed and precision in which information and knowledge could be forwarded, improved. The primordial methods of

K. Banerjee (✉) · H. Sharma · A. Sengupta
ITM University (Gwalior), Gwalior, Madhya Pradesh, India
e-mail: babun.banerjee@gmail.com

V. Lakshminarayanan and I. Bhattacharya (eds.), *Advances in Optical Science and Engineering*, Springer Proceedings in Physics 166,
DOI 10.1007/978-81-322-2367-2_58

using word of mouth and the old-fashioned mailing system generated opportunity for telegraph systems and telephones, the wired communication and later for the wireless communication as mobile phones. Wireless communication has numerous benefits over traditional wired network. Wireless communication and advanced circuit technology made it possible to develop inexpensive, tiny, low power and multi-functional sensing and communicating devices for various purpose [1]. These small sensing devices have the capabilities of sensing, computing and communicating with each other. These devices are called sensors. A sensor is a miniature device used to sense the atmosphere of its surroundings, collect data, and draw some significant deduction after processing the raw data. Later it can be used to identify some phenomena. The collection of these sensor nodes are called wireless sensor networks (WSN). Now a days WSN is used in various applications like defences, remote patient monitoring (Healthcare), forest fire prevention, wildlife monitoring, underwater habitat monitoring, sports, temperature sensing, environmental monitoring, and industrial monitoring, ecology [2] and so on. Sensor network consists of several (in fact a huge number of) sensor nodes and less number of gateway node. Gateways are considered as powerful in comparison with normal sensor node. Gateway nodes are often called as Base Station (BS). Every single node has its own sensor device, a micro-processor, memory unit and radio, so it has the ability of sensing environments, process data and communicate with others. Final outcome is submitted to base station. The data diffused from sensor to sensor or sensor and finally to BS. Conventionally communication is done wirelessly using radio frequency (RF) channel. But, modern free-space optics (FSO) communications via modulating retro reflectors has shown its potentiality, owing to necessity of lesser communications energy. In FSO, sender and receiver needs to be in line of sight (LOS). In this paper, we have assumed that hybrid RF/FSO is used for communication. In this hybrid method FSO is regularly used, but when sender and receiver are not in LOS, RF is used. It is verified that hybrid RF/FSO WSN lasts at least twice as long as RF WSN lasts. So hybrid RF/FSO WSN has become a great choice in terms of power conservation. But due to error prone nature and ever altering topology of a sensor node, it is easy to make a node malicious or deploy a malicious node in the sensor network. Security should be taken into consideration because most of sensor networks is inclined to various attacks [3] and it usually possess several mission-critical responsibilities [4].

So security is a great concern for any WSN as RF WSN as well as hybrid RF/FSO is prone to various attacks. Though some attacks (like jamming attack in physical layer) can be avoided by using hybrid RF/FSO WSN but, there are chances of other attacks. To counter security threat, a series of concrete action is needed. In this paper different attacks in hybrid RF/FSO are identified and some possible solutions are provided to prevent those attacks.

58.2 Security Goals

Several metrics such as authenticity, availability, confidentiality, integrity, nonrepudiation, freshness, forward secrecy and backward secrecy are essential goals to achieve in order to attain security. Those should be provides by any routing protocols of WSNs.

58.2.1 Authentication

Authentication can deliver trustworthiness of the message by recognizing its birthplace. Attackers in a sensor network can alter the contents of a packet as well as inject superfluous bogus packets in the network. The destination node must be able to identify the sender of a packet among original and attacker. Data authentication is a practice of ascertaining the identity of senders. Usually a message authentication code (MAC) is calculated to achieve authenticity [5].

58.2.2 Availability

Availability is the defined as whether a particular node is able to use the resources assigned for communication and whether the network is currently available for the messages to communicate. WSNs must be robust against attacks. If there are any attacks, the effect must be negligible and should not have any impact on overall network performance.

58.2.3 Confidentiality

Confidentially is the capability to conceal messages from an attacker so that any message transferred via the WSN remains hidden and confidential. The standard tactic for keeping sensitive data hidden is to encrypt the data with a secret key that only intended receivers have. Therefore confidentiality is achieved.

58.2.4 Integrity

Integrity ensures the reliability of the message. It also refers to the skill to make sure that a message has not been meddled with, altered or changed whereas communicating on the network. Even if the WSN has confidentiality measures in place, there is always a chance that the integrity can be compromised.

58.2.5 Forward Secrecy

After a sensor leaves a wireless sensor network, it should be unable to read any future messages.

58.2.6 Backward Secrecy

A recently added sensor node should be unable to intercept any old and previously transmitted message.

58.2.7 Data Freshness

Data freshness is a property that implies the data is recent or fresh, and it confirms that a node has not repeated any previous messages. It is imperative when there is a shared key in the network.

58.2.8 Non Repudiation

This is a desirable property. It means after sending a message, any sensor node cannot rebuff. This property is analogous to digital signature in cryptography.

58.3 Threat Model

Threat model is generally classified into two categories-based on device capability and based on attacker type or the location of attacker [6]. The earlier classification is a matter of the device capability of the attacker. This can also be segmented into mote-class attacker and laptop-class attacker. A mote-class attacker has access to very limited sensor nodes [3]. Those nodes and attacker node have same capability in terms of power, processing and hardware specifications. The laptop-class attacker has access to a robust device than its own. Those devices can be a laptop or like devices. So the attacker has advantages than general notes in regard to power, processing and communication. So a laptop-class attacker can be more treacherous than a mote-class attacker or an attacker with conventional low power sensor nodes. A usual low power sensor node can use to jam the radio link of the RF/FSO WSN, while a laptop-class attacker could jam the entire network by means of a stronger transmitter. A laptop-class attackers have a high bandwidth, minimal-latency communications channel, which is not available to mote-class attacker. The second grouping is based on attacker location. It is once more classified into insider and outsider attacks. An insider attack is taken place when an authorized sensor node within the network has gone bad. Compromised node is an example of an insider attack. Recognition and avoidance of this type of attack is tough. An outsider attack occurs when an attacker has no special access to the network. Attacker is positioned outside of the sensor network. It is again classified into passive and active attack [6]. Passive attackers are those which assemble sensitive data from the sensor network, for which privacy and confidentiality are gone astray. The goal behind these attacks is to steal owner data like a ghost. It does not temper the data. So detection is hard and prevention technique of this attack is more beneficial. An example of a passive attack is eavesdropping. On the other hand active attack is more expected to be discovered rather be prevented.

58.4 Attacks on RF/FSO WSN

There are so many causes for need of security in WSNs. First of all, WSNs have vulnerability to several attacks owing to the broadcasting nature of communication medium. Moreover, in some circumstances, nodes are positioned in a dangerous location. Attack can happen anywhere in the transmission. By using hybrid RF/FSO, we can prevent a few physical layer attacks. Various attacks in hybrid RF/FSO WSN like Denial of service, blackhole, sinkhole, HELLO flood, Sybil, wormhole, acknowledgement spoofing, rushing are discussed below:

58.4.1 *Denial of Service (DoS) Attack*

Denial of service (DoS) is caused by a malicious node and it degrades network performance [7]. By sending additional redundant packets, the informal DoS attack tries to use all the resources attainable to the target node. As a consequence, it hinders authorized nodes in network from getting admission or resources available to them [2]. DoS attack is projected to the opposition's effort to corrupt or annihilate a network, and furthermore for every events that decrease a network's capability to set up a service [8]. DoS attack has several variants.

58.4.2 *Selective Forwarding*

The routing has an autonomy which is the packet forwarding. Selective forwarding attack drops several packets [2]. Selective forwarding is a type of the DoS attack. In this attack compromised node forwards only few chosen packets and drops the rest.

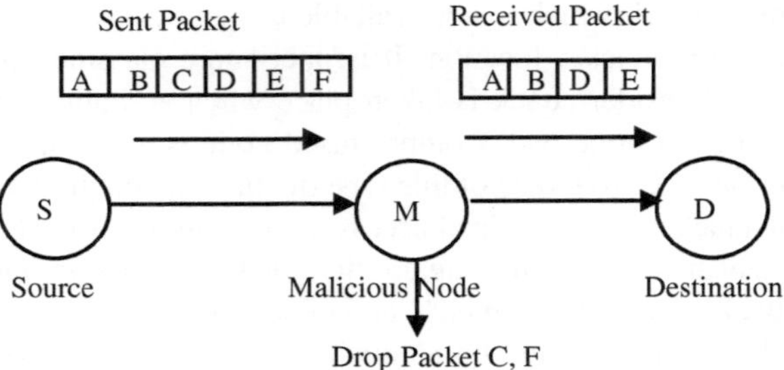

Fig. 58.1 Selective forwarding attack: here *S* is source node. *M* is the malicious node and *D* is the destination node. *S* sends six packets, but *M* drops two of them and *D* gets only four packets

Attacker needs to be present in the route of the data stream to inject selective forwarding attack. To do so, he can use can use sinkhole attack or Sybil attack. This attack is also called a greyhole attack. Selective forwarding is tricky to detect. Figure 58.1 demonstrates selective forwarding attack.

58.4.3 Wormhole Attack

Wormhole is a low-latency connection among two detached portions of the network via out-of-band channel. This means a message, which typically traverses several hops, traverses only solitary node. The time of the delivery can be meaningful in the routing scheme, particularly if the influenced message holds routing information. The attacker can send repeated packets through the wormhole to generate the illusion that two distant nodes are neighbors. Thus the node seems to be only a single hop from the base station. Wormhole attacks could probably be used in grouping with selective forwarding or eavesdropping. Discovery of wormhole can become very difficult when used in combination with the Sybil attack [3].

58.4.4 Sinkhole Attack

The objective of the sinkhole attack is to trap as much of the traffic as possible to the malicious node [9]. The core idea behind this attack is that the mischievous node attempts to look very eye-catching with respect to the routing algorithm. This goal can be accomplished without difficulty, for instance, an attacker can spoof the route advertisement or can create a high quality path to the BS using wormhole attack [2]. Sinkhole can be used for selective forwarding attack, which is very efficient and easy. Sinkhole attack can modify messages or can alter the hop count. Sinkhole attack can construct a loop in the network. Sensor networks are

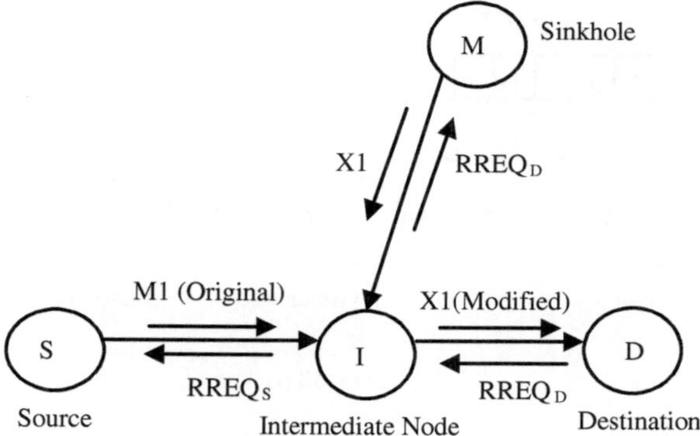

Fig. 58.2 Sinkhole attack: here *S* is source node, *D* is destination node, *I* is an intermediate node and *M* is the malicious node. *S* wishes to send message *M1* to *D* via in-between mode I. But *M* intercepts it and change the message to *X1*. Now *D* finally obtains message *X1*

predominantly vulnerable to sinkhole attacks is because of their dedicated communication type. As every packets share the identical destination (BS), a malicious node desires only to offer a lone excellent route to the base station with the purpose of influencing a huge number of nodes. Figure 58.2 shows sinkhole attack. In this example, sinkhole alter a message.

58.4.5 Blackhole Attack

A crucial alternative of selective forwarding attack is called a blackhole attack [10]. In such instance, all the packets are dropped by the malicious node. Yet a node acting like a black hole can be effortlessly spotted by the neighboring nodes, considered as dead and omitted from the routing path. Figure 58.3 demonstrates attack.

58.4.6 HELLO Flood Attack

It is one of the most primitive attacks in WSNs. In quite a few protocols, nodes advertise themselves to the neighbors by broadcasting the HELLO packets. Nodes which gets such packet determines, that the broadcasting node is its neighbor and is within the usual radio range. A laptop-class attacker can use an influential radio to send these HELLO packets to other nodes, which are at far more distant than the standard radio range from it. These nodes will direct their messages and will try to grasp the neighbor, which is not in the range of their radio. It can be prevented by blocking method [11]. HELLO flood attack is illustrated in Fig. 58.4.

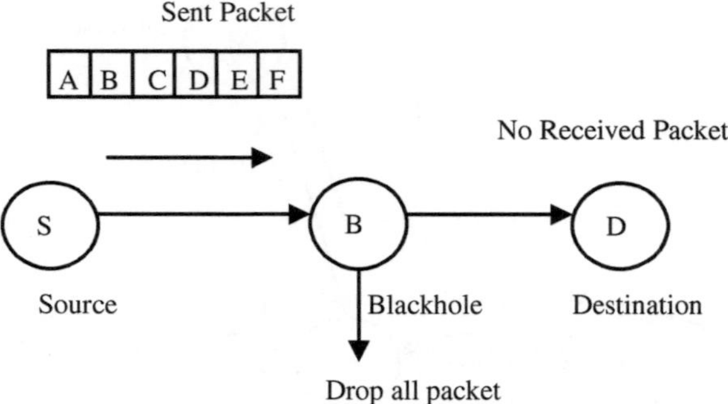

Fig. 58.3 Blackhole attack: Here *S* is source node. *M* is the blackhole node and *D* is the destination node. *S* sends six packets, but *M* drops all of them and *D* gets none

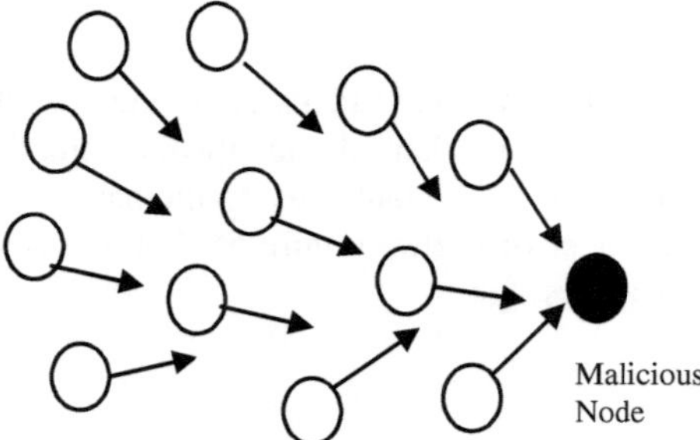

Fig. 58.4 HELLO flood attack: *black* node is malicious node and other are general sensor nodes. All other nodes are attracted to malicious node

58.4.7 Acknowledgement Spoofing

Acknowledgement spoofing is all about crafting delusion. An attacker spoofs link layer acknowledgements to convince the node that a dead neighbor of it is alive or certain weak link is dependable. The impact is alike to selective forwarding, selected packets are gone with great probability.

58.4.8 Sybil Attack

Here, the attacker presents a number of false identities to other nodes in the network. Sybil attack can meaningfully lessen network performance. Most of the routes, possibly will pass through the attacker, therefore Sybil attack can eradicate distributed storage, topology or even the strongest multipath routing algorithms [1]. Sybil attack is also a prominent threat for geographic routing. In geographic

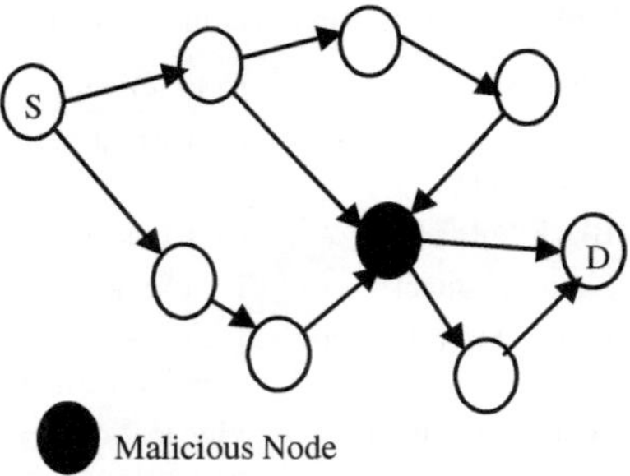

Fig. 58.5 Sybil attack: here *S* is source node and destination is *D*. Here all traffic passes through the malicious node

routing, nodes occasionally need to exchange coordinates. In Sybil attack (In geographic routing) attacker node can be virtually present at more locations at a time and consequently affecting the routing algorithm. Sybil attack is a grave threat not only for routing protocols, but also for other network algorithms for instance distributed storage or voting algorithm. Figure 58.5 shows elementary Sybil attack.

58.4.9 Rushing Attack

An attacker circulates route request and reply messages rapidly throughout the network. This overpowers any genuine future route request messages, i.e. nodes discard them, because nodes suppress other reproductions of a route request that they have by this time processed.

58.5 Defensive Measures

Different threats can be solved using different methods. Symmetric (DES) or asymmetric key cryptography can be used to secure communication [12]. In RF WSN, asymmetric key cryptography is not advisable due to high energy requirement. But, in RF/FSO WSN we can use asymmetric key cryptography as this type of WSN lasts twice than RF WSN [13]. Secure data aggregation is another promising solution to various threats. Secure localization also can be used as it targets to prevent the consequence of any type of attacks. Through access control WSNs can be secured as well. The problem of access control can be considered as controlling access to resources to restricted objects. But, commonly secure routing protocols are used for routing as well as securing the network. Possible solutions to those attacks are given in Table 58.1.

Table 58.1 Defensive techniques that can be used to prevent various attacks

Attacks	Countermeasures
Selective forwarding	Consistent network monitoring using source routing
Black hole attack	Constant network monitoring by means of source routing
Sinkhole attack	By checking the radio range frequently for the nodes
HELLO flood attack	Each neighbor node needs to authenticate with each other using identity verification protocol, blocking method
Wormhole attack	Physical observing of field devices and consistent monitoring of network using source routing. Monitoring system may perhaps use packet leash
Acknowledgement spoofing	By regular monitoring of network
Sybil attack	Rearranging of devices and changing session key
Rushing attack	By organized monitoring of network

58.6 Conclusion

In the earlier, there was a slight attention on WSNs security but, in the present day, need of security in WSNs has becomes more apparent since, WSNs are used much more than previous times. Some limitations make WSNs pretty much different from traditional wireless ad hoc networks. WSNs have some unique characteristics as well. Due to some of those unique characteristics WSNs are very much prone to different kind of attacks. In this paper, we have tried to pinpoint different potential attacks of hybrid RF/FSO WSN. Though some characteristics of this technique helps to overcome a few attacks, we cannot avoid all attacks. Hence routing protocol developers should aim to achieve security goals by preventing security threats.

References

1. Akkaya K, Younis M (2005) A survey on routing protocols for wireless sensor networks. Ad Hoc Netw 3:325–349
2. Ahmad Salehi S et al (2013) Security in wireless sensor networks: issues and challenges. In: Proceedings of the IEEE international conference on space science and communication, pp 356–360
3. Karlof C, Wagner D (2003) Secure routing in wireless sensor networks: attacks and countermeasures. In: Proceedings of the 1st IEEE international workshop on sensor network protocols and application, pp 113–127
4. Shi E, Perrig A (2004) Designing secure sensor networks. Wirel Commun Mag 11(6):38–43
5. Wang N et al (2006) Wireless sensors in agriculture and food industry—recent development and future perspective. Comput Electron Agric 50(1):1–14
6. Shashikala, Kavitha C Dr (2012) A survey on secured routing protocols for wireless sensor network. In: Proceedings of IEEE ICCCNT
7. Wood AD, Stankovic JA (2002) Denial of service in sensor networks. Computer 35(10):54–60

8. Dohler M (2008) Wireless sensor networks: the biggest cross-community design exercise to-date. Recent Pat Comput Sci 1(1):9–25
9. Bao F, Chen R et al (2012) Hierarchical trust management for wireless sensor networks and its applications to trust-based routing and intrusion detection. IEEE Trans Netw Serv Manage 9 (2):169–183
10. Yin J, Madria SK (2003) A hierarchical secure routing protocol against black hole attacks in sensor networks. In: Proceedings of IEEE CCT, pp 1282–1286
11. Zhang Y, Li X et al (2011) A hierarchy-based dynamic key management for clustered wireless sensor network. Energy Procedia 13:967–7974
12. Banerjee K, Sharma H, Chaurasia BK (2014) Secure communication for cluster based wireless sensor network. In: Proceedings of IEEE CICN, pp 867–871
13. Sivathasan S et al (2008) RF/FSO wireless sensor networks: a performance study. In: Proceedings of IEEE GLOBECOM, pp 1–5

Chapter 59
Review on Li-Fi Technology

Rajarshi Roy Chaudhuri, Kaustav Dutta and Archisman Saha

Abstract With advent of various communication technologies one can access the whole world at one go. The impact of internet on our day to day life has been so extensive that it is impossible to think of a day without it. It has become a fundamental requirement in our daily lives. Survey reports show that nearly 46 % of homes throughout the world have access to the internet. And the percentage is growing each day. With such a high demand there has been a looming Radio Frequency spectrum crisis, which paved the way of the invention of a new technology:-LI-FI. LI-FI, acronym of light fidelity, is a new wireless technology which has the ability to provide high speed internet connection within localized environment. Till today we are familiar with WI-FI which uses radio spectrum for communication. Even though it gives a speed of nearly 150 Mbps (as per IEEE802.11n), it isn't sufficient to satisfy all users. On the other hand LI-FI uses spectrum which comprises a wide range of frequencies, from the infrared through visible, down to the ultraviolet spectrum for communication which has the ability to produce a theoretical speed of 10 Gbps. It is not only confined to light-emitting diode (LED) or laser technology or to any specific receiving technique, LI-FI is a framework for all those technologies which provides new ways to all present as well as future services or applications.

R.R. Chaudhuri (✉) · K. Dutta · A. Saha
Department of Electronics and Communication Engineering, Institute of Engineering and Management, Y-12 Engineering Campus, Salt Lake, Sector V, Kolkata, India
e-mail: roychaudhurirajarshi@gmail.com

K. Dutta
e-mail: kaustavdutt@rediffmail.com

A. Saha
e-mail: archis.1992@gmail.com

V. Lakshminarayanan and I. Bhattacharya (eds.), *Advances in Optical Science and Engineering*, Springer Proceedings in Physics 166,
DOI 10.1007/978-81-322-2367-2_59

59.1 Introduction

In a very lucid language we can define LIFI as LIGHT based wireless technology (WIFI). The only difference of LIFI with WIFI is that instead of radio waves it takes the help of visible light spectrum to communicate and instead of modems, light-emitting diode (LED) lamps are used which are fitted with transmitter as well as receiver [1]. We are involved with light for millions of years but it has a very little or no impact on heath issues. For this reason as well as the confinement of the previous technology i.e. WIFI, has paved the way for the discovery of this technology. Data can be encoded in light, at a speed in which LEDs flicker. When it is off 0 is transmitted and when on 1 can be transmitted through it—thus giving rise to digital signals. The modulation intensity is so rapid that human eyes cannot detect it and LED appears to be constantly glowing [2].

Li-Fi, has already achieved lightning fast speed in the laboratories. Scientists at the Heinrich Hertz Institute situated in Berlin, Germany, have been able to obtain a data transfer speed rate of over 500 Mbps where they used a white-light LED for their experiment. Herald Haas, who is known to be the father of this technology, has set up a firm which sells a consumer visible light communication (VLC) transmitter. It has the ability to transmit data at about 100 Mbps which is much faster than most UK broadband connections [3].

Prof. Harald Haas, a pioneer in the field of optical wireless communications at the University of Edinburgh was the first person to demonstrate this technology. He has been able to show how a light-emitting diode bulb (LED) combined with signal processing technology successfully streamed a video to the computer. He was the first person to coin the term "Light Fidelity" or LI-FI. With the help of this technology, a one-watt white LED bulb has the ability to provide internet connectivity to four computers simultaneously. At the same time LI-FI holds the potential of providing a future where data for smart phones, tablets and laptops will be transmitted via light in a room. Operating on the principle of data transfer through light illumination, security is also ensured–as if you are unable to see the light; you can never access the data (Fig. 59.1) [4, 5].

59.2 Designing a LI-FI Based System

A LIFI product is mainly an assembly of four primary sub-components:

- Bulb
- Radio Frequency Power Amplifier (RFPA) circuit
- Printed Circuit Board (PCB), and
- An enclosure

The function of the Printed Circuit Board is to control all the electrical signals of the lamp and also contains the microcontroller which manages the different

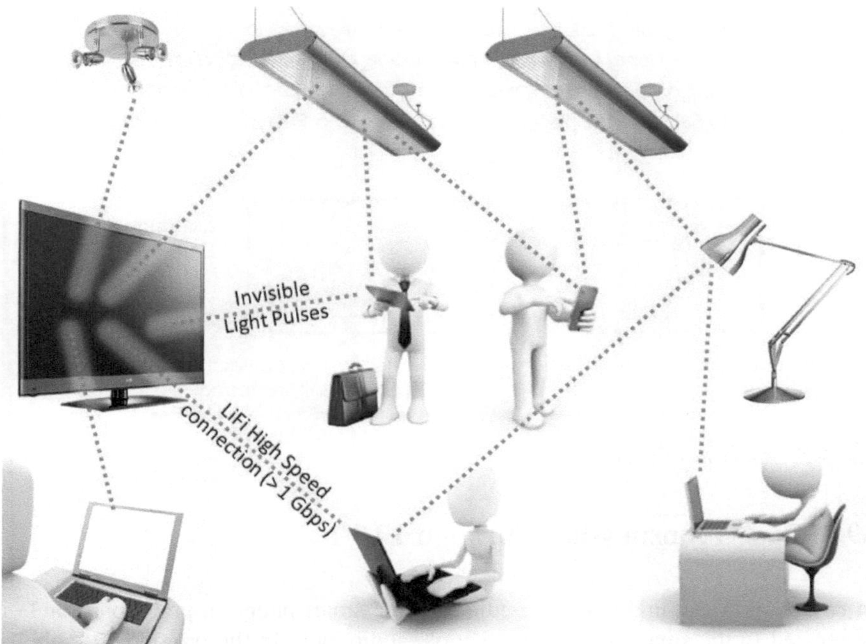

Fig. 59.1 A future Li-Fi environment

functions of the lamp. The solid-state Power Amplifier generates a radio frequency signal which is converted into an electric field. Due to high concentration of energy in the electric field the contents of the bulb gets transformed to a plasma state at the center of the bulb. This controlled plasma is responsible for the generation of an intense source of light. All these subcomponents are contained in an enclosure made up of aluminum.

59.2.1 Function of the Bulb Sub-assembly

The centre of a LIFI system consists of a bulb sub-assembly, in which a sealed bulb which is inserted in a dielectric material. This particular design is much more reliable than other well-known light sources which usually use degradable electrodes inside the bulb. The dielectric material has two functions; primarily it acts as a waveguide for the radio frequency energy radiated by the Power Amplifier and secondly it also helps in concentrating the electric field thus focusing energy in the bulb. This energy from the electric field is responsible for heating the material in the bulb readily and transforms it to a plasma state that has the power to emit light of high intensity (Fig. 59.2) [6].

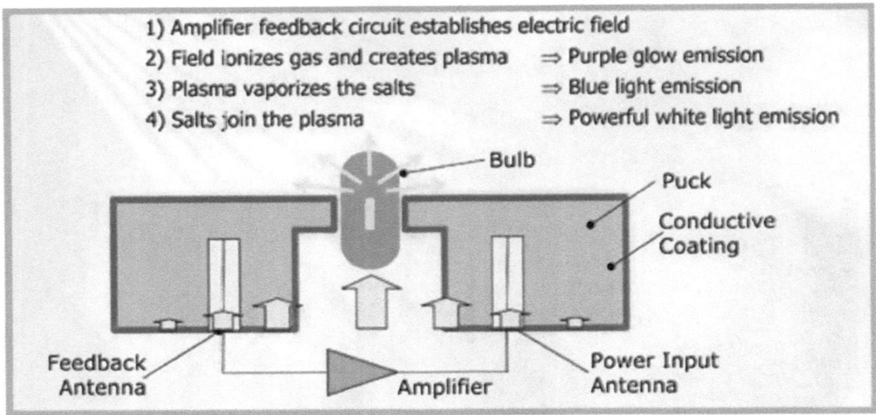

Fig. 59.2 Bulb sub-assembly

59.3 Data Transmission Through LI-FI

In case of transmitting a simple audio file in a smart phone; a protoboard table is connected to the smart phone via a simple audio jack. In the protoboard table the received audio signal is converted into an optical signal which is transmitted by a special emitter across the light spectrum generated by a LED. At the receiving end, a receptor which is located in a speaker captures the signal and converts it back to an audio signal which can be played by the speaker.

For wireless internet transmission, the principle remains the same but it makes use of a receptor device which is designed to be placed over a router. The router incorporates an LED lamp to transmit the data so that anyone falling within the range of light emitted by the LED lamp will be able to access the internet (Fig. 59.3).

59.4 Superiority of LI-FI Over Other Technology

1. Data transmission can be done at a very high speed, as high as 30 GB per minute [5].
2. Visible Spectrum is used in Li-Fi making it safer than radio frequency signals.
3. VLC can be used in aircrafts.
4. Li-Fi can be used for medical fields, and also in such areas where infrared, WI-FI, and Bluetooth technologies are banned [7].
5. Li-Fi has the potential to work under water.
6. There are about nineteen billion bulbs spread worldwide, they just need to get replaced with LED bulbs that transmit data [8].

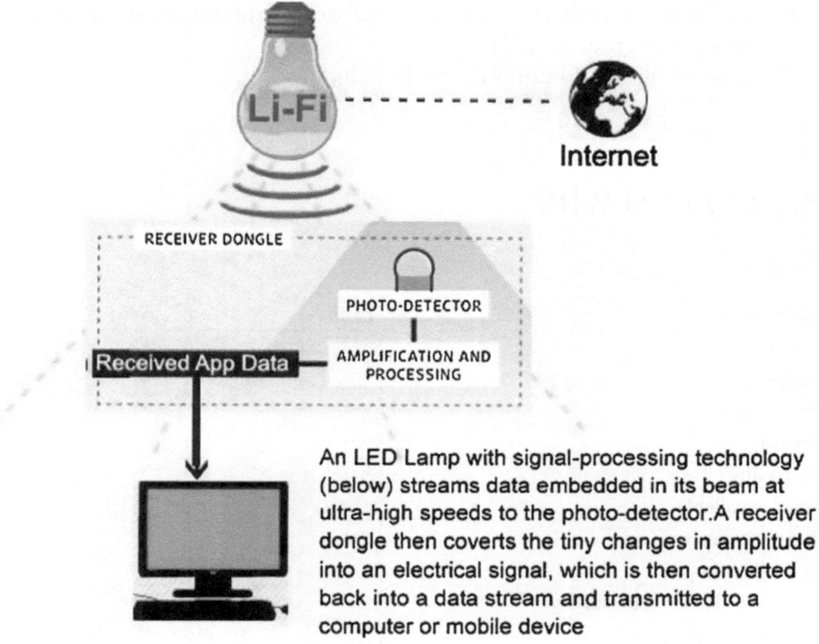

Fig. 59.3 This is how data is transmitted through a LED lamp

7. Security is an added benefit, since light does not penetrate through walls can also be used in streets for traffic control. Cars having LED based headlights and backlights can communicate with each other and prevent road accidents. Even the traffic lights can also communicate to the car with the help of this technology [9].
8. By implementing this technology worldwide every street lamp has the potential to be a free access point.
9. Li-Fi is the solution for the problem concerning the shortage of radio frequency bandwidth [10].

59.5 Limitations of LI-FI

1. One of the most basic problems is that light cannot pass through opaque objects, so if the receiver somehow gets blocked then the signal will immediately get cut out and transmission is disabled [11].
2. Two major issues that have to be kept to be kept in mind by the companies (while providing VLC services) are reliability and network coverage. External sources of light like sun rays, bulbs; and other opaque materials may interfere in

the transmission path and this may lead to interruption or disturbance in the communication process [12].
3. The cost of installation of VLC is quite high.

59.6 LI-FI V/S WIFI

Li-Fi/Wi-Fi comparison

Parameter	Li-Fi	Wi-Fi
Speed	***	***
Range	*	**
Data density	***	*
Security	***	**
Reliability	**	**
Power available	***	*
Transmit/receive power	***	**
Ecological impact	*	**
Device-to-device connectivity	***	***
Obstacle interference	***	*
Bill of materials	***	**
Market maturity	*	***

* low *** medium *** high

59.7 Conclusion

Light Fidelity or Li-Fi is one of the upcoming and growing technologies which have the ability to compete with various other existing technologies or being developed. Since in this technology light is the main source of transmission it will be very advantageous and can be implemented in various fields that is not possible with Wi-Fi and other related technologies. *The present research in VLC is focussed on finding a modulation scheme for IM/DD for point to point VLC links taking into account the fact that VLC would serve two purposes at the same time namely, illumination and gigabit wireless communication.* OFDM has proved to be the most suitable choice as a digital modulation scheme for use in Li-FI Technology. The Li-Fi technology based on VLC principles is advantageous over the Wi-Fi technology in terms of *spectrum availability, high efficiency, secure data transfer,* and, *high data rate transmission.* Owing to the high frequency range of visible light spectrum Li-Fi is a powerful wireless network solution to the looming RF-spectrum crisis. Bi-directional point to point communication using Li-FI system has been practically realizable. *At present work on an entire Li-Fi attocell networking solution is going on.* Li-Fi Technology will find huge applications in various industries, offices, hospitals and even in homes for high speed data transfer because of its safety features and its intelligent use of the visible light spectrum which otherwise is used in most cases to serve the purpose of illumination.

References

1. http://seminarprojects.com/s/seminar-report-on-lifi
2. Goyal M, Saproo D, Bhagashra A (2013) New Epoch of wireless communication: light fidelity. Int J Innovative Res Comput Commun Eng 1(2), April 2013
3. Thomson I (2013) Forget Wi-Fi, boffins' get 150Mbps Li-Fi connection from a light bulb: many (Chinese) hands make light work. The Register. Retrieved Oct 2013
4. Haas H (2011) Wireless data from every light bulb. TED Global. Edinburgh, Scotland
5. http://en.wikipedia.org/wiki/Li-Fi
6. www.lificonsortium.org
7. Visible-light communication: tripping the light fantastic: a fast and cheap optical version of Wi-Fi is coming. The Economist. Jan 2012. Retrieved Oct 2013
8. Lim I (2011) Li-Fi—internet at the speed of light. The gadgeteer, Aug 2011
9. LI-FI the latest technology in wireless IJCCE. 2(3), july 2012
10. http://dvice.com/archives/2012/08/lifi-ten-ways-i.php
11. http://beyondweblogs.com/what-is-li-fi-is-this-replacing-Wi-Fi
12. Smith T (2012) WTF is... Li-Fi? Optical data transfer's new leading light? The Register. Retrieved Oct 2013
13. Tsonev D, Sinanovic S, Haas H (2013) An IEEE standard for visible light communicationsvisiblelightcomm.com, dated April 2011. Complete modeling of nonlinear distortion in OFDM—based optical wireless communication. IEEE J Lightwave Technol 31 (18):3064–3076. doi:10.1109/JLT.2013.2278675

Part XIII
Optical Design

Chapter 60
Sub Aperture Polishing of Fused Silica Aspheric Surface Using Dwell Time Approach

Neeraj Pandey, A. Kumar, K.K. Pant, Vinod Kumar and A. Ghosh

Abstract Aspheric surfaces are the desired surfaces for an optical designer due to their ability to control aberration effects using less number of surfaces. Traditional full aperture polishing methods are not suitable for aspheric surface polishing. The present paper describes the sub aperture polishing of aspheric surface in Fused Silica using dwell time approach. The linear feed rate has been optimized to produce optical quality surface finish using less number of iterations. The surface characterization is done using stylus profilometer and optical profilometer.

60.1 Introduction

Spherical surfaces are not usually the optimal shape for lens surfaces or for mirrors for most optical applications. Optical systems can benefit tremendously if they can use aspherical surfaces—surfaces that are not spherical. The use of aspheres allows better quality images with the use of fewer elements. The application of aspheric surfaces is limited to a tiny fraction of optics because of the difficulty in fabricating and testing these surfaces. Optical designers always require less number of surfaces with more degree of freedom for removing the aberrations. Using aspheric surfaces, optical designer can design an optical system with less number of elements, small optical system and improved performance [1]. The sag equation of an aspheric surface is as following:

$$z = \frac{ch^2}{1 + \sqrt{1 - (1+k)c^2h^2}} + Ah^4 + Bh^6 + Ch^8 + Dh^{10}\ldots \quad (60.1)$$

where, c is the curvature = ($1/R$), k denotes the conic constant and A, B, C, D are 4th, 6th, 8th and 10th order deformation coefficients respectively.

N. Pandey (✉) · A. Kumar · K.K. Pant · V. Kumar · A. Ghosh
Instruments Research and Development Establishment, Dehradun 248008, India
e-mail: neerajpandey@irde.drdo.in; neerajspn@gmail.com

V. Lakshminarayanan and I. Bhattacharya (eds.), *Advances in Optical Science and Engineering*, Springer Proceedings in Physics 166,
DOI 10.1007/978-81-322-2367-2_60

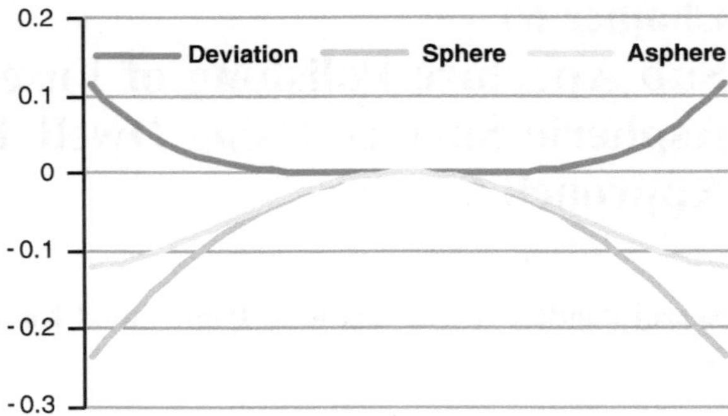

Fig. 60.1 Sag profile of fused silica asphere

In this paper, we present the polishing of Fused Silica aspheric surface using sub aperture polishing technique. A convex aspheric surface in Fused Silica material having base radius of curvature 73.117 mm and diameter 12 mm has been polished using sub aperture polishing technique.

The component details are as following:

Component Parameters

- Material—Fused Silica
- Final edge diameter = 12.000 mm
- Spherical radius $R1$ = 10.711 mm Convex
- Aspheric base radius $R2$ = 73.117 mm Convex
- Conic constant $k = 0$
- Aspheric terms:

$$A = 0.103578\text{e}{-03}, \quad B = -0.213284\text{e}{-06}$$

Figure 60.1 shows the sag profile analysis. The red curve shows the sphere, green shows the asphere and blue shows the departure of asphere from base sphere. The maximum departure at the edge is 50 μm. The sag analysis shows that the surface is convex at the center while it is being concave at the edges. The curvature is changing form center to edge and there is inflection point. This behavior of asphere creates problem in polishing as well as in testing of surface.

60.2 Sub Aperture Polishing Technique

Full aperture technique is best suited for spherical optics, while an aspheric surface has zonal departures in various zones. A full aperture polishing tool cannot correct aspheric surface therefore using a smaller sub aperture tool, we will get finer corrections over the various zones [2].

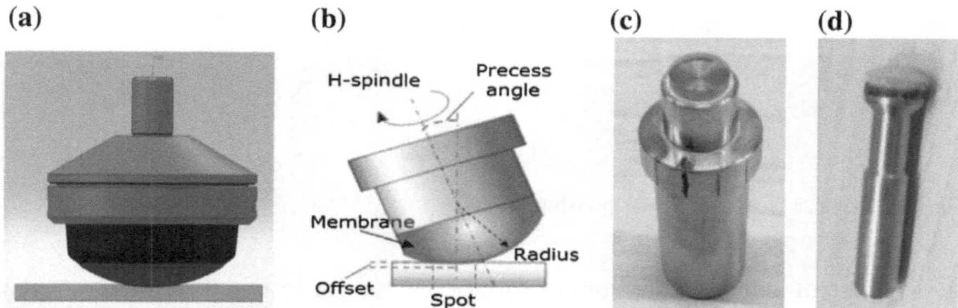

Fig. 60.2 **a** Sub aperture tool. **b** Sub aperture tool geometry. **c** Mounting tool. **d** Polishing tool

The spot size of a membrane based sub aperture tool can be controlled by the application of pressure. We used polyurethane (LP66) and Uninap cloth for various polishing cycles [3].

Sub aperture tool geometry is shown in Fig. 60.2b. We have developed a mounting tool in stainless steel material for easy mounting of aspheric lens and to remove centering error problem due to mounting. A sub aperture polishing tool is shown in Fig. 60.2d.

60.3 Polishing, Measurement and Successive Iterations

Material removal rate during the optical polishing can be defined by the Preston's equation [4] and it depends up on the tool pressure and relative velocity between tool and workspace.

$$\frac{\mathrm{d}m}{\mathrm{d}t} = c_{\mathrm{p}} P_0 V_{\mathrm{r}} \tag{60.2}$$

where dm/dt is the average thickness removal rate, P_0 is the applied pressure, and V_r is the average relative velocity of the polishing particle relative to the substrate. The Preston's constant (c_p) describes molecular level effects during the polishing process. The rate of removal increases linearly with pressure and velocity. By changing the relative velocity between polishing tool and substrate, one can achieve desired material removal at the substrate. The material removal can also be described in terms of linear feed rate and dwell time approach can be applied for material removal.

We have used raster polishing mode for polishing of Fused Silica aspheric surface for local corrections. In raster polishing mode, job axis (C-Axis) is stationary while tool is moving either along X-axis or Y-axis. During raster polishing, feed rate is variable as per the error profile of the surface. Hence material removal

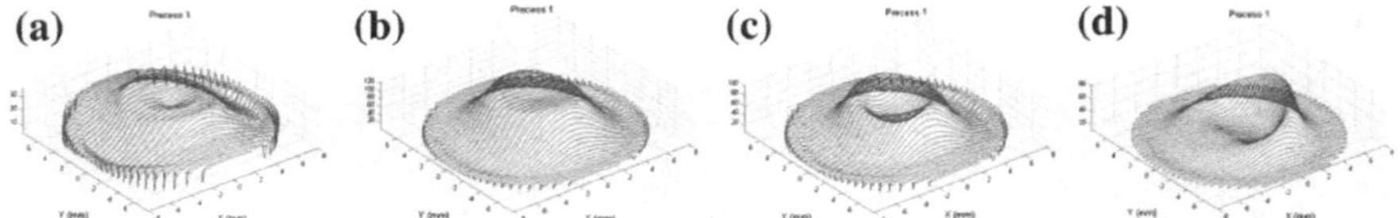

Fig. 60.3 Linear feed rate in four polishing runs

rate is different in different regions of surface and we get desired material removal at surface along different regions.

Polyurethane pad is used for pre-polishing due to fast cutting action. Raster mode has been chosen because grinded surface is very good and surface peak to valley (PV) is of the order of 1.5 μm. Surface profile measurements were done using Form Talysurf contact profilometer. Linear feed rate has been optimized as per the surface profile measurements and required corrections. Feed rate values between 20 and 80 mm/min provide good surface finish. Figure 60.3 shows the linear feed rate plots for various runs. The feed rate is moderated as per the required material removal at a particular place.

Figures 60.3 and 60.4 show the feed rate profiles and surface profiles in various polishing cycles. After first polishing cycle, the measured surface profile is shown in Fig. 60.4a. In this surface profile, more material has been removed at the central part as compared to material removed at the outer zonal part. In next polishing cycle, feed rate (Fig. 60.3b) is moderated such that high feed rate is at central part while lower feed rate at the outer zonal part. Hence material removal rate is small at the central part while it is more at the outer zonal part. Using this approach to moderate linear feed rate, the surface PV was reduced in successive polishing iterations.

Among four polishing runs, first two runs are raster with polyurethane LP 66 pad having tool radius 14 mm, while other two raster runs with Uninap cloth pad having tool radius 7 mm. In four consecutive polishing iterations, we achieve successive improvements on surface and finally we achieved surface PV 0.4 μm and roughness (Ra): 10 nm.

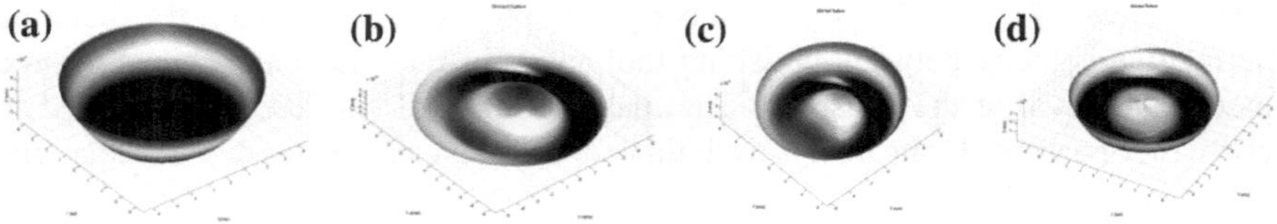

Fig. 60.4 Surface profile achieved in four polishing runs. **a** *PV* 1.3 μm, **b** *PV* 1 μm, **c** *PV* 0.8 μm, **d** *PV* 0.4 μm

60.4 Conclusion

Convex aspheric surface in Fused Silica material has been polished successfully using sub aperture polishing technique. The linear feed rate has been optimized for better results and surface accuracy. We have done four polishing runs having total polishing time 147 min. The finished surface has PV: 0.4 μm, RMS: 40 nm and Ra: 10 nm.

Acknowledgments Authors are thankful to Dr. S.S. Negi, Director, IRDE for his support and encouragement for the work. Authors are also thankful to Sh. S.P. Gaba, Scientist-'G', for his valuable suggestions.

References

1. Boothroyd G (2007) Handbook of lapping and polishing. CRC Press, Taylor & Francis Group, New York
2. Evans CJ (2003) Material removal mechanisms in lapping and polishing. CIRP Ann 52 (2):611–633
3. Bingham RG (2000) A novel automated process for aspheric surfaces. In: Proceedings of 45th annual meeting, vol 4093. Current developments in lens optical design and engineering, pp 445–448
4. Preston FW (1927) The theory and design of plate glass polishing machines. J Soc Glass Tech 11:214

Part XIV
Opto-Electronic Devices

Chapter 61
Implementation of Reed Muller Expansion Technique Using Mach-Zehnder Interferometer Based All Optical Reversible Gates

Ashis Kumar Mandal, Supriti Samanta, Goutam Kumar Maity and Nabin Baran Manik

Abstract From the last century the reversible logic has formed as an unconventional form of computing. It is relatively very new in the area of extensive applications in low power CMOS, quantum computing, digital signal processing (DSP), nanotechnology, communication, etc. In all-optical domain with the help of Feynman and Toffli gates we represent the implementation of Reed Muller transformation technique for any canonical SOP (sum of product) expression and here also we have introduced their principle of operations and used a theoretical model to complete the task. In the field of ultra-fast all-optical signa processing Mach–Zehnder interferometer (MZI), semiconductor optical amplifier (SOA)-based, has an important function. The different logical (realization of Boolean function) functions can be executed by this method with Feynman and Toffoli gates in the domain of reversible logic-based information processing.

A.K. Mandal (✉) · N.B. Manik
Department of Physics, Jadavpur University, Kolkata, West Bengal, India
e-mail: ashiskumarmandal7@gmail.com

N.B. Manik
e-mail: nabin.b.manik@gmail.com

S. Samanta
Department of Physics, Chaipat Girls' High School, Daspur, West Bengal, India
e-mail: supritisamanta28@gmail.com

G.K. Maity
Electronics and Communication Engineering, MCKV Institute of Engineering, Howrah, India
e-mail: goutam123_2005@yahoo.co.in

V. Lakshminarayanan and I. Bhattacharya (eds.), *Advances in Optical Science and Engineering*, Springer Proceedings in Physics 166,
DOI 10.1007/978-81-322-2367-2_61

61.1 Introduction

The logic blocks (gates like AND, OR, NAND) in our traditional computer are irreversible. After generating the output bits the blocks lose the input bits. The inputs of reversible gates are one-to-one correspondence with outputs. So, we can determine the inputs of such gates from its outputs. The computation with the reversible logic circuits gives us arbitrary small energy dissipation [1]. But, in conventional irreversible hardware computation does in energy dissipation due to information loss [2]. Many optical logic gates have been suggested to perform irreversible logic function [3]. A gate is universal if any logical reversible circuit can be designed using these gates. This paper presents a circuit for realization of universal FG and TG gates in all-optical domain. This paper is progressed in such way that in Sect. 61.2 Mach–Zehnder interferometer (MZI)-based optical switch is interpreted. All-optical circuit of MZI-based reversible FG and TG gates are discussed in Sects. 61.3 and 61.4. Reed-Muller theory and one solution of a problem are explained in Sects. 61.5 and 61.6. Corresponding simulation (by Matlab-7.0) results confirm gate's properties are also attached in Sect. 61.7 of this paper.

61.2 MZI-Based All-Optical Switch

When the incoming signal is to be switched into the interferometer, it is split between the arms of the interferometer. The incoming signal emerges from crossbar port in the absence of a control signal. The presence of control pulse changes the refractive index of the medium given by $\Delta n = n_2 I$, here, Δn is the change in the refractive index of the medium, n_2 the nonlinear refractive coefficient and I the intensity of the light incident on the medium. The incoming signal is switched over to bar port due to a change in the index adds a phase shift between the two arms of the interferometer. This switching process is based on cross phase modulation (XPM). The XPM used SOA-MZIis the great important interferometric structure due to its low-energy requirement, simplicity, compactness and stability [4–7]. Symmetric MZI-SOA all-optical switch is shown in Fig. 61.1a, b. Two semiconductor optical amplifiers (SOA-1 and SOA-2) are inserted in each arm of a MZ interferometer. It has two input ports (port-1 and port-2) and two output ports (port-3 or Bar-port and port-4 or Cross-port).

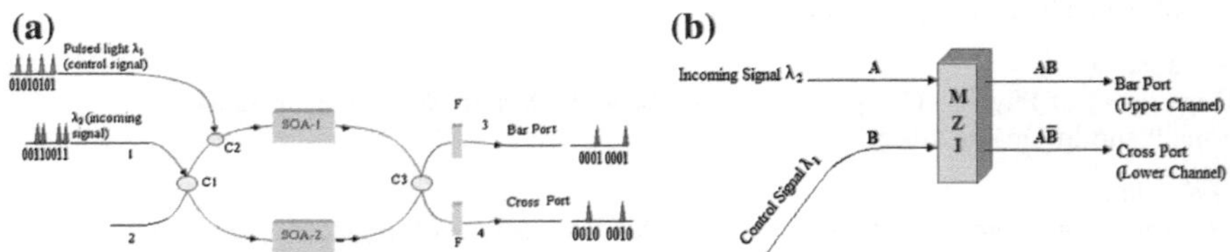

Fig. 61.1 **a** Symmetric MZI-SOA all-optical switch. **b** Schematic block diagram of MZI

The incoming signal pulse of wavelength λ_2 enters through port-1, is divided equally by the coupler C1 (50:50) and propagates simultaneously in the two arms and port-2 is kept open. At that time, through coupler C2 a pulsed signal of wavelength λ_1 enters to the upper arm in such a way that most power goes through upper arm. This pulsed signal saturates the SOA-1 and changes its refractive index, while the SOA-2 remains the unsaturated gain state. As a consequence, a differential phase shift can be gained between the data signal of two arms. So, light is present in the port-3 (bar port), as shown in Fig. 61.1a and no light is present in the port-4 (cross port). This is known as 'switched state'. Both SOA (SOA-1 and SOA-2) get the same unsaturated gain when control signal is absent. Then no light is present in the port-3 (bar port) but light is present in the port-4 (cross port). This is called 'no-switched state'. Optical filters (F) are inserted in front of the output ports for blocking the λ_1 signal (control signal). The MZ scheme is desirable for cross-gain saturation as it does not reverse the bit pattern and results in a higher on–off contrast because nothing exits from bar port during 0 bits. Schematic block diagram of MZI is shown in Fig. 61.1b.

61.3 MZI-Based Feynman Gate (FG)

Figure 61.2a (Schematic diagram), b (MZI-based circuit) show the Feynman gate. It is a 2*2 gate and is also called as Controlled NOT. It is widely used for fan-out purposes. This gate has the mapping of inputs (A, B) to outputs $P = A$, $Q = A \oplus B$. It can be formed using 2 MZI based all optical switch, 2 Beam Combiner and 3 Beam Splitter in all optical reversible computing. This Feynman gate can be constructed using 2 MZI based optical switches thus the optical cost of Feynman gate is considered as 2 and delay of the optical Feynman gate is considered as 1Δ (Table 61.1).

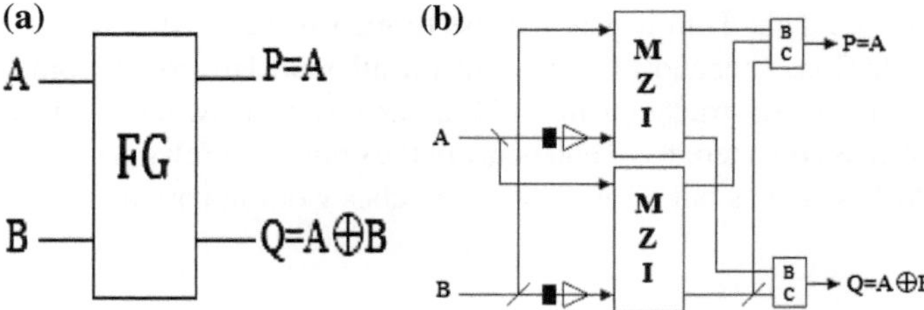

Fig. 61.2 **a** Schematic diagram. **b** MZI-based circuit. BC Beam Combiner, / *BS* Beam Splitter, ▷ *EDFA* Erbium Doped Fiber Amplifier, ■ *WC* Wavelength Converter

Table 61.1 Truth table of FG

A	*B*	*P*	*Q*
0	0	0	0
0	1	0	1
1	0	1	1
1	1	1	0

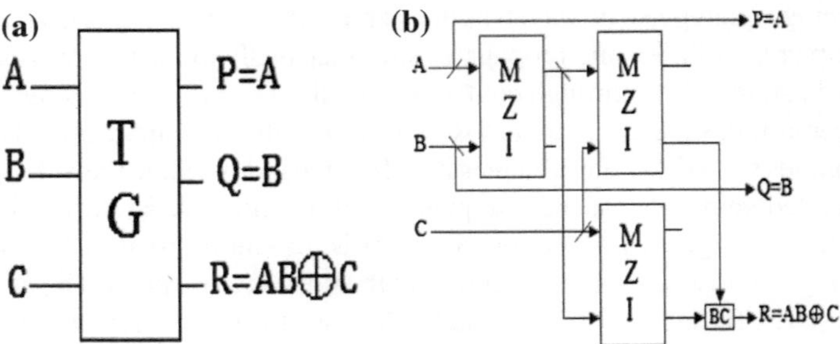

Fig. 61.3 **a** Schematic diagram. **b** MZI-based circuit. BC Beam combiner, / *BS* Beam splitter, ▷ *EDFA* Erbium Doped Fiber Amplifier, ■ *WC* Wavelength Converter

Table 61.2 Truth table of TG

A	*B*	*C*	*P*	*Q*	*R*
0	0	0	0	0	0
0	0	1	0	0	1
0	1	0	0	1	0
0	1	1	0	1	1
1	0	0	1	0	0
1	0	1	1	0	1
1	1	0	1	1	1
1	1	1	1	1	0

61.4 MZI-Based Toffoli Gate (TG)

Figure 61.3a (Schematic diagram), b (MZI-based circuit) show the Toffoli gate. It is a 3*3 gate and it has the mapping of inputs (*A*, *B*, *C*) to outputs $P = A$, $Q = B$, and $R = AB \oplus C$. This Toffoli gate can be formed using 3 MZI based all optical switch, 1 Beam Combiner and 4 Beam Splitter in all optical reversible computing. The Toffoli gate can be constructed using 3 MZI based optical switches thus the optical cost of this gate is considered as 3 and delay of the optical Toffoli gate is considered as 2Δ as two MZI switches out of three MZI switches work in parallel (Table 61.2).

61.5 Transformation of Any Canonical SOP Expression into an Expression Containing 'XOR' and 'AND' Operators Using Reed Muller Expression

The Reed-Muller expansion technique which will be discussed in this paper can be used very easily for this purpose. It can be written down the XOR-AND expression just by looking at the canonical SOP (sum of product) form of switching expression. This is true also for the reverse transformation.

The canonical SOP form of any N-variable function can be written as

$$F(X_0, X_1, \ldots, X_{N-1}) = a_0 m_0 + a_1 m_1 + \cdots + a^N_{2-1} m^N_{2-1}$$

where all $m_i (i = 0 \text{ to } 2^N - 1)$ are the minters. If $a_i = 1$ then m_i is present and if $a_i = 0$ then m_i is absent in expression. The variables within minters appear either in true form or in the complement form. The expression which will be obtained after simplification will be called a positive or true polarity Reed-Muller expression because the variables appear in the true form only. Now a 3-variable function *F*(*A*, *B*, *C*) which is written in canonical SOP form can be transformed into positive XOR-AND form using the following procedure:

$$\begin{aligned} F(A, B, C) &= a_0 m_0 + a_1 m_1 + a_2 m_2 + a_3 m_3 + a_4 m_4 + a_5 m_5 + a_6 m_6 + a_7 m_7 \\ &= a_0 (\overline{A}\,\overline{B}\,\overline{C}) \oplus a_1 (\overline{A}\,\overline{B} C) \oplus a_2 (\overline{A} B \overline{C}) \oplus a_3 (\overline{A} BC) \oplus a_4 (A \overline{B}\,\overline{C}) \\ &\quad \oplus a_5 (A \overline{B} C) \oplus a_6 (AB\overline{C}) \oplus a_7 (ABC) \end{aligned}$$

Using the property $m_i + m_j = m_i \oplus m_j$ for $i \neq j$.

$$= a_0(1 \oplus A)(1 \oplus B)(1 \oplus C) \oplus a_1(1 \oplus A)(1 \oplus B)C \oplus a_2(1 \oplus A)B(1 \oplus C) \oplus a_3(1 \oplus A) BC \oplus a_4 A(1 \oplus B)(1 \oplus C) \oplus a_5 A(1 \oplus B)C \oplus a_6 AB\,(1 \oplus C) \oplus a_7(ABC)$$

Replacing any complement variable X by $1 \oplus X$

$$= b_0 \oplus b_1 C \oplus b_2 B \oplus b_3 BC \oplus b_4 A \oplus b_5 AC \oplus b_6 AB \oplus b_7 ABC$$

After simplification where

$$\begin{aligned} b_0 &= a_0 \\ b_1 &= a_1 \oplus a_0 \\ b_2 &= a_2 \oplus a_0 \\ b_3 &= a_3 \oplus a_2 \oplus a_1 \oplus a_0 \\ b_4 &= a_4 \oplus a_0 \\ b_5 &= a_5 \oplus a_4 \oplus a_1 \oplus a_0 \\ b_6 &= a_6 \oplus a_4 \oplus a_2 \oplus a_0 \\ b_7 &= a_7 \oplus a_6 \oplus a_5 \oplus a_4 \oplus a_3 \oplus a_2 \oplus a_1 \oplus a_0 \end{aligned}$$

The coefficients b_0, b_1, b_2, … etc. may have values 0 or 1.
They indicate which of the product terms are absent or present (Fig. 61.4).

Fig. 61.4 Ladder of n-input TG (n-input AND)

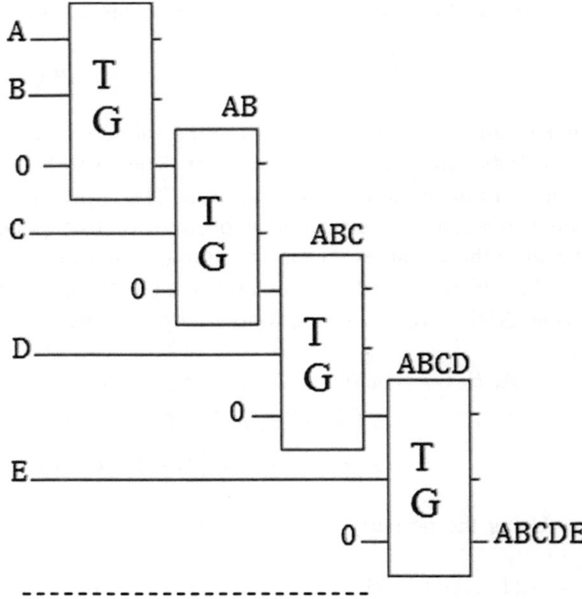

61.6 Transformation of the Function $F = \sum m\ (0, 2, 3, 5, 7)$ into XOR-AND Expression Using Reed Muller Expression Technique and Drawing of the Circuit Using FG-TG Gate Only (as Example of Problem)

The above function can be written as

$$F(X_2, X_1, X_0) = a_0m_0 + a_2m_2 + a_3m_3 + a_5m_5 + a_7m_7$$

where $a_0 = a_2 = a_3 = a_5 = a_7 = 1$ but $a_1 = a_4 = a_6 = 0$

The general XOR-AND expression for any 3 variable canonical SOP expression is

$$F(X_2, X_1, X_0) = b_0 \oplus b_1X_0 \oplus b_2X_1 \oplus b_3X_1X_0 \oplus b_4X_2 \oplus b_5X_2X_0 \oplus b_6X_2X_1 \oplus b_7X_2X_1X_0$$

where, the b-coefficients can be obtained from the a-coefficients using the triangle shown in Fig. 61.5. From this triangle we get,

$$b_0 = b_1 = b_3 = b_4 = b_7 = 1$$
$$b_2 = b_5 = b_6 = 0$$

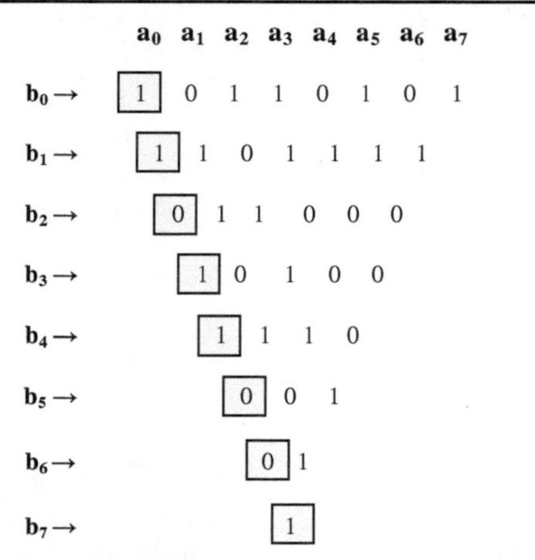

Fig. 61.5 The b-coefficients can be obtained from the a-coefficients using the triangle

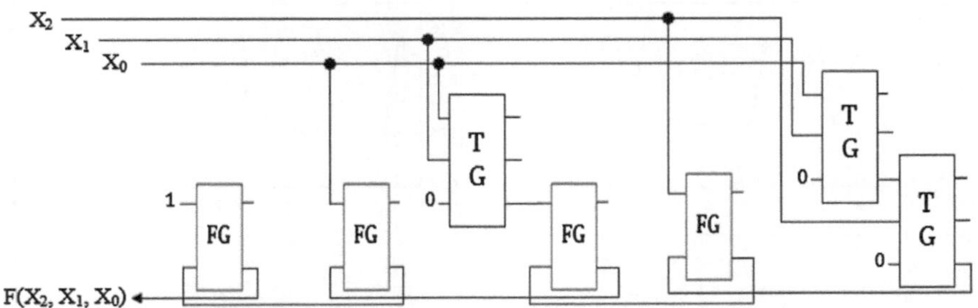

Fig. 61.6 Implemented circuit

Therefore, the canonical XOR-AND expression for the given function is

$$F(X_2, X_1, X_0) = 1 \oplus X_0 \oplus X_1X_0 \oplus X_2 \oplus X_2X_1X_0$$

The implemented circuit is shown in Fig. 61.6 with FG, TG and n-input TG (like n-input AND gate) (Table 61.3).

61.7 Simulation and Result

Here, '1' state refers the presence of light and '0' state refers absence of light. Simulation is done using Matlab-7. The vertical axis in Fig. 61.7a–e indicates power in dBm, while horizontal axis represents time scale in ps.

Table 61.3 Truth table of Fig. 61.6

X_2	X_1	X_0	$F(X_2, X_1, X_0)$
0	0	0	1
0	0	1	0
0	1	0	1
0	1	1	1
1	0	0	0
1	0	1	1
1	1	0	0
1	1	1	1

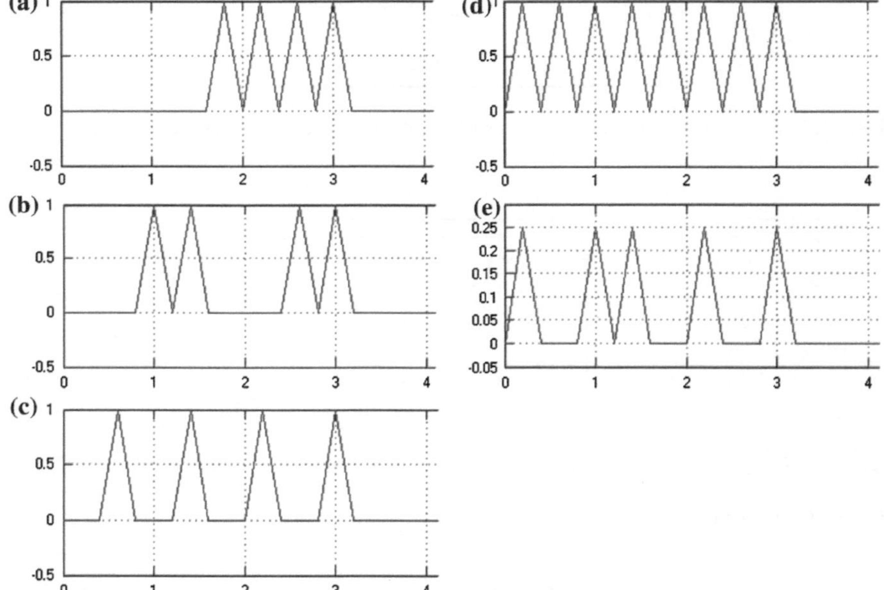

Fig. 61.7 **a** Input X_2, **b** Input X_1, **c** Input X_0, **d** Input '1', **e** Output $F(X_2, X_1, X_0)$

The timing instant of input X_2, X_1, X_0 and '1' of Fig. 61.6 for the occurrence of bit patterns are at 0.2, 0.6, 1.0, 1.4, 1.8, 2.2, 2.6, 3.0 ps. Upper four set wave-forms of Fig. 61.7a–e indicate the input bit sequences 00001111, 00110011, 01010101, 11111111 for the input variables X_2, X_1, X_0 and constant '1' respectively. And, the Fig. 61.7e wave-form indicates output bit sequences 10110101.

61.8 Conclusion

The all-optical scheme of Reed-Muller expansion technique is proposed and explained. It is important that the above explanations are based on simple model. Here, we have used MZI as it is often used in practice due to it can be easily integrated by using SiO_2/Si or InGaAsP/InP waveguides, resulting in a compact device [8]. Different logic functions in reversible system can easily be transformed into XOR-AND expression using Reed-Muller expansion technique and constructed circuit with these gates (FG and TG) in future.

References

1. Bennett CH (1973) Logical reversibility of computation. IBM J Res Dev 17:525–532
2. Landauer R (1961) Irreversibility and heat generation in the computational process. IBM J Res Dev 5:183–191
3. Kim J-Y, Kang J-M, Kim T-Y, Han S-K (2006) All-optical multiple logic gates with XOR, NOR, OR, and NAND functions using parallel SOA-MZI structures: theory and experiment. J Light Wave Technol IEEE 24(9):3392–3399
4. Agrwal GP (2001) Applications of nonlinear fibre optics. Academic Press, India [(An imprint of Elsevier, San Diego, USA) Chap. 3–4]
5. Schreieck R, Kwakernaak M, Jackel H, Gamper E, Gini E, Vogt W et al (2001) Ultafast switching dynamics of Mach-Zehnder interferometer switches. IEEE Photonics Technol Lett 13 (6):603–605
6. Agrwal GP (2002) Fibre-optic communication systems, 3rd edn. Wiley, Inc Publication, New York [Chap. 8 (WDM Components)]
7. Mandal AK, Maity GK (2014) An all-optical new universal gate using Mach-Zehnder interferometer. In: 2014 IEEE, sixth international conference on computational intelligence and communication networks (ICCICN 2014), pp 1044–1048 (978-1-4799-6929-6/14)
8. Leuthold J, Besse PA, Eckner J, Gamper E, D¨ ulk M, Fischer S et al (1999) All-optical Mach-Zehnder interferometer wavelength converters and switches with integrated data and control-signal separation scheme. J Lightwave Technol 17(6):1056–1066

Chapter 62
Design and Simulation of 1 × 4 Demultiplexer by Using 2D-Photonic Crystal Ring Resonator for ITU-T G.692.2 S+C Band CWDM System

Mayur K. Chhipa

Abstract In this paper, heterostructure-based demultiplexer (De-MUX) is designed for ITU-T G.694.2 Coarse Wavelength Division Multiplexing S+C Band system using Two-Dimensional (2D) Photonic Crystal Ring Resonator (PCRR) which is useful for Photonic-Integrated Circuits and designing in future optical networks.

62.1 Introduction

Photonic crystals (PCs) are artificial dielectric or metallic structures in which the refractive index changes periodically. Photonic crystals have attracted much interest due to their novel electromagnetic wave characteristics and important scientific and engineering applications. PhCs may inhibit the propagation of certain range of wavelengths in either one direction or in all directions, providing the possibility to confine and trap the light in a cage [1]. Their effect to the propagation of electromagnetic waves is similar to the periodic potential in a semiconductor crystal. Necessarily, photonic crystals contain internal regions of high and low dielectric constant that are repeated regularly. From the communication capacity point of view, the electron-based conventional communication system has shown the physical limitation due to a rapid growth of the internet and multimedia. If the information is processed as an optical signal itself, not an electrical signal, then the information processing speed will be increased and it can provide much more convenience. Therefore, the optical communication system will be expected that it will lead the future communication system by using a photonic device [2].

M.K. Chhipa (✉)
Department of Electronics and Communication Engineering,
Government Engineering College, Ajmer 305001, Rajasthan, India
e-mail: mayurchhipa1@gmail.com

V. Lakshminarayanan and I. Bhattacharya (eds.), *Advances in Optical Science and Engineering*, Springer Proceedings in Physics 166,
DOI 10.1007/978-81-322-2367-2_62

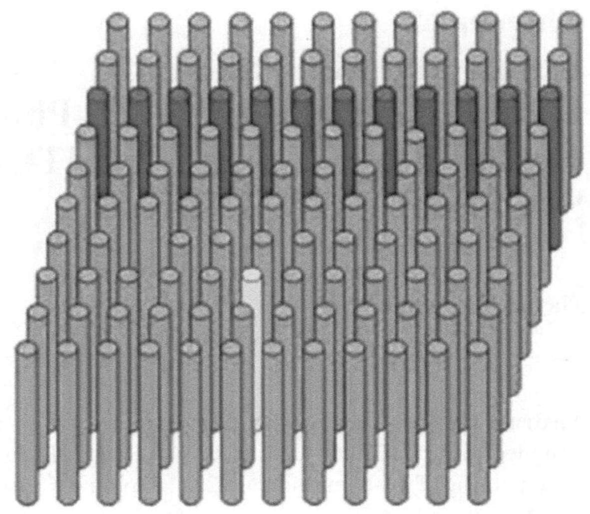

Fig. 62.1 2D-Photonic crystals structure with dielectric material rods in the air background

In networking, it is used to describe a device that receives a transmission of several signals over a single line and can properly decode the single line signal into multiple signals. The equipment used to demultiplex the signals into several sources which were multiplexed before is called the demultiplexer. Generally, demultiplexing components are realized using arrayed waveguide gratings, thin-film filters, or fiber Bragg gratings (FBG) [3]. Various concepts for realizing a demultiplexing component utilizing the extraordinary properties of PhCs have recently been proposed. These ideas are optical microcavities, directional couplers, multimode self-imaging waveguides, and super-prisms [4] (Fig. 62.1).

In the literature, Two-Dimensional (2D) PC-based ADFs are designed and analyzed using square lattice and triangular lattice. Among these, 2D square lattice PC-based WDDM provides effective confinement of light can easily control the propagation modes, easy fabrication, and simple structure than others. So far, the square lattice PC-based WDDM is designed by introducing point defects and/or line defects [5].

62.2 Simulation and Design

The Plane Wave Expansion (PWE) method is used to investigate the Photonic Band Gap (PBG) of photonic crystal structure to obtain a band diagram. The band diagram of structure is shown in Fig. 62.2 which has PBG for Transverse Electric (TE) mode. Hence, we are restricted for TE PBG, whose electric field is parallel to the rod axis. The PBG range for the structure is 0.57752 1/λ–0.823551 1/λ whose respective wavelength range extend from 1213 to 1714 nm where as PBG range of Region2 is 0.567149 1/λ–0.818245 1/λ whose respective wavelength ranges extends from 1241

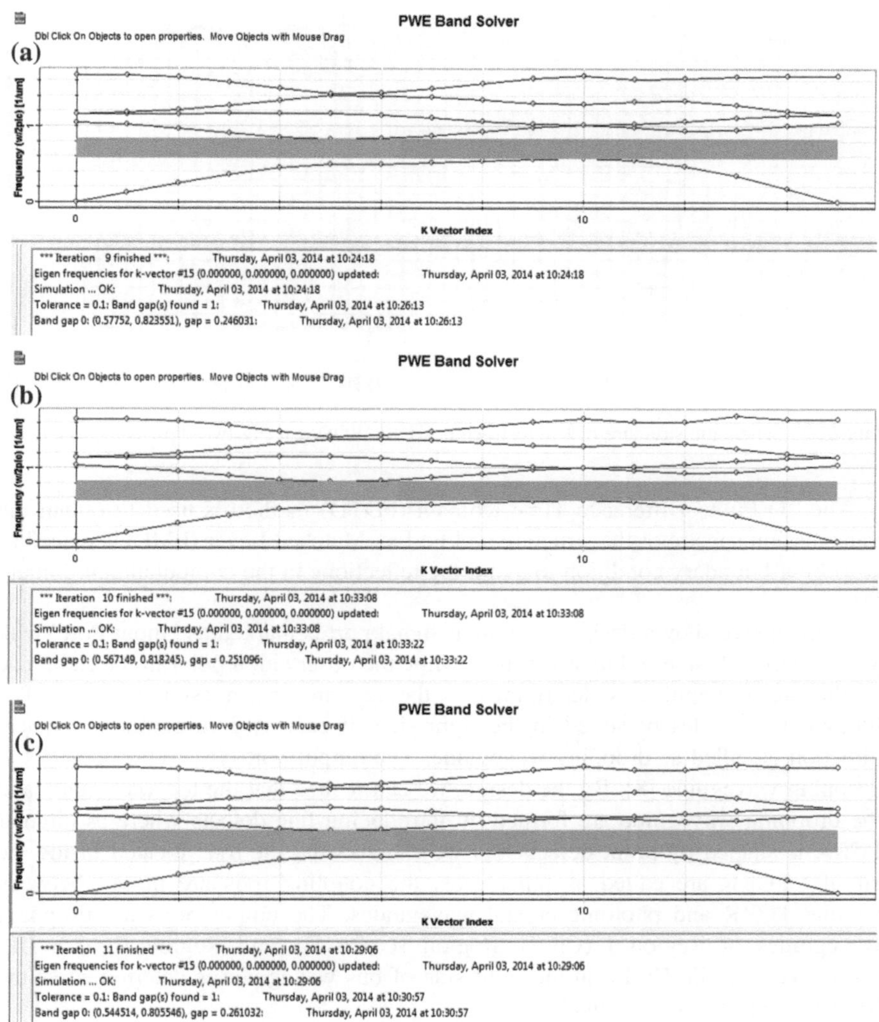

Fig. 62.2 Band gap diagram of 18 × 2I PC square lattice structure for the **a** Region 1, **b** Region 2, and **c** Region 3 (without introducing any defects)

to 1830 nm, and PBG range of Region3 is 0.544514 1/λ–0.805546 1/λ whose respective wavelength ranges 1263–1862 nm extends from, respectively.

The schematic structure of heterostructure-based WDDM using circular PCRR is shown in Fig. 62.3. The number of rods in *Z* and *X* direction is 21 and 54. The distance 0.540 nm is in between two adjacent rods, which is known as lattice constant and the radius of the Si rod is 0.1 μm. The structure is having refractive indices as 3.255 for first, 3.32 for second, and 3.47 for third region, respectively, refractive index of silicon dielectric material [6].

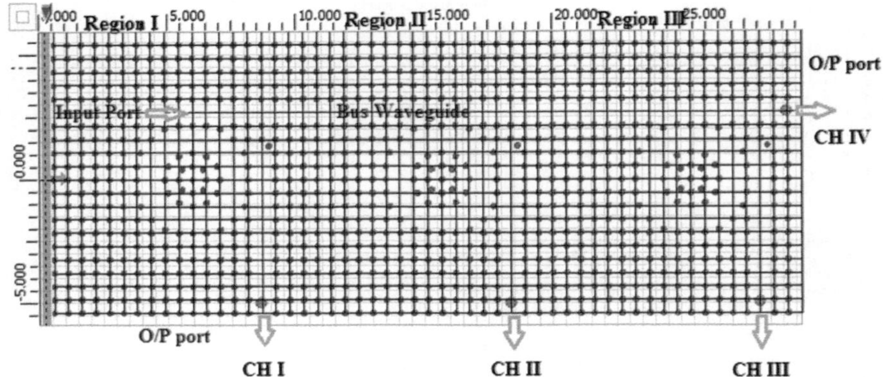

Fig. 62.3 Schematic structure of heterostructure-based WDDM using two-dimensional PCRR

The 2D Finite Difference Time Domain (FDTD) method is used to obtain the transmission spectra of the structure and Perfect Matched Layer (PML) is placed as absorbing boundary condition to avoid the reflections in the computational domain [7].

In the above shown designs, the input signal port is marked as "Input Port" with arrow in the left side of bus waveguide while the observation point in the right side of the bus waveguide is determined as the forward transmission terminal. The bottom waveguides poisoned in the right side of each ring resonator in vertical direction is called as dropping waveguide. The output port observation point is at dropping waveguide [8]. By the design, we can realize that the top waveguide and the dropping waveguide are formed by introducing line defects where as circular PCRR is shaped by point defects. As we can see that the rods located inside the circular PCRR are called as inner rods, the coupling rods are placed between circular PCRR and photonic crystal waveguides. The output ports are dropping waveguides of Region I (Ch I), Region II (Ch II), and Region III (Ch III), respectively, while Ch IV in the right side of bus waveguide is determined as the forward transmission terminal.

62.3 Simulation and Results

A Gaussian modulated continuous wave input signal, covering whole frequency or wavelength range of interest is placed into the input port [7]. The Discrete Fourier Transform (DFT) of the fields that are being calculated by 2D-FDTD method to measure the transmission spectra at output dropping waveguide at observation point.

Figure 62.4 shows the transmission spectra of heterostructure-based WDDM using circular PCRR [8] with the dropping efficiency about 88, 78, 72, and 98 %.

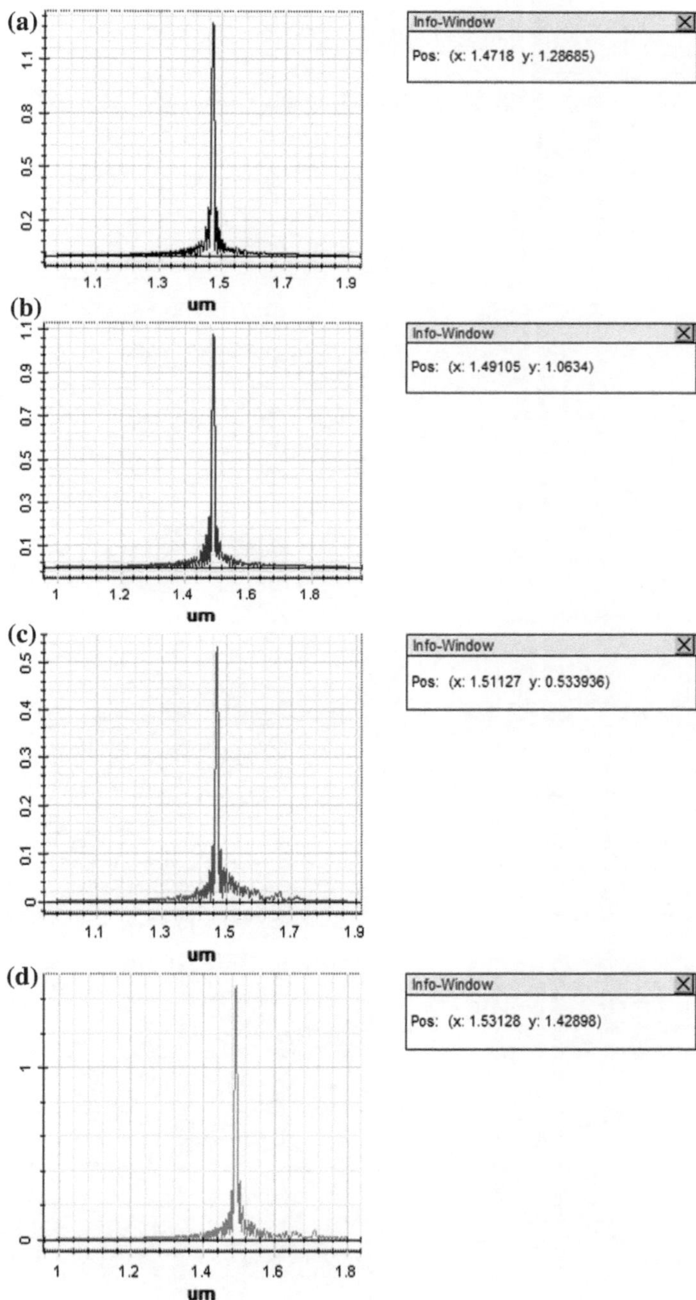

Fig. 62.4 Dropped wavelength at **a** Ch I 1471 nm, **b** Ch II 1491 nm, **c** Ch III 1511 nm, and **d** Ch IV 1531 nm of heterostructre-based WDDM

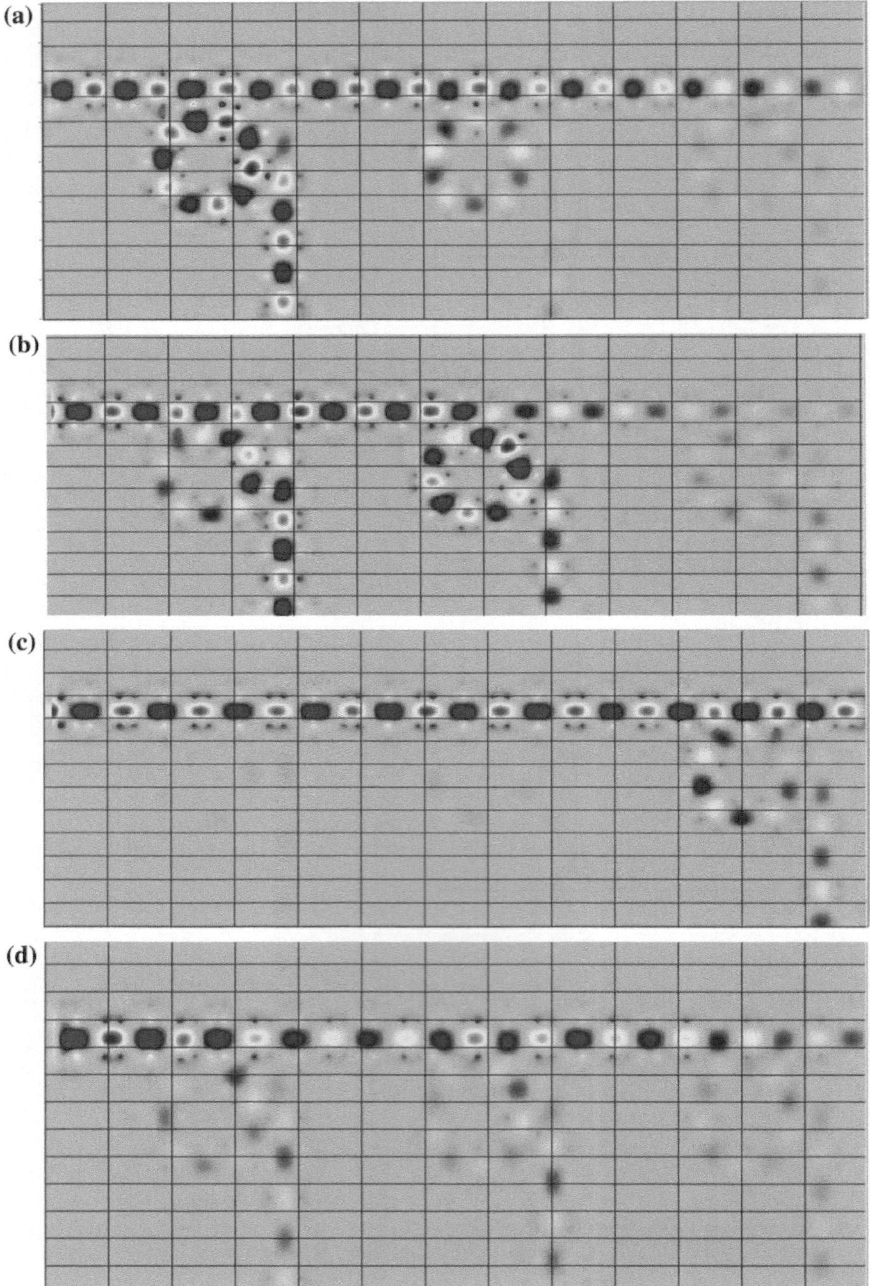

Fig. 62.5 Electric field pattern of the circular PCRR-based WDDM at **a** 1471 nm, **b** 1491 nm, **c** 1511 nm, and **d** 1531 nm wavelength

Figure 62.4 depicts the electric field pattern of "on" resonance at 1471, 1491, 1511, and 1531 nm wavelength. The coupling management at ports "Ch I", "Ch II", "Ch III", and "Ch IV" are obtained by conducting discrete fourier transform (DFT) of the fields that are calculated by 2D-FDTD method and the dropped wavelength could be shown at observation point 1 for 1471 nm, observation point 2 for 1491 nm and similarly observation point 3 for 1511 nm wavelength as shown in Fig. 62.4a–d of 1531 nm wavelength. The input and output signal power is recorded by observation points which are positioned at the input and output ports.

Figure 62.5a–d depicts the electric field pattern of "on" resonance at 1471, 1491, 1511, and 1531 nm, respectively.

62.4 Conclusion

The heterostructure-based WDDM is designed to guide four channels centered at 1471, 1491, 1511, and 1531 nm using three circular photonic crystal ring resonators. The designed structure is constructed by 2D pillar type photonic crystal in square lattice with three different refractive indices. The simulation shows 88, 78, and 72 % dropping efficiency of channel I, channel II, and channel III and 98 % at channel IV. All the simulation and design work is done on OptiFDTD licensed version software.

Hence, such kind of devices would be more useful for the realization of integrated optic circuits for CWDM/DWDM systems and, future access and metro networking applications.

References

1. John S (1987) Strong localization of photons on certain disordered dielectric super lattices. Phys Rev Lett 58:2486–2488
2. Joannopoulos JD, Meade RD, Winn JN (1992) Photonic crystals—molding the flow of light, 2nd edn, pp 2–5
3. ITU-T Recommendation G.694.2. Spectral grids for WDM applications: CWDM wavelength grid
4. Ghorbanpour H, Makouei S (2013) 2-channel all optical demultiplexer based on photonic crystal ring resonator. Front Optoelectron 6(2):224–227
5. Wang C-C, Chen L-W (2010) Channel drop filters with folded directional couplers in two-dimensional photonic crystals. Phys B 405:1210–1215
6. Selim R, Pinto D, Obayya SSA (2011) Novel fast photonic crystal multiplexer-demultiplexer switches. J Opt Quantum Electron 42(8):425–433
7. Qiu M, Jaskorzynska B (2003) Design of a channel drop filter in a two-dimensional triangular photonic crystal. IEEE J Appl Phys Lett 83(6):1074–1076
8. Robinson S, Nakkeeran R (2011) Add drop filter for ITU.T G.694.2 CWDM systems. In: Proceedings of 3rd IEEE conference ICECT

Chapter 63
Design of Tunable Wavelength Demultiplexer for DWDM Application Based on 1-D Photonic Crystal with KTP Defect

Sanjeev K. Srivastava, Raj Kumar Tomar, Sanjay Srivastava and S.P. Ojha

Abstract We theoretically investigate the tunable wavelength selective filter by using one-dimensional photonic crystal (1D PC) with potassium titanyl phosphate (KTP) as a defect layer For this we study the transmission characteristic of the multilayered structure composed of $(MgF_2/ZnS)^N(KTP)/(MgF_2/ZnS)^N$. In order to get the transmittance spectra of defect PC we apply transfer matrix method (TMM). We use KTP crystal as defect layer because it is an electro-optical material and its refractive index changes with the change in the electric field intensity. Due to change in the refractive index, defect modes of the PC structure can be tuned at the desired wavelength. From the analysis of the transmittance spectra it is found that the average change in the central wavelength of defect mode is 0.6 nm/ 0.01×10^{10} V/m. This property of tunable PC filter can be exploited in designing a wavelength demultiplexer for DWDM application in optical communication.

63.1 Introduction

During the last twenty-five years the optical devices based on photonic crystals have generated great interest. Photonic crystals (PCs) are new class of optical material having alternate layers of low and high refractive index materials with

S.K. Srivastava (✉) · R.K. Tomar
Department of Physics, Amity Institute of Applied Sciences,
Amity University, Noida, India
e-mail: sanjeev17th@yahoo.co.in

S. Srivastava
Department of Physics, MMV, Banaras Hindu University, Varanasi, India

S.P. Ojha
Department of Physics, Indian Institute of Technology,
Banaras Hindu University, Varanasi, India

V. Lakshminarayanan and I. Bhattacharya (eds.), *Advances in Optical Science and Engineering*, Springer Proceedings in Physics 166,
DOI 10.1007/978-81-322-2367-2_63

lattice parameters on the order of wavelength of optical wave. Due to their ability to control and manipulate the propagation of light they possess potential application in the field of photonics [1–5]. In these structures propagation of optical wave is completely prohibited in a certain range of frequency called photonic band gaps (PBGs). By introducing the defect into conventional 1d-PC, the defect mode (or tunneling mode) can be generated in the PBG. The defect in normal PC can be introduced by changing the thickness of the layer, inserting another material into the structure or removing a layer from normal PC structure. The frequency (or wavelength) of defect mode can be tuned by changing the operating temperature, by applying an external electric and/or magnetic field or by optical illuminations and leads to various photonic devices [5–10]. One of the important applications of tunable PC filter is employed in designing the dense wavelength division multiplexer DWDM system [11, 12]. DWDM is a device used to divide and combine different wavelength channels each carrying an optical data signal. In the present communication we propose the design of demultiplexer for DWDM application by using 1d-PC with KTP defect. Since KTP crystal is an electro optical material, its refractive index changes with the application of electric filed intensity. When the KTP material is used as defect layer between two Bragg mirrors one get the transmission mode at the desired wavelength.

63.2 Theoretical Development

Figure 63.1 represents the schematic view of 1D PC structure with KTP defect layer. The proposed PC structure has the form $(AB)^N D(AB)^N$ where A and B are the materials of low and high refractive indices having thickness d_1 and d_2. D represents the defect layer; N is the total number of layers.

A transmittance spectrum of the structure is obtained by applying the transfer matrix method (TMM) [13, 14]. The characteristics matrix corresponding to kth layer, for the TE (s-polarized) and TM (p-polarized) waves has the form

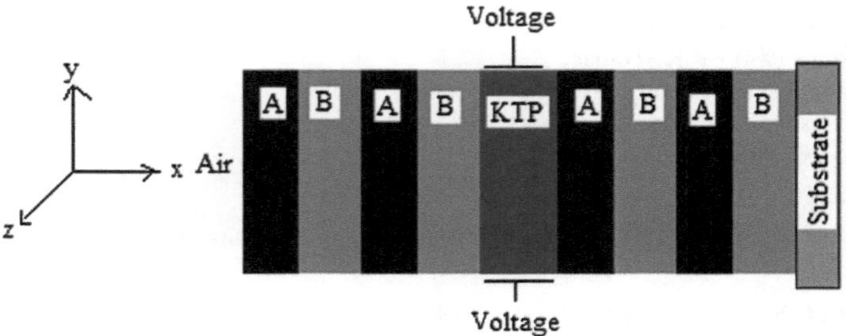

Fig. 63.1 Schematic representation of 1d PC structure with KTP defect

$$M_k = \begin{bmatrix} \cos \alpha_k & \frac{1}{iq_k} \sin \alpha_k \\ -iq_k \sin \alpha_k & \cos \alpha_k \end{bmatrix} \quad (1)$$

where $q_k = n_k \cos \theta_k$, ($k = 1, 2, 3$) for the TE wave and $q_k = \cos \theta_k / n_k$ for the TM wave, $\alpha_k = \frac{2\pi}{\lambda} n_k d_k \cos \theta_k$, θ_k is the ray angle inside the layer of refractive index n_k, d_k is the thickness of the layers and λ is the wavelength in the incidence medium. The total characteristics matrix for the N period of system is given by

$$M = (M_k)^N \quad (2)$$

The reflectance spectra of the proposed PC structure for s-polarized and p-polarized waves are given by [13, 14]

$$R(s,p) = \left| \frac{(M_{11} + q_t(s,p)M_{12})q_i(s,p) - (M_{21} + q_t(s,p)M_{22})}{(M_{11} + q_t(s,p)M_{12})q_i(s,p) + (M_{21} + q_t(s,p)M_{22})} \right|^2 \quad (3)$$

The values of q_i and q_t for s- and p-polarized waves are given as

$$q_i(s) = n_i \cos \theta_i; \quad q_t(s) = n_t \cos \theta_t \quad (4)$$

$$q_i(p) = \cos \theta_i / n_i \quad q_t(p) = \cos \theta_t / n_t \quad (5)$$

Here θ_i and θ_t angle of incidence in the incident medium and substrate having refractive indices n_i and n_t respectively. The transmittance (T) spectrum can be obtained by using the expression:

$$T(s,p) = 1 - R(s,p) \quad (6)$$

The refractive index of the KTP material exposed to radiation of intensity E at wavelength 532 nm, along the z-axis can be expressed as [15, 16]

$$n_3 = n_0 - \left(\frac{1}{2}\right) n_0^3 \gamma E, \quad (7)$$

where n_0 is the refractive index of the crystal when no intensity is applied to the crystal and has value 1.778. γ is the non-linear constant and has value in this case as 9.5×10^{-12} m/V. E is the intensity of applied electric filed and varies as $E = E_0 \sin \omega t$, here ω is the frequency of AC field applied along the z-axis [16] and its value is chosen as $\frac{\pi}{50}$ and the value of E_0 as 10^{10} V/m.

63.3 Results and Discussions

Schematic diagram of DWDM system is depicted in Fig. 63.2. Input source having ten different wavelengths is incident on PC1 at angle 15°. The first wavelength is transmitted through PC1 while the other wavelengths are reflected at incident angle 15°. The reflected wavelength in each case is incident at same angle on the next PC and this process progresses up to PC10. Each PC structure is kept on at different electric field intensity so that they can transmit a very narrow band of wavelength. The refractive indices of MgF_2 and ZnS are taken as 1.38(n_1) and 2.3(n_2) respectively and their thicknesses are taken as 280.8 nm (d_1) and 168.5 nm (d_2), which is in accordance with the quarter wave stack condition $n_1 d_1 = n_2 d_2 = \frac{\lambda_0}{4}$ where λ_0 (= 1550 nm). While the thickness of KTP layer (defect layer) is taken as 272.5 nm [0.606 of (d_1 + d_2)]. The transmittance spectra of air/$(MgF_2/ZnS)^8$(KTP)/$(MgF_2/ZnS)^7$/air at different value of electric field intensity and incidence angle 15° is shown in Fig. 63.3 for TE mode. From the analysis of the transmittance curve we observe a defect mode around the wavelength 1549.7 nm with line width of 0.3 nm. For DWDM system the width the dropping line has sufficient value. Thus, if input source having different wavelengths ranges from 1549 to 1556 nm is incident at an angle 15° on the suggested PC structure at E = 1 × 10^{10} V/m, then it will pass a single extremely narrow band of wavelength centered 1549.7 nm and all other wavelengths will be reflected. Now by changing the electric field (E.F.) intensity at

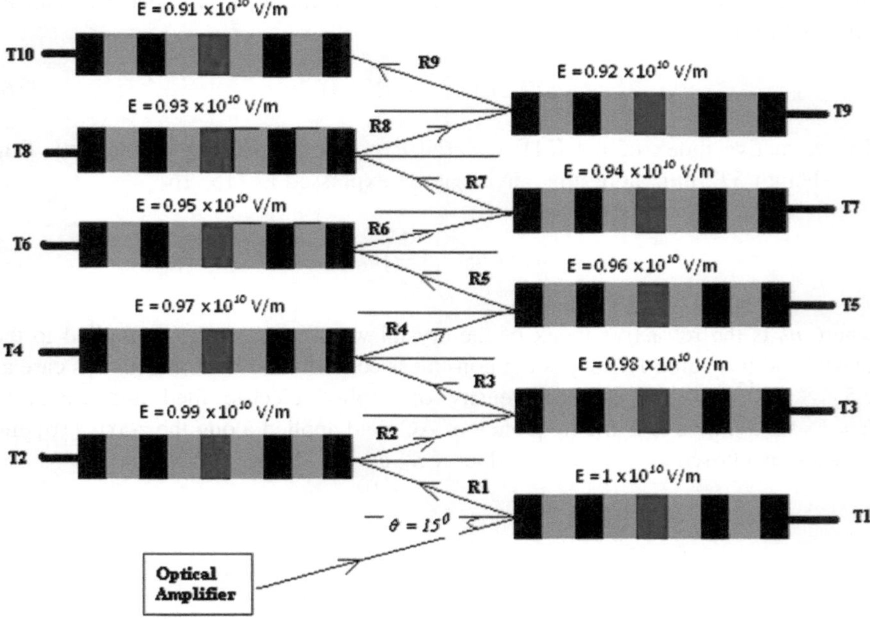

Fig. 63.2 Design of wavelength division demultiplexer based on 1d PC with KTP defect

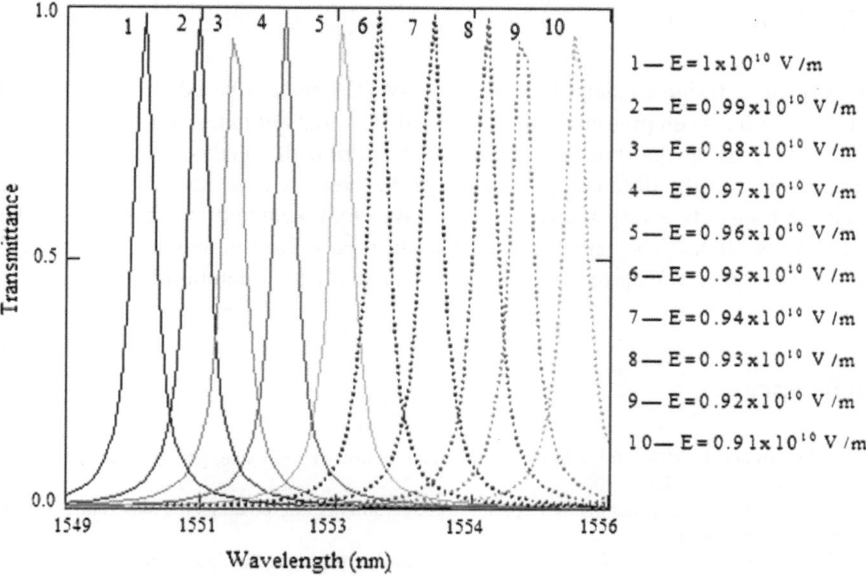

Fig. 63.3 Transmittance curve of PC structure with KTP defect at different intensity of E.F

Table 63.1 Central wavelength of transmission peak and the line width at the different value of electric field intensity

E.F. intensity (10^{10} V/m)	Central wavelength of transmission peak (nm)	Line width (nm)
1.0	1549.7	1549.5 − 1549.8 = 0.3
0.99	1550.4	1550.2 − 1550.5 = 0.3
0.98	1550.9	1550.7 − 1551.0 = 0.3
0.97	1551.6	1551.4 − 1551.7 = 0.3
0.96	1552.2	1552.1 − 1552.4 = 0.3
0.95	1552.7	1552.6 − 1552.9 = 0.3
0.94	1553.4	1553.3 − 1553.6 = 0.3
0.93	1554.1	1554.0 − 1554.3 = 0.3
0.92	1554.6	1554.4 − 1554.7 = 0.3
0.91	1555.3	1555.1 − 1555.4 = 0.3

various values such that 0.99×10^{10}, 0.98×10^{10}, 0.97×10^{10}, 0.96×10^{10}, 0.95×10^{10}, 0.94×10^{10}, 0.93×10^{10}, 0.92×10^{10} and 0.91×10^{10} V/m the transmission peak is shifted to different wavelengths which are centered at 1550.4, 1550.9, 1551.6, 1552.2, 1552.7, 1553.4, 1554.1, 1554.6 and 1555.3 nm respectively. The line width of each peak has the value of 0.3 nm, which is reported in Table 63.1.

63.4 Conclusion

In summary, design of demultiplexer for DWDM system based on 1d- PC with KTP defect has been presented and discussed. The transmittance spectrum has been obtained by using the transfer matrix method. From the analysis of the transmittance a curve at the different of values of E.F. intensity it is observed the first PC structure transmits a very narrow and of wavelength centered at a particular value and reflects others. The reflected wavelength in each case is incident on next PC, this process progress up to last PC and the system acts a demultiplexer.

References

1. Yablonovitch E (1987) Inhibited spontaneous emission in solid-state physics and electronics. Phys Rev Lett 58:2059–2062
2. John S (1987) Strong localization of photons in certain disordered dielectric superlattices. Phys Rev Lett 58:2486–2489
3. Weiss SM, Haurylau M, Fauchet PM (2005) Tunable photonic bandgap structures for optical interconnects. Opt Mater 27740–745
4. Srivastava SK, Ojha SP (2009) Broadband optical reflector based on Si/SiO_2 one-dimensional graded photonic crystal structure. J Mod Opt 56:33–40
5. Yuan K, Zheng X, Li C-L, She WL (2005) Design of omnidirectional and multiple channeled filters using 1d photonic crystal containing a defect layer with a negative refractive index. Phys Rev E71 066604:1–5
6. Suthar B, Bhargava A (2012) Temperature dependent tunable photonic channel filter. IEEE Photon Tech Lett 24:338–340
7. Hung H-C, Wu C-J, Chang S-J (2011) Terahertz temperature dependent defect mode in a semiconductor dielectric photonic crystal. J Appl Phys 110:093110-1-6
8. Srivastava SK (2014) Study of defect modes in one-dimensional photonic crystal structure containing high and low T_c superconductor as defect layer. J Sup Nov Magn 27:101–114
9. Skoromets V, Nmec H, Kadlec C, Fattakhova-Rohlfing D, Kužel P (2013) Electric filed tunable defect mode in one-dimensional photonic crystal operating in the terahertz range. Appl Phys Lett 102:241106-1-4
10. Tian H, Zi J (2005) One-dimensional tunable photonic crystals by means of external magnetic fields. Opt Comm 252:321–328
11. Habibiyan H, Ghafoori-Fard H, Rostami A (2009) Tunable all optical photonic crystal channel drop filter for DWDM system. J Opt A Pure Appl Opt 11:065102
12. Kumar A, Suthar B, Kumar V, Singh KS, Bhargava B (2012) Tunable wavelength demultiplexer for DWDM application using 1-D photonic crystal. Prog Electromag Res Lett 33:27–35
13. Yeh P (1988) Optical waves in layered media. Wiley, New York
14. Born M, Wolf E (1998) Principles of optics. Cambridge University Press, Cambridge
15. Li WC, Liu ZJ, Zhao HD, Sha XP, Li ZQ (2012) Characteristics of tunable micro cavity based on one-dimensional photonic crystal doping KTP as a defect layer. Opto Electron Lett 8:0021–24
16. Bordui PF, Jacco JC, Loiacono GM, Stolzenberger RA, Zola JJ (1987) Growth of large single crystals of KTP from high temperature solution using heat pipe based furnace system. J Crystal Growth 84:403–408

Chapter 64
Study and Implementation of White Power-LED Based Indoor Lighting Application for the Healthcare Sector

A. Chakraborty and R. Ganguly

Abstract With the current technological growth in the field of device fabrication, white power-LED's are available for solid state lighting applications. This is a paradigm shift from electrical lighting to electronic lighting. The implemented systems are showing some promise by saving a considerable amount of energy as well as providing a good and acceptable illumination level. However, the 'useful life' of such devices is an important parameter. If the proper device is not chosen, the desired reliability and performance will not be obtained. In the present work, different parameters associated with reliability of such LED's are studied. Four different varieties of LED's are put to test the 'useful life' as per IESNA LM 79 standard. From the results obtained, the proper LED is chosen for further application. Subsequently, lighting design is done for a hospital waiting room (indoor application) with 24 × 7 lighting requirements for replacement of existing CFLs there. The calculations show that although the initial cost is higher for LED based lighting, yet the savings on energy and replacement of the lamp results in a payback time of less than a year.

64.1 Introduction

LED's have long been used as indicators on electronic equipment. However in the last decade or so, their applications have spread into different domains like signalling, control panels etc. In the recent past, with the development of high-power white LED's, white power-LED applications in to the field of lighting have started. With this technology in place, there is a paradigm shift in the field of lighting. The concept of electronic lighting solutions with the help of power LED's are gradually penetrating different segments setting aside the conventional electrical lighting both in indoor and outdoor applications initiating the need to study their lifetimes [1, 2].

A. Chakraborty · R. Ganguly (✉)
Institute of Engineering and Management, Salt Lake, Kolkata, India
e-mail: rajivganguly72@gmail.com

V. Lakshminarayanan and I. Bhattacharya (eds.), *Advances in Optical Science and Engineering*, Springer Proceedings in Physics 166,
DOI 10.1007/978-81-322-2367-2_64

The constant and rapid technological growth in this area is pushing us towards an era of global lighting revolution. In line with the global development, there is also a substantial growth in the field of LED based lighting applications in the Indian subcontinent.

With the current technological growth, it is felt that appropriate time has come to shift from incandescent and fluorescent lighting technologies to white LED based lighting applications. It is understood that lighting is responsible for 19 % of global electricity use [3] and in India, one fifth of the electricity consumption is through lighting [4]. In India, the initiation towards implementing LED technology for lighting applications has already started in past few years. The primary areas of implementation are of course government organizations and big industry houses [5]. Some examples are:

- Haldia Development Authority in Haldia, West Bengal, India is saving around 70,000 kW hours of electricity and over US$9000 in energy bills each month since installing LED street lights.
- Kolkata Municipal Corporation in West Bengal, India boasted electricity savings of 52 % by switching from HPSV to LED lamps in a pilot project [5].

In addition to energy efficiency of LED's, there are other substantial benefits like—longer rated life than normal bulbs/CFL's, produces better quality of light per energy input, offers additional savings through reduced maintenance costs, reduction in air-conditioning cooling load, easy integration and control in smart buildings (e.g. digital load control in BMS), they do not contain mercury and hence is safer for the environment, they can be coupled to solar cells through converters as both are dc devices [6]. However, some basic issues need to be addressed for successful implementation of this technology. They are—adequacy of lighting when conventional lighting is replaced by LED lighting (lumen output), colour of light as good as conventional light or not (colour rendering index, CRI), cost involved in replacement from conventional to LED lighting, long-term cost savings due to LED replacements, time frame for this saving to happen (payback period), operating life and the frequency of the lamp replacement. In the present work, a majority of these issues are addressed with related studies on individual LED's as well as LED lighting systems used with luminaries in some chosen application.

64.2 LED Lighting Design Issues

On a broader classification, LED lighting systems can have two application domains—outdoor and indoor. The considerations for design vary to a great extent depending on the application. One of the important parameters is thermal management of the lighting system which varies considerably in these two cases and so on. Moreover, there are installations where replacement of an existing lighting

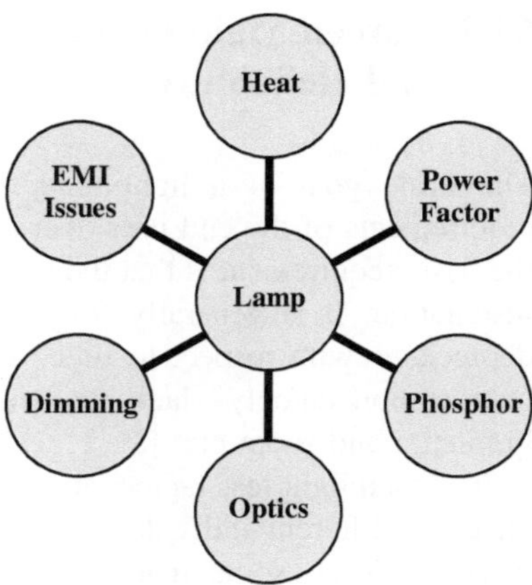

Fig. 64.1 Typical lamp design considerations

system (namely incandescent/halogen/CFL) may be done (retrofit systems) or it may be a fresh application development with new LED lamps. Accordingly, the designs will be different.

Figure 64.1 shows some of the design issues that needs to addressed in the design of lamp replacement. One of the basic requirements is that the design should be reliable and for this purpose, the following considerations are important—Power Management, Thermal Management, Optical Management and Assembly Integrity [7]. Figure 64.2 diagrammatically represents these considerations.

Failures within SSL luminaire may result from one or more of the above mentioned functional aspects of luminaire design and manufacturing namely—power management, thermal management, optical management and luminaire assembly integrity. Figure 64.2 provides an overview of the relationships between various components, materials and design elements [7].

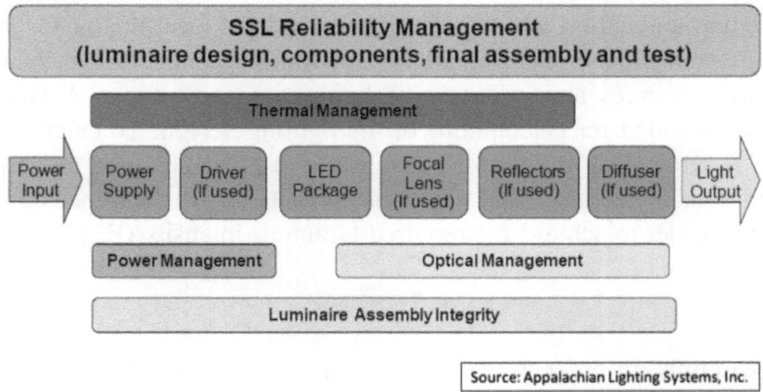

Fig. 64.2 LED lighting luminaire components and reliability considerations

64.3 Lumen Maintenance, Useful Life, Maintenance Factor and Reliability

The basic goal of a lighting system be it indoor or outdoor is to meet the requirements of the end user over the lifetime of the installation. For this purpose, the basic requirement is that the lighting should be adequate over the lifetime of the installation. It is generally found that the lumen output of any lighting system depreciates with respect to time. There are various factors leading to this lumen depreciation namely—lamp emissions, changing surface properties of the lamp, luminaire and room etc. [8].

The total light loss factors applied during the calculation process is the combined effect of different individual factors. The factors are either *recoverable or non-recoverable*. By some maintenance activities, some factors can be recovered. As for example, the dirt deposited on the lighting fixture may be cleaned periodically to recover the lumen depreciation. However, some other factors cannot be recovered over time. They result in permanent lumen depreciation from the system. As for example, the following non-recoverable changes occur in the individual power LEDs due to the passage of the current through it:

- *Degradation of the semiconductor material* with increase in the number of cracks, faults etc.
- *Degradation of the metal-semiconductor junctions* of the power LED device
- *Degradation of the phosphor*
- *Yellowing of the epoxy encapsulation*

So, the biggest design challenge for a power-LED based lighting system is to have the right lumen maintenance at all times considering recoverable and non-recoverable light-loss factors. It is worth mentioning here that, Narendran et al. [9] has defined 70 and 50 % levels of lumen maintenance for general lighting as well as for lighting of unimportant areas like staircase etc. respectively and currently that is the acceptable level. The EN12464-1 standard determines the required luminous intensity intended for a particular application which is the minimum light level that needs to be safeguarded independently of the degradation of luminous flux of the light sources, defective light sources and pollution of luminaries and spaces. A surplus must therefore be provided for calculations of the lighting design. To calculate this, the concept of *maintenance factor* (MF) is introduced [10]. This is a reduction factor which is combined with the initial luminous intensity (E_i) in the dimensioning of the installation in order to achieve the required luminous intensity (E_m):

$$E_m = E_i \times \dot{MF}\ldots \tag{64.1}$$

As for example, in order to maintain a luminous intensity of 250 lux, an initial luminous intensity of 357.14 lux needs to be maintained by only considering the useful life up to 70 % of lumen depreciation. For a complete design, however, the other factors (as mentioned below) contributing to lumen depreciation should also be considered and MF to be modified accordingly—lumen depreciation of the LED lamp, decline in output from the luminaire due to pollution, space pollution, frequency of lamp defects without lamps immediate replacement.

As seen in Fig. 64.2, the reliability of the system is otherwise a complex function and is dependent on different parameters associated with the system. The overall system reliability is as shown in 64.2 which suggests that the total reliability is a product of reliabilities of the individual components of the system namely—electrical, interconnections, LED, optical, thermal and mechanical aspects of the system [11].

$$R_{system} = R_{electrical} \times R_{connections} \times R_{LEDS} \times R_{optical} \times R_{thermal} \times R_{mechanical} \quad (64.2)$$

So, (64.2) clearly suggests that during the design of a lighting system using LEDs, each and every component of the system should adequately be taken care of as a lower reliability of any one of the aspects will affect the overall reliability of the system.

64.4 Experimental Procedure to Determine Useful Life of Power LED

From (64.2) we can see that there are several factors that contribute to the reliability of the LED lighting systems. However, the heart of the system is the power LED and its electrical driving circuitry. These two components are the major contributing factors to the overall system reliability. It is understood that in majority of the applications, constant current drivers are needed to drive power LED based lighting systems [12–14]. In the present case, a constant current driver of appropriate rating is being considered for the intended application. The selection of the power LED, therefore, plays a crucial and a pivotal role in the overall system design. [1, 2] indicate the different methods for estimating the useful life of LED, one by measuring the lumen depreciation characteristics of the LED and the other by measuring the junction temperature.

To understand the 'useful life' of power LED's, eight different variants of 1 W power LED's (two from standard manufacturers and the rest two from the commercially available varieties) are chosen. For each type of LED, two numbers of LED's are taken for averaging the results. They are put to useful life measurement by measuring the lumen depreciation characteristics as per [1].

The eight power LED's are connected in series and driven by the same constant current of 350 mA at 45° ± 1 °C temperature (tropical condition). This arrangement primarily maintains the same current through the LED's at all points of time during the experiment. The schematic diagram of the experimental arrangement is shown in Fig. 64.3. The light output is measured by suitable photodetector (PD). As per IESNA LM 79 standard, data for the first 1000 hours are not considered as this

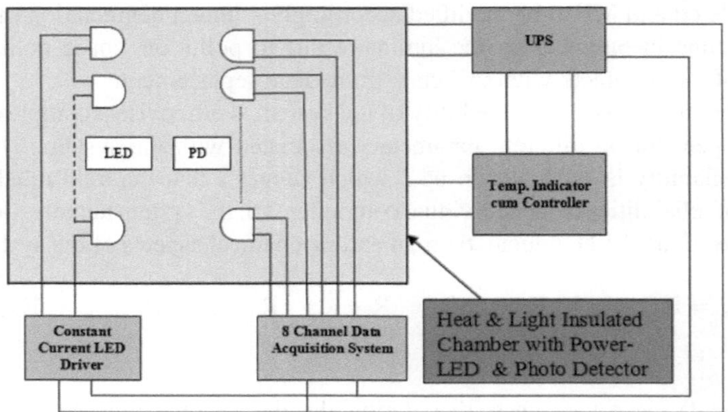

Fig. 64.3 Schematic arrangement to determine the useful life of power LED's

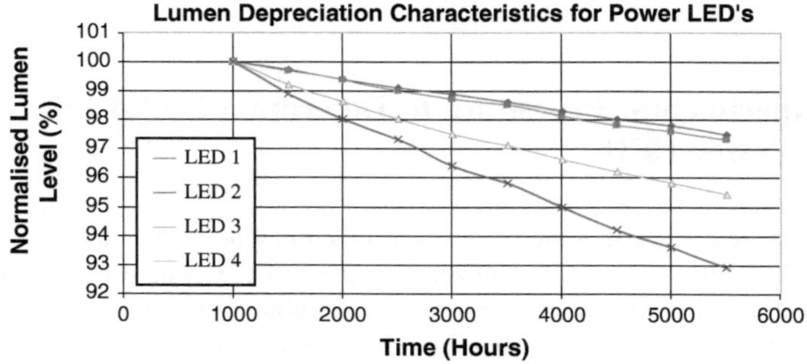

Fig. 64.4 Lumen depreciation characteristics for four sets of power LED's

period is required for stabilization of the light output and this period is called the seasoning period [1]. Lumen depreciation data is taken between 1000 to 5500 hours for all the LED's, stored in a data acquisition system, averaged and normalized considering the 100 % light level as the light output after 1000 h of operation [1]. Four sets of data available after the experiment is plotted in a graph as shown in Fig. 64.4 which shows the lumen depreciation characteristics upto 5500 hours. LED 1 and LED 2 are from standard manufacturers and LED 3 and 4 are from the commercially available variety.

64.5 Results Obtained and Inference

Figure 64.4 shows the lumen depreciation characteristics for the four sets of LED's. It is clear from the figure that LED's 1 and 2 which are from standard manufacturers' shows similar lumen depreciation characteristics. On the contrary, LED's 3

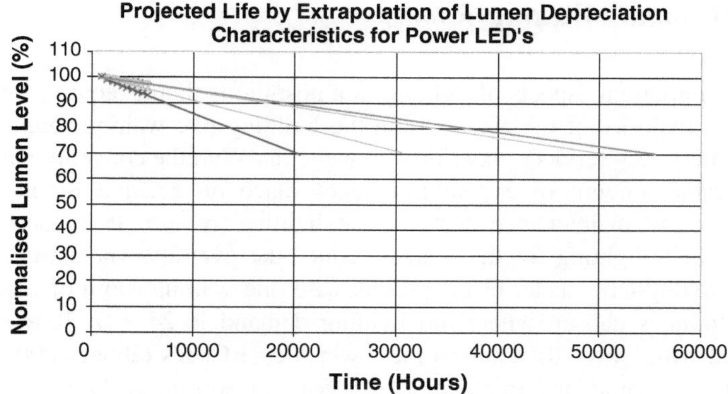

Fig. 64.5 Projected lumen depreciation characteristics for four sets of power LED's to estimate the useful life

and 4 are from the commercially available variety and they represent a faster lumen depreciation. The curves are extrapolated to estimate the expected useful life of the LED's as per IESNA LM-79 standard. It is shown in Fig. 64.5. It is clearly visible from Fig. 64.5 that for LED's 1 and 2, the expected useful life is more than 50,000 h. But for LED 3, it is around 30,000 h and for LED 4 it is about 20,000 h. So, it is understood from Fig. 64.5 that LED 1 and 2 are best suited for lighting applications at important places like office, showroom etc. with a good useful life (considering 70 % level). So, in the present work, LED's 1 and 2 from standard manufacturers are chosen for some intended application.

64.6 LED Lighting System Design

As mentioned earlier that there is a paradigm shift in the lighting sector with the introduction of power LED's into the lighting segment. This shift is primarily from electrical to electronic (solid state) lighting (SSL) which results in numerous advantages. In Sect. 64.1, some important issues are outlined that needs clarification for successful implementation of LED's for lighting applications.

It has been mentioned in [16] that there are numerous applications for LED lighting like—residential, commercial, entertainment and outdoor and infrastructure. The interesting aspect is that it is very convenient to use various intelligent features into these applications [15, 16]. In the following sections, a commercial application (indoor) is discussed namely—hospital (waiting room). This area requires lighting for a long period and is so chosen for discussion in the present case.

64.6.1 Hospital (Waiting Room)

One of the important aspects of lighting in a hospital is that it requires lighting for 24 × 7 in a majority of the areas within the building like waiting rooms, nursing stations, emergency services, exterior walkways etc. With the conventional lighting, a considerable amount of expenditure takes place on account of energy cost, replacement and maintenance costs of the lighting system. It is estimated that shifting to LED lighting for these areas reduce the overall ownership cost of the system in a long-term scale. In the present case, the waiting room of a hospital in Kolkata, India is chosen where the lighting demand is 24 × 7. It was initially illuminated with CFL's. But due to the lower life of CFL's (about 8000 h), lamps required to be maintained frequently (~1 year) or so due to 24 × 7 operational requirements. So, an alternative LED solution for the same is provided with LED's having a much higher useful life (>50,000 h). The following considerations are made in the design of the LED lighting solution [17]—efficiency, comfort and performance. Table 64.1 illustrates the parameters of the lamps (CFL and LED) and the related energy savings, payback period etc.

From Table 64.1, it is clear that although the initial investment is higher in case of LED based lighting solution (about $1000 in the present case), but the savings it produces in terms of lower energy costs, lower maintenance costs and longer operational lifetime, the lifetime cost/the ownership cost of the system over the entire lifetime is far lower as compared to the corresponding CFL based lighting solution. With the present LED lighting solution, the lamps are expected to run on an average of 5.7 years considering 70 % lifetime. Figure 64.7 shows the cost comparison of the two systems.

It is clear from Fig. 64.6 that the payback time can be obtained from the point of intersection of the two graphs (shown by the red vertical line touching the time axis) which is in this case is less than one year. So, the initial higher cost is justified. The

Table 64.1 Comparison of CFL and LED lighting for hospital (waiting room)

Description	CFL	LED lamps
Area under consideration	300 sq. m.	300 sq. m.
Number of lamps	100	100
Per lamp power consumption	24 W	10 W
Total power consumption	2400 W	1000 W
Per lamp cost with fixture	$8.00	$18.00
Total consumption per year (24 × 7) (kWh)	21,024	8760
Total electricity cost per year ($0.1 per kWh)	$2102	$876.00
Total lamp Cost (Initial investment)	**$800.00**	**$1800.00**
Maintenance and replacement cost after 1 year	$300.00	Nil
Total expenses after 1 year	**$3202**	**$2676**
Total expenses after 2 years	**$5604**	**$3552**
Total expenses after 3 years	**$8006**	**$4428**

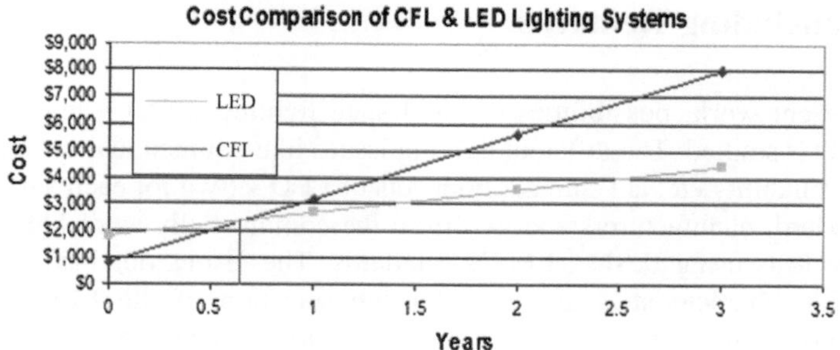

Fig. 64.6 Cost comparison of CFL and LED lighting systems

Fig. 64.7 Actual photograph of the implementation in the hospital waiting room

lux levels are measured using a portable luxmeter at different points of the room and shows agreeable results for replacement. Figure 64.7 shows the actual photograph of the hospital (waiting room).

64.6.1.1 Reduction in CO_2 Emission

It is seen from Table 64.1 that switching from CFL to LED based lighting solution, there is a reduction of 12,264 kWh of energy consumption annually for this installation. Considering a reduction of 0.42 kg of CO_2 per kWh [17], the net reduction of CO_2 emission will be 5150.88 kg per year for this installation which is having a positive impact on the environment.

64.7 Concluding Remarks

In the present work, possibilities of solid state lighting (SSL) using high power white LED is studied. The different lighting issues like—lumen maintenance, useful life, MF, reliability etc. are studied. Four sets of LED's (two for each variety)—two from standard manufacturers and two from the commercially available variety are put to life tests using IESNA LM-79 standards. The results obtained from lumen depreciation characteristics are used for the selection of the LED for further applications keeping in mind about the lumen maintenance, MF, reliability and other related issues. Subsequently, the lighting design is done for a hospital waiting room (24 × 7 operations) and is implemented after replacing the CFL lighting already existent there. The cost comparisons of both the systems are done and it is found that the payback period is less than a year. So, the cost of implementation is justified. The system is implemented for more than a year now and the performance is monitored at regular intervals which show acceptable lumen maintenance levels as well as power savings. The environmental impact by savings of CO_2 is also calculated. Based on the results obtained, further bigger implementation is planned at the outdoor conditions. Further study is being carried out to design and implement outdoor lighting systems.

References

1. Bhattacharya S, Chakraborty A, Ganguly R (2009) A novel technique for fast determination of 'useful life' of white phosphor-converted led by measuring lumen depreciation characteristics for different forward currents for solid state lighting applications. In: 18th international photovoltaic science and engineering conference and exhibition, Kolkata, India. 19–23 Jan 2009
2. Hong E, Narendran N (2004) A method for projecting useful life of LED lighting systems.In: Proceedings of SPIE, third international conference on solid state lighting 5187:93–99
3. www.cleanrevolution.org
4. http://www.ese.iitb.ac.in/~suryad/Lighting-CEP.pdf
5. www.theclimategroup.org/what-we-do/news-and-blogs/led-revolution-underway-in-indian-cities/
6. http://www.pcj.com/dnn/Portals/0/Documents/LED%20Lighting%20Technology.pdf
7. LED luminaire lifetime: recommendations for testing and reporting, 2nd edn. Next Generation Lighting Industry Alliance with the U.S. Department of Energy, June 2011
8. Lumen maintenance and light loss factors: consequences of current design practices for LEDs. Solid State Lighting Program, Building Technologies Program, Office of Energy Efficiency and Renewable Energy, U.S. Department of Energy, Sept 2013
9. Narendran et al. (2005) What is the useful life for white LEDs. J Illum Eng Soc 20(1):57–67
10. The maintenance factor for LED lighting. www.etaplighting.com
11. Lumileds P (2012) Evaluating the lifetime behaviour of LED systems. White Paper, WP15 20121218
12. Berman M (2012) What type of LED driver or power supply do we need? Available: http://us.tdk-lambda.com/ftp/other/what-led-driver-do-i-need.pdf
13. Why drive white LEDs with constant current. AN 3256. Available: http://www.maximintegrated.com/app-notes/index.mvp/id/3256

14. http://www.meanwell.com/webnet/search/seriessearch.html
15. http://www.ledlightingexplained.com/led-lighting-myths/led_lighting_Getting_It_ Right.pdf
16. Carner P (2012) Bringing intelligence to LED lighting applications. White Paper, Texas Instruments, Dallas. Available: www.ti.com/lit/wp/spry211/spry211.pdf
17. http://www.cogaprel.com/images/Healthcare_Brochure.pdf

Chapter 65
Characteristics of II–VI Quantum Dot Infrared Photo-Detectors

C.M.S. Negi, Dharmendra Kumar and Jitendra Kumar

Abstract We theoretically investigate the performance of a II–VI ZnCdSe/ZnSe-based quantum dot infrared photodetector (QDIP) utilizing intersubband hole transitions in the valence band of the QDs to absorb infrared radiation. The analysis starts with the computation of band structure via multi-band effective mass model based on the Luttinger-Kohn Hamiltonian with the inclusion of strain effects. The theoretical formulation is further used to determine the spectral responsivity and dark current characteristics of the QDIP.

65.1 Introduction

Infrared detectors are required in the various imaging applications, including medical diagnostics, astronomy, remote sensing, space-based surveillance, and thermal imaging [1]. Initially, infrared detectors were developed using narrow band gap semiconductors such as HgCdTe and InSb [2], which make use of interband transitions to detect the infrared radiation. Operating wavelength of these detectors can be modified by controlling the band gap of the semiconductor material via changing the alloys composition. However, the processing of narrow gap semiconductors creates problem in the epitaxial growth, non-uniformity, low yield and high cost [3]. Alternatively, intersubband transitions in the quantum hetero-structures have been utilized to detect the infrared radiation. The attractive feature of intersubband transition based detectors is that they do not require narrow band gap semiconductors. This provides flexibility to use any available wide band gap material to fabricate infrared detector. Moreover, quantum hetero-structure

C.M.S. Negi (✉)
Department of Electronics, Banasthali Vidyapith, Banasthali 304022, Rajasthan, India
e-mail: nchandra@banasthali.in

D. Kumar · J. Kumar
Department of Electronics Engineering, Indian School of Mines,
Dhanbad 826004, Jharkhand, India

V. Lakshminarayanan and I. Bhattacharya (eds.), *Advances in Optical Science and Engineering*, Springer Proceedings in Physics 166,
DOI 10.1007/978-81-322-2367-2_65

parameters such as dimensions and the band offset can be designed to tailor the spectral response hence Operating wavelength of these detectors.

In particular, intersubband optical transitions in self-assembled semiconductor quantum dots have attracted a great deal of attention because of their interesting optical and electronic properties resulting from the three-dimensional confinement of charge carriers. Due to the expected unique characteristics, QDs have shown there potential as a possible alternative to the HgCdTe and quantum well infrared detector (QWIP) for applications in infrared detection [4]. The three dimensions quantization of energy states in the QDs leads to several benefits in the quantum dot infrared photodetectors (QDIPs), such as, longer relaxation time of carriers, normal incidence operation, and lesser dark current over QWIPs [5].

Longer relaxation time of carriers and lesser dark current permit high operating temperatures. QDIPs were usually fabricated using self-assembled QDs made from III-V based material systems. Instead, in the last few years, II–VI ZnSe based material systems have been comprehensively studied for their possible use in infrared detection [6–8].

In this article, we therefore investigate a p-type ZnCdSe/ZnSe-based QDIP functioning through the intervalence band transitions as opposite to the intraband transitions between the different states of electrons energy levels in the conduction band.

The QDIP is studied with the $\mathbf{k} \cdot \mathbf{p}$ approach under the envelope function approximation [9]. The energy subbands and wavefunctions are obtained from the 4×4 Kohn-Luttinger Hamiltonian. Strain effects are incorporated through the addition of Bir-Pikus Hamiltonian to the general Luttinger Hamiltonian. Subsequently, we evaluate the dipole matrix elements for intraband transitions. These calculations are further utilized to determine the responsivity and dark current of the QDIP.

65.2 Theoretical Formulations

The QDIP structure under investigation is consists of layers of periodically distributed $Zn_{1-x}Cd_xSe$/ZnSe QDs. The QD is assumed to be disc shaped and modeled by the parabolic confinement potential along x- and y-directions (in-plane) and finite quantum well potential well along the z-axis (growth direction). The strength of confinement potentials are determined through the various physical parameters comprising valence band offset, in-built strain and size of the QDs. The energy and wavefunctions of the valence band states are determined by the $\mathbf{k} \cdot \mathbf{p}$ method represented by the Kohn-Luttinger Hamiltonian including strain effects [10]. For in-plane confinements the basis functions are in the form of the Hermit Gaussian functions. In z-direction the energies are determined by the transcendental solution of the Schrodinger equation and basis functions can be approximated as hard wall wavefunctions. The energy eigenvalues and wavefunctions computed after the numerical diagonalization of Kohn-Luttinger Hamiltonian are employed for the calculations of Fermi levels and optical matrix elements for intersubband transitions, which are further utilized to obtain the various parameters of QDIP including absorption coefficient [11]. The size inhomogeneity of the QDs which leads to the inhomogeneous broadening in the absorption spectra is accounted through the Gaussian broadening function. The quantum efficiency of the QDIP is determined through the knowledge of the absorption spectra.

Photocurrent generation in QDIP is related to the photo excitation of holes from the QDs followed by the absorption of infrared radiation. The photocurrent density in the QDIP is calculated by the following relationship [11]:

$$J_p = e\eta\Phi_i P_{\mathrm{QD}} \exp\left(\frac{e^2 P}{C_Q kT}\right) \Big/ \{p_0(P_{\mathrm{QD}} - \langle P \rangle)\} \tag{65.1}$$

where, J_p is the photocurrent density, e is the electronic charge, η is the quantum efficiency, Φ_i being the normally incident photon flux, C_Q is the quantum dot capacitance, T is the absolute temperature, k is the Boltzmann constant, P_{QD} is the maximum allowable number of holes inside the QDs, P is the average number of holes in the QDs, and p_o is the capture probability of uncharged QDs.

Photo emission and capturing process of holes modify the number of holes, hence the charge distribution inside the QDs. This leads the modification of local current density injected by the emitter contact and can be represented by the following equation [12]:

$$j = J_m \frac{\theta}{\langle P \rangle} \exp\left\{ q\left(V + V_D - \frac{\langle P \rangle}{P_{\mathrm{QD}}} V_{\mathrm{QD}} \right) \Big/ (M+1)kT \right\} \tag{65.2}$$

where, J_m is the maximum photocurrent that can be injected by the emitter contact, V is the bias voltage, $\theta = 1.11\left[\mathrm{erf}\left(0.84\, eL_Q\sqrt{\langle P \rangle \rho_Q^{\frac{3}{2}}/\varepsilon kT}\right)\right]^2 \varepsilon kT/\left(e^2\sqrt{\rho_Q}\right)$, $V_{\mathrm{QD}} = 2\pi eM(M+1)\rho_Q L(1-\delta)P_{\mathrm{QD}}/\varepsilon$, $V_D = 2\pi eM(M+1)\rho_D L/\varepsilon$, $\delta = 0.72\sqrt{2}/\pi ML\sqrt{\rho_Q}$, where $L_Q = 1/\sqrt{\rho_Q}$ is the lateral spacing and L is the spacing between the QDs layer along transverse directions, M represent the number of layers in the QDIP, ε is the permittivity of the semiconductor and ρ_D denotes the doping density of the acceptors. Next we determine the average number of holes $\langle P \rangle$ occupied by the QDs when infrared radiation is incident on the QDIP by equalizing (65.1) and (65.2) which balances the capture into and emission process of holes from QDs. Substituting the value of $\langle P \rangle$ in (65.1) provides the average photo current density of the QDIP. Responsivity is determined by the equation $J_p/(\hbar\omega\Phi_{in})$. The terminal current under dark condition in the QDIP can be calculated by considering that the dark current in QDIP originates mainly due to the thermionic emission and quantum tunneling of holes in the QDs. The final expression of the average dark current density takes the form [13]:

$$J_d = e\rho_Q(G_{\text{tun}} + G_{\text{th}})/p_c \quad (65.3)$$

Here, $G_{\text{tun}} = G_{0t} \exp\left(4/3\sqrt{2em_t^* \phi_B^3}/h\text{F}\right) \exp(-\Delta\varepsilon/kT) \exp(A)$, $G_{\text{th}} = G_0 \exp(-\epsilon_q/kT) \exp(A)$, In these equations G_{tun} and G_{th} represents the rate of tunneling and thermionic emission of holes, respectively, m_t^* is effective transport mass of hole that accounts the contribution from both heavy hole and light hole, ϕ_B symbolizes height of the potential barrier from the topmost filled energy level in the valence band, F denotes the applied electric field, $\Delta\varepsilon$ calculates energy difference from the ground state to the top most filled energy level, ϵ_q represents ionization energy from the ground state of QDs and $A = \left(\pi\hbar^2 \langle P \rangle / m_t^* kTa_Q^2\right)$, and a_Q is the QD radius. The average number of holes occupied by the QDs while no light is incident on the QDIP can be determined by equalizing the (65.3) and (65.2). Substituting the value of average number of holes in (65.3) estimates the dark current in QDIP.

65.3 Results and Discussions

We now study the performance characteristics of QDIP using the formulation explained in the Sect. 65.2. The material parameter used for this analysis are taken from the [14, 15], in addition, the parameters for $Zn_{1-x}Cd_xSe$ are calculated using the linear interpolation of the CdSe with ZnSe parameters. Structural parameters used in these calculations are provided in Table 65.1.

Figure 65.1 shows the theoretical spectral responsivity curve of a QDIP as a function of strain (ε) at an applied bias of 1 V. Strain can be considered as a lattice disparity between the QD and barrier material, hence it can be modified by varying the composition of the QD material. The main peak in all curves is attributed to the heavy hole (hh) ground state to the light hole (lh) first excited state. The strain shifts the main peak towards higher energy (lower wavelength). The effect of strain on the valence band states is more complex, the hydrostatic strain shifts the hh and lh levels by the same amount and biaxial strain increase the separation among the hh and lh states, so that, increase in the strain leads to the increased energy separation between hh and lh, thus leads to the shift in the main peak of the responsivity spectra towards shorter wavelengths. Figure 65.1 also shows that, strain can be used to tune the peak response wavelength of the detector from 4.5 to 11 μm.

Figure 65.2 shows the theoretical dark current-voltage characteristics of a QDIP as a function of temperature. As anticipated, dark current density raises rapidly with

Table 65.1 Structural parameters for QDIP

P_{QD}	ρ_Q	ρ_D	L	a_Q	h_Q (height of QDs)
8	5×10^{10} cm^{-2}	3×10^{10} cm^{-2}	80 nm	10 nm	5 nm

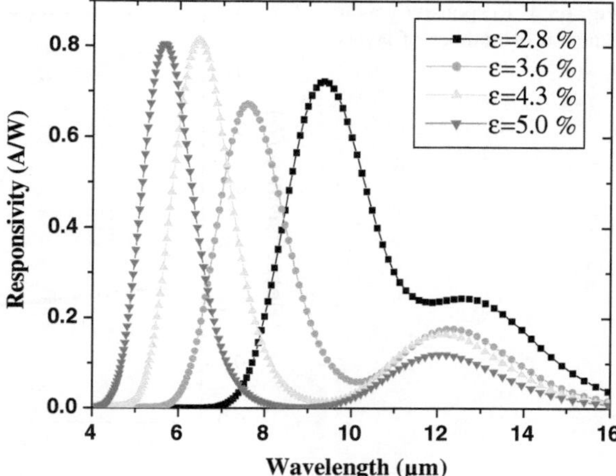

Fig. 65.1 Spectral responsivity of the detector as a function of strain

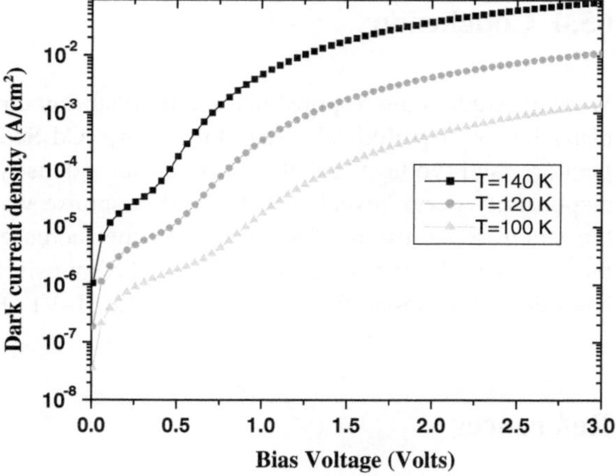

Fig. 65.2 Dark current Characteristics of QDIP as a function of applied voltage and temperature

the temperature and voltage. The temperature and applied bias dependence on the dark current originates from the thermionic emission and field assisted tunneling of holes in the QDs.

Figure 65.3 displays the effect of number of layers and applied voltage on the responsivity of the QDIP. The decrease in responsivity with the number of layer is due to the reduction of number holes occupied by the QDs. The increase of responsivity with the applied voltage in attributed to the lowering of the barrier and increase in the number of holes occupied by the QDs.

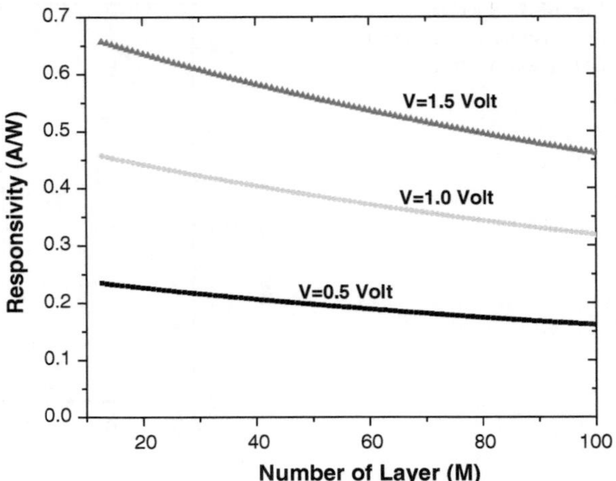

Fig. 65.3 Responsivity as a function of number of layers

65.4 Conclusion

We have studied the responsivity and dark current characteristics of QDIP designed using layers of periodically distributed $Zn_{1-x}Cd_xSe/ZnSe$ QDs. The effect of temperature and voltage on the dark current density has been studied. Spectral responsivity curve reveal that the peak response wavelength depends strongly on the strain. Responsivity decreases with the increase in the number of layers and increase with the applied voltage. Spectral responsivity curve demonstrate the broad band detection possibility of the wide band II–VI QDIPs.

References

1. Rogalski A (2009) Infrared detectors for the future. Acta Phys Pol, A 116:389–405
2. Rogalski A (2005) HgCdTe infrared detector material: history, status and outlook. Rep Prog Phys 68:2267
3. Levine BF (1993) Quantum-well infrared photodetectors. J Appl Phys 74:R1–R81
4. LaoY-F, Wolde S, UnilPerera AG, Zhang YH, Wang TM, Liu HC, Kim JO, Schuler- Sandy T, Tian Z-B, Krishna SS (2013) InAs/GaAs p-type quantum dot infrared photodetector with higher efficiency. Appl Phys Lett 103(241115):1–4
5. Negi CMS, Kumar D, Gupta SK, Kumar J (2013) Theoretical analysis of resonant cavity p-type quantum dot infrared photodetector. IEEE J Quantum Electron 49, 839–845
6. Ravikumar AP, Alfaro-Martinez A, Chen G, Zhao K, Tamargo MC, Gmachl CF, Shen A (2012) ZnCdSe/ZnCdMgSe quantum well infrared photodetector. Opt Express 20:22391–22397
7. Ravikumar AP, Garcia TA, Jesus JD, Tamargo MC, Gmachl CF (2014) High detectivity short-wavelength II-VI quantum cascade detector. Appl Phys Lett 105(6):061113(1–4)
8. Ravikumar AP, Chen G, Zhao K, Tian Y, Prucnal P, Tamargo MC, Gmachl CF, Shen A (2013) Appl Phys Lett 102:161107

9. Kumar J, Kapoor S, Gupta SK, Sen PK (2006) Theoretical investigation of the effect of asymmetry on optical anisotropy and electronic structure of Stranski-Krastanov quantum dots. Phys Rev B 74:115326(1–10)
10. Gupta SK, Kapoor S, Kumar J, Sen PK (2007) Strain induced effects on optical properties of magnetized Stranski-Krastanov quantum dots. Nanotechnology 18:325402(1–7)
11. Negi CMS, Kumar J (2014) Analysis of electromagnetically induced transparency-based quantum dot infrared photodetectors. J Opt Soc Am B 31:2121–2130
12. Ryzhii V, Khmyrova I, Pipa V, Mitin V, Willander M (2001) Device model for quantum dot infrared photodetectors and their dark-current characteristics. Semicond Sci Technol 16:331–338
13. Martyniuk P, Rogalski A (2009) Insight into performance of quantum dot infrared photodetectors. Bull Polish Acad Sci Tech Sci 57:103–116
14. Kapoor S, Kumar J, Sen PK (2010) Magneto-optical analysis of anisotropic CdZnSe quantum dots. Physica E 42:2380–2385
15. Lozykowski HJ, Shastri VK (1991) Excitonic and Raman properties of ZnSe/$Zn_{1-x}Cd_xSe$ strained layer quantum wells. J Appl Phys 69:3235

Chapter 66
Smartphone Based Platform for Colorimetric Sensing of Dyes

Sibasish Dutta and Pabitra Nath

Abstract We demonstrate the working of a smartphone based optical sensor for measuring absorption band of coloured dyes. By integration of simple laboratory optical components with the camera unit of the smartphone we have converted it into a visible spectrometer with a pixel resolution of 0.345 nm/pixel. Light from a broadband optical source is allowed to transmit through a specific dye solution. The transmitted light signal is captured by the camera of the smartphone. The present sensor is inexpensive, portable and light weight making it an ideal handy sensor suitable for different on-field sensing.

66.1 Introduction

The demand of mobile phone devices in the global market has been skyrocketing and reached over 7 billion across the globe by the end of 2013 [1]. Since the inception of first mobile phone in the year 1973 by Motorola incorporation, there has been an unprecedented growth in cellphone technology which led to addition of more and more features in the cellphone. Integration of different built-in facilities such as wi-fi, bluetooth, megapixel cameras, ambient light sensor driven by various application software have made it a ubiquitous device in day-to-days life of modern civilization. With the advent of technological breakthrough there have been dramatic improvements in smartphone technology which led to increase in RAM, faster CPU and greater storage capacity. There are different mobile platforms in the market like Android (Google), iOS (Apple), RIM (Blackberry) and windows (Microsoft Inc), which provide plethora of opportunities to develop any kind of applications in

S. Dutta (✉) · P. Nath
Applied Photonics and Nanophotonics Laboratory, Department of Physics, Tezpur University, Napaam, Assam 784028, India
e-mail: siba@tezu.ernet.in

P. Nath
e-mail: pnath@tezu.ernet.in

V. Lakshminarayanan and I. Bhattacharya (eds.), *Advances in Optical Science and Engineering*, Springer Proceedings in Physics 166,
DOI 10.1007/978-81-322-2367-2_66

accordance with users' need and get it installed in the smartphone itself. Such flexibilities have given birth to Google playstore which is a storehouse of android apps. Taking advantage of the present demand of mobile phone, different research groups are currently engaged in development of various handy and portable optical sensing devices integrated with smartphone for carrying different sensing operations [2–7].

In the present work we have exploited the camera of the smartphone for developing a portable colorimetric sensor to measure absorption band of coloured dyes. Absorption band of a dye is an identity of that particular dye as different coloured dyes occupy different absorption band of a broadband light source.

66.2 Sensor Set-up and Experimental

Figure 66.1 shows the smartphone integrated optical sensor. Light from a broadband source falls on a pinhole. The pinhole restricts the amount of light from entering into the system. The light coming out of the pinhole is allowed to fall on a collimating lens. The collimating lens makes the light beam parallel with respect to the optical axis. The collimated light beam is allowed to pass through a quartz cuvette filled with the dye sample solution. The transmitted light beam from the solution falls on a cylindrical lens which makes the collimated light beam into a focused line beam. The focused line beam falls on a transmission grating that disperses the beam into its constituent components. The dispersed spectrum falls on the CMOS image sensor which can be captured by the smartphone. Because of the presence of internal filter in the camera module of the smartphone, our designed sensor is capable of capturing spectrum in the visible domain spanning from 400 to 700 nm. imageJ software (a freely downloadable software) has been used for

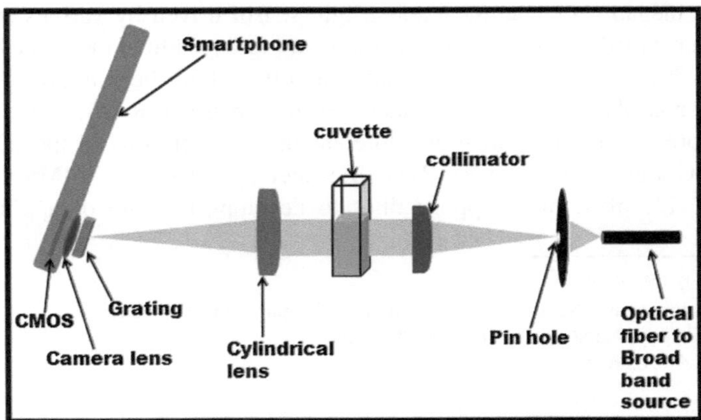

Fig. 66.1 Schematic diagram of the proposed sensor

plotting transmission intensity versus pixel unit of the captured image spectrum. By using MATLAB program pixel scale has been converted into wavelength scale.

We have followed the procedure adopted by Gallegos et al. [8] to convert the pixel scale into wavelength scale. To do so, two diode laser pointers (Green of 532 nm and Red of 655 nm) were used to illuminate the pinhole and the focused line beam of each laser source is captured by the camera of the smartphone. The two diode lasers occupy specific position in the visible domain and using this information the pixel resolution of the designed sensor is found to be 0.345 nm/pixel.

66.3 Sample Preparations

Different dyes namely Safranine O, Coomassie Brilliant blue R 250 and Malachite Green have been procured from SRL India while Rhodamine 6G and Sunset Yellow were procured from Himedia. 0.001 g of each dye is dissolved in double distilled water to prepare a concentration of 0.001% dye sample solution.

66.4 Results and Discussions

Quartz cuvette of 5 mm path length filled with 3 mL dye solution is introduced in the optical path of the sensor. The collimated light beam passes through the sample solution and the transmitted light beam is captured by the camera of the phone.

To have a very good resolution the camera must be locked at autofocus mode that ensures focusing at infinite distance. Figure 66.2a, b show the captured RGB image spectrums of the bare broadband source and image spectrum of Malachite Green dye. From the image spectrum we can visually see the absorption band of the specific dye. Figure 66.3 shows the transmission spectra of each of the dye sample as captured by our smartphone sensor. The figure also includes captured RGB spectra of the dyes as displayed on the screen of the smartphone.

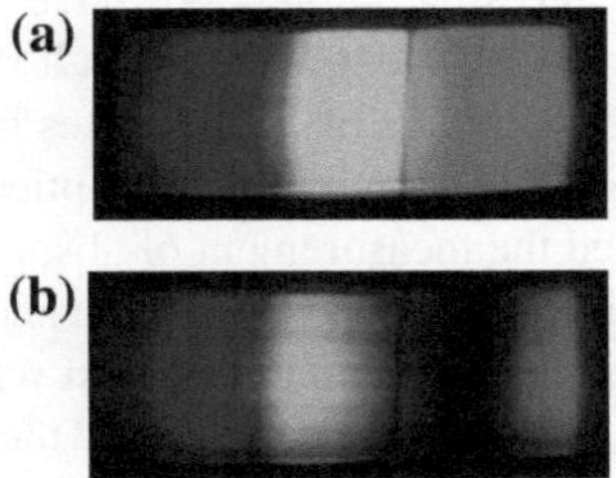

Fig. 66.2 Transmission RGB spectra of **a** broadband source and **b** Malachite green dye

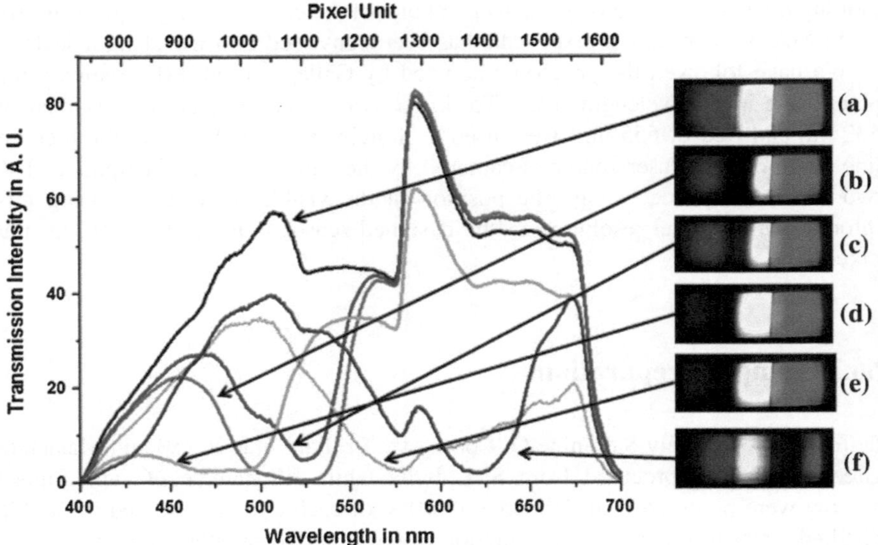

Fig. 66.3 Sensor characteristic response for *a* bare spectrum; *b* Rhodamine 6G; *c* Safranine O; *d* Sunset yellow; *e* Coomassie brilliant blue; *f* Malachite green

From the transmission spectroscopic plot it can be inferred that the dyes Rhodamine 6G, Safranine O, Sunset yellow, Coomassie brilliant blue R 250 and Malachite green have absorption band in the wavelength range 480–540, 500–530, 460–530, 560–615 and 600–630 nm respectively. In order to claim the firm validation of the data obtained with our sensor, we have taken the transmission spectra of these dyes using a standard spectrometer (SHIMADZU 2450). Figure 66.4 shows the transmission spectra of the dye samples taken with the standard spectrometer.

By Comparing the data obtained with our designed sensor with that of the standard spectrometer it can be inferred that our proposed colorimetric smartphone sensor provides highly reliable and almost accurate data. The present sensor works in direct coupling mode in which the collimated light beam is allowed to transmit through dye sample and the modulated transmitted signal from this sample is captured by the camera of the smartphone. We have repeated the investigations several times and we observe consistent results for all the considered samples. We can also measure the absorption bands of colored dyes in evanescent mode through incorporation of a right angled glass prism in the optical set-up.

Recently we have reported the measurement of absorption bands of colored dyes by evanescent wave coupled spectroscopy [9]. The evanescent field generated at the base of a right angled isosceles prism can be coupled with the dye solution and the resulting modulated light signal could be captured by the camera of the smartphone. The disadvantage of this technique is that the glass prism has to be cleaned properly after measuring the absorption band of each sample thus makes the process time

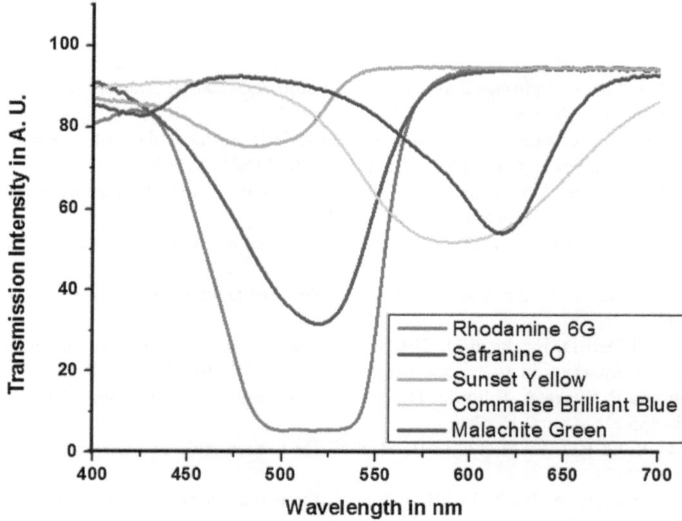

Fig. 66.4 Characteristic transmission spectra of the dyes obtained using a standard spectrometer

consuming. To reduce the cost further we can replace the broadband light source with a white LED source. Using an LED will add another advantage as we can increase or decrease its intensity in accordance with the pinhole size. Alternatively we can also use sunlight as the source which will make our designed sensor truly portable and suitable for sensing in outdoor environment. The presence of existing mobile network system can be used to send the data anywhere in the world which will make sensing investigation suitable in remote areas where carrying of a bulky laboratory device is an ardent task.

66.5 Conclusion

We have demonstrated the use of smartphone as an optical detection tool form measurement of absorption bands of colored dyes in visible domain. With integration of some simple laboratory optical components we have demonstrated the use of smartphone as a visible spectrometer with pixel resolution of 0.345 nm/pixel. The present smartphone sensor is capable of performing visible spectroscopy at level close to a commercial spectrometer. Since total cost involvement of the designed sensor being less than $150 excluding the smartphone and the sensor being compact, light weight and portable, we envision that the present sensor can be a very good alternative for performing visible spectroscopy experiments in near future.

References

1. Ozcan A (2014) Mobile phones democratize and cultivate next-generation imaging, diagnostics and measurement tools. Lab Chip 14(17):3187–3194
2. Preechaburana P, Gonzalez MC, Suska A et al (2012) Surface plasmon resonance chemical sensing on cell phones. Angew Chem Int Ed 51(46):11585–11588
3. Zhuo Y, Hu H, Chen WL et al (2014) Single nanoparticle detection using photonic crystal enhanced microscopy. Analyst 139(5):1007–1015
4. Tseng D, Mudanyali O, Oztoprak C et al (2010) Lensfree microscopy on a cellphone. Lab Chip 10(14):1787–1792
5. Oncescu V, Mancuso M, Erickson D (2014) Cholesterol testing on a smartphone. Lab Chip 14 (4):759–763
6. Oncescu V, O'Dell D, Erickson D (2013) Smartphone based health accessory for colorimetric detection of biomarkers in sweat and saliva. Lab Chip 13(16):3232–3238
7. Preechaburana P, Suska A, Filippini D (2014) Biosensing with cell phones. Trends Biotechnol 32(7):351–355
8. Gallegos D, Long KD, Yu H et al (2013) Label-free biodetection using a smartphone. Lab Chip 13(11):2124–2132
9. Dutta S, Choudhury A, Nath P (2014) Evanescent wave coupled spectroscopic sensing using smartphone. IEEE Photonics Technol Lett 26(6):568–570

Part XV
Opto-Electronic Materials

Chapter 67
Intrinsic Localized Modes in Metamaterials

Bijoy Mandal, Arindam Biswas, Swarup Samanta and A.K. Bhattacharjee

Abstract Metamaterials are also important nonlinear optical materials for applications in split-ring-resonator (SRR) in the antenna arrays. By using a Klein-Gordon approach, first classical breathers are shown. Then, by using a periodic boundary condition, quantum breathers are also shown in terms of TPBS parameters against coupling in the SRR system. An important variation of the TPBS parameters with coupling is observed. Finally, in a non-periodic boundary condition approach, the temporal variation of number of quanta is shown to derive the time of redistribution of quanta.

67.1 Introduction

The study of electrodynamics of substances with simultaneous negative values of magnetic permittivity (μ) and dielectric permittivity (ε) is quite interesting indeed. This was first formulated in 1968 by Veselago [1]. An unusual property of negative refraction in substances was predicted with both negative ε and μ. However, such substances do not occur in nature. A natural comes to our mind: in what sort of materials can these properties be observed? This question was answered by Smith et al. [2] based on some theoretical work of Pendry et al. [3, 4]. The fabrication of "metamaterial" (MM), i.e., an artificial structure consisted of metallic material wires that are responsible for the negative permittivity and metallic "split ring resonator" (SRR) responsible for the negative permeability. The SRRs may have different

B. Mandal
Department of CSE, NSHM Knowledge Campus, Durgapur, West Bengal, India

A. Biswas (✉) · S. Samanta
Department of ECE, Hooghly Engineering and Technology College, Hooghly 712103, India
e-mail: mailarindambiswas@yahoo.co.in

A.K. Bhattacharjee
Department of ECE, NIT, Durgapur, West Bengal, India

V. Lakshminarayanan and I. Bhattacharya (eds.), *Advances in Optical Science and Engineering*, Springer Proceedings in Physics 166,
DOI 10.1007/978-81-322-2367-2_67

shapes. The properties of optical and electrical for a metamaterial may be harnessed by the use of SRRs that has application in antennas in proper way. However, very high accuracy is required for the fabrication of SRRs. MMs also shows large magnetic response relatively at THz frequency unlike natural materials. So, their THz application makes significant advances.

For metamaterials, here the treatment focuses on a simple model system of a one-dimensional array of loosely coupled SRRs. The formulation is accomplished using a Taylor-expansion in the x-dimension. The magnetization of the metamaterial [5] is considered to study the field behavior in SRRs. The dynamical equation thus obtained gives a reliable solution for device application as supported by the numerical solutions. The modulation of the amplitude of solitonic waves is not carried out in the present work, which is a subject for our future work. For different cases of the system, soliton solutions for the Klein-Gordon equation for both bright and dark types are derived. Despite voluminous work done on SRRs based metamaterials, no derivation of classical dynamics from Hamiltonian of the system has been attempted to the best of our knowledge. Hence, an attempt is made here in this direction in the present work. Moreover, for QBs under both periodic and non-periodic boundary conditions, the variation of different TPBS parameters and 'time of redistribution' of the number of quanta with coupling are studied within the SRR system.

67.2 Theoretical Development

67.2.1 Classical Breathers

Klein-Gordon equation as we get the following Klein-Gordon dynamical equation:

$$\frac{\partial^2 q}{\partial \tau^2} - \text{ab}\frac{\partial^2 q}{\partial x^2} + b\left[q - \frac{\alpha}{3\varepsilon_l}q^3\right] - b\Lambda\Omega\sin(\Omega\tau) + \gamma\frac{\partial q}{\partial \tau} = 0 \tag{67.1}$$

Here, q_i is the polarization of the ith domain and ab denotes the interaction or coupling parameter incorporating the lattice distance term. To prepare the equation for numerical solution we introduce the auxiliary variable: $Q_i = \underset{i}{\dot{q}} = \frac{\partial q_i}{\partial \tau}$. This reduces the second order (67.1) to the first order system:

$$\dot{q} = Q \tag{67.2}$$

$$\dot{Q} = abDq - \gamma Q + b\left[q - \frac{\alpha}{3\varepsilon_l}q^3\right] + b\Lambda\Omega\sin(\Omega\tau) \tag{67.3}$$

The vectors q and Q represent the unknown charge and its time derivative in the domains that are arranged along the x-axis, and the term $b\left[q - \frac{\alpha}{3\varepsilon_l}q^3\right]$ should be interpreted as a nonlinear vector whose ith component is $b\left[q_i - \frac{\alpha}{3\varepsilon_l}q_i^3\right]$, where the coefficients of the anharmonic part are very important in deciding the formulation of discrete breathers. The domain of interest for the Klein-Gordon equation is typically the real line $x \in (-\infty, \infty)$. Therefore, there will use the Hermite and sinc second derivative matrices for D that is a banded matrix. Now, having selected an appropriate banded matrix D, it remains to solve our system of the first order differential equations that can be written as:

$$\begin{pmatrix} \dot{q} \\ \dot{Q} \end{pmatrix} = \begin{pmatrix} I & 0 \\ -\gamma & abD \end{pmatrix}\begin{pmatrix} Q \\ q \end{pmatrix} + b\begin{pmatrix} 0 \\ \left(q - \frac{\alpha}{3\varepsilon_l}q^3\right) \end{pmatrix} + \begin{pmatrix} 0 \\ U \end{pmatrix} b\Lambda\Omega\sin(\Omega\tau) \quad (67.4)$$

Here, U is a column vector whose elements are unity. We have used well-known 4th order Runge-Kutta method for the system (67.4). Next, we consider the DBs subjected to losses due to damping and the external driving force, i.e. for dissipative DBs. Now let us deal with QBs in terms of TPBS parameters in a periodic boundary condition.

67.2.2 Quantum Breathers

67.2.2.1 Periodic Boundary Condition Approach

Equation (67.1) show a general treatment of the dynamics mode in the array, for particularly modes that are localized over a small number of domains in that particular array strongly. Here, α is considered as the nonlinearity parameter. For next extended modes that are localized and very slowly range over a numbers of consecutive domains. Now, the discrete Hamiltonian (67.1) can be split as:

$$\tilde{H} = H_0 + H_1 \quad (67.5)$$

Then, a general 'basis' can be written for n particles. The numerical analysis was done with Fourier Grid Hamiltonian method [4] along with 1000 grids for 0.006 spacing to calculate various eigenvalues and eigenvectors. We are restricted to ourselves for two phonon states. Since on the working temperature, the number of phonon is very small. For reducing the computer memory requirement, we can take the advantage of translational invariance by periodic Bloch wave formulation, as shown in [4]. For a two-phonons case, the non-zero hopping coefficients (D_{mn}) are: $D_{01} = D_{10}, D_{12} = D_{21}$. Equation (67.6) show energy gap between the single phonon continuums along with a bound state is given by:

$$E_g = E_2 - E_0 - 2(E_1 - E_0). \tag{67.6}$$

where E_0, E_1 and E_2 are three eigenvalues at different points of wave vector (k) that are calculated from the computation to generate $E(k)$ versus k curve, that produce the signature of quantum breathers (QB) in terms of two-phonon bound state. The width of the single-phonon in the eigenspectrum is given by the magnitude of 4σ, where σ is expressed as:

$$\sigma = -\frac{\lambda}{2} D_{01}^2 \tag{67.7}$$

where λ is an interaction term, and D_{01} represents the coefficient for zero to single phonon generation. The variation of the single phonon spectrum width (W_{ph}) represents (through $D_{01} = D_{10}$) the creation of a new phonon or annihilation of an existing phonon. Again, the hopping coefficient f or a single phonon to become a two-phonon bound state is given by:

$$\mu = -\frac{\lambda}{2} D_{01} D_{12} = -\frac{\lambda}{2} D_{10} D_{21}. \tag{67.8}$$

All the above calculations were done for 51 sites or domains and $\lambda = 12$. More data points could be used in our present simulation, but here we are primarily focused to study nonlinearity or impurity induced critical behavior of discrete breather motion and its quantum origin. To treat the problem analytically we take the help of second-quantization method. By quantizing the Hamiltonian in (67.1) in a number conserving quantized form with $N \to \infty$ [6] with $\lambda_1 = \frac{\lambda}{(\lambda-\alpha)}$, $\eta = \frac{2\alpha}{(\lambda-\alpha)}$ and $E = 0$ in (67.1) leads to the equation for two-phonon amplitudes as:

$$\hat{H} = \sum_n a_n^+ a_n + \frac{3}{8} \eta a_n^{+2} a_n^2 + \frac{\lambda_1}{2} \{a_n^+ (a_{n+1} + a_{n-1}) + h.c.\} \tag{67.9}$$

67.2.2.2 Non-periodic Boundary Condition

The general Hamiltonian for the Klein-Gordon equation for order parameter (y_n) at nth site is written as:

$$H = \sum_n \frac{p_n^2}{2m} + \frac{A}{2} y_n^2 + \frac{B}{4} y_n^4 + k(y_n - y_{n-1})^2. \tag{67.10}$$

The first term is momentum at nth site (p_n), the second and third terms are nonlinear potential formulation and the last term contains an interaction constant (k). From the above (67.8), after deducting the classical equation of motion and rescaling of time, we get:

$$\tilde{H} = \sum_n \frac{1}{2}p_n^2 + \frac{1}{2}y_n^2 + \eta y_n^4 + \lambda(y_n - y_{n-1})^2. \qquad (67.11)$$

where $\eta = \frac{B}{4A}\lambda = \frac{k}{2A}$, A and B are two constants. Now, let us use creation and annihilation Bosonic operators at the nth site and the above Hamiltonian (67.9) is quantized. The non-number conserving methods for four sites and an arbitrary number of particles are shown in an important work done by P[7]. However, the method which is presented above that gives a generalized idea for solving the system for arbitrary number of particles on arbitrary number of sites. So, it is clearly distinguished from the other investigations in our method.

After second quantization, a general 'basis' is then created. For the characterization of quantum discrete breathers, we need to make the Hamiltonian time-dependent. We can take the help of temporal evolution for number of bosons at each site of the system: $\langle n_i \rangle(t) = \langle \Psi_t | \hat{n}_i | \Psi_t \rangle$. Using ith eigenstate of the Hamiltonian, we make it time dependent as follows:

$$|\Psi_i(t)\rangle = \sum_i b_i \exp(-iE_i t/\hbar)|\psi_i\rangle. \qquad (67.12)$$

where ψ_i and E_i is the ith eigenvector and eigenvalue respectively, and t is time. The Planck's constant (h) is taken as unity and $b_i = \langle \psi_i \mid \psi(0) \rangle$ for each site i and for a given range of t, where $\psi(0)$ stands for the initial state.

67.3 Results and Discussion

In Although Lazarides and coworkers [7] and others have shown 3-D pictures of classical breathers, but that were based on non-linear Shrodinger equation formalism. However, in the present work, we have derived a non-linear Klein-Gordon equation, where coupling or interaction between different SRR elements is also very important. From the spatial discretization by using spectral collocation method, it is interesting to see the overview of K-G breathers in 3-D pictures. With linear permittivity = 2.0, coupling = 0.05, the classical breathers are shown for zero damping in Fig. 67.1 and that for non-zero damping (γ = 0.20) is shown in Fig. 67.2. It is seen from Fig. 67.1 that the breathers are symmetric while the effect of damping is noticeable to some extent in Fig. 67.2, where relatively high value of damping is applied. It should be mentioned that a damping value up to 0.18 does not disturb the breather profile to a significant extent, and from a value of 0.20, it starts showing the effect of damping. In metamaterials, the value of damping is generally not considered very high [5] so that breather oscillations could be considered stable.

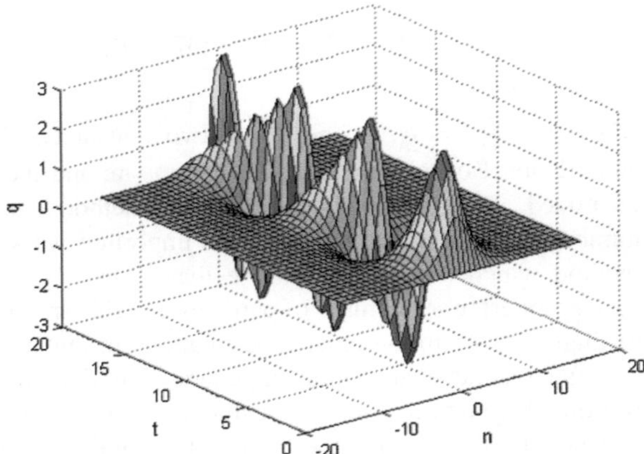

Fig. 67.1 Dissipative breathers for λ = 0.05, $\varepsilon_l = 2$ and $\gamma = 0$. Symmetric breathers are observed without any damping

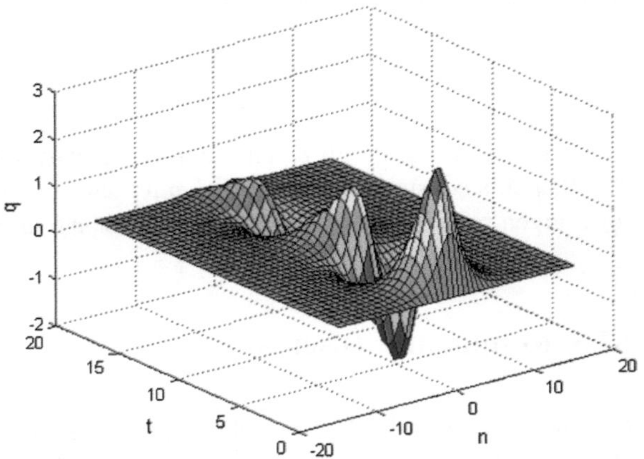

Fig. 67.2 Dissipative breathers for λ = 0.05, $\varepsilon_l = 2$ and $\gamma = 0.20$. The effect of damping is not appreciable even at such a high value

We first show a typical eigenspectrum of two-phonon bound state (TPBS) by using the formalism, as described in [8] for a periodic boundary condition in terms of using a Bloch function by using Fourier Grid Hamiltonian method. For a very low value of linear dielectric constant of 0.002 with focusing nonlinearity $\alpha = +1$ and a small value of interaction constant of 0.05, it is noted from Fig. 67.3 that a continuum of states with single particle with a breather band with two phonon bound state or quantum breather state can also be observed.

For a non-periodic boundary condition, a typical simulation for a smaller value of interaction or coupling is shown below in Fig. 67.4

Here, the critical time for redistribution ($t_{\rm re}$) is around 19.20, i.e. where two quanta meet or tend to meet. It is known that by engineering the geometry of the SRR assembly, the interaction between SRR elements can be varied.

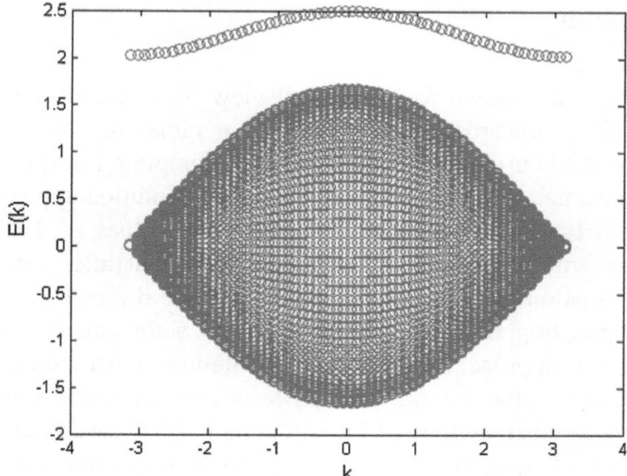

Fig. 67.3 Eigenspectrum for λ = 0.05, ε_l = 0.002 and α = +1. The continuum represents single phonon continua and the quantum breather band or two-phonon bound state in the upper branch of the spectrum

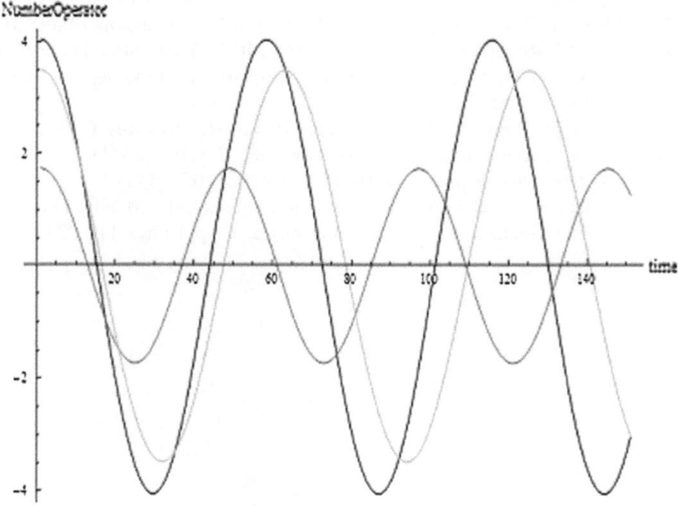

Fig. 67.4 For metamaterials with SRR assembly, 8 particles on 3 sites with $|\Psi(0)\rangle = |7, 1, 0\rangle$, ε_l = 2, α = +1, $\lambda' = 0.01$ and $\Lambda = 0$ in (67.12), (t_{re} = 19.20)

67.4 Conclusion

Classical breathers are shown for an overall view in 3 dimensions that does not show any sensitivity towards higher damping, or rather the breather oscillations seem to be quite stable up to a quite high value of damping. For quantum breathers in SRR based metamaterials, in a periodic boundary condition approach the second quantization with bosonic operators gives rise some values of TPBS parameters against coupling within the SRR system that show that after smaller values of coupling, the formation quantum breathers seems to be difficult thereby increasing both single-phonon hopping energy and hopping coefficient. For a non-periodic boundary condition approach, the temporal evolution spectra show the quantum breathers life time in femtosecond. This piece of information is considered useful for a future study in this new field of investigation in SRR based metamaterials and other application of quantum breathers in such important nonlinear optical materials.

References

1. Veselago VG (1968) Sov Phys Usp 10:509
2. Smith DR, Padilla WJ, Vier DC, Nemat-Nasser SC, Schultz S (2000) Phys Rev Lett 84:4184
3. Pendry JB, Holden AJ, Roberts DJ, Stewart WJ (1998) J Phys Condens Matter 10:4785
4. Pendry JB, Holden AJ, Roberts DJ, Stewart WJ (1999) IEEE Trans Micr Theory Tech 47:2075
5. Schulman LS, Tolkunov D, Mihokova E (2006) Structure and time-dependence of quantum breathers. Chem Phys 322:55–74
6. O'Brien S, McPeake D, Ramakrishna SA, Pendry JB (2004) Phys Rev B 69:241101
7. Giri P, Choudhary K, Sengupta A, Bandyopadhyay AK, Ray PC (2011) Discrete breathers in nonlinear $LiNbO_3$-type ferroelectrics. J Appl Phys 109:054105–054112
8. Biswas A, Choudhary K, Bandyopadhyay AK, Bhattacharjee AK, Mandal D (2011) Quantum pinning transition due to charge defect in ferroelectrics. J Appl Phys 110:024104–024111

Chapter 68
Optically Enhanced SnO_2/CdS Nanocomposites by Chemical Method and Their Characterization

Vijay Kumar, P. Rajaram and Y.C. Goswami

Abstract SnO_2/CdS heterostructures nanocomposites were synthesized using newly develop ultrasonic sol-gel method. Stannous chloride, cadmium chloride ($CdCL_2 \cdot 2H_2O$) and thiourea were used as Sn, Cd and S precursors respectively and Ethylene glycol was used as a complexing agent. The samples were characterized by XRD, SEM, EDX, optical studies. All the XRD peaks are identified for SnO_2 however a slight shift is observed with addition of CdS. EDX confirms the presence of Sn, Cd and S in the samples. AFM and SEM studies also confirm the nanofibers structures with roughness 2.9136 nm and conversion of hollow tubes into nanofibers. The UV–Vis spectrum of the nanostructures displays a new absorption band range lies in the range between 450–530 nm compared with the bare SnO_2 hollow tubes. The strong emission peak is observed at 375 nm in UV region for all the samples and intensity of emission become weaker due to incorporation of CdS nanoparticles. Addition of CdS introduces effective charge separation in the heterostructures which controlled the intensity of photo luminescence makes them suitable for optoelectronic applications.

68.1 Introduction

Nanocrystals in the form of tubes, wires, belts, and dots represent a building block that can be used to assemble a range of electronic, photonic and sensing devices. The size, shape and phase of inorganic nanocrystals are key elements to vary the optical, electrical properties [1]. Consequently, the controlled fabrications with

V. Kumar (✉) · Y.C. Goswami
School of Physical Sciences, ITM University, Turari, Gwalior, M.P 474001, India
e-mail: vijaynadda83@gmail.com

Y.C. Goswami
e-mail: y_goswami@yahoo.com

P. Rajaram
SOS Physics, Jiwaji University, Gwalior, M.P 474001, India

V. Lakshminarayanan and I. Bhattacharya (eds.), *Advances in Optical Science and Engineering*, Springer Proceedings in Physics 166,
DOI 10.1007/978-81-322-2367-2_68

functional properties are required to meet the ever-increasing demands. Metal oxide systems made up of SnO_2 are of great interest owning to high chemical stability and good flexibility in fabrication. Tin oxide is a wide band gap (3.6 eV) at 300 K and n-type semiconductor material which has been widely used as gas sensors, photocatalysts and anticoating statistics [2–5]. However lot of research is required to improve the sensitivity and selectivity of these oxides. The application of these nanomaterials may be restricted by their own physical and chemical characteristics like bandgap, electrical, optical properties, gas sensitivity and chemical stability. It is reasonable to suggest modification of SnO_2 nanostructures. Hence for improvement, another material such as CdS is coated on the SnO_2 core. CdS is an n-type (2.42 eV) material with high light- collecting ability in the visible region and a potential to accept or release electron. Therefore SnO_2/CdS heterostructure is a type-II material in which the lower band gap (CdS) is deposited around a core of SnO_2 which has a higher band gap. The photogenerated electrons confined to the core and hole is located in the shell. SnO_2/CdS has a potential application such as gas sensor [6] and optoelectronic devices [7]. SnO_2/CdS heterostructure prepared by various method like sonochemical deposition [6], dip-coating [8], solution synthesis [9] and colloidal approach [10] etc. In the present work, we have reported type-II SnO_2/CdS heterostructures grown by ultrasonically assisted sol-gel method.

68.2 Experimental Detail

The SnO_2/CdS heterostructure nanocomposites were prepared in two stages. In the first stage transparent sol of SnO_2 nanoparticles were prepared using continuous ultrasonication of $SnCl_2 \cdot 2H_2O$ in ethylene glycol at 70 °C for 3 h with few drops of HCL to prevent hydrolysis. In the second stage cadmium chloride and thiourea were added and stirrered for another few hours. A transparent yellowish sol was obtained, after 24 h that converted into gel. Spin coater was used to obtained films. All the samples were characterized by various techniques. UV–Vis spectra were carried by Perkin Elmer λ-25 instrument in the lab, X-ray diffractogram were obtained in the 2θ ranging from 20 to 80° with Cu Kα radiation of wavelength 1.546 Å using Bruker D8 Advanced XRD, Scanning electron micrographs (SEM) and EDX curves were obtained by using SX100 (Cameca). AFM micrographs were obtained by using digital Instrumentation—Inc, USA.

68.3 Results and Discussion

Figure 68.1a and b, shows the X-ray diffraction pattern of bare SnO_2 and SnO_2/CdS heterostructures grown by ultrasonic sol-gel method. All the peaks are indentified for (110), (101) and (220) tetragonal phase of SnO_2 [JCPDS Card no. 41–1445]. Slight shift towards lower angle in 110 peaks is observed for SnO_2/CdS sample

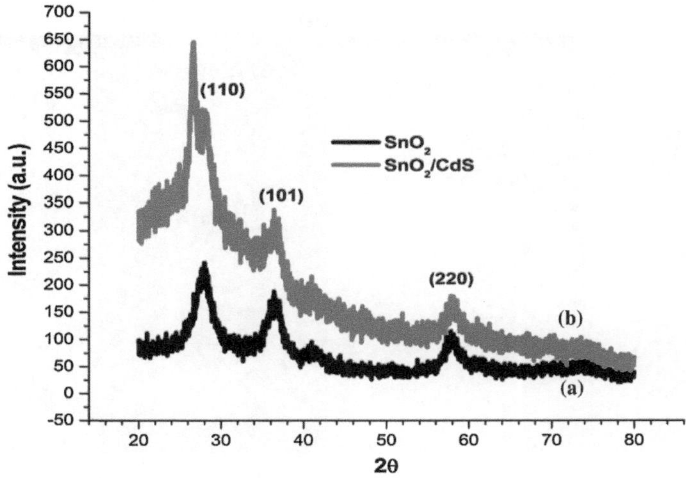

Fig. 68.1 X-ray diffractograms for **a** bare SnO_2 and **b** SnO_2/CdS nanostructures

Fig. 68.2 Scanning electron micrograph of **a** bare SnO_2 and **b** SnO_2/CdS nanostructures

which may be due to tensile strain reported in [11] as the SnO_2 has much smaller lattice constant than CdS shell. It resulted in possibility of core/shell nanostructures. No peak for CdS shell is observed which might be occur due to deposition of very thin layer of CdS on SnO_2.

SEM of SnO_2 and SnO_2/CdS (Inset EDX spectra) shown in Fig. 68.2. SEM studies show large randomly oriented grains and their conversion into nanofibers. EDX spectra shown in inset represent the presence of Cd and S compositions on the SnO_2 hollow nanostructures.

Figure 68.3a, b show the 2 dimensional and 3 dimensional AFM micrographs of (a) bare SnO_2 and (b) SnO_2/CdS nanostructures. Bare SnO_2 shows the hollow

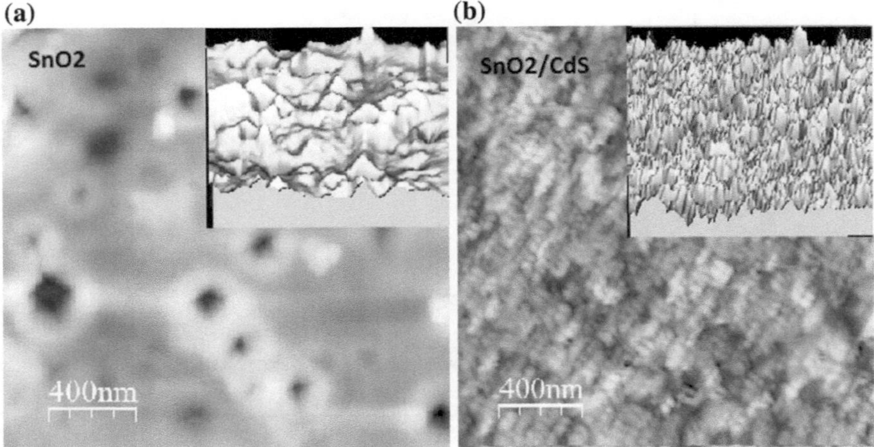

Fig. 68.3 2D and 3D atomic force micrographs of **a** bare SnO_2 and **b** SnO_2/CdS core/shell nanostructures

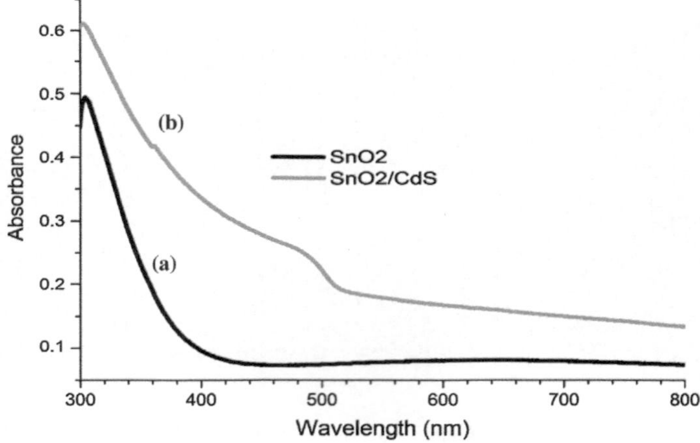

Fig. 68.4 Optical absorbance spectra of (**a**) bare SnO_2 and (**b**) SnO_2/CdS core/shell nanostructures

nanostructures with roughness 7.7308 nm [12] that converted into nano fibers with roughness 2.9136 nm on addition of CdS. It also indicates the formation of nanocomposites.

Figure 68.4 shows the optical absorption studies of bare SnO_2 and SnO_2/CdS nanostructures absorbed mainly UV-light, whereas SnO_2/CdS films show enhanced absorption at wavelength greater than 400 nm. Absorption edge lies in the range 450–530 nm wavelength which correspond to the band gap energy of CdS nanoparticles whereas CdS is a visible-light photo sensitizer for SnO_2. Such observations make them suitable for optoelectronic devices where more visible region can be utilized.

68.4 Conclusions

SnO_2/CdS heterostructure composites were obtained newly developed by ultrasonically sol-gel method. Slight shift in X-ray peaks are observed in SnO_2 due to CdS indicates the formation of core/shell structures. AFM studies observed the roughness of SnO_2/CdS decreases with depositing CdS on SnO_2 hollow nanostructures was obtained. SEM and EDX studies also confirm the nanofibers nanostructures. EDX confirms the presence of Sn, Cd and S respectively. The UV–Vis spectra of the nanostructures displays a new absorption band range lies in the range between 450–530 nm compared with the bare SnO_2 hollow tubes makes them suitable for various optoelectronic devices.

Acknowledgments The authors are thankful to MPCST and AICTE for funding this work. The authors are also thankful UGC-DAE consortium for scientific research Indore India for providing XRD and AFM facility and IIT Roorki for providing SEM and EDX facility.

References

1. Alivisatos AP (1996) Semiconductor clusters, nanocrystals, and quantum dots. Science 271:933–937
2. Stergiopoulos T, Arabatzis IM, Cachet H, Falaras P (2003) Photoelectrochemistry at SnO_2 particulate fractal electrodes sensitized by a ruthenium complex: solid-state solar cell assembling by incorporating a composite polymer electrolyte. J Photochem Photob A: Chem 155:163–170
3. Bergeron BV, Marton A, Oskaam G, Meyer GJ (2005) Dye-sensitized SnO_2 electrodes with iodide and pseudohalide redox mediators. J Phys Chem B 109:937–943
4. Han S, Jang B, Kim T, Oh SM, Hyeon T (2005) Simple synthesis of hollow tin dioxide microspheres and their application to lithium-ion battery anodes. Adv Funct Mater 15:1845–1850
5. Goswami YC, Kumar V, Rajaram P (2014) Sonochemical growth of SnO_2 hollow nanostructures on cellulose acetate flexible substrates. Mater Lett 128:425–428
6. Gao T, Wang T (2004) Sonochemical synthesis of SnO_2 nanobelt/CdS nanoparticle core/shell heterostructures. Chem Commun 22:2558–2559
7. Sun H, Kang SZ, Mu J (2009) Preparation and optoelectronic property of a SnO_2/CdS nanocomposite based on the flowerlike clusters of SnO_2 nanorods. J Dispersion Sci Technol 30:384–387
8. Nasr C, Hotchandani S, Kim WY, Schmehl RH, Kamat PV (1997) Photoelectrochemistry of composite semiconductor thin films. Photosensitization of SnO_2/CdS coupled nanocrystallites with a ruthenium polypyridyl complex. J Phys Chem B 101:7480–7487
9. Yan S, Hu D, Hu F, Wu J, Huang N, Xiao Z (2011) Solution-based synthesis of SnO_2 nanoparticle/CdS nanowire heterostructures. CrystEngComm 13:4580–4585

10. Kennedy R, Martini I, Hartland G, Kamat P (1997) Capped semiconductor colloids: synthesis and photochemistry of CdS capped SnO_2 nanocrystallites. J Chem Sci 109:497–507
11. Wang K, Chem JJ, Zhou WL, Zhang Y, Yan YF, Pern J, Mascarenhas A (2008) Direct growth of highly mismatched type II ZnO/ZnSe core/shell nanowire arrays on transparent conducting oxide substrates for solar cell applications. Adv Mater 20:3248–3253
12. Goswami YC, Kumar V, Rajaram P, Ganesan V, Malik MA, O'Brien P (2014) Synthesis of SnO_2 nanostructures by ultrasonic-assisted sol–gel method. J Sol-Gel Sci Technol 69:617–624

Chapter 69
Temperature Effect on Optical Gain of CdSe/ZnSe Quantum Dots

Dharmendra Kumar, C.M.S. Negi and Jitendra Kumar

Abstract In this work, theoretical formulation for the computation of electronic structure and gain spectra of CdSe/ZnSe quantum dot has been developed in the framework of multi-band k.p. approach. The optical gain of CdSe/ZnSe quantum dot has been determined. The gain is found to increase with carrier density but it is found to decrease as we increase the temperature.

69.1 Introduction

The study of temperature dependent optical gain in self-assembled CdSe/ ZnSe quantum dot (QD) structure has paid lot of attention since last decade. It is essential to understand fundamental behaviour of optoelectronic devices such as QD laser diode [1] and quantum dot infrared photodetector (QDIP) [2], because the number of electrons in the confined states of QDs determines the characteristics of such devices. Valerini et al. [3] have shown that the increase in temperature reduces photoluminescence (PL) intensity and the emission energy is red shifted, in another study it has been found that due to increase in temperature output intensity decreases [4].

In this research work, we theoretically explore the gain characteristics of self-assembled CdSe/ZnSe QD modelled by parabolic confinement potential along x–y plane, while in the growth direction (z-axis) confinement potential is modelled using finite well potential which arises due to the band offsets. We report the high dependence of the gain spectra on the carrier density and temperature of the QD. The four bands Luttinger Hamiltonian is numerically diagonalized using harmonic oscillator basis to determine the energy eigenvalues and eigen vectors.

D. Kumar (✉) · J. Kumar
Department of Electronics Engineering, Indian School of Mines,
Dhanbad, Jharkhand 826004, India
e-mail: dharmendra0127@gmail.com

C.M.S. Negi
Department of Electronics, BanasthaliVidyapith, Banasthali, Rajasthan 304022, India

V. Lakshminarayanan and I. Bhattacharya (eds.), *Advances in Optical Science and Engineering*, Springer Proceedings in Physics 166,
DOI 10.1007/978-81-322-2367-2_69

The calculated eigenvectors and eigenvalues are further employed to calculate the dipole matrix elements for interband transitions which is followed by the calculation of the optical gain of QD.

69.2 Theoretical Discription

In this work, a disc shape QD having elliptical cross section is modelled by a parabolic confinement potential $V_{x,y}(x,y)$ in the x–y plane and confinement along z-direction is considered to be finite quantum well. The growth axis of quantum well is considered with z-axis.

The parabolic confinement $V_{x,y}(x,y)$ is given by

$$V_{x,y} = \frac{1}{2}\alpha_x x^2 + \frac{1}{2}\alpha_y y^2 \tag{69.1}$$

The parameter $\alpha_{x,y}$ decides the strength of confinement. The eigenenergies and wavefunctions of valence band (VB) and conduction band (CB) have been calculated under effective mass approximation. For VB we have used 4×4 Luttinger Hamiltonian to find electronic structure.

The envelope function for the electrons in the CB has been calculated by solving Schrödinger equation in parabolic confinement potential along in-plane and solving transcendental equation for finite well potential confinement along growth direction. The discrete energy values for the CB can be obtained as

$$E_C = \left[\left(n_x + \frac{1}{2}\right)\hbar\omega_x + \left(n_y + \frac{1}{2}\right)\hbar\omega_y + E_{ez}\right] \tag{69.2}$$

here, n_x and n_y represent the quantum numbers for the in-plane confinement, $\omega_{x,y} = \sqrt{\alpha_{x,y}/m_e}$, m_e denotes the effective mass of electron and E_{ez} being the energy due to z-confinement.

To determine the complex VB structure, we have used 4×4 Luttinger Hamiltonian under effective mass approximation given as [5, 6]

$$\begin{bmatrix} H_{hh} & S & R & 0 \\ S^* & H_{lh} & 0 & -R \\ R^* & 0 & H_{lh} & S \\ 0 & -R^* & S^* & H_{hh} \end{bmatrix} \tag{69.3}$$

We can expand the Hamiltonian on harmonic oscillator basis function to get eigen energies of heavy holes and light holes as [7]

$$E_{hh} = \left[E_{z,hh} + \left(n_x + \frac{1}{2}\right)\hbar\omega_{hh}^{x} + \left(n_y + \frac{1}{2}\right)\hbar\omega_{hh}^{y}\right]\delta_{n_x,n_x'}\delta_{n_y,n_y'} \tag{69.4}$$

$$E_{lh} = \left[E_{z,lh} + \left(n_x + \frac{1}{2}\right)\hbar\omega_{lh}^{x} + \left(n_y + \frac{1}{2}\right)\hbar\omega_{lh}^{y}\right]\delta_{n_x,n_x'}\delta_{n_y,n_y'} \tag{69.5}$$

where, $\omega_{hh(lh)}^{x,y} = \sqrt{\alpha_{x,y}/m_{hh(lh)}}$, and $E_{hh(lh)}^{z}$ denotes the discrete energy of holes along growth axis, we express the isotropic frequency as ω_0 when $\alpha_x = \alpha_y$ while dissimilarity arising due to anisotropy can be define as $\partial\omega$. The expression for the frequencies can be represented as:

$$\omega_{hh,lh}^{x,y} = \omega_0 \pm \delta\omega\sqrt{\gamma_1 \pm \gamma_2} \tag{69.6}$$

The anisotropy parameter is A $= \delta\omega/\omega_0$, in the above equation $\gamma_1, \gamma_2, \gamma_3$ are the Luttinger parameter.

The resulting Luttinger Hamiltonian matrix of size 100 × 100 is diagonalized numerically to compute energy eigenvalues and eigen vectors [8]. The computation error is less than 0.1 %. Subsequently we obtain the linear optical gain by calculating the dipole matrix elements using the solution of (69.3).

69.2.1 Gain Formula

The expression for the linear optical gain of QD can be derived by using density-matrix theory approach and can be represented as [9]

$$g(E) = \frac{2\pi e^2\hbar N_D}{cn_r\varepsilon_0 m_0^2}\sum_{c,v}\frac{\left|P_{cv}^{\sigma}\right|^2(f_c - f_v)}{E_{cv}}B_{cv}(E - E_{cv}) \tag{69.7}$$

where, c signifies the discrete energy state in the CB, v denote the discrete energy levels in the VB, f_c and f_v are the Fermi distribution function for the discrete state in CB and VB, respectively of the QD, n_r stands for the refractive index of QD material, E_{cv} represents the interband transition energy, N_D is the density of carriers in QD, $\hbar$ being the reduced Planck's constant, c denotes the light speed, e being the electron charge and B_{cv} characterize the branding in the gain spectra, for single QD it can be written as [10]

$$B_{cv}(E - E_{cv}) = \frac{\hbar\Gamma_{cv}/\pi}{(E - E_{cv})^2 + (\hbar\Gamma_{cv})^2} \tag{69.8}$$

where, $\hbar\Gamma_{cv}$ symbolizes the homogeneous broadening owing to the interband scattering and related to the relaxation time τ by $\Gamma_{cv} = 1/\tau$.

The transition matrix element for interband transitions can be evaluated using relation

$$\left|P_{cv}^{\sigma}\right|^{2} = \left|I_{c,v}\right|^{2} M^{2}, \tag{69.9}$$

I_{cv} denote the overlap integral between hole and electron envelope wave functions, and it can be obtained by the k.p method. M is the dipole moment element can be written as

$$M^{2} = \frac{m_0^2}{12m_e^*} \frac{E_g\left(E_g + D\right)}{\left(E_g + \frac{2D}{3}\right)} \tag{69.10}$$

here m_e^* denotes the effective mass of electron, E_g is the energy band gap, and D signifies the energy of the spin orbit interaction in the QD.

Electron density N and hole density P ($\simeq$ N) for intrinsic material in the QD are related to quasi Fermi levels (E_{fc} and E_{fv}) as [11].

$$\mathrm{N} = \sum_i \frac{2}{\left[1 + \exp\left(\frac{E_{ci} - E_{fc}}{\mathrm{KT}}\right)\right] V} \tag{69.11}$$

$$\mathrm{P} = \sum_i \frac{2}{\left[1 + \exp\left(\frac{E_{fv} - E_{hi}}{\mathrm{KT}}\right)\right] V} \tag{69.12}$$

The origin of the energy states is taken at the bottom of the CB, E_{ci} and E_{hi} are the quantized energy level, V is the Volume of the elliptical QD, and $i = n_x, n_y, m_s$.

69.3 Result and Discussion

This section discusses the effect of carrier density, and temperature on the gain spectra of CdSe/ZnSe QD. For numerical calculation the material parameters for bulk CdSe are: $E_g = 1.75$ eV [12], $n_r = 2.54$ [13], $m_e = 0.13m_0$ [14], Luttinger parameters $\gamma_1 = 2.1$, $\gamma_2 = \gamma_3 = 0.55$ [15], the band offset at the interface of the hetro-junction is 0.2782 eV for VB and 0.1498 eV for CB.

Figure 69.1a displays gain spectra for different injected carrier densities and temperature within the dot. As the carrier density is increased, the quasi-Fermi level will shift towards higher energy states, thus in turn, enhancing the gain.

In Fig. 69.1b displays the temperature effect on the gain spectra for the QD at an injected carrier density of $N = 4 \times 10^{24}$ m^{-3}. It can be seen that as the temperature is increased, the peak of the gain curve gets red shifted and also the peak value of the gain is reduced. With increase in temperature, the effective band gap decreases,

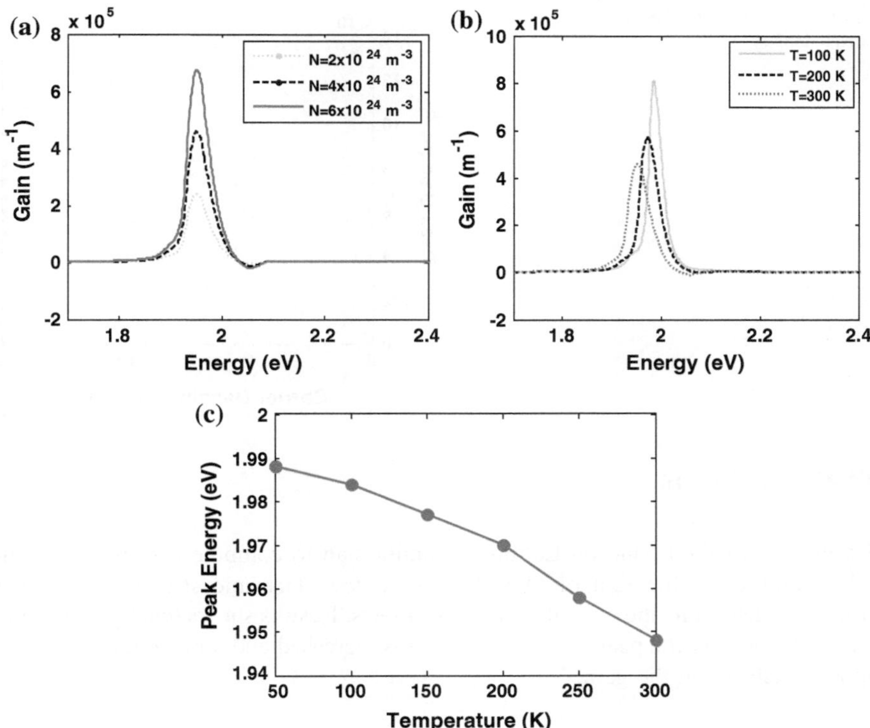

Fig. 69.1 **a** Gain spectra versus transition energy for different carrier density (*N*) for height of QD = 2 nm and radius = 6 nm. **b** Gain spectra versus transition energy for different Temperature (*T*). **c** Peak energy versus Temperature (*T*)

thereby increases the height of the QD and hence its confinement is lowered which results in the red shift. It is well known that as temperature increases, the carriers spread in QDs, charge carrier trapping on QDs increases the probability of non-radiative recombination increases, thereupon, the peak gain will decrease.

Figure 69.1c shows the variation of the peak energy at which the gain peak occurs as a function of temperature. D. Valeriniet.al., have reported the similar results for colloidal CdSe/ZnS core/shell QD [3].

The peak gain values as a function of carrier density for the QD are plotted in Fig. 69.2 for height of the quantum dot (L_c) taken to be 2 nm at T = 300 K. The peak value of the gain varies linearly with carrier density (*N*).

The peak gain is $47\,\mathrm{m}^{-1}$ for lower carrier density and the peak gain is $1.249 \times 10^6\,\mathrm{m}^{-1}$, for higher *N*. Increase in *N* increases peak value of gain but it saturates at higher *N*. This may be very significant in order to reduce the threshold carrier density for CdSe/ZnSe QD lasers.

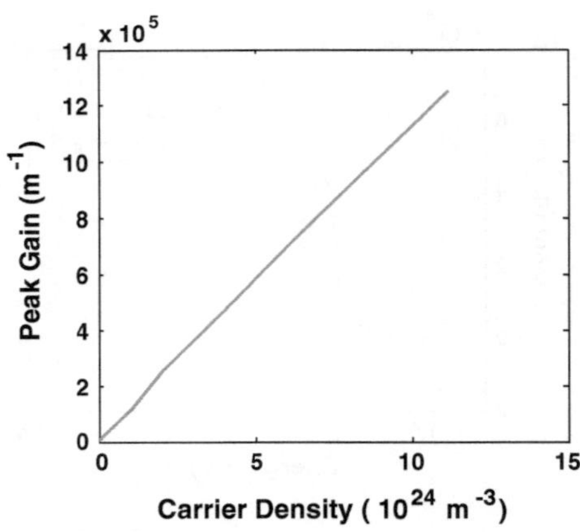

Fig. 69.2 Peak gain versus carrier density

69.4 Conclusion

We use k.p method band on Luttinger Hamiltonian to compute the energy eigenvalues and wave function of QD. The computed electronic structure is further utilized to determine the optical properties of QDs. Results shows that by increasing the carrier density the peak value of the gain is increased and temperature will have opposite effects on the gain.

References

1. Liu H et al (2011) Long-wavelength InAs/GaAs quantum-dot laser diode monolithically grown on Ge substrate. Nat Photonics 5:416–419
2. Negi CMS, Kumar D, Gupta SK, Kumar J (2013) Theoretical analysis of resonant cavity p-type quantum dotinfrared photodetector. IEEE J Quantum Electron 49:839–845
3. Valerini D et al (2005) Temperature dependence of the photoluminescence properties of colloidal CdSe/ZnS core/shell quantum dots embedded in a polystyrene matrix. Phy Rev B 71 (1–6):235409
4. Ranjbaran A (2012) Temperature effects on output characteristics of quantum dot white light emitting diode. Front Optoelectron 5(3):284–291
5. Chuang SL (2009) Physics of photonic devices. Wiley, Hoboken
6. Luttinger JM (1956) Quantum theory of cyclotron resonance in semiconductors: general theory. Phys Rev 102:1030–1041
7. Kumar J, Kapoor S, Gupta SK, Sen PK (2006) Theoretical investigation of the effect of asymmetry on optical anisotropy and electronic structure of Stranski-Krastanov quantum dots. Phys Rev B 74(1–10):115326
8. Kumar D, Negi CMS, Gupta SK, Kumar J (2013) Effect of shape anisotropy and size on electronic structure of CdSe/ZnSe quantum dots. IEEE Trans Nanotechnol 12:925–930
9. Sugawara M, Mukai K, Nakata Y, Ishikawa H, Sakamoto A (2000) Effect of homogeneous broadening of optical gain on lasing spectra in self-assembled InxGa1-xAs/GaAs quantum dot lasers. Phy Rev B 61(11):7595–7603

10. Sakamoto A, Sugawara M (2000) Theoretical calculation of lasing spectra of quantum-dot lasers: effect of homogeneous broadening of optical gain. IEEE Photonics Technol Lett 12:107–109
11. Asada M et al (1986) Gain and the threshold of three-dimensional quantum-box lasers. IEEE J Quantum Electron 22:1915–1921
12. Kostić R et al (2011) Nonlinear absorption spectra for intersubband transitions of CdSe/ZnS spherical quantum dots. J Nanophotonics 5:051810
13. Bahae MS, Hagan DJ, Strylandn EWV (1991) Dispersion of bound electronic nonlinear refraction in solids. IEEE J Quantum Electron 27:1296–1309
14. Adachi S (2005) Properties of Group-IV, III–V and II–VI Semiconductors. Wiley, New Jersey
15. Ekimov AI, Hache F et al (1993) Absorption and intensity-dependent photoluminescence measurements on CdSe quantum dots: assignment of the first electronic transitions. Opt Soc Am B 10:100–107

Chapter 70
Nonlinear Optical Characterization of Borotellurite Glass of Composition $0.1BaO\text{-}0.4TeO_2\text{-}0.5B_2O_3$ by Z-Scan Method

Anil Kumar, Devendra Mohan, A. Ghosh and A.K. Gupta

Abstract Present work focuses on the study of nonlinear optical properties of borotellurite glass of composition $0.1BaO\text{-}0.4TeO_2\text{-}0.5B_2O_3$. Third order optical nonlinear coefficients of borotellurite glass have been investigated by using the standard z-scan method using 100 femtosecond Ti: sapphire laser at 1550 nm wavelength with repetition rate of 1 kHz. Nonlinear refractive index n_2 was estimated to be 2.59×10^{-17} cm^2/W. Nonlinear absorption coefficient β was found to be 0.3189×10^{-12} cm/W. Nonlinear susceptibility $\chi^{(3)}$ has been calculated to be $\sim 1.8329 \times 10^{-14}$ esu.

70.1 Introduction

Single beam Z scan method [1–3] is a standard tool for determining nonlinear parameters of various materials because of its simplicity, sensitivity, accuracy and the ease of separation between nonlinear refraction and nonlinear absorption. In Z-scan technique the sample moves across the focal point of laser beam along the direction of propagation of the laser beam. Assuming Gaussian beam optics, this experiment allows an intensity scan of the irradiated sample, and provides information about the nonlinearity in the sample. This results due to change in the size of the incident beam on the sample which is minimum at the focus and then increases again on crossing the focus. The purpose of the experiment is to find the deterministic variation in transmission as the incident intensity changes by translation along the z-axis. The change in the transmittance of the focusing Gaussian beam in a medium is recorded as a function of position of medium.

A. Kumar (✉) · A. Ghosh · A.K. Gupta
Instruments Research and Development Establishment (IRDE), Dehradun 248008, India
e-mail: bansal_anil@irde.drdo.in

A. Kumar · D. Mohan
Guru Jambheshwar University of Science and Technology, Hisar 125001, India

V. Lakshminarayanan and I. Bhattacharya (eds.), *Advances in Optical Science and Engineering*, Springer Proceedings in Physics 166,
DOI 10.1007/978-81-322-2367-2_70

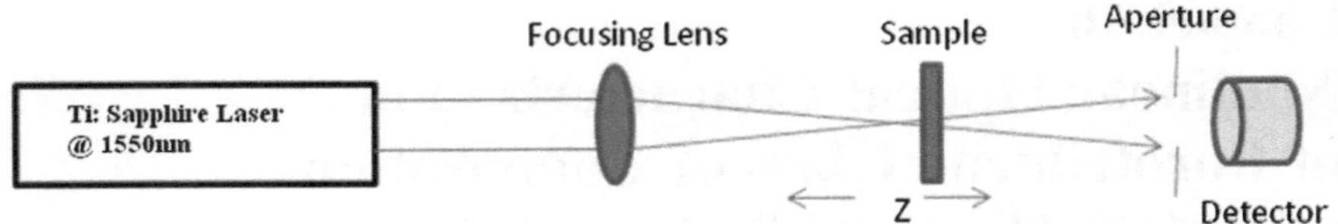

Fig. 70.1 Schematic diagram of Z-scan experiment (aperture will be removed in open aperture Z-scan experiment)

The experimental setup is hereby shown in Fig. 70.1. A 100 fs titanium sapphire laser at 1550 nm wavelength with repetition rate of 1 kHz is employed to provide exciting pulse. Two micro-joule energy pulses were used to excite the sample. The sample of $0.1BaO\text{-}0.4TeO_2\text{-}0.5B_2O_3$ is transparent for the IR radiation at 1550 nm and the linear refractive index of the sample was 1.3536.

70.2 Results and Discussion

The nonlinear refractive index was measured using the closed aperture z-scan configuration. Figure 70.2 shows the measured normalized transmittance as a function of z-position. The theoretical relation of *T*-*z* for the closed aperture case is given by

$$T = 1 + \frac{4\Delta\Phi_0(z/z_0)}{\left[(z/z_0)^2 + 9\right]\left[(z/z_0)^2 + 1\right]} \tag{70.1}$$

The solid curve in Fig. 70.2 shows theoretical value of *T* at corresponding *z*. Following standard relations are utilized in calculation of nonlinear refractive index,

Fig. 70.2 Closed aperture Z-Scan

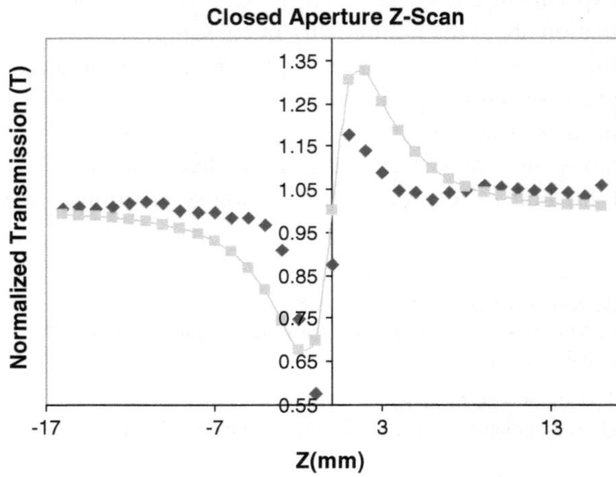

$$\Delta T_{p-v} = 0.406(1 - S)^{0.25}\Delta\Phi_0 \tag{70.2}$$

$$\Delta\Phi_0 = \frac{2\pi}{\lambda} n_2 I_0 L_{\text{eff}} \tag{70.3}$$

where $\Delta T_{p\text{-}v}$ is the transmittance difference (peak to valley) from the closed aperture scan, $\Delta\Phi_o$ is the on-axis nonlinear phase shift and S is the aperture transmittance given by $S = 1 - \exp(-2r_a^2/\omega_a^2)$, where r_a is the aperture radius and ω_a is the beam radius at the aperture in linear region when kept in far field. $I_0 = 2P/\pi\omega_o^2$ is the peak intensity of the laser beam at focus ($z = 0$) and $L_{\text{eff}} = (1 - e^{-\alpha L})/\alpha$ is the effective length of the sample with α as the linear absorption coefficient and L is the sample thickness. λ = 1550 nm, is the wavelength of interacting laser beam. From the best fit of (70.2) to the experimental data in Fig. 70.2, the value of $\Delta\Phi_0$ can be obtained as 0.633. The estimated value of n_2 for $0.1BaO\text{-}0.4TeO_2\text{-}0.5B_2O_3$ sample was $\sim 2.59 \times 10^{-17}$ cm²/W.

Open aperture z-scan was used for calculating nonlinear absorption coefficient β. Figure 70.3 shows the normalized transmission T as a function of the distance z-along the lens axis in the far field. The theoretical relation between T and z for the open aperture configuration is given by,

$$T = 1 - \frac{q_0}{2\sqrt{2}(1 + z^2/z_0^2)} \tag{70.4}$$

where $q_0 = \beta I_0 L_{\text{eff}}$ with z_0 as the Rayleigh range.

From the best fit of (70.4) to the experimental data in Fig. 70.3, the value of q_0 can be obtained as 0.54. Hence by using all known parameters we have calculated the nonlinear absorption coefficient β of $0.1BaO\text{-}0.4TeO_2\text{-}0.5B_2O_3$ and is found to be 0.3189×10^{-12} cm/W.

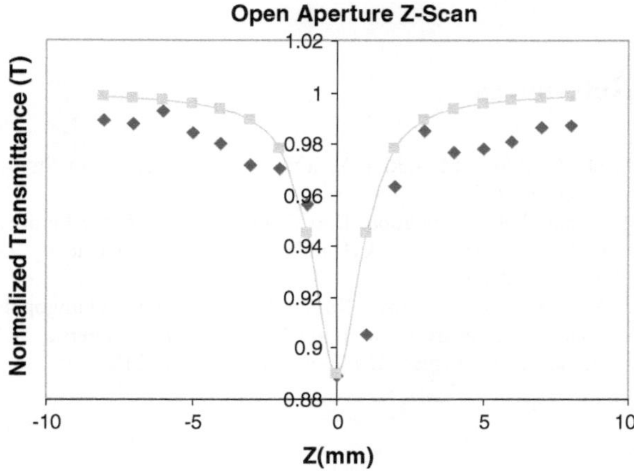

Fig. 70.3 Open aperture Z-scan

The imaginary and real part of third order nonlinear optical susceptibility $|\chi^{(3)}|$ was estimated from the calculated values of n_2 (closed aperture scan) and β (open aperture scan) using following relations,

$$\mathrm{Re}[\chi^{(3)}] = 10^{-4}\frac{\varepsilon_0 n_0^2 c^2}{\pi} n_2 \tag{70.5}$$

$$\mathrm{Im}[\chi^{(3)}] = 10^{-2}\frac{\varepsilon_0 n_0^2 c^2 \lambda}{4\pi^2} \beta \tag{70.6}$$

$$\chi^{(3)} = \sqrt{(\mathrm{Re}\chi^{(3)})^2} + \sqrt{(\mathrm{Im}\chi^{(3)})^2} \tag{70.7}$$

where ε_0 is the permittivity of free space, c is speed of light in vacuum and n_0 is the linear refractive index of the sample under experiment. The third order nonlinear susceptibility calculated for $0.1BaO$-$0.4TeO_2$-$0.5B_2O_3$ using above relations was $\sim 1.8329 \times 10^{-14}$ esu.

70.3 Conclusion

We have investigated the nonlinear optical coefficients of $0.1BaO$-$0.4TeO_2$-$0.5B_2O_3$ with fs titanium sapphire laser at 1550 nm wavelength. Nonlinear refractive index n_2 was calculated and found to be 2.59×10^{-17} cm^2/W. Third order nonlinear susceptibility $\chi^{(3)}$ was also calculated and found to be 1.8329×10^{-14} esu. Nonlinear absorption coefficient, β was calculated to be 0.3189×10^{-12} cm/W.

Acknowledgments Authors are thankful to Prof. A.K. Kar of Heriot Watt University, Edinburgh, UK for providing the experimental support for this work. Authors are also thankful to Dr. S.S. Negi, director IRDE and Sh. S.P. Gaba, Scientist 'G' for their support and encouragement for the work.

References

1. Sheik- Bahae M, said AA, wei TH, Hagan DJ, Van Stryland EW (1990) IEEE J Quantum Electron 26:760
2. Kumar RSS, Venugopal Rao S, Giribabu L (2007) Femtosecond and nanosecond nonlinear optical properties of alkyl phthalocyanines studied using Z-scan technique. Chem Phys Lett 447:274–278
3. Mathews SJ, Chaitanya Kumar S, Giribabu L, Venugopal Rao S (2007) Large third-order optical nonlinearity and optical limiting in symmetric and unsymmetrical phthalocyanines studied using Z-scan. Opt Commun 280:206–212

Chapter 71
Microwave Assisted Synthesis of Highly Luminescent ZnS Nanostructures Using Zinc Dithiocarbazic Complex Chemical Route

Ranjana Sharma, Bhoop Singh, Vijay Kumar, Y.C. Goswami, Rajeev Singh and D. Kumar

Abstract Bis(parahydroxyacetophenonebenzyldithiocarbazate) zinc(II) and bis (acetophenonebenzyldithiocarbazate) zinc(II) complexes have been synthesized. Zinc sulphide nanostructures were obtained from both complexes at different deposition temperatures using microwave assisted chemical route. The surface morphology of as-prepared ZnS nanoparticles were obtained using X-ray diffraction (XRD), transmission electron microscopy (TEM) and high transmission electron microscopy (HRTEM). Particles with spherical shape are shown in the TEM images of these samples. The optical properties of the particles studied by UV–Vis and photoluminescence (PL) spectroscopy showed evidence of quantum confinement.

71.1 Introduction

Semiconductor nanoparticles exhibit various attracting properties that differ them from those of corresponding bulk materials [1]. Confinement of electrons and holes in a small volume and large number of atoms on the surface is comparable to that inside the particles are responsible for this specific behavior. Surface of these nanostructures are found to be very reactive because of missing neighboring atoms and are bound by weaker forces [2]. Surface phenomena like sublimation,

R. Sharma (✉) · B. Singh · R. Singh
Department of Chemistry and Environmental Science, ITM-GOI, Sithouli, Gwalior 474001, MP, India
e-mail: ranjanagoswami1978@gmail.com

V. Kumar · Y.C. Goswami
School of Physical Sciences, ITM University, Turari, Gwalior 474001, MP, India

D. Kumar
PG Department of Chemistry, SMS Govt Science College, Gwalior 474001, MP, India
e-mail: rajeevs_rathore@rediffmail.com

V. Lakshminarayanan and I. Bhattacharya (eds.), *Advances in Optical Science and Engineering*, Springer Proceedings in Physics 166,
DOI 10.1007/978-81-322-2367-2_71

desorption, adsorption, chemical reaction, etc. occur due to increase in surface particles effects. II–VI semiconductor nanostructures have attracted special attention to research community due to their luminescence, optical and electronic properties. II–VI nanostructures have very significant importance because of their fundamental role not only in basic research but also in technological applications [3, 4]. Synthesis and characterization of these II–VI semiconductors are recent area of research. Zinc sulphide (ZnS) is a direct band gap semiconductor material with bandgap 3.7 eV. It gives better chemical stability as compared to other chalcogenides. It exhibits wide optical transparency from the visible light deep infrared region and have wide applications in photoluminescence (PL), electroluminescence (EL) and optical sensing devices. Recently developed ZnS nanoparticles have very good potential for applications in nonlinear optical devices and fast optical switches [5, 6].

ZnS nanostructures can be synthesize buy various methods [7–11]. These methods include vapor-liquid-solid (VLS) methods, chemical vapor deposition (CVD), thermal evaporation and liquid phase colloidal syntheses in aqueous or non-aqueous media. Capping is required for stabilizing the nanostructures. Generally it is done by with organic materials in most of the chemical methods The sol–gel technique is one of the most common methods for preparing semiconductor nanoparticles. Organometallic compounds are recently reported to synthesize the nanostructures which decompose cleanly at the reaction temperature giving rise to the desired active species [12]. The metal coordinating groups donate electron to metal atoms and thus prevent aggregation and further growth and the surfactant molecules bind to the particle surfaces in a manner that dictates the crystallographic axis along which growth will take place and helps in the shape selective synthesis. A single-source precursor route of synthesizing nanocrystals has some advantages over a multiple precursor technique. It is one-step synthesis technique and mixing of materials at the molecular level prevents formation of mixed phases or in the end product. In this paper, we present a new microwave synthesis route for preparation of ZnS nanoparticles using single molecular precursor.

71.2 Experimental

The synthesis of ZnS nanoparticles consists of three steps: synthesis of Schiff bases, synthesis of single molecular precursor and synthesis of nanoparticles. S benzyl dithiocarbazate is prepared in the lab by adding hydrazine and KOH and mixing CS_2 dropwise. KOH is dissolved in the specific ration with alcohol water mixture. The mixture was cooled in ice. The hydrazine hydrate was added slowly. Finally A solution of CS_2 was added in alcohol dropwise. The yellow oily layer was then separated using a separating funnel. Mixture was kept in the ice bath and benzyl chloride was added slowly with mechanical stirring. The product was filtered out and washed and recrystallized form methanol. Schiff base was prepared by the simple condensation reaction between carbonyl compounds parahydroxyacetophenone and acetophenone with s benzyl dithiocarbazate.

71.2.1 Synthesis of Single Molecular Precursor

Schiff base was dissolved in the ethanol and this mixture was gradually added to the ethanolic solution of zinc acetate and zinc chloride with the constant stirring of mixture (in the ratio 2:1). White solid mass of the complexes was filtered off washed with the ethanol, finally dried and used for synthesis of ZnS nanostructures.

71.2.1.1 Deposition of ZnS by Microwave

In reaction 10 mg of precursor was dissolved in 8 mL of oleylamine and hrs and 1 gm TOPO after which 0.1 mL of dilute HC1 was added to the mixture. The reaction mixture was then stirred vigorously while the temperature was elevated to either 120 or 175 °C and maintained at that value for varied amounts of time. After cooling to room temperature, the white precipitate of ZnS was collected and washed by centrifugation with water and ethanol. The obtained product was thoroughly washed with distilled water and then dried. The sample was characterized by Optical spectrophotometer, Photoluminescence X-ray diffraction (XRD) and transmission electron microscopy (TEM). Optical characterizations was performed in the PC ray research Centre for Advanced studies a. Infrared spectra were recorded on a Perkin Spectrum Two, ATR instrument (6000–550 cm^{-1}, resolution 2 cm^{-1}). Melting points were recorded Melting Point Apparatus. XRD studies were performed on a Bruker AXSD 8 diffractometer using CuKα in the range 20–70° in a step size of 0.05 with a varying count rate depending upon the sample. TEM images were collected on Philips CM200 transmission electro microscope using an accelerating voltage 200 kV. TEM samples were prepared by putting a drop of suspension of the sample in ethanol on carbon coated copper grid.

71.3 Results and Discussion

Figure 71.1 shows X-ray diffractograms for ZnS nanostructures by different complexes. Line broadening of the XRD peaks is observed indicates the nanoscale range materials which consist of particles. All the diffraction peaks are identified and indexed to cubic phase (sphalerite) of ZnS (JCPDS: 77-2100). Additional peaks for Zn metal are also observed with precursor I whereas absent or give less precursor II. For ZnS 1 dominant peak is (111) however for ZnS (2) dominant peak is (200). The peaks for ZnS (1) are sharper than ZnS (2) indicates better crystallinity [11].

Optical transmission spectra for ZnS particles obtained with precursor I is shown in Fig. 71.2. Very high Transmission 96–98 % in visible region is observed. Absorption edge shift towards low wavelength as compare to spectra for bulk also confirms the formation of nano structures. Sharp absorption edge occurs indicates good quality of the particles with uniform structures [7].

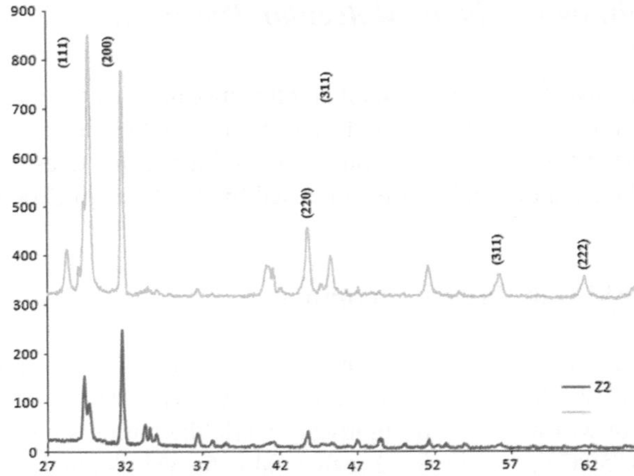

Fig. 71.1 X ray Diffractograms for ZnS (1) and ZnS (2)

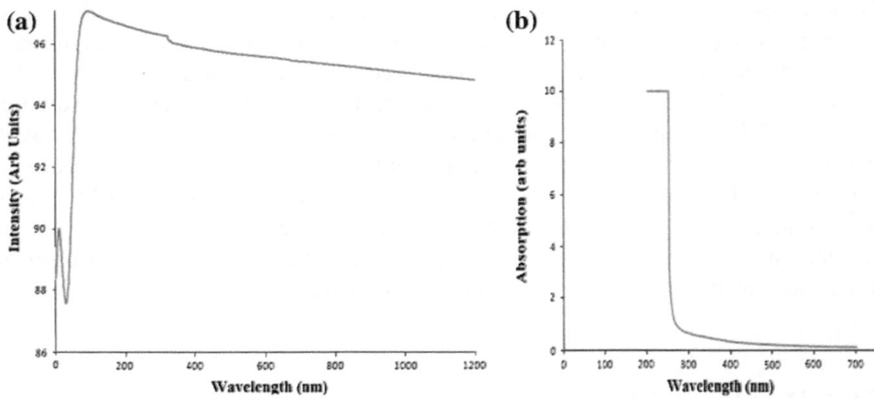

Fig. 71.2 **a** Transmission and **b** absorption spectrum for ZnS nanostructure

Figure 71.3 show PL spectra of ZnS nanocrystals obtained from precursor (I) and (II) respectively. The excitation was given at 280 nm. Large Pl peak is observed around 490 nm with ZnS obtained form precursor I whereas ZnS obtained with precursor II give comparatively smaller peak. However at 650 all ZNS gives the peak at the same place [8]. Good pl spectra I visible region makes them suitable for optoelectronic applications. In order to study the size, shape, and crystal structure of the clusters TEM and HREM images were taken and shown in Fig. 71.4a and b. Overview of images do not provide much information other than the particles are uniform in size and tend to agglomerate. Lattice fringes in HRTEM image shows the good quality ZnS nanostructures.

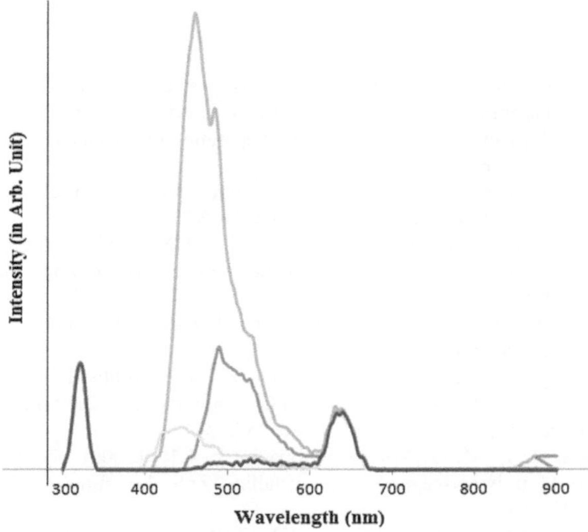

Fig. 71.3 Photoluminescence spectra of ZnS nanostructures

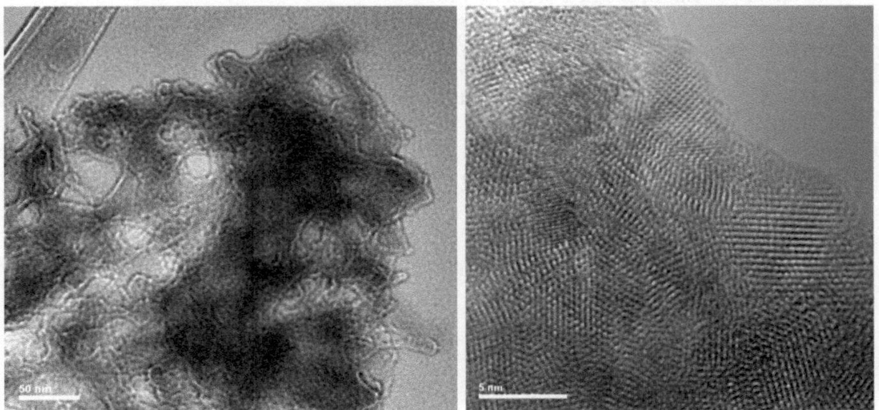

Fig. 71.4 TEM and HRTEM spectra of ZnS nanostructures

71.4 Conclusions

Good quality ZnS nanostructures can be obtained by microwave assisted single molecular precursor method. The obtained particles are nano crystalline with good transmission and Pl spectra suitable for various optoelectronic devices.

Acknowledgments The authors are thankful to MPCST for funding this work.

References

1. Chandran A, Francis N, Jose T, Georg KC (2010) Synthesis structural characterization and optical bandgap determination of ZnS nanoparticles. Acad Rev 17(1–2):17–21
2. Xu JF, Ji W, Lin JY, Tang SH, Du YW (1998) Preparation of ZnS nanoparticles by ultrasonic radiation method. Appl Phys A 66:639–641
3. Bodo B, Singha R, Das SC (2012) Structural and optical properties of chemically synthesized ZnS nanostructures. Int J Appl Phys Math 2(4):271–287
4. Willander M, Klason P, Yang LL, Al-Hilli SF, Zhao QX, Nur O (2008) ZnO nanowires: chemical growth, electrodeposition, and application to intracellular nano-sensors. Phys Status Solidi C 5:3076–3083
5. Jassby D, Wiesner M (2011) Characterization of ZnS nanoparticle aggregation using photoluminescence. Langmuir 27(3):902–908
6. Israr MQ, Sadaf JR, Yang LL, Nur O, Willander M, Palisaitis J, Persson POÅ (2009) Trimming of aqueous chemically grown ZnO nanorods into ZnO nanotubes and their comparative optical properties. Appl Phys Lett 95:073114
7. Maity R, Maiti UN, Mitra MK, Chattopadhyay KK (2006) Synthesis and optical characterization of polymer-capped nanocrystalline ZnS thin films by chemical process. Physica E 33:104–109
8. Cheng X, Zhao Q, Yang Y, Tjong SC, Li RKY (2010) A facile method for the synthesis of ZnS/polystyrene composite particles and ZnS hollow micro-spheres. J Mater Sci 45:777–782
9. Yuan JL, Kajiyoshi K, Yanagisawa K, Sasaoka H, Nishimura K (2006) Fabrication of silica nanocoatings on ZnS-type phosphors via a sol–gel route using cetyltrimethylammonium chloride dispersant. Mater Lett 60:1284–1286
10. Shafiq I, Sharif A, Sing LC (2009) ZnS_xSe_{1-x} nanowire arrays with tunable optical properties grown on ZnS nanoribbon substrates. Phys E 41:739–745
11. Wang C, Ao YH, Wang PF, Zhang SH, Qian J, Hou J (2010) A simple method for large-scale preparation of ZnS nanoribbon film and its photocatalytic activity for dye degradation. Appl Surf Sci 256:4125–4128
12. Barrelet CJ, Wu Y, Bell DC, Lieber CM (2003) Synthesis of CdS and ZnS Nanowires Using Single-Source Molecula Precursors. J Am Chem Soc 125:11498–11499

Chapter 72
Optoelectronics of Cu^{2+}-Doped TiO_2 Films Prepared by Sol–Gel Method

S. Rai and Pranab J. Dihingia

Abstract Both undoped and Cu^{2+}-doped TiO_2 films have been prepared by sol-gel dip-coating method and characterized by UV-Vis, FTIR, and PL spectroscopy. Also, SEM images of the prepared films were to taken to analyze the surface morphology. Finally, the optoelectronic characterization of the prepared Cu^{2+}-doped films have been performed using *I–V* measurements under two different light conditions. The surface plasmon resonance (SPR) bands of copper have been observed between 500 and 600 nm. Also, the variation of current under white light illumination was found to much more prominent at a particular voltage than that of dark current.

72.1 Introduction

Transition metal ions such as copper (Cu^{2+}) also significantly influences the optoelectronic properties of certain semiconductors such as TiO_2. TiO_2 itself is a transition metal oxide associated with interesting optical and electronic properties. Cu^{2+}-doped TiO_2 is a good absorber of sunlight for solar-energy conversion. Cu^{2+} doping of TiO_2 creates an extra state in the forbidden gap of TiO_2, which gives two extra peaks in the absorption spectrum of the oxide semiconductor [1]. It has been found that the presence of Cu^{2+} in the TiO_2 lattice shifts the absorption edge of TiO_2 towards visible region [2], which, in turn, is responsible for visible region photo-catalytic action of Cu^{2+}-doped TiO_2 nanomaterials. This red-shift of the

S. Rai (✉)
Department of Physics, Mizoram University, Aizawl 796004, Mizoram, India
e-mail: srai.rai677@gmail.com

P.J. Dihingia
Department of Physics, Dibrugarh University, Dibrugarh 786004, Assam, India

V. Lakshminarayanan and I. Bhattacharya (eds.), *Advances in Optical Science and Engineering*, Springer Proceedings in Physics 166,
DOI 10.1007/978-81-322-2367-2_72

absorption edge of Cu^{2+}-doped TiO_2 nanoparticles from UV to visible region is due to the charge transfer between the dopant and the conduction band (valence band) of TiO_2 or a *d-d* transition in the crystal field [3].

In this work, TiO_2 films, both undoped and Cu^{2+}-doped have been prepared by sol–gel dip-coating method at room temperature. The prepared films have been analyzed for optoelectronic properties by UV-Vis, FTIR, and current voltage (*I–V*) measurements. The structural analyses of the films have been done with XRD, SEM, and TEM.

72.2 Experimental

The materials used in the preparation of copper doped TiO_2 films are: (a) titanium isopropoxide (Acros Organics, 98+ %) as TiO_2 precursor, (b) isopropanol (Rankem, assay = 99 %), (c) acetic acid (Acros Organics, 99.5 % pure), (c) copper(II) chloride (CDH, 99 %), (d) distilled water, (e) ultrasonically cleaned optically flat glass slides, (f) TCO-coated glass slides. Titanium isopropoxide (TIPO) was dissolved in isopropanol with continuous stirring. Addition of acetic acid helps in dissolving TIPO. The stirring was done for 45 min-1 h. A viscous sol was obtained. In this sol, cleaned glass slides (both plain and TCO-coated) were dipped to get precursor gel films. These films when heated to 65 °C and above, crystalline TiO_2 films were formed. For copper doping, an appropriate amount of $CuCl_2$ was dissolved in isopropanol prior to the addition of TIPO. Then, the same procedure as above was repeated for copper doped TiO_2 films. The UV-Vis absorption spectra of the prepared films were taken with the help of a Scinco PD array UV-Vis spectrophotometer (S-3100). The FTIR spectrum of a TiO_2 film was recorded with the help a of Shimadzu IRPrestige-21 FTIR spectrophotometer. The XRD and SEM-EDS analysis of the films were done with a Bruker AXS D8 Advance X-ray diffractometer of SAIF, Kochi, India (CuK_α radiation of wavelength 1.5406 Å) and a JEOL JSM-6390LV scanning electron microscope with a JEOL JED-2300 energy dispersive spectrometer of SAIF, Kochi, India respectively. TEM images of the samples were recorded with a JEOL JEM-2100 transmission electron microscope with resolution ranging from 1.9 to 1.4 Å at SAIF, North-Eastern Hill University, Shillong (India).

72.3 Results and Discussion

72.3.1 *FTIR Spectra*

Å The FTIR spectrum of undoped TiO_2 powder is shown in Fig. 72.1. The undoped crystalline powder was obtained from the same sol as used for dip-coating to obtain

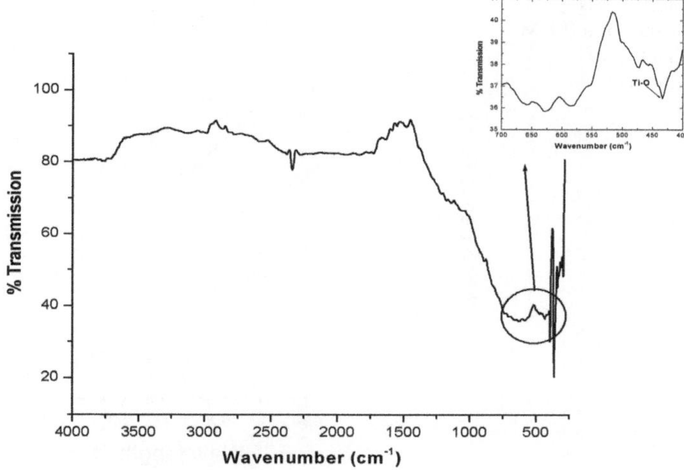

Fig. 72.1 FTIR of undoped crystalline TiO_2 powder

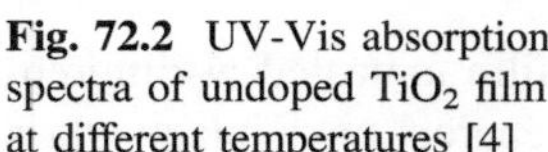
Fig. 72.2 UV-Vis absorption spectra of undoped TiO_2 film at different temperatures [4]

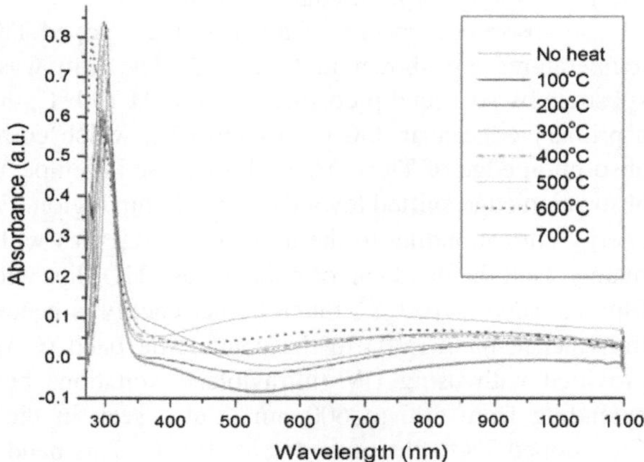

the TiO_2 films. The band at around 433 cm^{-1} can be assigned as due to Ti-O bonds in the TiO_2 crystalline structures [5] (Fig. 72.2).

72.3.2 UV-Vis Absorption Spectra

From the UV-Vis absorption spectra of undoped TiO_2 film at different temperatures it can be inferred that the 600 nm visible region absorption band appears because of the formation of defects associated with oxygen vacancies that originate from the reduction of TiO_2 [6] after heat treatment at 700 °C. Also, the absorption edge for

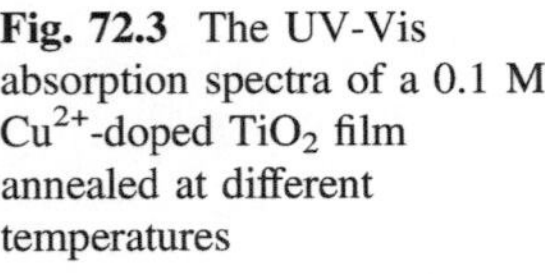

Fig. 72.3 The UV-Vis absorption spectra of a 0.1 M Cu^{2+}-doped TiO_2 film annealed at different temperatures

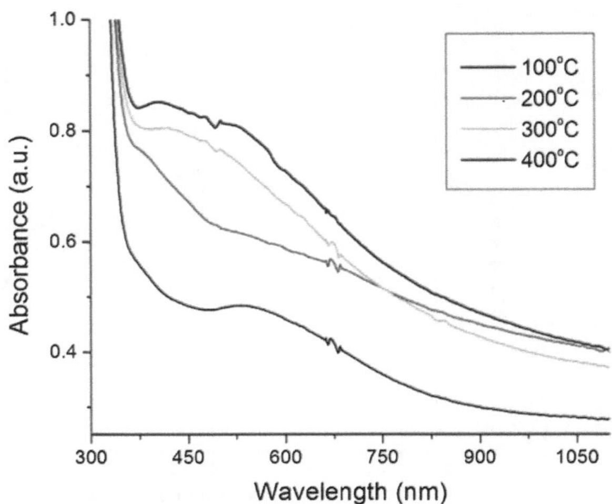

the TiO_2 film annealed at 200 °C lies near about at 330 nm. It was found that the energy of the absorption edge is about 3.761 eV.

The absorption spectra of a 0.1 M Cu^{2+}-doped TiO_2 film annealed at different temperatures are shown in Fig. 72.3. The film was prepared on a plane glass substrate by sol–gel dip-coating method. At 100 °C, the absorption spectrum shows a primary edge near 350 nm (3.546 eV), which corresponds to the characteristic absorption edge of TiO_2 [7]. With increase in temperature from 100 to 400 °C, the absorption edge shifted towards 350–380 nm region. For this temperature range, the energy corresponding to the absorption edge lies within 3.546–3.266 eV approximately. But, the band gap of bulk anatase TiO_2 lies within 3.1–3.2 eV. Thus, for the film annealed at 100 °C, much higher energy is needed to excite an electron from the valence band (VB) to the conduction band (CB) of TiO_2. This can only be provided with using UV (ultraviolet) excitations below 350 nm. A broad band extending from 500 to 600 nm is also seen in the absorption spectrum of the Cu^{2+}-doped TiO_2 film annealed at 100 °C. This band should not be confused with the absorption band obtained in case of undoped TiO_2 film (Fig. 4.3 in Chap. 4) annealed at 700 °C, where it appeared only due to the formation of defects associated with oxygen vacancies that originated from the reduction of TiO_2 after annealing at 700 °C.

The band between 500 and 600 nm can be explained with the help of surface plasmon resonance (SPR) absorption of metal nanoparticles. The resonance condition is established when the frequency of the photons matches the natural frequency of the surface (valence) electrons oscillating against the restoring force of the positive nuclei. In fact, the optical properties of metal nanostructures in the

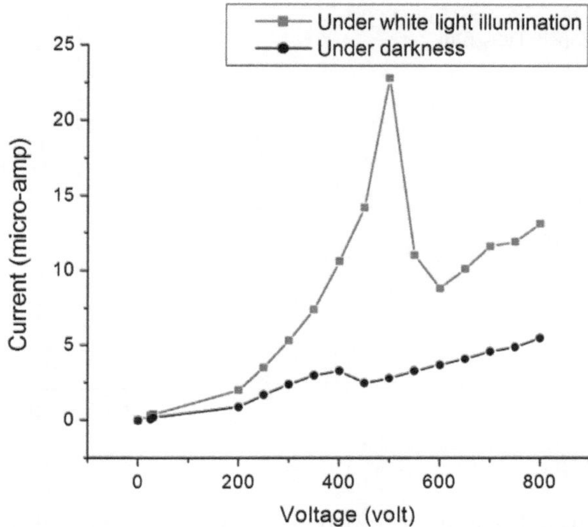

Fig. 72.4 The current-voltage (*I–V*) characteristics of a 0.01 M Cu^{2+}-doped TiO_2 film prepared on TCO-coated glass slide (annealed at 400 °C)

visible region are dominated by the surface plasmon (SP) absorption [8]. It has been reported that copper nanoparticles show an SPR band in the 573–600 nm range. Also, some studies show that the SPR band of copper may appear even at 562 nm [9], while some others have reported that it may lie between 560 and 582 nm also. The position of the SPR band is dependent on certain parameters such as solvent used in the synthesis, capping agent, etc. The obtained band between 500 and 600 nm can be easily assigned as either due to the SPR absorption of copper or due to the *d-d* transitions of Cu^{2+}. The increase in bandwidth of the SPR may be related to the decrease in size of the nanoparticles (NPs). The observed SPR band indicates that copper NPs have been formed in the TiO_2 matrix and it strongly depends on the dielectric constant of the TiO_2 matrix. Though the film at 100 °C exhibits a diffuse absorption band in the visible region, yet it is difficult to obtain good *I–V* characteristics of such a film under visible light conditions since photo-generation of electron-hole pairs does not take place in TiO_2 below 350 °C. The SPR band of copper seems to get red-shifted with increase in annealing temperature. At 400 °C, this band becomes prominent and centred at 582 nm. There is also a small band centred at 404 nm, which is due to the defect states introduced in the TiO_2 lattice due to annealing.

I–V characteristics of a 0.01 M Cu^{2+}-doped TiO_2 film prepared on a TCO-coated glass and annealed at 400 °C is shown in Fig. 72.4. The characteristics were obtained under both dark and white light illumination conditions. The white light was obtained from an LED torch (300 lm). The film was prepared on a TCO-coated

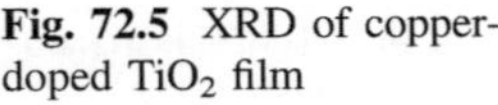

Fig. 72.5 XRD of copper-doped TiO_2 film

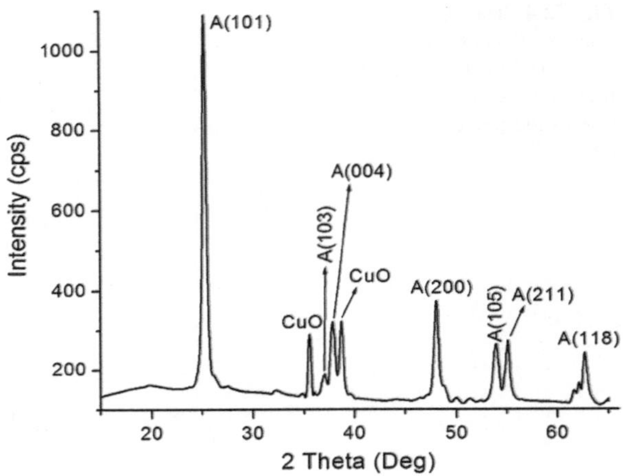

glass because the resistivity of TiO_2 is very high even at elevated temperatures (2.5×10^4 Ω cm at 700 °C and 10^{12} Ω cm at 25 °C). The room temperature during these measurements was 20 °C. It was observed that both dark current and current under illumination increased up to a certain value of applied voltage. But the variation of current under illumination was much more than that of dark current.

72.3.3 XRD and SEM Analyses of Copper-Doped TiO_2 Film

The X-ray diffractogram of a copper (Cu)-doped TiO_2 film is shown in Fig. 72.5. Signatures of crystalline anatase (A) phases of TiO_2 and CuO can be clearly seen from the diffractogram.

The SEM images of the films reveal roughly spherical particles with few clusters. The corresponding EDS shows the presence of the elements Ti, Cu, and O (Fig. 72.6).

72.3.4 TEM Images of the Copper-Doped TiO_2 Film

The film material on the glass slide was subjected to TEM analysis. Some oval and roughly spherical particles of less than 20 nm in size can been seen in the TEM image (Fig. 72.7).

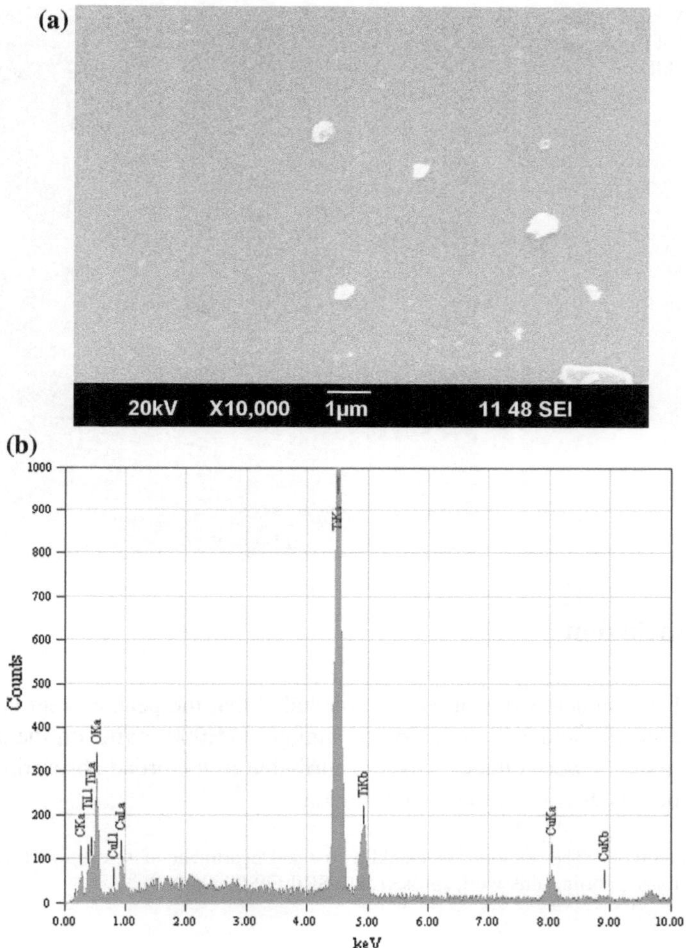

Fig. 72.6 **a** SEM image of Cu-doped TiO_2 film, **b** the corresponding EDS of copper-doped TiO_2 film

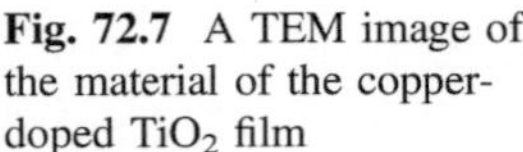

Fig. 72.7 A TEM image of the material of the copper-doped TiO_2 film

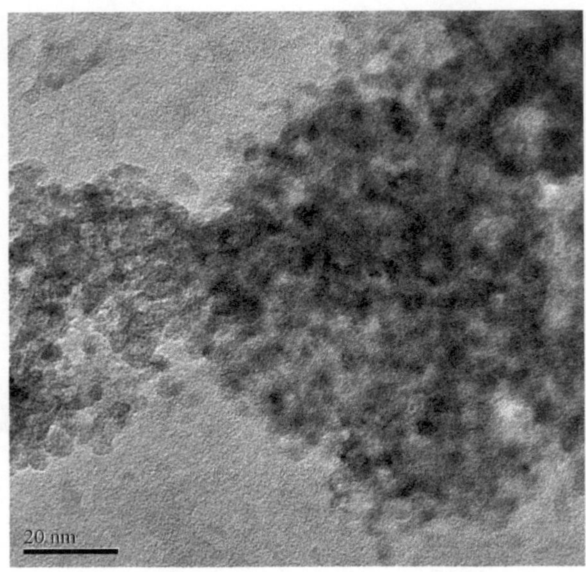

72.4 Conclusion

From the I–V characteristics, it can be concluded that the peak current is expected to increase steeply under illumination of sunlight (75,000 lm/m^2 on the surface of earth) at a much lower voltage. This can attributed to the broad absorption band of copper doped TiO_2 film in the visible region.

Acknowledgements The authors are thankful to the Department of science and Technology, New Delhi for supporting this work (project no.: SR/LOP-0039/2010).

References

1. Blondeau G, Froelicher M, Froment M, Hugot-Le Goff A, Zerbino J (1979) Influence of copper addition on the optical properties of TiO_2. J Electrochem Soc 126:1592–1596
2. Vidyasagar CC, Naik YA, Venkatesha TG, Manjunatha P (2012) Sol–gel synthesis using glacial acetic acid and optical properties of anatase Cu–TiO_2 nanoparticles. J Nanoeng Nanomanufact 2:91–98
3. Subramanian V, Wolf EE, Kamat PV (2004) Catalysis with TiO_2/gold nanocomposites. Effect of metal particle size on the fermi level equilibration. J Am Chem Soc 126:4943–4950
4. Dihingia PJ, Rai S (2012) Synthesis of TiO_2 nanoparticles and spectroscopic upconversion luminescence of Nd^{3+}-doped TiO_2-SiO_2 composite glass. J Lumin 132:1243–1251
5. Yang LL, Lai YS, Chen JS, Tsai PH, Chen CL, Chang J (2005) Compositional tailored sol–gel SiO_2–TiO_2 thin films: crystallization, chemical bonding configuration, and optical properties. J Mater Res 20:3141–3149

6. Kuznetsov VN, Serpone N (2006) Visible light absorption by various titanium dioxide specimens. J Phys Chem B 110:25203–25209
7. Jiménez González AE, Gelover Santiago S (2007) Structural and optoelectronic characterization of TiO_2 films prepared using the sol–gel technique. Semicond Sci Technol 22:709–716
8. Zhang JZ (2009) Optical properties and spectroscopy of nanomaterials. World Scientific Publishers, Singapore, pp 207–208
9. Dang TMD, Le TTT, Fribourg-Blanc E, Dang MC (2011) Synthesis and optical properties of copper nanoparticles prepared by a chemical reduction method. Adv Nat Sci: Nanosci Nanotechnol 2(1):015009

Chapter 73
Growth and Characterization of Nanocrystalline CuInSSe Thin Films by Spray Pyrolysis

Vipin Shrotriya and P. Rajaram

Abstract Nanocrystalline CuInSSe thin films were grown on glass substrates using the spray pyrolysis technique. The CuInSSe films were co-deposited from an aqueous solution containing $CuCl_2$, $InCl_3$, thiourea and SeO_2. EDC was used as a complexing agent and films were deposited at the constant temperature 300 °C. XRD, Scanning Electron Microscope (SEM) and Energy Dispersive Analysis of X-Ray (EDAX) were used to characterize the samples and the results indicate that the films are single phase, p-type in conductivity and having the chalcopyrite structure. Optical studies show that the optical gap values are 1.23 and 1.39 eV for films grown from spray solutions having S/Se ionic ratios 0.3 and 0.7 respectively. Using the Scherer formula the average crystallite size of the films was found to be in the range 5–11 nm.

73.1 Introduction

$CuInS_2$ and $CuInSe_2$ play an important role for thin film solar cell applications having electrical conversion efficiency about 5–12.5 % [1–3] and 15.4 % [4] respectively. $CuInS_2$ has a band gap 1.53 eV, which is higher than the optimum gap of 1.2 eV and $CuInSe_2$ has a band gap 1.02 eV which is lower than the optimum gap. By alloying sulphur with $CuInSe_2$ to form CuInSSe, the band gap may be increased from 1 to 1.2 eV resulting in an increase in the open circuit voltage and conversion efficiencies in their solar cells. Very few researchers have reported on the synthesis of CuInSSe thin films using various methods [5–7]. Spray pyrolysis is a simple and low cost technique not requiring sophisticated instrumentation for large scale commercialization [8]. In this technique, solutions of the constituents are mixed together and converted into fine droplets by forcing the liquid mixture through a nozzle. The reactants in the droplets are pyrolyzed on a heated substrate.

V. Shrotriya (✉) · P. Rajaram
School of Studies in Physics, Jiwaji University, Gwalior 474011, MP, India
e-mail: vipinshrotriya@gmail.com

V. Lakshminarayanan and I. Bhattacharya (eds.), *Advances in Optical Science and Engineering*, Springer Proceedings in Physics 166,
DOI 10.1007/978-81-322-2367-2_73

A pyrolysis reaction leads to the deposition of films of the desired compound while other products evaporate as gaseous species. In this work we present our work on the fabrication and characterization of nanocrystalline CuInSSe thin films grown using spray pyrolysis.

73.2 Experimental Details

CuInSSe thin films were prepared on glass substrates using the spray pyrolysis method. $CuCl_2.2H_2O$ (CDH), $InCl_3$ (CDH), $NH_2.CS.NH_2$ (CDH) and SeO_2 (CDH) dissolved in distilled water were used as the starting solutions. A few mgs of ethylene diamine dihydrochloride (EDC) was added to the solution to prevent the precipitation of the selenium salt. The atomic ratio of Cu: In: (S + Se) in the solution was kept at 1:1:4, and the S/Se ratio was kept at 0.3 and 0.7. The excess thiourea plus SeO_2 were taken to compensate for the loss of S and Se during the spray process. The pH of the solution was maintained at 10 by adding a few drops of ethanolamine. Compressed air was used as a carrier gas in the spray process. The spray rate was kept between 3 and 5 ml/min and the deposition time was 30 min for all the films grown. During deposition of CuInSSe thin films, substrate temperatures between 250 and 350 °C were tried and after a few initial trials, the substrate temperature was fixed at 300 °C. After growth, the CuInSSe thin films were annealed in air at 300 °C for half an hour. A Philips D8 advance X-ray diffractometer employing Cu-Kα radiation of wavelength 1.5406 Å was used for structural studies. The Scanning Electron Microscope (SEM) and Energy Dispersive Analysis of X-Ray (EDAX) studies were carried out using a Philips FE-Quanta 200 microscope and optical transmission spectra were obtained using a UV-VIS Spectrophotometer (Shimadzu UV-2450).

73.3 Results and Discussion

73.3.1 X-ray Diffraction

Figure 73.1 shows the XRD patterns of CuInSSe thin films on glass slides for two different compositions. The patterns show that the grown films are single phase, polycrystalline and having a preferred (112) orientation. The (112) plane for pure $CuInSe_2$ occurs at a two theta value of 26.69° whereas that for pure $CuInS_2$ occurs at 28.24° [9]. We can see that for our films, as the (S/Se) concentration is increased, the (112) peak shifts towards higher angles as shown in the inset. The main peaks at 27.2°, 44.9° and 53.17° 2θ indicate that the synthesized material is CuInSSe (JCPDS 036-1311). No peaks were found from any other phases. The crystallite size calculated (for the (112) peak) using the Debye Scherer formula is found to be 5–11 nm.

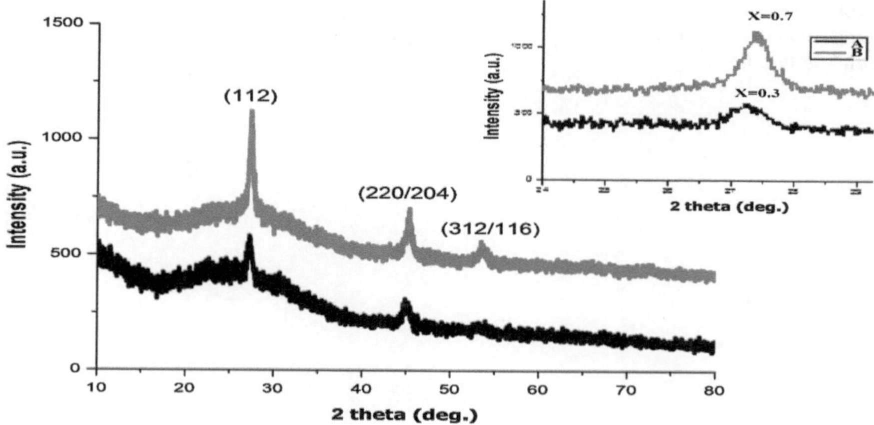

Fig. 73.1 XRD patterns of annealed CuInSSe thin films prepared from solutions with (*A*) S/Se = 0.3 and (*B*) S/Se = 0.7. The inset shows the shift in (112) peak with increasing 'S' content

73.3.2 SEM and EDAX Studies

Figure 73.2 shows the SEM micrographs and EDAX spectra of CuInSSe thin films. From the micrographs we can say that the surface is covered with very fine particles (~15–20 nm) for S/Se = 0.3 while for S/Se = 0.7 the particles are much bigger (~50–60 nm). The EDAX compositions of the grown films are shown in Table 73.1 and it can be seen that both the films are slightly non-stoichiometric.

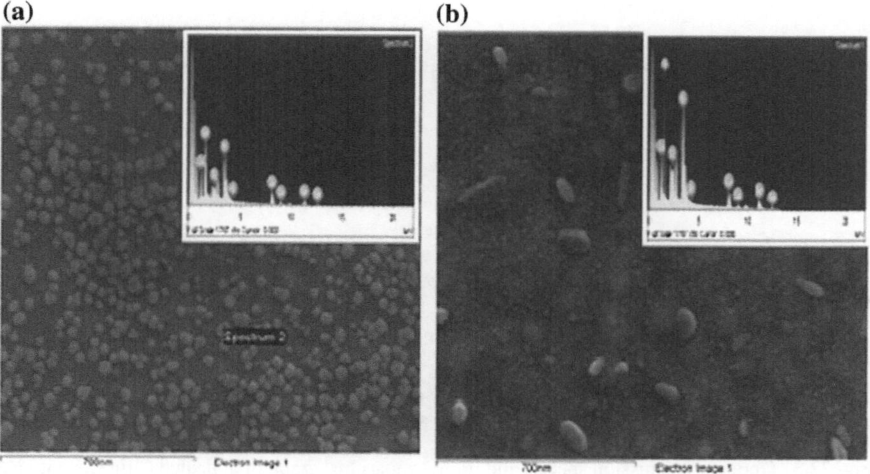

Fig. 73.2 SEM micrographs and EDAX spectra of annealed CuInSSe thin films prepared from solutions with **a** S/Se = 0.3 and **b** S/Se = 0.7

Table 73.1 EDAX compositions of the annealed CuInSSe thin films

Element	Atomic percentage at S/Se = 0.3	Atomic percentage at S/Se = 0.7
S K	7.19	11.05
Cu K	24.19	17.50
Se L	37.51	38.40
In L	31.11	33.05

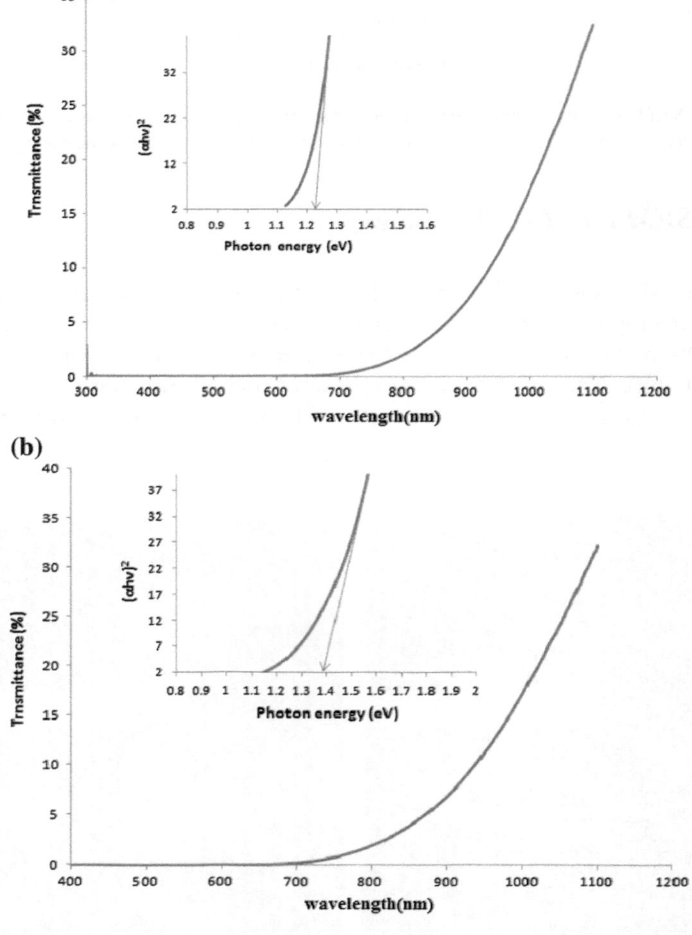

Fig. 73.3 Optical transmission spectra and $(\alpha h\upsilon)^2$ versus $h\upsilon^2$ graph of annealed CuInSSe with **a** S/Se = 0.3 and **b** S/Se = 0.7

This suggests that SeO_2 used for the Se ions is not giving the desired composition, and either has to be replaced or given further treatments. The Hot probe method shows that the deposited thin films have p-type conductivity.

73.3.3 *Optical Transmission Studies*

To study the optical characteristics of the films, transmittance spectra were recorded in the range 350–1100 nm. The presence of direct band gap was determined from $(\alpha h\upsilon)^2$ against $h\upsilon$ plots, obtained from the transmittance spectra. The band gap value of the films is found to be 1.23 and 1.39 eV with the concentration of S/Se = 0.3 and 0.7 respectively. Figure 73.3 shows the transmission spectra of CuInSSe thin films.

73.4 Conclusions

In conclusion, polycrystalline CuInSSe thin films with S/Se = 0.3 and 0.7 were obtained on glass substrates using the spray pyrolysis technique. The grown thin films are *p*-type, and single phase with the chalcopyrite structure. The band gap value of the films is found to be 1.23 and 1.39 eV for the ionic bath compositions S/Se = 0.3 and 0.7 respectively.

Acknowledgments This work is being funded by the Madhya Pradesh Council of Science and Technology (MPCST), Bhopal, India through research project No. 1078/CST/R&D/2012 for which Authors are thankful. One of the Authors, V. Shrotriya is grateful to MPCST for the JRF. The Authors are also thankful to IIC, IIT Roorkee for providing XRD and SEM/EDAX facilities.

References

1. Mitchell KW, Pollock GA, Mason AV (1988) 7.3 % efficient $CuInSe_2$ solar cell. In: Proceeding of the 20th IEEE photovoltaic specialists conference, IEEE, New York, p 1542. doi:10.1109/PVSC.1988.105967
2. Goossens A, Hofhuis J (2008) Spray deposited $CuInS_2$ solar cells. Nanotechnol 19:424018–424025. doi:10.1088/0957-4484/19/42/424018
3. Braunger D, Hariskos D, Walter T, Schock HW (1996) An 11.4 % efficient polycrystalline thin film solar cell based on $CuInS_2$ with a Cd-free buffer layer. Sol Energy Mater Sol Cells 40 (2):97–102. SSDI 0927-0248(95)00069-0
4. Hedstrom J et al (1993). ZnO/CdS/Cu(In,Ga)Se_2 thin film solar cells with improved performance. In: Proceedings of the 23rd IEEE photovoltaic specialists conference, pp 364–371
5. Subbaramaiah K, Sundara Raja V (1992). Structural and optical properties of spray-deposited $CuIn(S_{1-x}Se_x)_2$ thin film. Thin Solid Films 208:247–251
6. Hou WW, Bob B, Li SH, Yang Y (2009). Low-temperature processing of a solution-deposited CuInSSe thin-film solar cell. Thin Solid Films 517:6853–6856. doi:10.1016/j.tsf.2009.06.032

7. Chavhan S, Sharma R (2006) Growth, structural and optical properties of non-stoichiometric $CuIn(S_{1-x}Se_x)_2$ thin film deposited by solution growth technique for photovoltaic application. J Phys Chem Solids 67:767–773. doi:10.1016/j.jpcs.2005.11.013
8. Bates CW, Nelson KF, Raza SA, Mooney JB, Recktenwald JM, Macintosh L, Lamoreaux R (1982) Spray pyrolysis and heat treatment of $CuInSe_2$ for photovoltaic applications. Thin Solid Films 88:279. doi:10.1016/0040-6090(82)90058-X
9. Kazmerski LL, Ayyagari MS, Anvor GA, White FR, Merrill AJ, Merrill AJ (1976) Electron and X-ray diffraction analysis of ternary compound (I-II-VI_2) thin films. Thin Solid Films 37:323

Chapter 74
Performance Enhancement of Joint Fractional Correlator for Digital Holography Based Three-Dimensional Object Recognition Using Wavelet Filter

Dhirendra Kumar and Naveen K. Nishchal

Abstract In this paper, performance of a joint fractional correlator for digital holography based three-dimensional (3D) object recognition has been enhanced using wavelet filter. Digital Fresnel holograms of 3D objects have been optically captured and numerically reconstructed. The correlation has been carried out by comparing the two 3D images reconstructed from corresponding digital holograms with the help of joint fractional correlator. The reconstructed objects are multiplied with appropriately scaled Mexican-hat wavelet filter in Fresnel domain to improve the performance of the fractional correlator. The performance measure parameters; discrimination ratio and peak-to-correlation energy have been computed to show that the filter enhances the correlation output.

74.1 Introduction

Three-dimensional (3D) object recognition has been a challenge in the field of optical and digital signal processing. Many techniques for 3D recognition have been demonstrated by researchers [1]. Digital holography based 3D object recognition [1, 2] has provided a good option since digital holograms capture all the information of a 3D object into a two-dimensional (2D) complex fringe pattern [3–7]. Digital holography offers flexibility for improving the reconstructed wavefront [6, 7]. The techniques based on digital holography have been efficiently used for 3D object recognition [8–11]. Millan [1] has discussed the important problems of time-variant information, multidimensional image processing, and color content in 3D correlation. An overview of 3D recognition based on matching the holographic representations of 3D objects through correlation has been presented by Tsang et al. [2]. Seifi et al. [8] proposed 3D object recognition by directly matching the diffraction patterns, which reduces the computational cost by recording the diffraction patterns in

D. Kumar · N.K. Nishchal (✉)
Department of Physics, Indian Institute of Technology Patna, Patliputra Colony, Patna 800013, India
e-mail: nkn@iitp.ac.in

V. Lakshminarayanan and I. Bhattacharya (eds.), *Advances in Optical Science and Engineering*, Springer Proceedings in Physics 166,
DOI 10.1007/978-81-322-2367-2_74

low dimensional space. Nelleri et al. [9, 10] proposed 3D object recognition with the help of wavelet matched filter to enhance the discrimination ability of the correlation scheme. Javidi et al. [11–14] proposed different technique for 3D recognition in which the complex distribution of the 3D object is recorded in single Fresnel plane using phase-shifting digital holography and a digital hologram of 3D reference pattern acts as a correlation filter. A 2D correlator may also be used to identify a target in 3D space with the help of different perspectives of an object [15].

The fractional Fourier transform (FRT) based correlator; joint fractional correlator [16–19] has been widely used for correlation purpose. Zhu et al. [18] proposed a fractional correlator for recognition of transparent 3D objects, which performs better than a joint Fourier transform correlator. Further, improvement in the performance of the joint fractional correlator has been reported [19]. In this paper, performance of the joint fractional correlator for digital holography based 3D object recognition has been improved by using a wavelet filter. The digital Fresnel holograms of real 3D objects have been optically recorded and reconstructed by numerical algorithm. The correlation operation has been performed using joint fractional correlator by comparing the two 3D images which were reconstructed from respective digital holograms. Improvement in the performance of the fractional correlator has been shown when the reconstructed objects are multiplied with appropriately scaled Mexican-hat wavelet filter (MWF) before the correlation operation is performed. The performance measure parameters; discrimination ratio (DR) and peak-to-correlation energy (PCE) have been computed that shows enhancement in the correlation results.

74.2 Theory

74.2.1 Joint Fractional Correlator

The FRT of function $g(x, y)$ with order μ may be written as,

$$G(\xi, \eta) = \Im^{\mu}[g(x, y)] \tag{74.1}$$

Here (x, y) and (ξ, η) represent coordinates in input and fractional domains, respectively. In a joint fractional correlator, target object, $g(x, y)$ and reference object, $g_r(x, y)$ are fractional Fourier transformed with order μ. The joint fractional power spectrum is obtained by multiplication of target signal with complex conjugate of reference signal. The correlation coefficient is obtained by taking inverse fractional Fourier transform of the joint fractional power spectrum, as shown in (74.2).

$$C(x, y) = \Im^{-1}[\Im^{\mu}\{g(x, y)\} \times [\Im^{\mu}\{g_r(x, y)\}]^*] \tag{74.2}$$

74.2.2 Wavelet Filter

The MWF has been used for feature extraction of 3D objects. The MWF is the second derivative of the Gaussian function. The 2D MWF is represented by (74.3).

$$w(x, y) = \left[1 - \left(x^2 + y^2\right)\right] \exp\left(-\frac{x^2 + y^2}{2}\right) \tag{74.3}$$

The MWF in the Fourier domain is represented by (74.4)

$$W(u, v) = s^2\left(u^2 + v^2\right) \exp\left(-2\pi^2 s^2\left(u^2 + v^2\right)\right) \tag{74.4}$$

Here, (x, y) and (u, v) denote input and Fourier domain coordinates, respectively and s represents scale factor. An improved correlation results for 3D recognition is achieved by filtering the complex valued 2D information reconstructed from a digital hologram in the Fourier domain.

74.2.3 Experimental Results and Discussions

The experimental arrangement for recording digital Fresnel holograms of real 3D objects has been shown in the Fig. 74.1. A 35 mW He–Ne laser (make: Research Electro-Optics, USA) having 632.8 nm wavelength acts as a source of coherent light. The incident light beam is divided into two parts with a beam splitter: reference beam and object beam. The object beam is the light beam which is scattered from the real 3D object. The obtained interference pattern is captured by a 5 Megapixel CMOS camera (make: The Imaging Source, Germany; Model: DFK 23GP031) having number of pixels 2592 × 1944 with 2.2 µm square pixel. Figure 74.2a, b show the recorded reference and target digital Fresnel holograms, respectively. The 3D images are numerically reconstructed. The reconstructed 3D images of reference and target objects have been shown in Fig. 74.3a–d. Figure 74.3c, d are obtained after

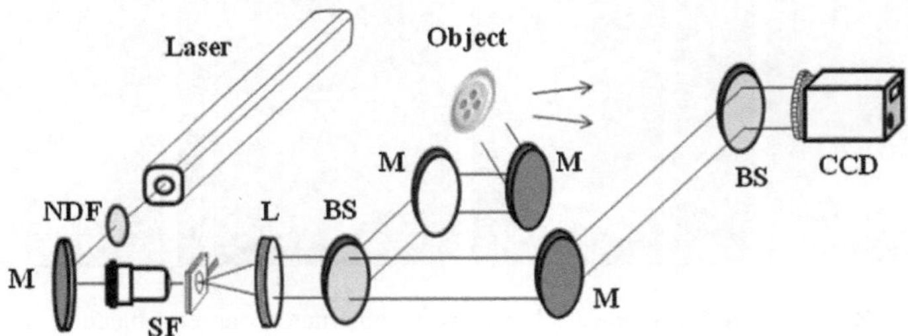

Fig. 74.1 Optical set-up for recording digital holograms of real 3D objects. *NDF* neutral density filter, *SF* spatial filter, *L* lens. *BE* beam expander, *BS* beam splitter, and *M* mirror

application of wavelet filter on the reconstructed 3D objects shown in Fig. 74.3a, b. The correlation output as shown in Fig. 74.4 has been obtained using joint fractional correlator. Figure 74.4a, b show autocorrelation and crosscorrelation outputs, respectively without using the wavelet filter. Autocorrelation and crosscorrelation obtained after the application of wavelet filter on the reconstructed reference and target objects have been shown in Fig. 74.4c, d, respectively.

74.2.4 Performance Measure

The numerical study has been carried out on MATLAB (R2010 b) platform. The performance measure parameters; DR and PCE, which have been defined in (74.3) and (74.4) for joint fractional correlator are computed with the correlation output obtained from both the processes; with and without application of wavelet filter on the reconstructed reference and target objects. The obtained values have been given

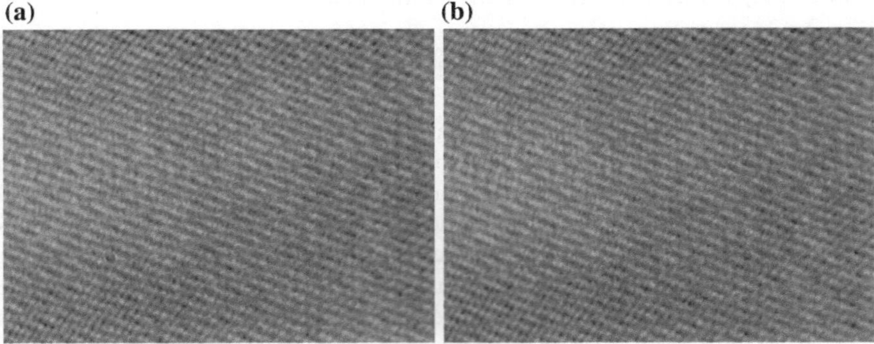

Fig. 74.2 Experimentally recorded digital Fresnel hologram of **a** reference and **b** target 3D object

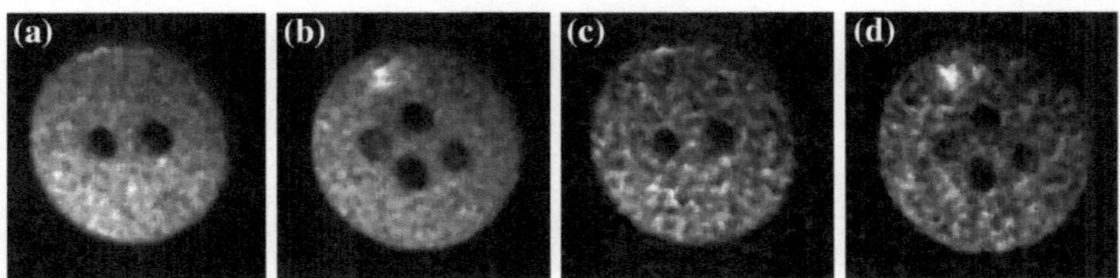

Fig. 74.3 **a** Reconstructed reference object, **b** reconstructed target object, **c** reconstructed reference object after application of wavelet filter, and **d** reconstructed target object after application of wavelet filter

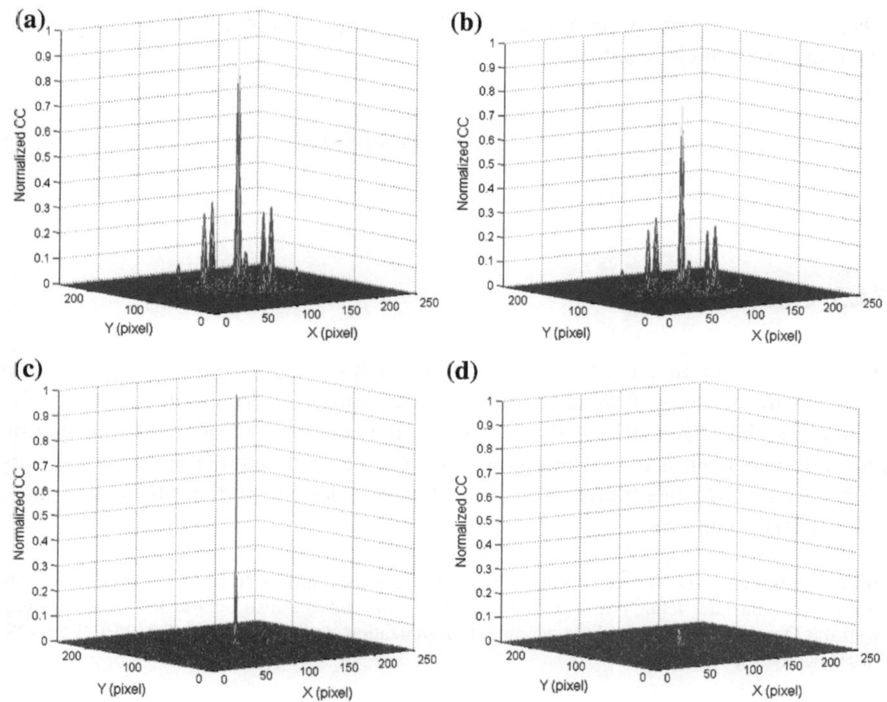

Fig. 74.4 Correlation results. **a** Auto-correlation and **b** cross-correlation when wavelet filter is not applied; **c** auto-correlation and **d** cross-correlation when wavelet filter is applied to joint fractional correlator

in Table 74.1. With the calculated values it can be inferred that the performance of the correlator enhances when a wavelet filter is used.

The DR is defined as,

$$\mathrm{DR} = 1 - \frac{\text{highest peak due to cross-correlation}}{\text{auto-correlation peak height}} \tag{74.5}$$

The PCE is defined as,

$$\mathrm{PCE} = \frac{|c(0,0)|^2}{E_y} \tag{74.6}$$

Table 74.1 Performance measure parameters for joint fractional correlator obtained for the two processes

Process	DR	PCE
Without wavelet filter	0.233	0.064
With wavelet filter	0.928	0.414

where $c(0,0)$ denotes the correlation peak intensity. E_y represents the correlation plane energy, which is defined as,

$$E_y = \int \int_{-\infty}^{\infty} c(x, y)\,\mathrm{d}x\mathrm{d}y \tag{74.7}$$

74.3 Conclusion

An improvement in the performance of the joint fractional correlator has been shown with the help of wavelet filtering. An improved correlation results for 3D recognition has been achieved by filtering the complex valued 2D information reconstructed from a digital hologram in the Fourier domain. Preliminary results with a single 3D object have been presented. We need to further study with large and different data sets.

Acknowledgments D. Kumar sincerely acknowledges the funding from the University Grants Commission through letter No. F.2-10/2011(SA-I). The authors also acknowledge the funding from the Defense Research and Development Organization, under Grant No. ERIP/ER/1200428/M/01/1473.

References

1. Millan MS (2012) Advanced optical correlation and digital methods for pattern matching-50th anniversary of Vander Lugt matched filter. J Opt 14(10):103001
2. Tsang PWM, Poon TC, Liu JP, Situ WC (2014) Review of holographic-based three-dimensional object recognition techniques. Appl Opt 53(27):G95–104
3. Schnars U, Jueptner W (2005) Digital holography: digital hologram recording, numerical recording, and related techniques. Springer, Heidelberg
4. Yamaguchi I, Zhang T (1997) Phase-shifting digital holography. Opt Lett 22(16):1268–1270
5. Awatsuji Y, Fujii A, Kubota T, Matoba O (2006) Parallel three-step phase-shifting digital holography. Appl Opt 45(13):2995–3002
6. Dong YC, Wu J (2010) Space-shifting digital holography with dc term removal. Opt Lett 35 (8):1287–1289
7. Cuche E, Marquet P, Depeursinge C (2000) Spatial filtering for zero order and twin image elimination in digital off-axis holography. Appl Opt 39(23):4070–4075
8. Seifi M, Denis L, Fournier C (2013) Fast and accurate 3D object recognition directly from digital holograms. J Opt Soc Am A 30(11):2216–2224
9. Nelleri A, Gopinathan U, Joseph J, Singh K (2005) Wavelet based three-dimensional object recognition using single off-axis digital Fresnel hologram. Proc. SPIE 5827:30–37
10. Nelleri A, Joseph J, Singh K (2006) Recognition and classification of three-dimensional phase objects by digital Fresnel holography. Appl Opt 45(17):4046–4053
11. Javidi B, Tajahuerce E (2000) Three-dimensional object recognition by use of digital holography. Opt Lett 25(9):610–612

12. Frauel Y, Javidi B (2001) Neural network for three-dimensional object recognition based on digital holography. Opt Lett 26(19):1478–1480
13. Tajahuerce E, Matoba O, Javidi B (2002) Shift-invariant three-dimensional object recognition by means of digital holography. Appl Opt 40(23):3877–3886
14. Javidi B, Kim D (2005) Three-dimensional-object recognition by use of single exposure on-axis digital holography. Opt Lett 30(3):236–238
15. Rosen J (1998) Three-dimensional joint transform correlator. Appl Opt 37(32):7538–7544
16. Jin W, Yan C, Ma L, Ye H, Wang H (2006) Joint extended fractional Fourier transform correlator. Opt Commun 268(1):34–37
17. Jin SI, Lee S-Y (2002) Joint transform correlator with fractional Fourier transform. Opt Commun 207(1–6):161–168
18. Zhu B, Zhao H, Liu S (2002) Three-dimensional transparent objects recognition based on the digital holography and fractional correlator. Optik 113(5):209–212
19. Bhagatji A, Nishchal NK, Gupta AK, Tyagi BP (2008) Extended fractional wavelet joint transform correlator. Opt Commun 281(1):44–48

Part XVI
Quantum Optics and Information Processing

Chapter 75
Efficient, High Power, Low Spectral Distortion and ASE Free Amplification of Mode Locked Yb-doped Fiber Laser

P.K. Gupta, P.K. Mukhopadhyay, C.P. Singh, A.J. Singh, S.K. Sharma, K.S. Bindra and S.M. Oak

Abstract Power amplification of all-fiber Yb-doped double clad fiber (Yb DCF) amplifier seeded by all-normal dispersion mode locked Yb-doped fiber oscillator is reported. 21.8 W of amplified average power at 39 MHz repetition rate was obtained at pump power of 35 W corresponding to $\sim$62 % pump to signal conversion efficiency. The single pulse energy of the amplified mode locked pulses was $\sim$560 nJ. The spectrum of the pulses shows no serious distortion and ASE contribution on power amplification.

75.1 Introduction

All-normal dispersion (ANDi) Yb-doped mode locked fiber lasers [1–4] have generated a great deal of interest due to simplicity in design, configurability in all-fiber components and their suitability for high energy/high power scaling. In most of the ANDi oscillator, the output was taken either before or at the nonlinear polarization rejection (NPR) port to extract maximum pulse energy, although, the spectral profile was much structured and distorted further on amplification. Recently, we have shown that by extracting the pulses after the NPR port in ANDi Yb-doped fiber laser can provide a clean seed source which is considerably resistant to distortion on amplification as compared to that from the NPR-port and can be compressed into femtosecond regime without any significant side-lobes in the pulse profile [5]. In this paper, we report power amplification of pulses extracted after the NPR port from mode locked Yb-doped ANDi laser. An output port after the NPR port is implemented by introducing a polarizing beam splitter followed by a half wave plate (HWP) in the oscillator. Mode locked pulses obtained from this output

P.K. Gupta · P.K. Mukhopadhyay (✉) · C.P. Singh · A.J. Singh · S.K. Sharma · K.S. Bindra · S.M. Oak
Solid State Laser Division, Raja Ramanna Centre for Advanced Technology, Indore 452013, India
e-mail: pkm@rrcat.gov.in

V. Lakshminarayanan and I. Bhattacharya (eds.), *Advances in Optical Science and Engineering*, Springer Proceedings in Physics 166,
DOI 10.1007/978-81-322-2367-2_75

port were amplified into Yb-doped double clad fiber (Yb DCF) amplifier. Around 21.8 W average amplified signal power with 35 W coupled pump power was measured at fundamental repetition rate of 39 MHz.

75.2 Experimental Set-up

The schematic of the experimental setup is shown in Fig. 75.1a. It consists of an oscillator, a preamplifier and a power amplifier segment. The oscillator is configured in ANDi configuration containing a single mode Yb-doped fiber pumped in-core by a single mode fiber coupled laser diode at 975 nm, a couple of single mode fibers (SMF), a pair of fiber pigtailed collimators (COL1 and COL2) and polarizing beam splitters (PBS1 and PBS2), HWP, polarization insensitive isolator (ISO), a band pass filter with 10 nm transmission bandwidth at 1060 nm and wavelength division multiplexers (WDM1 and WDM2). The WDM1 protects the pump diode from backward propagating ASE and WDM2 combines pump and signal beam. The total length of the resonator is ~5.2 m including around 20 cm free space between collimator pairs and around 5 m long SMF's making the total group velocity dispersion (GVD) of the cavity ~0.11 ps^2. PBS1 acts as NPR-port and PBS2 along with HWP acts as a variable output coupler (OP-port). The mode locking in the oscillator set-up could be easily obtained by biasing the polarization

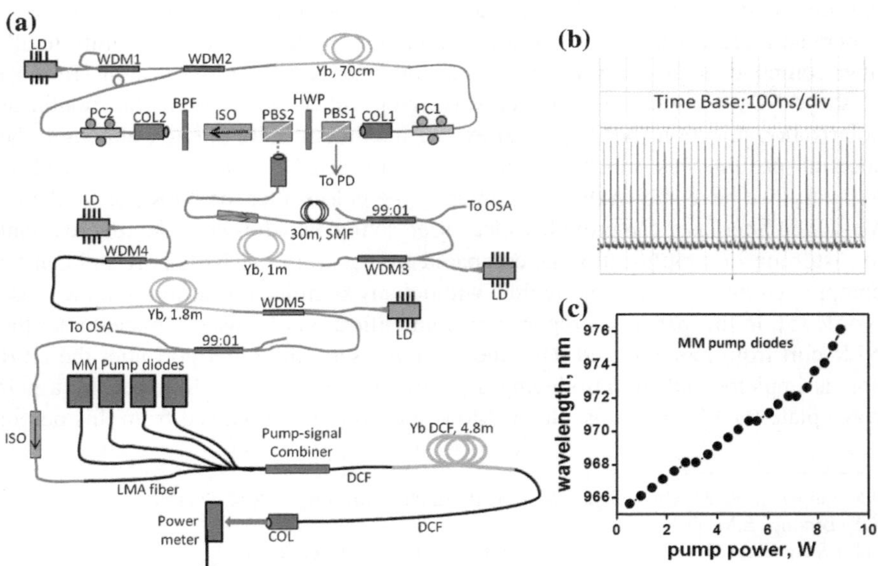

Fig. 75.1 **a** Schematic of Yb-doped mode locked fiber amplifier set-up, **b** a typical mode locked pulse train and **c** MM pump diode characterization

controllers and once it is mode locked, it remains so for a long time and is self-starting. The mode locked pulse train from NPR-port is detected through photo diode at the oscilloscope (Fig. 75.1b shows a typical mode locked pulse train) and that from OP-port is coupled into fiber pigtailed collimator for seeding the amplifier. This seed signal is stretched into 30 m long single mode fiber to counter the effect of nonlinearity by reducing the peak power of mode locked pulses and an isolator is placed to protect the oscillator from any feedback. A 99:01 fiber coupler is implemented to monitor the spectral profile of the mode locked pulses from the oscillator (OP-port).

The pre-amplifier consists of two stages. In the first stage, a single mode Yb-doped fiber (YDF, length 1 m, mode field diameter 6 μm) was pumped in-core by co-propagating pump beam from a single mode fiber coupled laser diode (FCLD). After this, an isolator and an ASE reduction filter (not shown in schematic) was spliced to protect the oscillator from any feedback and to reduce contribution of ASE in the amplified signal, respectively. In the second stage, 1.8 m long YDF is pumped in-core by co-propagating and counter-propagating pump beams from FCLD's. The output of the pre-amplifier is taken from the signal port of WDM5. A 99:01 fiber coupler is spliced to monitor the spectral profile (1 % port) after pre-amplification and rest of the signal (99 % port) is fed into power amplifier followed by an isolator.

The power amplifier segment consists of 4.8 m long Yb DCF (LMA YDF-20/130-VIII, Nufern, USA), 4-nos of high power multimode fiber (core diameter 105 μm, N.A. 0.15) coupled laser diodes at 976 nm and a 6:1 multimode pump combiner (MPC) with signal feed through. The MPC combines the pump beam from the laser diodes and delivers the combined diode beam to its output port (DCF, core/clad diameter 20/125 μm, core N.A. 0.08). Input end of the signal port of MPC (core diameter: 20 μm) is connected to the preamplifier segment and its output port is spliced with the Yb DCF. Each multimode fiber coupled laser diode was characterized for its drift in emission wavelength with output power. It was found that the emission wavelength of laser diodes increases with output power (Fig. 75.1c) and at the maximum output power of ~9 W; the emission wavelength is ~976 nm which is close to the peak of the absorption spectrum in Yb-doped fiber. The length of the Yb DCF was chosen to have maximum absorption of the pump beam. Total 35 W of pump power is coupled to the inner clad of Yb DCF. The other end of Yb DCF was spliced with a high power fiber pigtailed collimator. This completes the design of oscillator, pre-amplifier and power amplifier set-up.

75.3 Results

The oscillator could be mode locked easily by orienting the polarization controllers as the pump power coupled into Yb-doped fiber exceeds ~200 mW. It is pertinent to note that the output coupling at the OP-port can be varied over a wide range with the help of the HWP placed between the PBS1 and PBS2, and can be set to an optimum value. For zero OP coupling, the output occurs mainly through the NPR-port and with an optimum OP-coupling (~30 %), the output occurs mainly through the OP-port. We denote these two modes of operation as "NPR-setup" and "OP-setup" respectively. A maximum average power ~80 mW and 130 mW at 39 MHz repetition rate could be extracted from these output ports respectively under stable mode locking operation. Spectral profiles of the mode locked pulses obtained from these output ports are shown in Fig. 75.2a, b respectively. The spectral profile of the mode locked pulses from the NPR-setup is quite structured with usual 'hare-head' shape and have a root mean square width ($\Delta\lambda_{rms}$) ~5.54 nm. In contrast to this, the spectral profile from the OP-setup is comparatively smoother and is dome shaped with $\Delta\lambda_{rms}$ ~4.8 nm. This difference in spectral profile is due to strong pulse shaping by the band pass filter and NPR of heavily chirped mode locked pulses. In heavily chirped mode locked pulses, the nonlinear chirp is placed at the wings of the pulses which is rejected at the NPR-port and hence the spectral profile at the NPR-setup is structured whereas that from the OP-setup is relatively smoother. In Fig. 75.2, spectral profile is also shown at logarithmic scale in the insets. It can be seen clearly that in either case, the ASE is peaked more than 30 dB below the signal peak and hence the contribution of ASE is negligible.

The low power amplification characteristics of the mode locked pulses obtained from these two respective output ports was studied [5]. Around 10 mW of signal coupled into 1 m long Yb-doped fiber was amplified to an average power of 250 mW. The pre-amplification of seed pulses from the NPR-port exhibits distortion in its spectral profile (Fig. 75.3a) and develops side lobes on compression into grating pair (Fig. 75.3b). On the other hand, seed pulses from the OP-port nearly retain their spectral profile on amplification (Fig. 75.3c) and the temporal profile after compression is clean and free from any side lobes (Fig. 75.3d). The compressed pulse durations (FWHM) were measured to be 141 fs and 156 fs respectively. Also, the nonlinearity induced spectral broadening is more significant in the NPR-port amplified spectra as compared to that in OP-port amplified spectra. Moreover, a considerable amount of ASE (~3 %) can be seen in Fig. 75.3a while

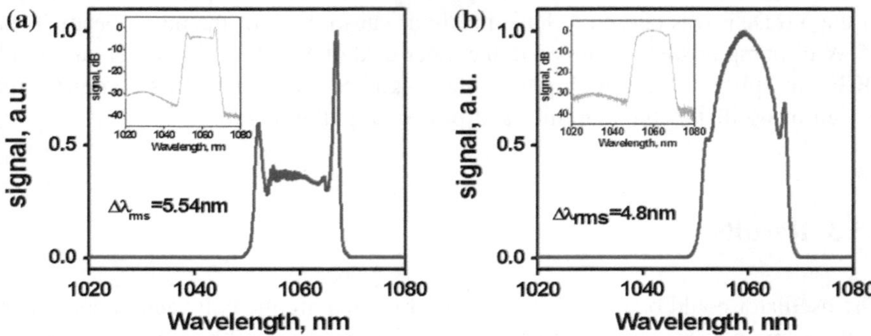

Fig. 75.2 Spectral profile of the mode locked pulses from **a** NPR-setup and **b** OP-setup respectively. Inset shows respective spectral profile at logarithmic scale

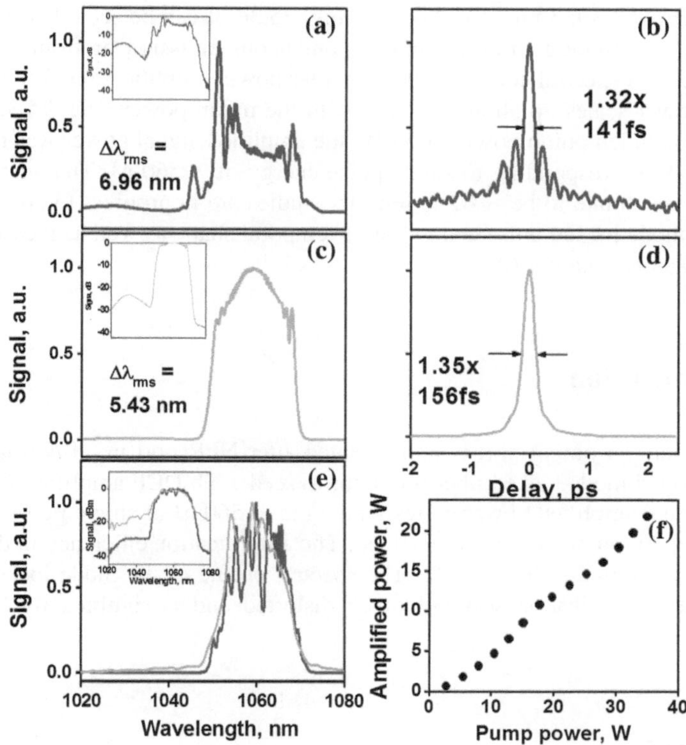

Fig. 75.3 **a** Spectral profile of the pre-amplified pulses seeded from the NPR-port. **b** Its AC trace after compression. **c** Spectral profile of the pre-amplified pulses seed from OP-port. **d** Its AC trace. **e** Spectral profile of the mode locked pulses before (*blue color*) and after (*red color*) power amplification. Inset shows respective spectral profiles at logarithmic scale. **f** Power amplification characteristics in Yb DCF (color figure online)

that in Fig. 75.3c it is below 0.5 %. It is to be noted that the same pre-amplifier set-up was used for the amplification of both the seed pulses. Thus, the seed pulses from the OP-port are considerably resistant to distortion in its spectral as well as temporal shape on amplification and hence further amplification studies are made only with the seed pulses from the OP-port. As shown in Fig. 75.1, before power amplification, the pre-amplified mode locked pulses were seeded further for small scale amplification in another 1.8 m long single mode Yb-doped fiber amplifier segment pumped in-core from co-propagating and counter-propagating pump beams to enhance the signal strength. Since several ASE reduction filters and isolators are placed at different amplification stages which have considerably large insertion losses, the amount of input signal to the power amplifier segment was ~80 mW. Figure 75.3e, f show the power amplification characteristics in Yb DCF amplifier. The recorded spectral profiles of the mode locked pulses before power amplification (blue color, $\Delta\lambda_{rms}$ ~5.17 nm) and after power amplification (red color, $\Delta\lambda_{rms}$ ~6.3 nm) are shown in Fig. 75.3e. It can be seen that there is no significant contribution from amplified spontaneous emission or from Stimulated Raman induced spectral components even after power amplification. The amplified average power scales up almost linearly with the pump power (Fig. 75.3f). At the maximum coupled pump power of 35 W, the amplified signal power was measured to be 21.8 W corresponding to single pulse energy of ~560 nJ. The amplification efficiency was found to be ~62 %. Further studies are in progress to compress the amplified mode locked pulse and to see its temporal shape as well as measure pulse width on autocorrelation (AC).

75.4 Conclusion

In conclusion, an output port is implemented after NPR port in all-normal dispersion Yb-doped mode locked fiber oscillator to seed a Yb DCF amplifier. More than 20 W average amplified power corresponding to ~560 nJ of single pulse energy at 39 MHz repetition rate has been obtained. The amplification efficiency in the power amplifier was found to be ~62 %. The spectral profile of the mode locked pulses after power amplification was not much distorted and is significantly free from ASE.

References

1. Chong A, Buckley J, Renninger W, Wise F (2006) All-normal-dispersion femtosecond fiber laser. Opt Exp 14(21):10095–10100
2. Chong A, Renninger W, Wise F (2007) All normal dispersion femtosecond fiber laser with pulse energy above 20 nJ. Opt Lett 32(16):2408–2410
3. Kiew K, Renninger W, Chong A, Wise F (2009) Sub 100 fs pulses at Watt level powers from a dissipative soliton fiber laser. Opt Lett 34(5):593–595
4. Mukhopadhyay PK, Özgören K, Budunoğlu İL, İlday FÖ (2009) All-fiber low-noise high-power femtosecond Yb-fiber amplifier system seeded by an all-normal dispersion fiber oscillator. IEEE J Sel Top Quant 15(1):145–152
5. Mukhopadhyay PK, Gupta PK, Bindra KS, Oak SM (2013) Note: amplification characteristics of all-normal-dispersion mode-locked Yb-doped fiber laser: influence of input pulse shape. Rev Sci Instrum 84:076107

Chapter 76
Theoretical Analysis of Direct Transition in SiGe/GeSn Strained Quantum Well Structure by Finite Difference Method

Prakash Pareek and Mukul K. Das

Abstract This paper focuses on the potential of compressive strained group IV quantum well (QW) structure in photo sensitive devices specially photodetectors. Direct absorption characteristic in compressive strained SiGe/GeSn QW is determined with the help of Schrödinger equation. The Schrödinger equation for Γ-valley, Heavy hole band and Light hole band is solved by finite difference method (FDM). Due to compressive strain, heavy hole (HH) band pushes upwards and light hole (LH) band lowers on addition of Sn concentration. Absorption coefficient of QW is evaluated by using Fermi's golden rule. Result shows that significant absorption is obtained in short wave infrared range. HH to conduction band transition provides greater absorption than LH to conduction band transition. Result also shows that compressive strained GeSn become direct band gap in nature for Sn concentration greater than 0.15 (15 %).

76.1 Introduction

Photonic devices have witnessed a widespread uses in recent years owing to their unique properties that put them on a higher level as compared to their electronic counterparts. Choice of semiconductor material for optoelctronic devices is an important point of discussion. Direct band gap material is preferred for efficient radiative transition. In direct band gap semiconductors, valence band maximum and the conduction band minimum occur at the same level and an upward or downward transition does not require a change in momentum. Thus the probability of radiative recombination is very high in direct band gap semiconductor. They can show strong

P. Pareek (✉) · M.K. Das
UGC SAP Research Laboratory, Department of Electronics Engineering, Indian School of Mines, Dhanbad 826004, India
e-mail: ppareek1@gmail.com

M.K. Das
e-mail: das.mk.ei@ismdhanbad.ac.in

V. Lakshminarayanan and I. Bhattacharya (eds.), *Advances in Optical Science and Engineering*, Springer Proceedings in Physics 166,
DOI 10.1007/978-81-322-2367-2_76

emission or absorption at optical communication wavelength ($\sim$1.55 µm). So photonic technology relies heavily on direct band gap III–V materials like GaAs [1]. Unfortunately, GaAs cells are extremely expensive to manufacture (gallium is rarer than gold) and they produce toxic wastes, and arsenic is poisonous [1]. This fact forces the concerned researchers to work towards a new group of semiconductor materials which can act as a replacement for III–V group.

Recently, there has been a great deal of interest among researchers on the design and analysis of photo sensitive devices based on Tin (Sn) incorporated Gr-IV alloys. Incorporation of proper amount of α-Sn into Ge is the most promising route for engineering the electronic and optical properties of such devices. Alloying Ge with α-Sn can effectively reduce the direct band gap of Ge more than its indirect band gap and, hence, a direct band gap GeSn alloy can be realized [2, 3]. For strain free $Ge_{1-x}Sn_x$ alloys, the indirect band gap to direct band gap transition occurs for Sn concentration $x = 0.9 - 0.11$. By introducing tensile strain in GeSn, the direct band gap can be obtained at lower concentration of Sn ($x = 0.4 - 0.8$) as compared to relaxed GeSn [3, 4]. It has also been found in recent times that the direct band gap can be obtained by introducing compressive strain in $Ge_{1-x}Sn_x$ alloys for $x > 0.15$ [5, 6]. Despite of larger requirement of Sn concentration for converting it from indirect to direct bandgap as compared to tensile strained GeSn alloys, compressive strained GeSn alloys have generated a great deal of interest among researchers. Because, the compressive strain allows the use of strain relaxed Ge/SiGe buffer layer for the growth of $Ge_{1-x}Sn_x$ layer. This enables the defect free epitaxial growth and also enables to achieve the amount of Sn incorporation beyond its solid solubility limit of 1 % [3]. Moreover, the inclusion of compressive strain in the GeSn multiple-quantum well (QW) laser can effectively modify the valence band structure to reduce the threshold carrier density. Compressive strain in the GeSn wells makes the top valence band heavy hole (HH) like, resulting in a dominant transverse electric (TE) gain in laser [7].

Compressive strained GeSn can also be utilized for short wave infrared photodetector [8], which is very attractive for several applications like in the the field of spectroscopic sensing which relies on the strong gas/liquid/solid absorption in SWIR range [9]. Accurate physics based modeling for compressive strained GeSn alloy based structure is required for re-engineering the electronic, optical properties of the structure before fabrication. In this work theoretical analysis for direct transition in a compressive strained SiGe/GeSn QW structure is done. Finite difference method (FDM) is used to solve the Schrödinger equation in the well for Γ valley conduction band, HH band and Light hole (LH) band separately.

76.1.1 Theoretical Analysis

The QW structure considered in our analysis is explained as follows. A 20 Å thick Ge_xSn_{1-x} layer is sandwiched between two relaxed $Si_{0.1}Ge_{0.9}$ barriers each of width 60 Å to form a single QW. In order to facilitate the analysis for direct inter-band transition, width of the well is selected in such a way that there exists only one bound state for each of the conduction and valence band in the well. The Schrödinger equation with effective mass approximation and considering the strain effect is considered in our analysis and is given as [10]:

$$\left[\frac{-\hbar^2}{2}\frac{\partial}{\partial z}\frac{1}{m_p}\frac{\partial}{\partial z}+\frac{\hbar^2 K_t^2}{2m_p}+V_p(z)\right]\psi = E\psi \tag{76.1}$$

where, $\hbar$ is planck's constant, z is position variable, K_t is the transverse wave vector, ψ is wave function, E is Eigen energy, m and V are respectively the effective mass and band profile. Suffix p stands for type of band e.g., $p = c$ for Γ conduction band, $p =$ hh for HH valence band, $p =$ lh for LH valence band. As the Γ valley is of only interest here, the transverse wave vector, K_t is taken as zero.

The equation is solved using FDM to obtain Eigen energies and wave function in the well. The whole region of interest is divided into N number of small elements of equal width, s and the equation is solved for each of the elements [11]. The next step is to obtain the matrix representing the Hamiltonian operator, H which is given by (76.2)

$$H = -\frac{\hbar^2}{2m_p}\frac{\partial^2}{\partial z^2}+V_p(z) \tag{76.2}$$

FDM basically converts Schrodinger equation which is a differential equation into a difference equation by standard finite difference technique as shown in (76.3). By substitution of (76.3) and (76.4) in (76.2), expression for matrix H is obtained after simplification as shown in (76.5)

$$\left(\frac{\partial^2}{\partial z^2}\right)_{z=z_N} \rightarrow \frac{1}{s^2}\left[\psi(z_{N+1}) - 2\psi(z_N) + \psi(z_{N+1})\right] \tag{76.3}$$

$$V_p(z)\psi(z) \rightarrow V_p(z_N)\psi(z_N) \tag{76.4}$$

$$H = [V_{pN} + 2t_0]\delta_{a,b} - t_0\delta_{a,b+1} - t_0\delta_{a,b-1} \tag{76.5}$$

where $t_0 = \frac{\hbar^2}{2m_p s^2}$, $\delta_{a,b}$ is Kronecker delta which is expressed as $\delta_{a,b} = \left\{\begin{matrix} 1; & a = b \\ 0; & a \neq b \end{matrix}\right\}$.

Following boundary conditions (76.6) were used for position dependent band profile, V_p (Z) ($p = c$ for Γ conduction band, $p =$ hh for HH valence band and $p =$ lh for LH valence band) within and outside the well are used in process of above solutions.

$$V_c(z) = \left\{ \begin{array}{ll} 0; & |z| \leq \frac{W}{2} \\ \Delta E_c; & |z| > \frac{W}{2} \end{array} \right\}; \quad V_{\mathrm{hh}}(z) = \left\{ \begin{array}{ll} 0; & |z| \leq \frac{W}{2} \\ -\Delta E_{\mathrm{hh}}; & |z| > \frac{W}{2} \end{array} \right\};$$
$$V_{\mathrm{lh}}(z) = \left\{ \begin{array}{ll} 0; & |z| \leq \frac{W}{2} \\ -\Delta E_{\mathrm{lh}}; & |z| > \frac{W}{2} \end{array} \right\} \tag{76.6}$$

where, W is well width. ΔE_c, $\Delta E_{v_{\mathrm{hh}}}$ and $\Delta E_{v_{\mathrm{lh}}}$ are conduction band, HH band and LH band offsets respectively for strained $Ge_{1-x}Sn_x$ layer on relaxed $Si_{0.1}Ge_{0.9}$. ΔE_c, $\Delta E_{v_{\mathrm{hh}}}$ and $\Delta E_{v_{\mathrm{lh}}}$ were calculated from reported empirical relationships [6]. Effective masses (m_c, m_{hh} and m_{lh}) were calculated by using strain theory and linear interpolation [6]. The material parameters for SiGe and GeSn alloys were calculated by using linear interpolation with bowing effect [3, 6]. Now, the total solution is obtained in the form of Hamiltonian matrix of the order of N. The permissible Eigen energies within the well are obtained from the Hamiltonian matrix, H and the corresponding wave functions are obtained from the Eigen vectors.

After evaluating the Eigen energies for conduction band and valence band, absorption coefficient is evaluated for QW with the help of Fermi golden rule and using the following mathematical expression [10] (76.7), for absorption coefficient α.

$$\alpha(\hbar\omega) = \frac{\pi q^2}{n_r c \varepsilon_0 m_0^2 \omega} \sum_{n,m} \left|I_m^n\right|^2 \hat{e} \cdot p_{cv}\, \rho_r. \quad \text{Heaviside}\,(\hbar\omega - E_0 + E_{c1} - E_{v1})\,(f_v - f_c) \tag{76.7}$$

where, ω is photon frequency, q is electronic charge, c is speed of light, n_r is refractive index of well, ε_0 is static dielectric constant, m_0 is electron rest mass, E_0 is direct band gap of $Ge_{1-x}Sn_x$ which is calculated by linear interpolation of direct band gap of Ge and Sn considering bowing effect [3]. E_{c1} and E_{v1} are bound state Eigen energies in Γ conduction band and valence band respectively, I is overlap integral of the conduction subband n wave function and valence subband m wave function, p_{cv} is momentum matrix element between conduction band and valence band (v = hh for HH band, v = lh for LH band) for $K_t = 0$, e is polarization unit vector and ρ_r is reduced joint density of states in QW. QW assumed to be undoped in this work, only valence band is filled with carriers so carrier probability occupancy of conduction band is taken as zero ($f_c = 0$, $f_v = 1$). Gaussian line shape function with broadening of 70 meV is considered in this analysis in place of Heaviside line shape function to account for inhomogeneity in GeSn alloy [12].

76.2 Results and Discussion

Eigen energies and wave functions for Γ valley of conduction and valence band are obtained from Hamiltonian matrix as mentioned in the previous section and are shown in Fig. 76.1a–d for different values of Sn concentration, x. Figure 76.1 shows

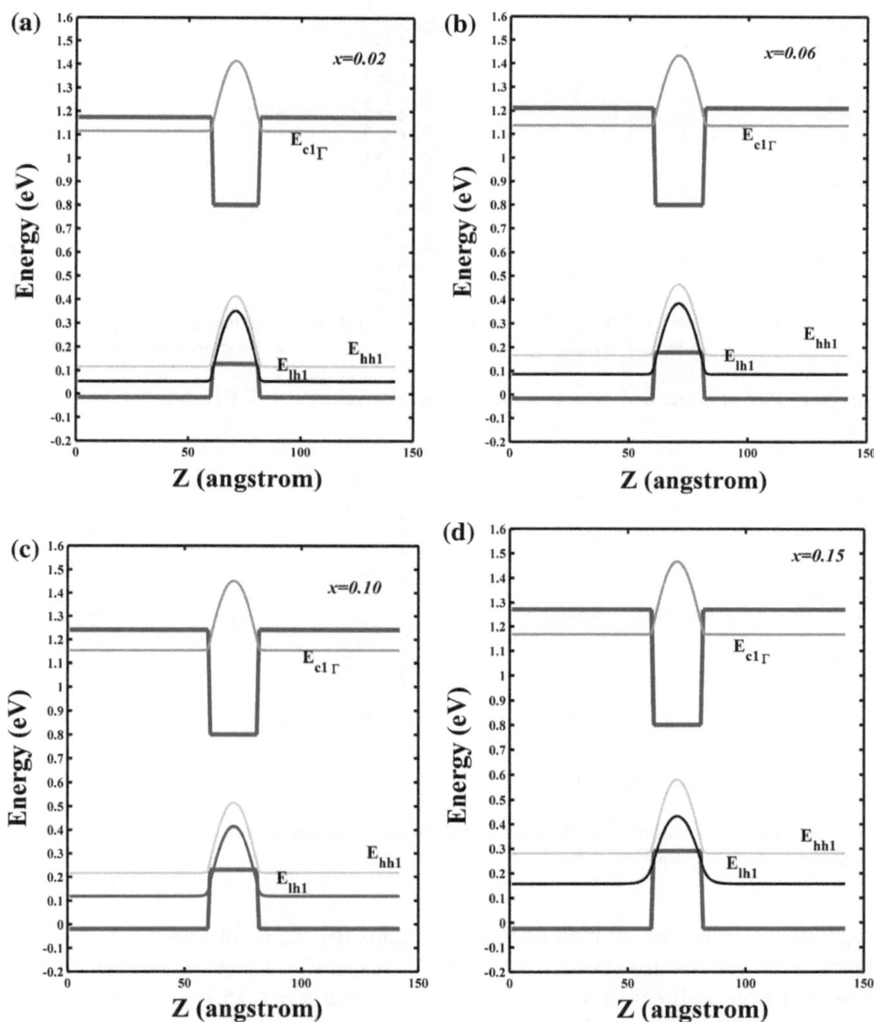

Fig. 76.1 Plot of Eigen energies and wave functions for different values of x. **a** $x = 0.02$, **b** $x = 0.06$, **c** $x = 0.10$, **d** $x = 0.15$

that with increasing x, the HH band shifts upwards and LH band shifts downwards and hence, the separation between HH band Eigen energy (E_{hh1}) and LH band Eigen energy (E_{lh1}) increases. This is due to the presence of compressive strain in GeSn layer and this causes reduction of required transition energy for direct transition from HH to conduction band. So, x has effect on the direct band gap of the material.

To clearly understand the effect of x on the band gap, direct band gap energies, E_g (Γ-hh) and E_g (Γ-lh) were plotted as a function of x and is shown in Fig. 76.2a, b

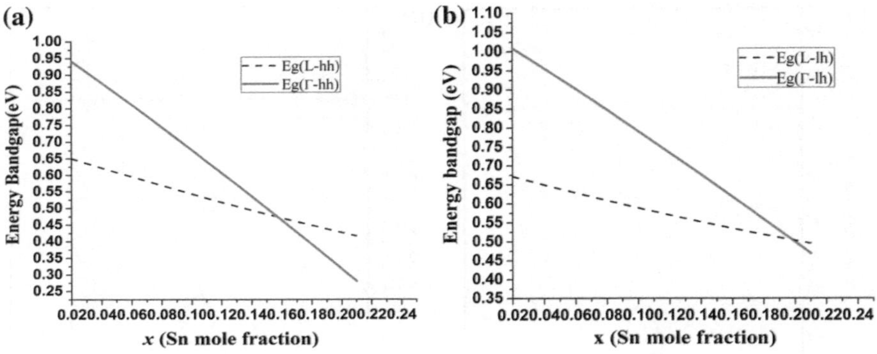

Fig. 76.2 Plot of direct and indirect band gap energies versus x. **a** Γ-HH, **b** Γ-LH

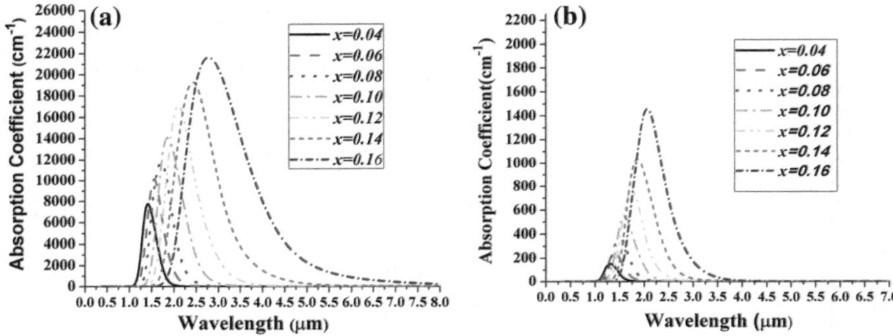

Fig. 76.3 Plot of absorption coefficient (α) for direct transition versus wavelength for different values of x. **a** HH- Γ transition, **b** LH- Γ

respectively. Reported indirect energy band gaps [6], E_g (L-hh) and E_g (L-lh), as a function of x have also been plotted in the respective figure. It is seen that the transition from indirect to direct band gap occurs at $x = 0.15$ in case of E_g (Γ-hh) and at $x = 0.20$ in case of E_g (Γ-lh). Absorption coefficient, α is determined using (76.7) and is shown in Fig. 76.3a, b respectively for the direct HH-conduction band LH-conduction band transition. Results indicate that HH–CB transition provides larger absorption than that of LH-conduction band transition due to reduction of transition energy for HH-conduction band transition. The absorption is red shifted on increasing Sn content in case of HH–CB transition and peak absorption lies in the range of 2–4 μm. Thus due to significant absorption at SWIR wavelength, this structure can be utilized for photodetection applications.

76.3 Conclusion

Potential of compressively strained Group IV QW structures in photonic applications is explored in this work. SiGe/GeSn QW structure investigated for photo detection applications by examining its direct transition characteristics. Direct transition characteristic of strained SiGe/GeSn is analyzed by solving Schrödinger equation by FDM method. The compressive strain in GeSn well makes the top valence band HH like which lead to higher absorption coefficient for HH-conduction band (Γ) transition than that of LH-conduction band (Γ) transition. The simulated results also indicate that for $x > 0.15$, GeSn well becomes direct band gap in nature. The peak absorption lies in short wave infrared range, so this QW structure can be used for SWIR photodetection. In future, GeSn based integrated photodetectors can become an attractive approach for monolithically integrated SWIR photodetectors. In addition, this QW structure is required to be theoretically investigated for laser operation.

References

1. Bhattacharya P (1994) Semiconductor optoelectronic devices. Prentice Hall India, USA, pp 2–7
2. Soref RA, Perry CH (1991) Predicted bandgap of the new semiconductor SiGeSn. J Appl Phys 69:539
3. Kouvetakis J, Menedez J, Chizmeshya AVG (2006) Tin based group IV semiconductors: new platforms for opto and micro electronics and silicon. Ann Rev Mat Res 36:497–554
4. D'Costa VR (2006) Optical critical points of thin-film Ge_{1-y} Sn_y alloys: a comparative Ge_1 $_{-y}Sn_y/Ge_{1-x}Si_x$ study. Phys Rev B 73:125207
5. Yahyaoui N, Sfina N, Lazzari J-L, Bournel A, Said M (2014) Band engineering and absorption spectra in compressively strained $Ge_{0.92}Sn_{0.08}$/Ge (001) double quantum well for infrared photodetection. Phys Status Solid C. doi:10.1002/pssc.201400054
6. Yahyaoui N, Sfina N, Lazzari J-L, Bournel A, Said M (2014) Wave-function engineering and absorption spectra in Si0.16Ge0.84/Ge0.84Sn0.16/Si0.16Ge0.84 strained on relaxed Si0.10Ge0.90 type I quantum well. JAP 115:033109
7. Chang G-E, Chang S-W, Chuang S-L (2010) Strain-balanced GezSn1-z-SixGeySn1-x-y multiple-quantum-well lasers. IEEE J Quant Electron 46:1813–1820
8. Gassenq A, Gencarelli F Van Campenhout J, Shimura Y, Loo R, Narcy G, Vincent B, Roelkens G (2012) GeSn/Ge heterostructure short-wave infrared photodetectors on silicon. Opt Express 20:27297–27303
9. Crowder JG, Smith SD, Vass A, Keddie J (2006) Infrared methods for gas detection. Mid Infrared Semicond Optoelectron 118:595–613 (Springer series in optical Science)
10. Chuang SL (1995) Physics of optoelctronic devices. Wiley, New York, pp 124–388
11. Dutta S (2005) Quantum transport: atom to transistor. Cambridge University Press, New York, pp 33–49
12. D'Costa VR, Fang Y, Mathews J, Roucka R, Tolle J, Menedez J, Kouvetakis J (2009) Sn alloying as a means of increasing the optical absorption of Ge at the C-and L-telecommunication bands. Semicond Sci Technol 24(115006):8

Chapter 77
Can Photons Influence Effective Mass?

B. Chatterjee, K. Sarkar and K.P. Ghatak

Abstract An attempt is made to study the effective electron mass (EEM) in opto electronic materials within the framework of $\vec{k} \cdot \vec{p}$ formalism in the presence of intense light waves which changes the dispersion relation fundamentally. It is found taking n-type Mercury Cadmium Telluride and Indium Gallium Arsenide Phosphide lattice matched to Indium Phosphide that the EEM exhibits increasing dependency with wavelength as the later one changes from red to violet together with the fact with increasing light intensity the same mass is being enhanced with different numerical values. The variations of EEM with light intensity and wavelength reflect the direct signature of intense light waves on the band structured dependent properties of opto electronic materials. In the absence of external photo excitation all the results get transformed to the well-known expressions as given in the literature and thus confirming the compatibility test.

77.1 Introduction

In recent years, with the advent of nano science and photonics, there has been extensive interest in investigating the various optoelectronic processes in optoelectronic nano structured compounds [1]. Incidentally, all the studies in the literature have been made with the fundamental constraints that the energy band structure will not at all change in the presence of very large electric field as available in nano-devices and this assumption is basically wrong. The physical

B. Chatterjee
Department of Electronics and Communication Engineering, Institute of Engineering and Management, Salt Lake, Kolkata, India

B. Chatterjee · K. Sarkar · K.P. Ghatak (✉)
Department of Basic Science and Humanities,
Institute of Engineering and Managemant, Kolkata,
West Bengal 700091, India
e-mail: kamakhyaghatakcu@gmail.com

V. Lakshminarayanan and I. Bhattacharya (eds.), *Advances in Optical Science and Engineering*, Springer Proceedings in Physics 166,
DOI 10.1007/978-81-322-2367-2_77

properties of opto-electronic compounds in the presence of intense photo excitations which change the basic dispersion relation have relatively less investigated in the literature [2]. In the present paper an attempt is made to investigate the effective electron mass (EEM) in opto-electronic materials under intense photo-excitation on the basis of newly derived electron wave vector-energy dispersion relation in the present case.

The EEM being inversely proportional to the carrier mobility occupies a central position to control the analysis of modern low dimensional photonic devices under different operating conditions [2]. The high value of the Fermi energy in low band gap semiconductors influences the EEM when it is energy dependent. For degenerate semiconductors, electrons at the Fermi level will participate in carrier transport and affect all the physical properties. The dispersion relation determines the Fermi energy through the carrier statistics and therefore, these two features ultimately control the variation of the EEM in degenerate compounds with respect to carrier concentration and other external variables in general. It appears from the literature that various carrier energy spectra have been proposed [3] which in turn creates the interest in studying the EEM in such materials under different external conditions [4, 5].

In this paper, we shall use n-$Hg_{1-x}Cd_xTe$ and n-$In_{1-x}Ga_xAs_yP_{1-y}$ lattice matched to InP as examples of opto electronic compounds respectively. It is important to note that the ternary alloy n-$Hg_{1-x}Cd_xTe$ is a classic narrow-gap semiconductor and is a vital optoelectronic compound because of the fact that its energy band gap can be altered to cover a spectral range from 0.8 to over 30 μm by adjusting the values of the alloy composition. The n-$Hg_{1-x}Cd_xTe$ finds wide uses in infrared detector compounds and photovoltaic detector arrays in the 8–12 μm wave bands. The above applications have generated an $Hg_{1-x}Cd_xTe$ technology for the commercial production of very high mobility single crystals, with specially prepared cleaned surface layers and the same compound is extremely suitable for studying the narrow sub-band physics since the relevant spectrum constants are within easy experimental reach. In addition, the quaternary materials are used extensively in high mobility transistors, nano structured optoelectronics, visible hetero-structure compounds lasers, and infrared light emitting diodes, lasers for various fiber optic systems, avalanche photo-detectors, tandem solar cells, detectors in optical fiber communications and long wavelength light sources respectively. Besides, new types of low dimensional optical devices can be fabricated successfully from various quaternary systems. Sections 77.2 and 77.3 contain theoretical background and results and discussion as pertinent to this paper.

77.2 Theoretical Background

The electron energy spectra in the present case can be written as [2]

$$\frac{\hbar^2 k^2}{2M_c} = \beta_{11}(E, \lambda_0) \tag{77.1}$$

$$\frac{\hbar^2 k^2}{2M_c} = \tau_{11}(E, \lambda_0) \tag{77.2}$$

$$\frac{\hbar^2 k^2}{2M_c} = \rho_{11}(E, \lambda_0) \tag{77.3}$$

Use of (77.1)–(77.3) lead to expression of the EEM in the present case as,

$$M^*(E_{FL}, \lambda_0) = M_c \beta'_{11}(E_{FL}, \lambda_0) \tag{77.4}$$

$$M^*(E_{FL}, \lambda_0) = M_c \tau'_{11}(E_{FL}, \lambda_0) \tag{77.5}$$

and

$$M^*(E_{FL}, \lambda_0) = M_c \rho'_{11}(E_{FL}, \lambda_0) \tag{77.6}$$

The electron concentration can be expressed as

$$n_0 = \left(3\pi^2\right)^{-1} \left(\frac{2M_c}{\hbar^2}\right)^{3/2} [M_1(E_F, \lambda_0) + N_1(E_F, \lambda_0)] \tag{77.7}$$

$$n_0 = \left(3\pi^2\right)^{-1} \left(\frac{2M_c}{\hbar^2}\right)^{3/2} g_v[M_2(E_F, \lambda_0) + N_2(E_F, \lambda_0)] \tag{77.8}$$

and

$$n_0 = \left(3\pi^2\right)^{-1} \left(\frac{2M_c}{\hbar^2}\right)^{3/2} g_v[M_3(E_F, \lambda_0) + N_3(E_F, \lambda_0)] \tag{77.9}$$

where, $M_1(E_F, \lambda_0) \equiv [\beta_{11}(E_F, \lambda_0)]^{3/2}$, $N_1(E_F, \lambda_0) = \sum_{r=1}^{s'} L(r) M_1(E_F, \lambda_0)$, $L(r) \equiv \left[2(k_B T)^{2r}(1 - 2^{1-2r})\xi(2r)\right]\left(\frac{\partial^{2r}}{\partial E^{2r}}\right)\Big|_{E=E_F}$, $M_2(E_F, \lambda_0) \equiv [\tau_{11}(E_F, \lambda_0)]^{3/2}$, $N_2(E_F, \lambda_0) = \sum_{r=1}^{s'} L(r) M_2(E_F, \lambda_0)$, $M_3(E_F, \lambda_0) \equiv [\rho_{11}(E_F, \lambda_0)]^{3/2}$ and $N_3(E_F, \lambda_0) = \sum_{r=1}^{s'} L(r) M_3(E_F, \lambda_0)$.

In the absence of light waves and for two band model of Kane the EEM and n_0 assume the well-known forms as given in [5].

77.3 Result and Discussions

Figures 77.1 and 77.2 exhibit the plots of EEM as functions of light intensity for n-$Hg_{1-x}Cd_xTe$ and n-$In_{1-x}Ga_xAs_yP_{1-y}$ lattice matched to InP respectively on the basis of the dispersion relations of Kane and parabolic energy bands in the presence of intense light waves. The Figs. 77.3 and 77.4 exhibit the dependence of the EEM on the wavelength of incident light for the said materials respectively.

From the plot (a) of Fig. 77.2 one can infer that the EEM for n-$Hg_{1-x}Cd_xTe$ in the presence of light waves increases with increasing I, in a more or less linear fashion in accordance with the perturbed three band model of Kane whereas from plot (c), one observes that the EEM on the basis of the perturbed two band model of Kane is greater as compared with plot (a) in the whole range of I. For low values of I, the curves (a) and (c) exhibits converging tendency. It is important to note that with respect to light intensity, the numerical values of the EEM in the absence of light waves, in the case of n-$Hg_{1-x}Cd_xTe$ are greater for both types of band models [curves (b) and (d)] as compared with that of (a) and (c) when $I \neq 0$.

It appears from the plots (a) and (c) of Fig. 77.1 that the EEM increases with I for both types of perturbed band models with different small curvatures and the difference in numerical values of EEM increases with increasing in I, although they converge to a particular value for $I = 10^{-6}$ nWm^{-2}. The curves (b) and (c) for

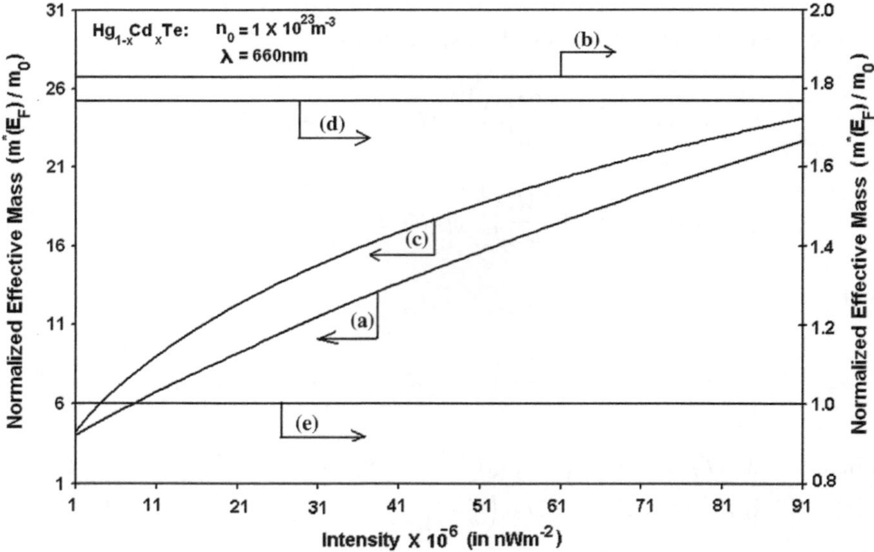

Fig. 77.1 Plot of the normalized EEM as a function of light intensity for n-$Hg_{1-x}Cd_xTe$ in the presence of intense photo excitation in which the curves a and c represent the three and two band models of Kane respectively. The curves b and d exhibit the same variation in the absence of external photo-excitation. The curve e in this case exhibits the parabolic energy band model both in the presence and in the absence of the external photo-excitation

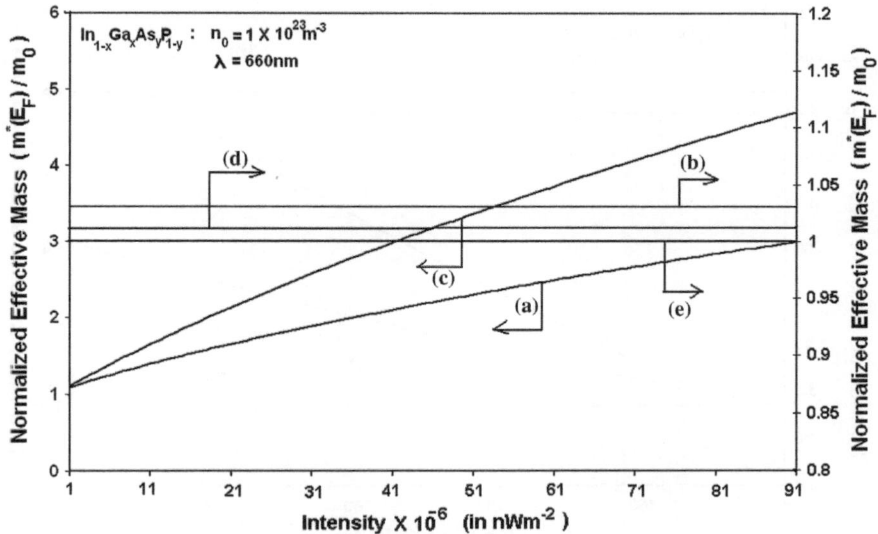

Fig. 77.2 Plot of the normalized EEM as a function of light intensity for $In_{1-x}Ga_xAs_yP_{1-y}$ lattice matched to InP in the presence of light waves in which the curves *a* and *c* represent the three and two band models of Kane respectively. The curves *b* and *d* exhibit the same variation in the absence of external photo-excitation. The curve *e* in this case exhibits the parabolic energy band model both in the presence and in the absence of the external photo-excitation

n-$Hg_{1-x}Cd_xTe$ of Fig. 77.3 increase with increasing λ and after intersection they exhibit the wide difference with each other. The spin-orbit splitting decreases the EEM remarkably for relatively higher values of λ.

From Fig. 77.3 it appears that the EEM in this case is much less when compared with the same as given *I*, in the presence of external intense light waves. From plots (a) and (c) of Fig. 77.4, we observe for $In_{1-x}Ga_xAs_yP_{1-y}$ lattice matched to InP that both the curves differ widely under photo-excitation with respect to λ in the whole range of wave length as considered in Fig. 77.4. As λ increases, from plots (a) and (c) of Fig. 77.3 we observe that the difference also increases. The influence of Δ on the EEM in the absence of external photo-excitation for three and two band models of Kane is very small as evident from the curves (b) and (d). It appears that the EEM increases as the wavelength shifts from violet to red. Figures 77.3 and 77.4 make the prominent signature of light immediately apparent since the EEM depends strongly on *I* and λ for the three and the two band model of Kane which is in direct contrast with that for the bulk specimens of the said compounds in the absence of external intense light waves. The variations of the EEM in Figs. 77.3 and 77.4 reflect the direct signature of the light wave on the band structure dependent optical properties of optoelectronic materials in general in the presence of external photo-excitation and the photon assisted transport for the corresponding nano structured photonic devices. The numerical values of the EEM in the presence of the light waves are larger than that of the same in the absence of light wave for both the three

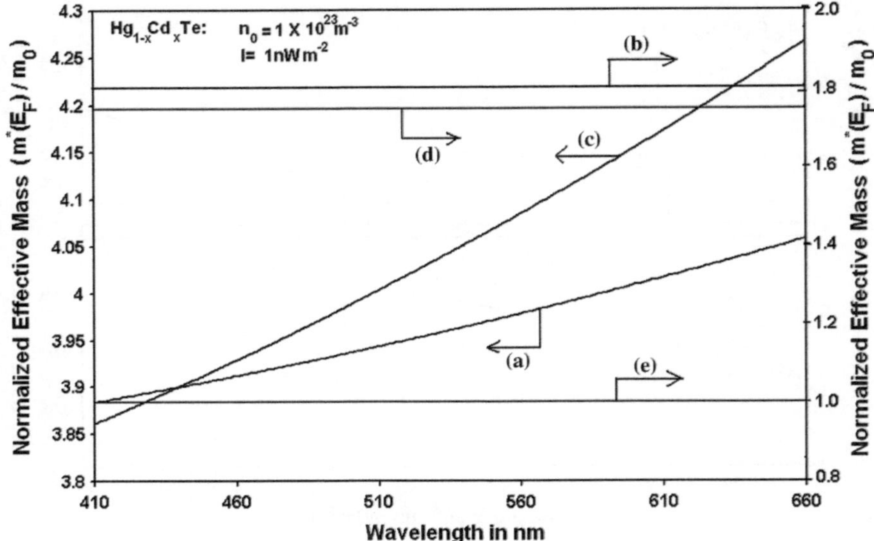

Fig. 77.3 Plot of the normalized EEM as a function of wavelength for n-$Hg_{1-x}Cd_xTe$ in the presence of light waves in which the curves *a* and *c* represent the three and two band models of Kane respectively. The curves *b* and *d* exhibit the same variation in the absence of external photo-excitation. The curve *e* in this case exhibits the parabolic energy band model both in the presence and in the absence of the external photo-excitation

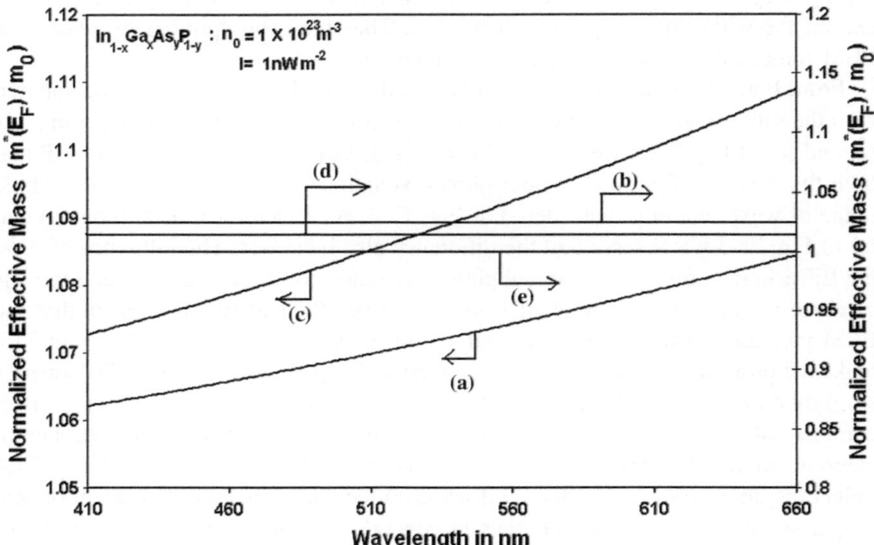

Fig. 77.4 Plot of the normalized EEM as a function of wavelength for $In_{1-x}Ga_xAs_yP_{1-y}$ lattice matched to InP in the presence of light waves in which the curves *a* and *c* represent the three and two band models of Kane respectively. The curves *b* and *d* exhibit the same variation in the absence of external photo-excitation. The curve *e* in this case exhibits the parabolic energy band model both in the presence and in the absence of the external photo-excitation

and the two band model of Kane. It is apparent that the EEM attains the maximum numerical value for ternary compounds and the minimum value for quaternary materials.

It is noteworthy that the basic (77.1) covers many compounds with different dispersion relations. Under certain limiting conditions, all the results of the EEM for different materials having various band structures lead to the well-known expression of the EEM having parabolic energy bands. The mathematical compatibility of our formulation is not only being manifested by this limiting test but also exhibits the fact that our simplified formalism is a rather generalized one, since we can obtain the corresponding well known results as given in the literature for semiconductors having parabolic energy bands as a consequence of our formulation. It is worth remarking that our formalism is not at all related to the Density-of-States technique as generally used in the literature [6]. The density-of-state function can be obtained from the carrier energy spectrum, but not vice versa. Therefore, we can infer that our investigation is more fundamental than those of the existing works due to the fact that the Boltzmann transport equation, which eventually controls the transport mechanism in nano devices, can be solved if and only if the carrier dispersion relations for different optical materials is known. The theoretical results as given in this paper would be rather useful in investigating various other experimental data related to this phenomenon. Finally, we can write that the results of this paper can be used to investigate the magnetic susceptibilities, the gate capacitance of MOS devices, the nonlinear response of nonlinear optical materials, the generalized Raman gain, the diffusion constants of the minority carriers, the Opto-thermal effects, the carrier statistics, the mobilities under different mechanisms of scattering, the Moss-Burstien shift, the plasma frequency, the quantized opto-Hall effect, the reflection coefficient, and other different opto-transport coefficients of modern low dimensional photonic devices operated in the presence of intense external photo excitation.

Acknowledgments The authors are grateful to Prof. G. Mugeraya, Director of National Institute of Technology, Agartala, Tripura, India and Prof. S. Chakrabarti, Director of Institute of Engineering and Management, Salt Lake, Kolkata, India respectively for their keen interest in the work and inspiration.

References

1. Chakrabarti S, Chakraborty M, Ghatak KP (2015) Einstein and his photoemission from heavily doped quantum wells. Rev Theor Sci (in press)
2. Ghatak KP, Chakrabarti S, Sen SK, Chakraborty S, Singh LS (2014) Einstein relation in heavily doped quantum wires superlattices with graded interfaces: simplified theory and suggestion for experimental determination. Adv Sci Eng Med 6(9):1042–1057
3. Bhattacharya S, Ghatak KP (2013) Fowler-Nordheim field emission: effects in semiconductor nanostructures. Springer series in soild state sciences, vol 170. Springer, Germany

4. Chakbarti S, Chatterjee B, Debbarma S, Ghatak KP (2015) Two dimensional effective electron mass at the fermi level in quantum wells of III-V, ternary and quaternary semiconductors. J Nano Sci Nanotechnol 15:1–12
5. Chatterjee B, Chakrabarti S, Chakraborty M, Ghatak KP (2014) Heavily doped semiconductors and the diffusion coefficient to mobility relationship. Rev Theor Sci 3:428–530
6. Schulman JN, Chang YC (1981) New method for calculating electronic properties of superlattices using complex band structures. Phys Rev B 24:4445–4448

Chapter 78
Oscillator Strength and Absorption Cross-section of Core-Shell Triangular Quantum Wire for Intersubband Transition

Arpan Deyasi and N.R. Das

Abstract Oscillator strength and absorption cross-section of core-shell triangular quantum wire is computed for intersubband optical transition between ground state and first excited state. Kane type band nonparabolicity of first order is considered to study the shift of cross-section peak with incident radiation, and also of the oscillator strength with structural parameters. Results are compared with that obtained from parabolic overestimation. Results show that both oscillator strength and peak magnitude of absorption cross-section decreases with increasing dimension, and the rate is significant when nonparabolic dispersion relation is taken into account. Findings are important for designing optical emitter/detector using core-shell triangular quantum wire.

78.1 Introduction

Electronic and photonic properties of semiconductor quantum structures have attracted the interest of both theoretical and experimental researchers in the last few years, precisely due to the rapid growth of fabrication procedure of lower dimension devices. Quantum well, wire and dot are the prime candidates among nanostructures, where carrier motion can be confined along one, two or three dimensions, leading to quantized energy states. Different geometrical shapes [1, 2] and material compositions [3, 4] are already implemented to design semiconductor quantum heterostructures for various applications like resonant tunneling diode [5], quantum cascade laser [6], transistor [7], optical emitter [8, 9], optical detector [10],

---

A. Deyasi (✉)
Electronics and Communication Engineering, RCC Institute of Information Technology, Kolkata 700015, India
e-mail: deyasi_arpan@yahoo.co.in

N.R. Das
Institute of Radio Physics and Electronics, University of Calcutta, Kolkata 700009, India
e-mail: nrd@ieee.org

V. Lakshminarayanan and I. Bhattacharya (eds.), *Advances in Optical Science and Engineering*, Springer Proceedings in Physics 166,
DOI 10.1007/978-81-322-2367-2_78

resonator [11], memory element [12] etc. Application-specific use of nanostructures requires precise computation of energy states, and thus choice of numerical method plays a key role to evaluate eigen energies for different devices [13, 14]. Recently, core-shell structure becomes an important candidate due to its inherent advantage of wider tuning range of emission/absorption spectra [15], higher absorption cross-section [16]. These structures have also improved oscillator strength compared to conventional quantum dot structures [17], which determines the probability of optical transitions between any two subbands [18]. Hence investigation of oscillator strength and absorption cross-section becomes very important for use of the heterostructure in photonic applications.

Oscillator strength of T-shaped quantum wire is computed for different material parameters [19], and also for self-assembly quantum dots [20] in presence of electric field. It is also computed for cylindrical quantum wire for field induced transition [21], and different potential configuration [22] Holovatsky [23] calculated it for elliptic nanotube and spherical dot [24]. In the present paper, authors computed oscillator strength for core-shell quantum wire of triangular geometry for the optical transition between ground and first excited states, and corresponding absorption cross-section is evaluated. $InAs/Ga_{0.47}In_{0.53}As$ is the material composition considered for simulation along with the incorporation of first-order band nonparabolicity of Kane-type, and simulated result is compared with that obtained for parabolic dispersion relation with similar geometrical dimensions and material composition. The parabolic assumption is nothing but an oversimplified mathematical consideration. Structural parameters are varied to observe the change of the parameters along with a wider range of incident radiation, and peak of the cross-section is also measured. Results are important to understand the probability of optical transitions in the structure, which plays key role in determining its candidature for optical transmitter/detector application.

78.2 Mathematical Modeling

We consider the schematic structure of triangular core-shell quantum wire as shown in Fig. 78.1, where electrons are confined in z and y directions.

The time-independent Schrödinger equation considering Kane-type nonparabolicity of first-order for wavefunction ψ in y–z plane in Cartesian system can be written as

$$\begin{aligned} &-\frac{\hbar^2}{2}\left\{\frac{\partial}{\partial y}\left[\frac{1}{m^*}\frac{\partial}{\partial y}\psi\right]+\frac{\partial}{\partial z}\left[\frac{1}{m^*}\frac{\partial}{\partial z}\psi\right]\right\} \\ &-\gamma\frac{\hbar^4}{4}\left\{\frac{\partial}{\partial y}\left[\frac{1}{m^*}\frac{\partial}{\partial y}\psi\right]+\frac{\partial}{\partial z}\left[\frac{1}{m^*}\frac{\partial}{\partial z}\psi\right]\right\}^2+V(y,z)\psi=E\psi \end{aligned} \tag{78.1}$$

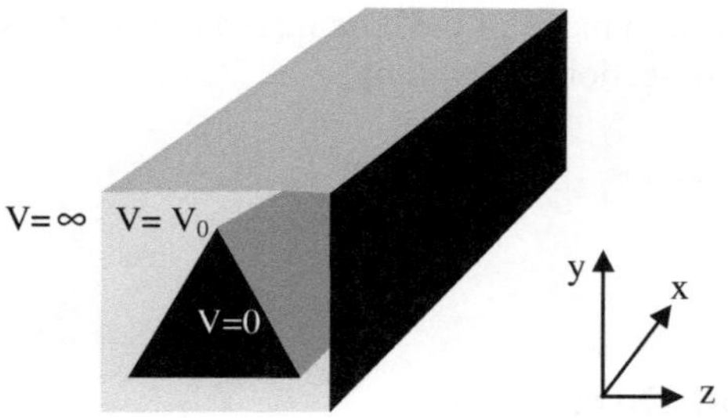

Fig. 78.1 Schematic structure of triangular core-shell quantum wire with Cartesian coordinate

where γ is the coefficient of band nonparabolicity. Introducing finite difference technique, (78.1) can be written as

$$\begin{aligned}
&-\frac{\hbar^4\gamma}{4m^{*2}(\Delta z)^4}[\psi(i+2,j)+\psi(i-2,j)-4\psi(i+1,j)-4\psi(i-1,j)+6\psi(i,j)]\\
&-\frac{\hbar^4\gamma}{4m^{*2}(\Delta y)^4}[\psi(i,j+2)+\psi(i,j-2)-4\psi(i,j+1)-4\psi(i,j-1)+6\psi(i,j)]\\
&-\frac{\hbar^4\gamma}{4m^{*2}(\Delta z)^2(\Delta y)^2}[\psi(i+1,j)+\psi(i-1,j)-2\psi(i,j)]\cdot[\psi(i,j+1)+\psi(i,j-1)-2\psi(i,j)]\\
&-\frac{\hbar^2}{2m^*(\Delta z)^2}[\psi(i+1,j)+\psi(i-1,j)-2\psi(i,j)]\\
&-\frac{\hbar^2}{2m^*(\Delta y)^2}[\psi(i,j+1)+\psi(i,j-1)-2\psi(i,j)]+V\psi(i,j)=E\psi(i,j)
\end{aligned} \tag{78.2}$$

where $i(j)$ is the no. of points along $z(y)$ axis, and $\Delta z(\Delta y)$ is dimension of the smallest grid along $z(y)$ direction. Solution of (78.2) gives eigenstates, from which intersubband energies ($\Delta E_{j,i}$, which denotes the intersubband transition energy form ith state to jth state) can be estimated. The efficiency of transition is determined by the parameter oscillator strength, which is given by

$$f_{ji}=\frac{2m^*\Delta E_{j,i}}{\hbar^2}|\langle j|z|i\rangle|^2 \tag{78.3}$$

where we assume the transition form quantum level 'i' to quantum level 'j'. Absorption cross-section can be computed from the knowledge of oscillator strength as

$$\sigma_{ji}=\frac{q^2T}{2\varepsilon_0 m^* nc}f_{ji}\left[1+T^2\left(\frac{E-\Delta E_{j,i}}{\hbar}\right)^2\right]^{-1} \tag{78.4}$$

where 'n' is the refractive index of the core material, and 'T' is the relaxation time. Peak absorption cross-section si given by

$$\sigma_{ji} = \frac{q^2 T}{2\varepsilon_0 m^* nc} f_{ji} \tag{78.5}$$

78.3 Results and Discussions

Using (78.3), oscillator strength of the structure is computed and plotted with width and length of the core region for the optical transition between ground state (E_{111}) and next higher excited state (E_{112}). In Fig. 78.2, it is observed that with increase of dimension for core layer, oscillator strength monotonically decreases, whereas the rate is almost negligible when parabolic dispersion relation is considered. This is due to the fact that for nonparabolic band structure, the dipole matrix element decreases rapidly whereas for parabolic band, transition energy decreases at a faster rate. But the matrix element is related with oscillator strength with square form, and hence the rate of change for the former is comparatively higher. Also with increase of core layer width, quantum confinement decreases, which causes reduction of eigenenergies, hence oscillator strength decreases.

In Fig. 78.3, absorption cross-section is plotted with incident radiation for two different core layer dimensions considering both types of dispersion relations. From the plot, it may be observed that with increase of dimension of either core width or height, cross-section peak appears at lower magnitude of incident radiation. The shift of peak position is larger for parabolic band compared to the nonparabolic one. As cross-section attains the peak corresponding to the transition energy between the two quantum levels, so the peak appears at the higher transition energy for lower width (/height) and goes to the lower value with increase of thickness. Again, since

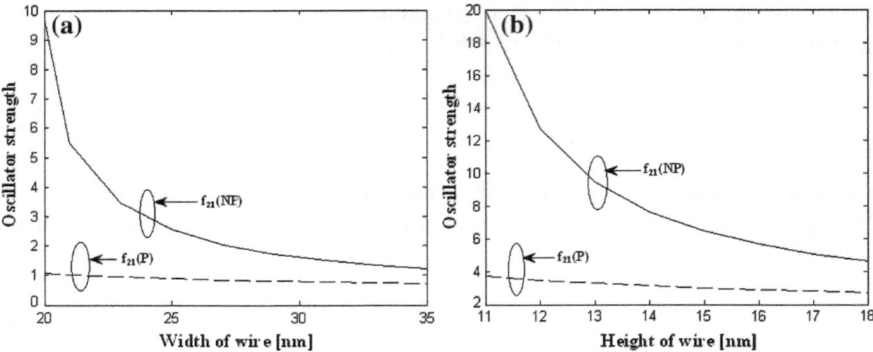

Fig. 78.2 Oscillator strength with **a** width of the core region for nonparabolic and parabolic band structures; **b** height of core region for nonparabolic and parabolic band structures

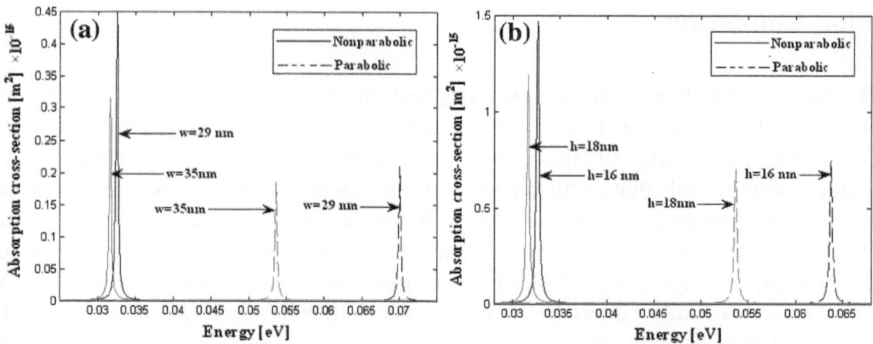

Fig. 78.3 Absorption cross-section with incident radiation for **a** two different core widths considering parabolic and nonparabolic band structures; **b** two different core heights considering parabolic and nonparabolic band structures

the transition energy increases due to nonparabolicity, so the peak appears at lower magnitude of incident radiation. But the shift of transition energy is very small with change of layer thickness in either of the quantized direction, so the shift of cross-section is almost negligible when nonparabolic dispersion relation is taking into account.

In Fig. 78.4, peak position of absorption cross-section is plotted with core layer dimensions. From the plot, it may be observed that with increase of width (/height), cross-section peak decreases. This is because higher core dimension decreases quantum confinement, which reduces energy. This effectively reduces oscillator strength. Since peak of the cross-section is directly proportional to oscillator strength, so peak magnitude reduces in a similar fashion to the oscillator strength. The rate is higher for nonparabolic band structure, and is almost negligible for parabolic overestimation.

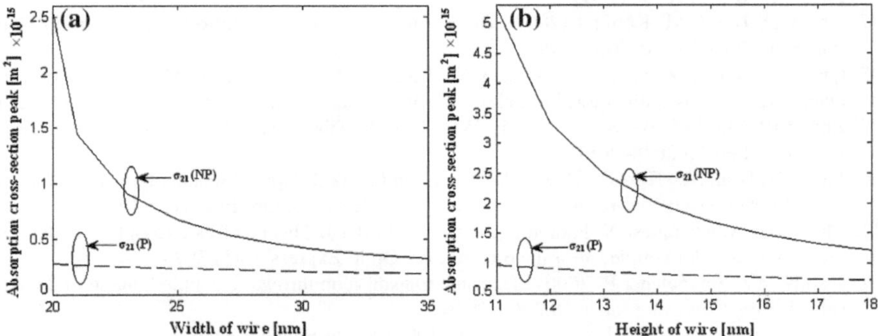

Fig. 78.4 **a** Absorption cross-section peak with width of core region for nonparabolic and parabolic band structures; **b** Absorption cross-section peak with width of core region for nonparabolic and parabolic band structures

78.4 Conclusion

Oscillator strength and absorption cross-section of core-shell triangular quantum wire is computed as function of structural parameters and with incident radiation. Band nonparabolicity of first order is considered for near accurate result. Computation reveals that oscillator strength rapidly decreases with dimension of the wire for nonparabolic semiconductor, whereas it remains almost constant when simplified parabolic band structure is considered. Hence it significantly depends on the bandgap of the material. Peak of absorption cross-section appears when incident radiation numerically equals to the intersubband transition energy, and magnitude of peak decreases with increasing dimension. Peak magnitude is higher for non-parabolic band structure. Result suggests that absorption is highly selective in core-shell structure as the cross-section profile may nearly be approximated to the Dirac-delta function.

References

1. Kim DH, You JH, Kim JH, Yoo KH, Kim TW (2012) Electronic structures and carrier distributions of T-Shaped $Al_xGa_{1-x}As/Al_yGa_{1-y}As$ quantum wires fabricated by a cleaved-edge overgrowth method. J Nanosci Nanotechnol 12:5687–5690
2. Khordad R (2012) Quantum wire with parallelogram cross-section: optical properties. J Theor Appl Phys 6(19):1–7
3. Wong BM, Leomnard F, Li Q, Wang GT (2011) Nanoscale effects on heterojunction electron gases in GaN/AlGaN core/shell nanowires. Nano Lett 11:3074–3079
4. Jia G, Wang Y, Gong L, Yao J (2011) Heterostructure type transformation of ternary $ZnTexSe_{1-x}$/ZnSe core-shell quantum dots. Digit J Nanomater Biostruct 6:43–53
5. Joel D, Singh MR (2010) Resonant tunneling in photonic double quantum well heterostructures. Nanoscale Res Lett 5:484–488
6. Allen SS, Richardson SL (1994) Theoretical investigations of resonant tunneling in asymmetric multibarrier semiconductor heterostructures in an applied constant electric field. Phys Rev B 50:11693–11700
7. Urban D, Braun M, König J (2007) Theory of a magnetically controlled quantum-dot spin transistor. Phys Rev B 76:125306
8. Qian F, Gradecak S, Li Y, Wen CY, Lieber CM (2005) Core/Multishell nanowire heterostructures as multicolor, high-efficiency light-emitting diodes. Nano Lett 5:2287–2291
9. Zhang JP, Chu DY, Wu SL, Ho ST, Bi WG, Tu CW, Tiberio RC (1995) Photonic-wire laser. Phys Rev Lett 75:2678–2681
10. Park BH, Baek SD, Kim JY, Bae J, Han H, Kwon O (2002) Optical sensing by using photonic quantum ring lasers and resonance-enhanced photodetectors. Opt Eng 41:1339–1345
11. Majumdar A, Manquest N, Faraon A, Vuckovic J (2010) Theory of electro-optic modulation via a quantum dot coupled to a nano-resonator. Optic Express 18(5):3974
12. Ivanov MV, Schmelcher P (2006) Electronic transmission through a coupled quantum dot and ring. J Phys Condens Matter 18:2963–2976
13. Ogawa M, Kunimasa T, Ito T, Miyoshi T (1998) Finite-Element analysis of quantum wires with arbitrary cross sections. J Appl Phys 84:3242–3249
14. Gangopadhyay S, Nag BR (1997) Energy levels in finite barrier triangular and arrowhead-shaped quantum wires. J Appl Phys 81:7885–7889

15. Balet LP, Ivanov SA, Priyatinski A, Achermann M, Kilmov VI (2004) Inverted core/shell nanocrystals continuously tunable between Type-I and Type-II localization regimes. Nano Lett 4:1485–1488
16. Nanda J, Ivanov SA, Htoon H, Bezel I, Piryatinski A, Tretiak S, Kilmov VI (2006) Absorption cross sections and Auger Recombination lifetimes in inverted core-shell nanocrystals: implications for lasing performance. J Appl Phys 99:034309
17. Bhattacharyya S, Das NR (2012) Effect of electric field on the oscillator strength and cross-section for intersubband transition in a semiconductor quantum ring. Phys Scr 85:045708
18. Yariv A (1975) Quantum electronics. Wiley, New York
19. Szymanska MH, Littlewood PB, Needs RJ (2001) Excitons in T-shaped quantum wires. Phys Rev B 63:205317
20. Bloom AR, Mirin RP, Silverman KL (2008) Reducing the oscillator strength in semiconductor quantum dots with a lateral electric field. In: Conference on lasers and electro-optics, San Jose
21. Wu S, Tomic S (2012) Exciton states and oscillator strengths in a cylindrical quantum wire with finite potential under transverse electric field. J Appl Phys 112(3):033715
22. Tshipa M (2014) Oscillator strength for optical transitions in a cylindrical quantum wire with an inverse parabolic confining electric potential. Indian J Phys 88(8):849–853
23. Holovatsky VA, Voitsekhivska OM, Gutsul VI (2008) Optical oscillator strengths for the electron quantum transitions in elliptic nanotubes. Rom J Phys 53(7–8):833–840
24. Holovatsky VA, Bernik I, Voitsekhivska OM (2014) Oscillator strengths of quantum transitions in spherical quantum dot $GaAs/Al_xGa_{1-x}As/GaAs/Al_xGa_{1-x}As$ with on-center donor impurity. Acta Phys Pol A 125(1):93–97

Author Index

V. Lakshminarayanan and I. Bhattacharya (eds.), *Advances in Optical Science and Engineering*, Springer Proceedings in Physics 166,
DOI 10.1007/978-81-322-2367-2

Subject Index

V. Lakshminarayanan and I. Bhattacharya (eds.), *Advances in Optical Science and Engineering*, Springer Proceedings in Physics 166,
DOI 10.1007/978-81-322-2367-2